COLORANTS AND AUXILIARIES

ORGANIC CHEMISTRY AND APPLICATION PROPERTIES

Volume 1 – Colorants

Colorants and auxiliaries

Organic chemistry and application properties

Volume 1 – Colorants

Edited by John Shore

BTTG–Shirley, Manchester, England

1990

Society of Dyers and Colourists

Published by the Society of Dyers and Colourists, PO Box 244, Perkin House, 82 Grattan Road, Bradford, West Yorkshire BD1 2JB, England, on behalf of the Dyers' Company Publications Trust.

Typeset by the Society of Dyers and Colourists and printed by Staples Printers Rochester Ltd

ISBN 0 901956 51 1

Dyers' Company Publications Trust

The Dyers' Company Publications Trust was instituted by the Worshipful Company of Dyers of the City of London in 1971 to encourage the publication of textbooks and other aids to learning in the science and technology of colour and coloration and related fields. The Society of Dyers and Colourists acts as trustee to the fund, its Textbooks Committee being the Trust's technical subcommittee.

Contributors

Geoffrey Hallas
Senior lecturer, Department of Colour Chemistry and Dyeing, University of Leeds

(the late) **Francis Jones**
Formerly senior lecturer, Department of Colour Chemistry and Dyeing, University of Leeds

David Patterson
Formerly senior lecturer, Department of Colour Chemistry and Dyeing, University of Leeds

John Shore
Dyeing & Finishing R & D Colour Services, BTTG–Shirley, Manchester

C Vivian Stead
Formerly of ICI plc, Colours and Fine Chemicals Division, Manchester

Preface to Volume 1

This textbook is a further addition to the series on colour and coloration technology initiated by the Textbooks Committee of the Society of Dyers and Colourists under the aegis of the Dyers' Company Publications Trust Management Committee, which administers the trust fund generously provided by the Worshipful Company of Dyers.

The major objective of this series of books, produced at irregular intervals over the last fifteen years, has been to establish a coherent body of explanatory information on the principles and application technology of relevance for students preparing to take the Associateship examinations of the Society. This particular book is directed specifically to the subject areas covered by Section A of Paper B: the organic chemistry and application of dyes and pigments and of the auxiliaries used with them in textile coloration processes. It is hoped that qualified chemists and colourists interested in the properties of colorants and their auxiliaries will also find this useful as a work of reference. For several reasons it is convenient to publish the material in two separate volumes: 1. Colorants, 2. Auxiliaries.

This first volume of the book concentrates on the chemical characteristics of dyes and pigments, with emphasis on attempts to interpret their colouring and fastness properties in terms of the essential structural features of colorant molecules. Some important intermediates and their uses in the synthesis of major chromophores are also indicated. The major theme of this volume, however, is the matrix of chemical classes and application ranges around which the *Colour Index International* is assembled. Some minor anomalies or inconsistencies in this matrix are noted but the fundamental value of this unique index to colorant manufacturers and users is universally recognised.

The authors and editor of this volume are greatly indebted to our referees for valuable comments and suggestions that have improved the content of our contributions. Our grateful thanks are due to Paul Dinsdale, the editor of the Society, and to Dr Jean Macqueen, the latter for her patient efforts in copy editing our original scripts into their published form without losing any significant points that we wished to express. We also thank Elaine Naylor and Jane Roberts for their very diligent work in keying the text and preparing the illustrations for publication from our original typescripts.

It is particularly sad to record that our mutual friend and co-author, Frank Jones, did not live long enough to see his chapter on metal-complex dyes published; I trust that we have preserved his original style intact.

JOHN SHORE

Contents

Chapters in Volume 2

CHAPTER 1

Historical development and classification of colorants

John Shore

1.1 HISTORICAL INTRODUCTION

Synthetic dyes and pigments have furnished the colorant user with a colourful palette for over a century. Many different substrates – natural or synthetic polymer structures of all kinds – can be dyed, printed or pigmented in virtually any colour to commercially acceptable standards of fastness, at a price. The history of synthetic colorants, their remarkable origin in research on coal-tar fractions in the mid-nineteenth century [1–4] and their rapid growth in volume and variety, especially during the first seventy years of their existence [5–10], have been well documented. Their development has been a notable achievement of organic chemists, who have had to overcome successive challenges to their ingenuity [11].

The important naturally occurring substrates of pre-industrial societies (cotton, linen, silk, wool, leather, paper, wood) share certain similarities, since they are all essentially saccharidic or peptidic polymers. They could thus be coloured using a relatively short range of dyes and pigments, perforce of natural origin also. An early objective of planned research on synthetic colorants, therefore, was to replace the leading natural extracts (alizarin and indigo) by their synthetic equivalents [12,13]. Simultaneously with this diligent and ultimately successful effort, other chemists were discovering totally new chromophores unknown in nature: azine, triarylmethane and other dyes from arylamine oxidation, azo colorants from the diazo reaction, and eventually azo–metal complexes and phthalocyanines. Building on success with indigo and anthraquinone derivatives, the systematic approach led on to related but new chromophores with outstanding properties: vat dyes and novel pigments.

Linked to this research by a common interest in certain versatile intermediates and a similar urge to extend the limited range of natural substrates, a new breed of organic chemist, the polymer specialist, was vigorously developing novel regenerated and synthetic fibres, plastomers, elastomers and resins. Most of these differed markedly in structure and properties from natural polysaccharides or polypeptides. Particularly during the second quarter of this century, urgent demands arose for special new colorants and application techniques designed to colour these substrates. Disperse dyes for ester fibres [14–17], basic dyes for acrylics [18,19] and pigments for the mass coloration of fibres and plastics [20,21] are typical examples of the response of the colour chemist. Natural fibres too have benefited from this broadening wave of research: reactive dyes for cellulosic and protein fibres [22–28] and fluorescent whiteners for undyed textiles, paper and detergent formulations [29–32] were discoveries stemming essentially from this exceptionally active period.

In recent years the emergence of an unknown substrate has become a rare event. The rate of introduction of radically new colorants, auxiliary products and processes has fallen markedly. An increasingly adverse balance has arisen between the escalating costs of the research effort together with more stringent hazard testing on the debit side, as against the diminishing value of marginal technical or economic improvements to existing ranges of colorants on established substrates. The pathways of colorant research have begun to turn away from conventional outlets for dyes and pigments towards more esoteric applications in electronic [33,34] or medical [35] fields.

1.2 EARLY ATTEMPTS TO CLASSIFY DYES AND PIGMENTS

The main objective of this chapter is to outline the principal system by which colorants are classified, namely the widely accepted *Colour Index* classification. After tracing the earlier developments from which this system has evolved [36], the distribution of existing dyes and pigments among the various classes listed therein will be introduced. Finally each of these classes will be discussed in turn, illustrated by structural formulae of typical commercial dyes.

The earliest comprehensive alphabetical listing of synthetic colorants and chemicals used in the coloration industry [37] was published in 1870, although it derived from previous texts on natural products. Certain later authors persisted with this alphabetical approach, but the beginnings of systematic classification based on chemical structure, with grouping within each structural class according to hue, were first seen in the 1880s [38]. These improved systems may be exemplified by that of Richardson [39] who listed the commercial names, chemical names and notional constitutions of about 100 'coal-tar dyes' in hue order. It is interesting that 50% of them were acid dyes and 20% basic dyes, and that 40% were placed in the 'red' category.

Undoubtedly the most successful of these early systems was the Schultz and Julius classification [40], later immortalised as 'Schultz's *Farbstofftabellen*' [41]. During the next fifty years seven editions of this standard reference index were published, and the number of chemical entries rose from about 280 in the First Edition (1888) to nearly 1500 in the Seventh (1931–32). The later editions of this work pioneered many of the features eventually adopted in the preparation of the *Colour Index* Second Edition.

The Society of Dyers and Colourists embarked on the production of the First Edition of its *Colour Index* in 1921 but production as a series of monthly issues was not completed until three years later. There were over 1200 entries for synthetic organic colorants, as well as smaller sections on natural dyes and inorganic pigments. The Society had intended to publish additions in 1928 but thereafter updating was discontinued [36]. By 1945 the need for a Second Edition had become urgent, and preparations for this began in collaboration with the American Association of Textile Chemists and Colorists. The four-volume Second Edition published in 1956–58 contained almost 3600 dyes of different constitution. A particularly useful innovation was the separate listing of commercial names (31 500) under equivalent C.I. generic names (4600 entries).

Since publication of the Second Edition, the *Colour Index* has become established as the leading reference work for the classification of colorants. A completely revised five-volume Third Edition was published in 1971. The growth in recorded information over the preceding period is indicated by the listing of approximately 7900 generic names and 36 000 commercial names; three supplementary volumes and numerous additions and amendments have taken the current listings of the *Colour Index* well above these levels. A comprehensive review of the information and a different approach to the problems of revision and updating, taking advantage of developments in data storage and retrieval technology, is planned for future versions of the *Colour Index* [36].

1.3 *COLOUR INDEX* CLASSIFICATION

Most of the organic dyes and pigments in the *Colour Index*, including many of those not assigned a specific chemical constitution number, are placed in one of the 25 structural classes according to their chemical type (where this is known). The largest class, azo colorants, is subdivided into four sections depending on the number of azo groups in the molecule. Metal-complex azo colorants [42,43] are designated separately by description but are classified together with their unmetallised analogues in the same generic class. Excluding the colorant precursors, such as azoic components and oxidation bases, as well as the sulphur dyes of indeterminate constitution, almost two-thirds of all the organic colorants listed in the *Colour Index* belong to this class, one-sixth of them being metal complexes. The next largest chemical class is the anthraquinones (15% of the total), followed by triarylmethanes (3%) and phthalocyanines (2%). No other individual chemical class accounts for more than 1% of the *Colour Index* entries.

The distribution of each chemical type between the major application groups of colorants is far from uniform (see Table 1.1). Stilbene and thiazole dyes are almost invariably direct dyes. Acridines and methines are usually basic dyes, whereas nitro, aminoketone and quinophthalone derivatives are often disperse dyes. Metal-complex azo and formazan types are mainly acid dyes but phthalocyanines are important for reactive dyes. Indigoids are predominantly vat dyes but anthraquinones retain considerable significance also as acid, disperse or reactive dyes. The data in this table are given in percentages because the actual numbers of dyes recorded gradually increase as new products are added. They relate to all those dyes listed where the chemical class is known, including products no longer in commercial use.

TABLE 1.1

Percentage distribution of each chemical class between major application ranges

	Distribution between application ranges/%								
Chemical class	Acid	Basic	Direct	Disperse	Mordant	Pigment	Reactive	Solvent	Vat
Unmetallised azo	20	5	30	12	12	6	10	5	
Metal-complex azo	65		10				12	13	
Thiazole		5	95						
Stilbene			98					2	
Anthraquinone	15	2		25	3	4	6	9	36
Indigoid	2					17			81
Quinophthalone	30	20		40				10	
Aminoketone	11			40	8		3	8	30
Phthalocyanine	14	4	8		4	9	43	15	3
Formazan	70						30		
Methine		71		23		1		5	
Nitro, nitroso	31	2		48	2	5		12	
Triarylmethane	35	22	1	1	24	5		12	
Xanthene	33	16			9	2	2	38	
Acridine		92		4				4	
Azine	39	39				3		19	
Oxazine		22	17	2	40	9	10		
Thiazine		55			10			10	25

There are in all nineteen generic name or application ranges of products listed in the *Colour Index*. C.I. Acid dyes constitute the largest range with currently over 2000 entries, 55% of them still in commercial use. Next come C.I. Direct dyes (1450 entries, 40% still active) and C.I. Disperse dyes (about 1250 entries, 60% active). C.I. Reactive dyes (75% active), C.I. Basic dyes, C.I. Solvent dyes and C.I. Pigments (all 60% active) continue to progress, but C.I. Vat dyes (45% active) and C.I. Mordant dyes (33% active) are apparently in decline [44]. C.I. Sulphur dyes can be regarded as a distinct chemical class as well as an application range, although a few vat dyes are manufactured in a similar way. Essentially C.I. Food dyes and C.I. Leather dyes are selections from the larger ranges of textile dyes and several application types are represented in the C.I. Natural dyes and pigments category. The groupings according to generic name also include colorant precursors (C.I. Azoic Components and C.I. Developers, C.I. Ingrain dyes, C.I. Oxidation Bases) and uncoloured products (C.I. Fluorescent Brighteners and C.I. Reducing Agents) associated with textile coloration. In spite of the inclusion of these somewhat anomalous categories, there has been little enthusiasm on the part of the publishers of the *Colour Index* to extend the coverage to auxiliary products in general, or to follow up earlier proposals for a separate publication in this area [36].

The use of *Colour Index* generic names for commercial dyes and pigments is now widespread among trade organisations, producers and user firms throughout the world. These designations avoid the duplication implicit in quoting the numerous commercial names from individual suppliers and facilitate comparisons between chemically similar products. Systematic chemical names are often complicated, liable to confusion and frequently not revealed by manufacturers for commercial reasons. Warnings against assuming that dyes or pigments with same C.I. generic name will necessarily behave identically when subjected to application or fastness tests have been sounded in recent years, however [45,46]. Disperse dyes and pigments are particularly prone to such anomalies because their performance is so dependent on physical form (see sections 2.4.1, 2.5, 2.7.1 and 3.1.5), but the presence of impurities can influence the behaviour of any pair of commercial colorants containing the same major component.

1.4 COLOUR AND CHEMICAL STRUCTURE

Before moving on to a description of the individual chemical classes and application ranges of dyes and pigments, it is necessary first to trace briefly the developments in understanding of the relationship between colour and chemical constitution. This subject has been reviewed most thoroughly elsewhere [47–50] and the intention here is only to outline the basic principles so that the reader can appreciate the need for such a variety of structural types of colorant. The often-conflicting requirements of colour and application characteristics form a major part of the subject matter in succeeding chapters.

The first general theory relating depth of colour to molecular structure was made by Witt (1876), who recognised that all dyes then known contained aryl rings bearing unsaturated groups, such as >C=O, –N=O or –N=N–, which he termed 'chromophores'. Intense colour could be developed from such a chromophoric grouping or 'chromogen' by attaching weakly basic substituents such as –OH or $-NH_2$ groups, called 'auxochromes', to the aryl ring. Subsequent developments recognised the particularly strong chromophoric effect of charged centres, as in the triarylmethane basic dyes which lack conventional uncharged chromophores, as well as the bathochromic (colour-enhancing) influence of pairs of auxochromes located in the 2,5-positions relative to the chromophore.

The next step was the quinonoid theory of Armstrong (1888). Although no longer tenable, it did help to explain the intense colour of the triarylmethanes and related basic dyes so prevalent at that time. Improved understanding of the interaction of visible and u.v. radiation with organic structures aroused interest in the tautomeric capabilities of dye molecules, such as benzenoid–quinonoid tautomerism in azo (1.1), anthraquinone (1.2) and triarylmethane (1.3) systems. Watson and Meek (1915) suggested that the oscillation between such alternative arrangements corresponded to the reversal of double and single bonds along the conjugated chain of the molecule: the longer the chain, the slower the period of vibration. This accorded qualitatively with the relation between conjugation and absorption.

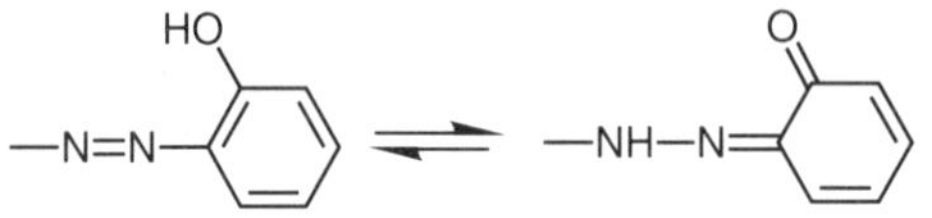

1.1

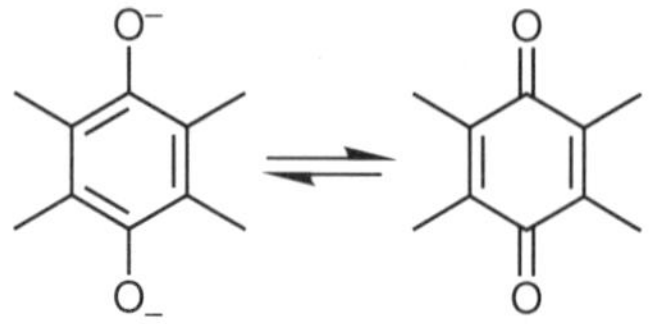

1.2

1.3

Mathematical elucidation of the principles of quantum mechanics in the 1920s, leading to the concept of molecular orbitals and a much more precise understanding of the nature of chemical bonding in the 1930s, provided the first opportunity for chemists to develop qualitative predictive methods of relating colour to chemical constitution. The equations involved are, however, so complicated that approximation methods had to be employed, particularly for calculating transition energies in organic molecules. Three distinct theoretical approaches were adopted: the valence bond, the molecular orbital and the free electron methods.

Valence bond theory was initially the most widely accepted approach, probably because it depended on familiar concepts of mesomeric effects in conjugated systems. The technique was never successful for quantitative calculation of the absorption spectra of dyes, however, because of the difficulties encountered when introducing the numerous canonical structures necessary for computational precision. The molecular orbital approach developed more slowly but was ultimately much more successful than valence bond theory. Dewar (1950) was able to predict absorption maxima for various types of cyanine dyes in excellent agreement with experiment. These results had an accelerating effect on the development of molecular orbital theory and an improved approach, the Pariser–Parr–Pople (or PPP) method, established itself during the next few years as the most useful and versatile technique for colour prediction, especially when the microelectronics revolution provided facilities for overcoming the complexity of the necessary calculations. The free electron (FEMO) theory had its origin in work on the conduction electrons of metals in the 1940s, when several workers independently recognised the close analogy between these and the delocalised π-electrons of polyene dyes. The method has since been extended to many other classes of dyes, notably by Kuhn [51] in the 1950s, but it has not found general acceptance for spectroscopic calculations, since it lacks ready adaptability in its present form [50].

1.5 CHEMICAL CLASSES OF DYES AND PIGMENTS

A brief description of each class of colorant will be given in this chapter, in order to show how they contribute to the overall distribution outlined in Table 1.1. For further details the reader is referred to the later Chapters 4 to 7. Many of the lesser-known chemical classes are more fully described in Chapter 6.

The order of discussion of the various chemical classes included here differs somewhat from that in the *Colour Index*. Thiazole dyes are dealt with immediately after the stilbene class because both of these, like the polyazo types, contribute notably to the range of direct dyes. The anthraquinone, indigoid, quinacridone, quinophthalone and aminoketone classes form another series with certain structural similarities and important applications in vat or disperse dyes and pigments. Phthalocyanine and formazan are stable metal-complex chromophores. The remaining seven categories included are already less important and still declining in commercial significance. The arylmethane, xanthene, acridine, azine, oxazine and thiazine chromophores share a limited degree of resemblance in structural terms.

1.5.1 Azo colorants

The presence of one or more azo (–N=N–) groups, usually associated with auxochromic groups (–OH or –NH–), is the characteristic feature of this class. Hydroxyazo dyes exhibit benzenoid–quinonoid tautomerism with the corresponding quinonehydrazones [52,53]. At least half of all commercial azo colorants belong to the monoazo subclass, that is, they have only one azo group per molecule. This proportion is considerably higher among the metal-complex azo dyes. Direct dyes represent the only application range where monoazo compounds are relatively unimportant; disazo and trisazo dyes are preferred in order to confer higher substantivity for cellulose. The numerous ways in which diazo and coupling components can be used to assemble azo dyes and pigments for many purposes are discussed in Chapter 4. Yellow azo chromophores are occasionally linked to blue anthraquinone or phthalocyanine structures in order to produce bright green dyes or pigments.

1.5.2 Thiazole dyes

The characteristic chromophore of this class is the thiazole ring itself, normally forming part of a 2-phenylbenzothiazole grouping. Most are yellow direct dyes of the azophenylthiazole (1.4) type, but a minority are simple basic dyes with an alkylated thiazolinium group (1.5), such as Thioflavine TCN (C.I. Basic Yellow 1). The thiazole ring enhances substantivity for cellulose and thus has been incorporated into certain anthraquinonoid and sulphurised vat dyes. Several important blue basic dyes are 2-phenylazo derivatives of 6-alkoxybenzothiazolinium compounds (1.6).

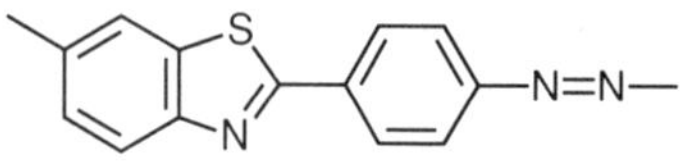

1.4

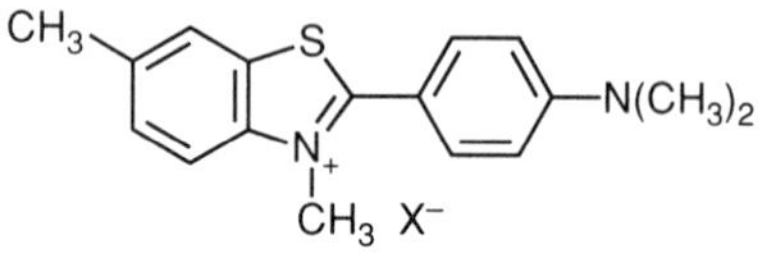

1.5

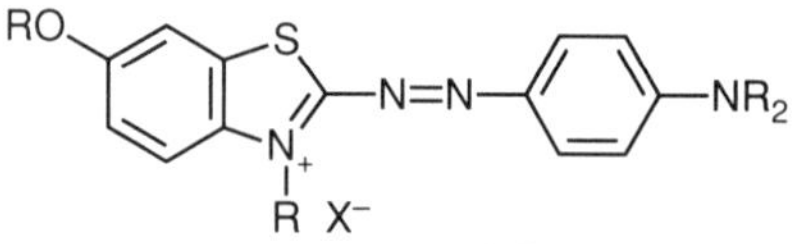

1.6

1.5.3 Stilbene dyes and fluorescent brightening agents

Stilbene dyes are mixtures of indeterminate constitution resembling polyazo direct dyes in their application properties. They result from the alkaline self-condensation of 4-nitrotoluene-2-sulphonic acid (1.7) or its initial condensation product 4,4′-dinitrostilbene-2,2′-disulphonic acid (1.8; X = NO_2), either alone or with various arylamines. The characteristic chromophores are azo- or azoxy-stilbene groupings (1.9). As with sulphur dyes, the C.I. generic names of stilbene dyes refer not to specific chemical entities but to mixtures of related compounds with closely similar dyeing and fastness properties. Almost all of them are yellow to brown direct dyes for cellulosic fibres and leather.

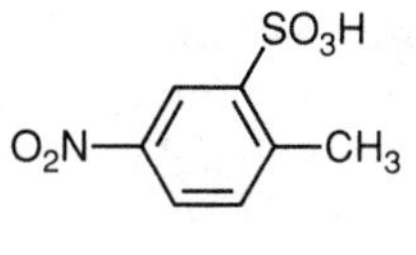

1.7

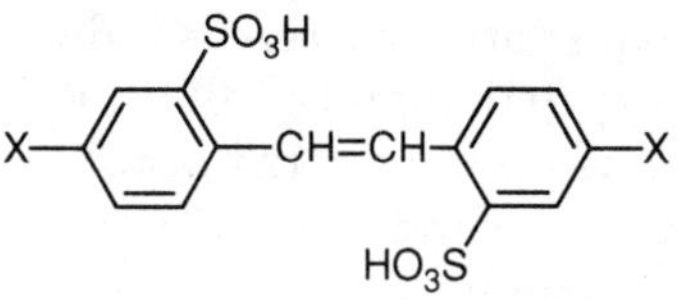

1.8

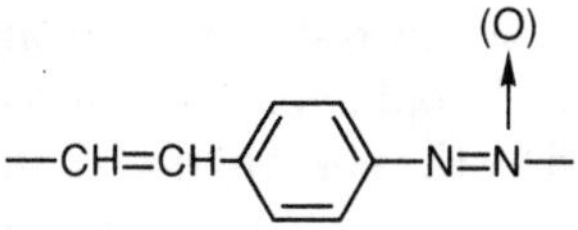

1.9

Approximately 75% of fluorescent brightening agents belong to the stilbene class. These are almost invariably derived from 4,4′-diaminostilbene-2,2′-disulphonic acid (1.8; X = NH_2), often condensed with cyanuric chloride to take advantage of the further contribution of the *s*-triazine rings to substantivity for cellulose.

1.5.4 Anthraquinone colorants

Strictly speaking, the characteristic chromophore (1.10) of these should be anthraquinone itself, but the term 'anthraquinonoid' is frequently extended, in the *Colour Index* as elsewhere, to include other polycyclic quinone structures. These are often synthesised from anthraquinone derivatives and most of them, including dibenzpyrenequinone (C.I. Vat Yellow 4), pyranthrone (C.I. Vat Orange 9), isoviolanthrone (C.I. Vat Violet 10) and violanthrone (C.I. Vat Blue 20), are strongly coloured even in the absence of auxochromes. Indanthrone (1.11; C.I. Vat Blue 4), the first polycyclic vat dye to be discovered, resulted from an unsuccessful attempt to link two anthraquinone nuclei via an indigoid chromophore. Polycyclic vat dyes and the many derivatives of anthraquinone applicable as acid, basic, disperse, mordant, reactive and vat dyes are discussed in Chapter 6.

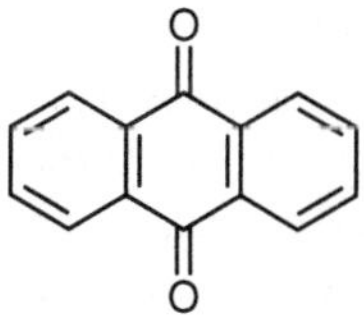

1.10

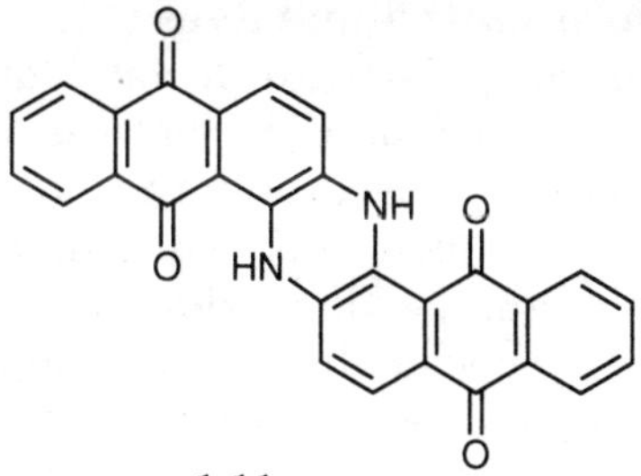

1.11

1.5.5 Indigoid colorants

This substantial class of vat dyes and pigments has declined gradually in importance relative to the anthraquinone derivatives. The still pre-eminent representative is indigo (1.12; C.I. Vat Blue 1), the dye most consistently in demand of all time. Originally obtained from natural sources, indigo was probably the decisive impetus for the early development of the synthetic dye industry [11]. In indigo and thioindigo (1.13; C.I. Vat Red 41) the chromophoric system is symmetrical. These dyes can exist in both *cis* and *trans* forms; the latter is the more stable form, and predominates in the solid state. Unsymmetrical indigoid dyes are also known, in which the two halves of the molecule united by the central C=C bond differ in substitution pattern, hetero-atom or orientation of the hetero ring. An example is Ciba Violet A (CGY), the monobrominated indole–thianaphthene analogue (1.14).

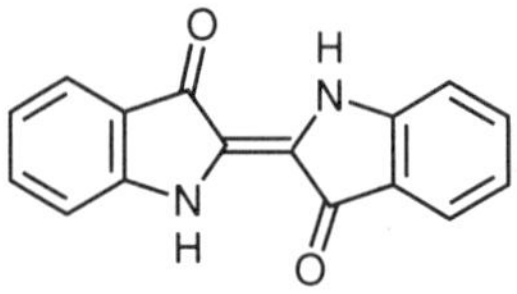

1.12

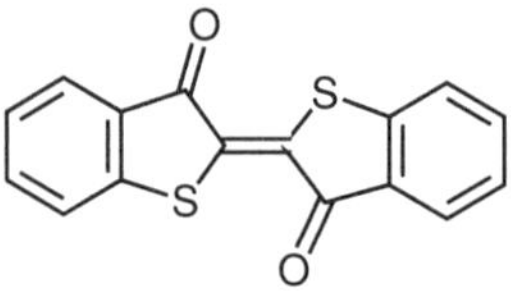

1.13

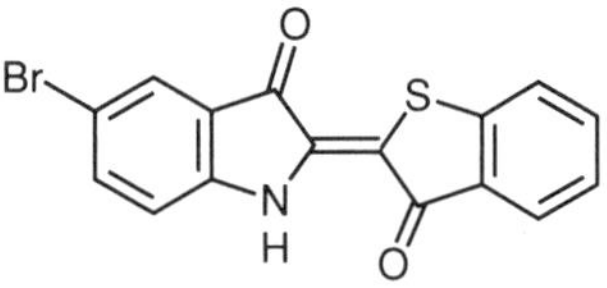

1.14

1.5.6 Quinacridone pigments

This chromophore (1.15) is somewhat reminiscent of the indigoid and anthraquinonoid types but it has not yielded useful vat dyes. Bluish-red pigments of the quinacridone class are especially important in violet and magenta colours or for deep reds in admixture with inorganic scarlets.

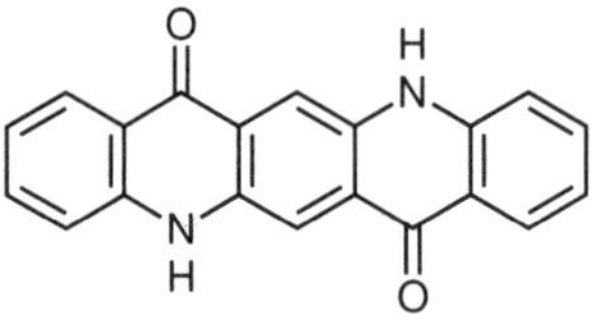

1.15

1.5.7 Quinophthalone dyes

The name of this structural class ('quinoline') in the *Colour Index* is not ideal because quinoline derivatives feature in other related classes, such as the methine basic dyes with a quinolinium cationic group. The class is more precisely associated with quinophthalone (1.16), the characteristic chromophore derived by condensation of quinoline derivatives with phthalic anhydride. This small class of yellow compounds contributes to the disperse, acid, basic and solvent ranges of dyes.

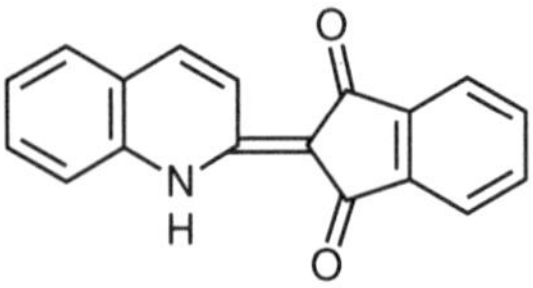

1.16

1.5.8 Aminoketone and hydroxyketone dyes

This small group of hydroxyquinone (1.17), arylaminoquinone (1.18) and aminonaphthalimide (1.19) derivatives has contributed a few members of some application ranges, mainly disperse yellow and red-brown vat dyes, but there are superior alternatives available from the major chemical classes.

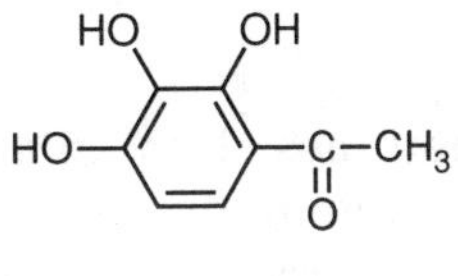

1.17

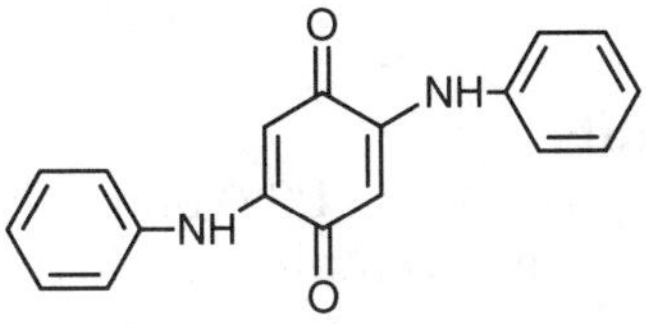

1.18

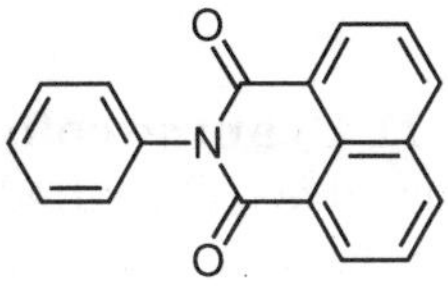

1.19

1.5.9 Phthalocyanine colorants

Substituted derivatives of metal-free phthalocyanine (C.I. Pigment Blue 16) and a series of metal complexes, notably copper phthalocyanine (C.I. Pigment Blue 15) (1.20; M = Cu), contribute brilliant blue and green colours to several application ranges [54]. Pigments and reactive dyes are especially dependent on this chemical class, but examples exist in all the important ranges except disperse dyes. These colorants are discussed together with the metal-complex azo dyes in Chapter 5.

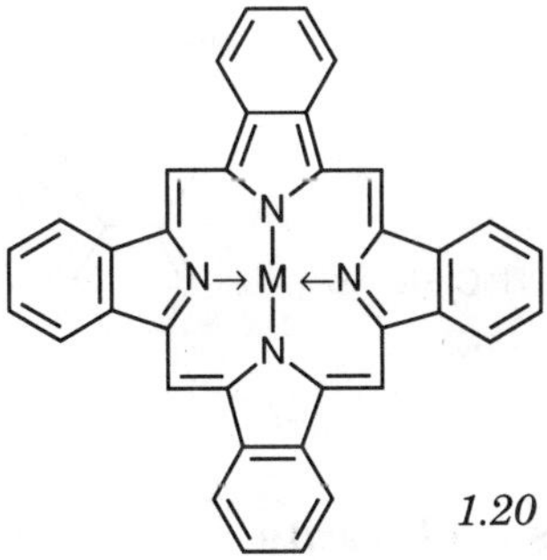

1.20

1.5.10 Formazan dyes

This small class of blue copper-complex dyes has made a significant contribution to the acid and reactive ranges in recent years (see sections 5.4.2, 5.4.3 and 7.5.8). The chromophore is the tricyclic chelate grouping illustrated (1.21), with the coordinated metal atom in the central position as in the metal phthalocyanines and 1:2 metal-complex azo dyes.

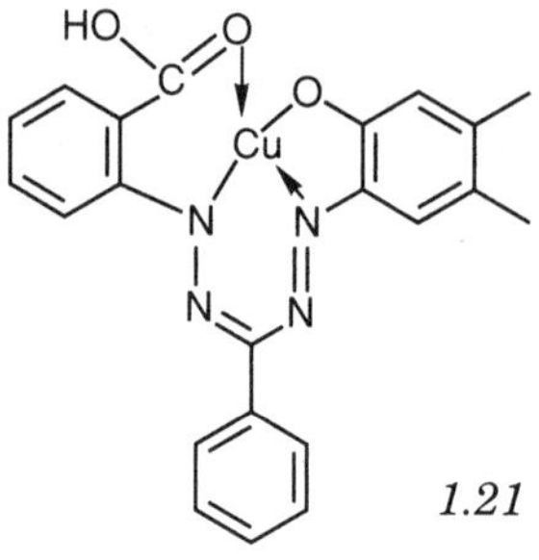

1.21

1.5.11 Cyanine colorants

These dyes, classified as methines (1.22) and polymethines (1.23) in the *Colour Index*, characteristically contain a conjugated system through one or more methine (–CH=) groups terminating with heterocyclic atoms, usually nitrogen as in the quinoline (1.22) and trimethylindoline (1.23) types. The most important function of cyanine colorants is as sensitisers in photography [55]. One or more methine groups may be replaced by nitrogen atoms, as in the azacyanines. Many methine compounds intended for textiles are yellow or red basic dyes, but uncharged yellow methine disperse (1.24) and azamethine chromium-complex solvent-soluble dyes (1.25) are also significant.

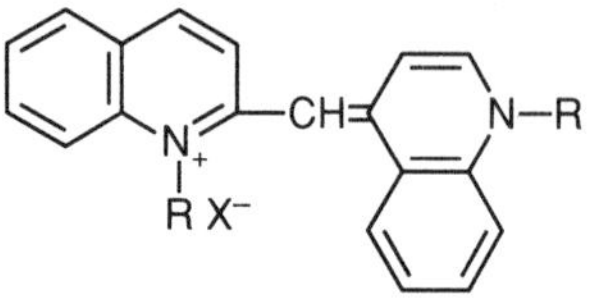

1.22

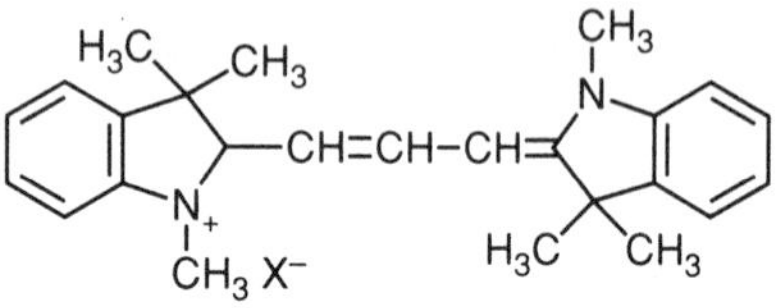

1.23

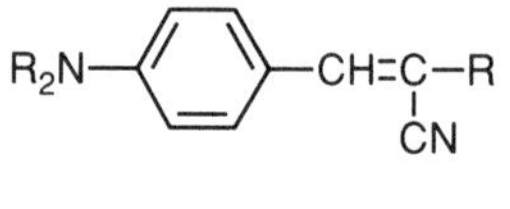

1.24

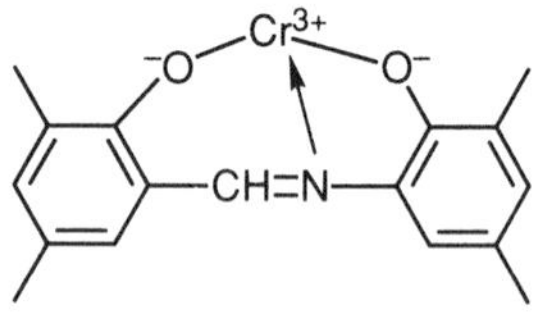

1.25

1.5.12 Nitro and nitroso colorants

Nitro dyes exhibit benzenoid–quinonoid tautomerism (1.26) and their colour is attributed mainly to the *o*-quinonoid form, since this can be stabilised by hydrogen bonding. The tautomeric *o*-nitroso-phenols or -naphthols readily form chelate complexes with metals. A few yellow nitro disperse dyes, including C.I. Disperse Yellow 1 (1.26), and brown acid dyes remain of significance; the remaining nitro colorants and nitroso compounds, such as C.I. Mordant Green 4 (1.27) and its iron complex (C.I. Pigment Green 8), are however mainly of historical interest.

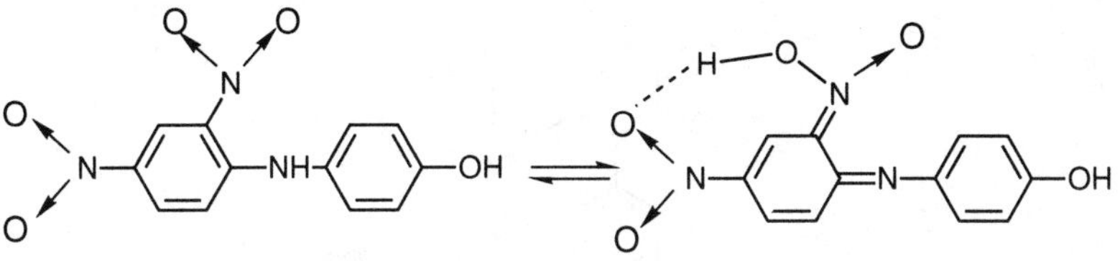

1.26

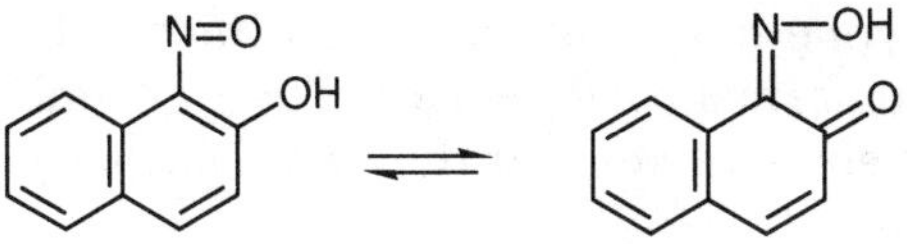

1.27

1.5.13 Diphenylmethane and triarylmethane colorants

Although treated as separate classes in the *Colour Index*, these structural types are closely related and the few diphenylmethane dyes like auramine (C.I. Basic Yellow 2) (1.28) are now of little practical interest. Commercial usage of the triarylmethane dyes and pigments has also declined considerably in favour of the major chemical classes. They were formerly noteworthy contributors to the acid, basic, mordant and solvent ranges, primarily in the violet, blue and green sectors. Numerous structural examples are given in the *Colour Index*; the terminal groupings can be amine/quinonimine, as in auramine and crystal violet (1.29; C.I. Basic Violet 3), hydroxy/quinone, or both. The aryl nuclei are not always benzenoid.

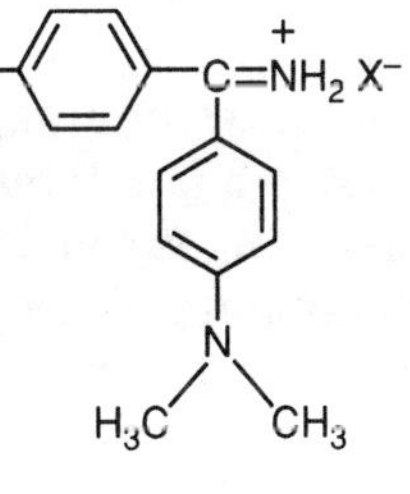

1.28

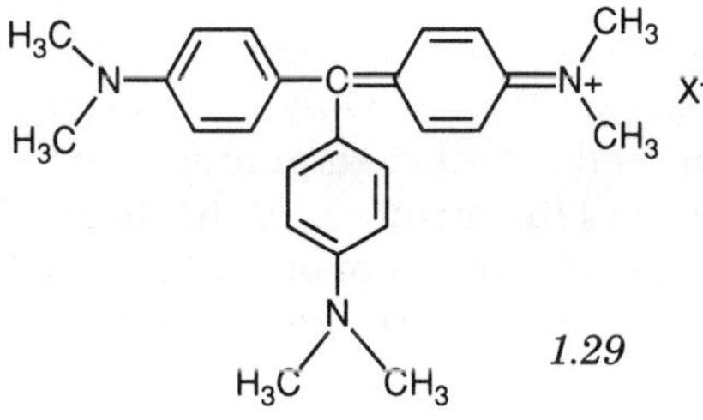

1.29

1.5.14 Xanthene colorants

Also structurally related to the triarylmethanes is the xanthene chromophore (1.30), in which two of the aryl nuclei are linked by oxygen to form a pyrone ring. Similar terminal groupings (amino, hydroxy, or both) are usually present. They have mainly contributed red members to the acid, basic, mordant and solvent ranges, but there has been a marked decline in their commercial importance.

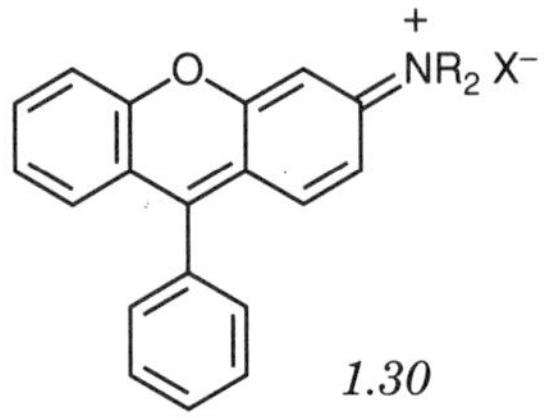

1.30

1.5.15 Acridine dyes

Acridine derivatives, such as acriflavine (1.31), can be regarded as relatives of the diphenylmethane class in which the two benzene nuclei are linked by nitrogen to form a pyridine ring. Their insignificance nowadays resembles that of their relatives, but they were formerly useful mainly as orange or yellow basic dyes [56].

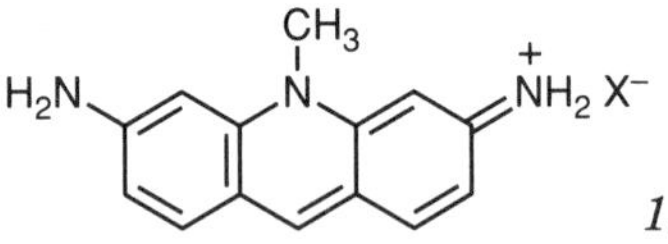

1.31

1.5.16 Azine, oxazine and thiazine colorants

These three classes are treated separately in the *Colour Index*, but it is useful to compare them in view of their structural similarity. Their chromophoric groupings differ only in the bridging link of the central pyrazine (1.32), oxazine (1.33) or thiazine (1.34) ring. Azine dyes from a wide variety of structural subclasses (quinoxalines, eurhodines, safranines, aposafranines, indulines, nigrosines) are illustrated in the *Colour Index*, but their commercial importance, mainly as red to blue basic dyes, blue acid dyes and blue or black solvent-soluble dyes, has declined. Thiazine dyes, such as methylene blue (1.34; C.I. Basic Blue 9), were never of much real significance. Only the oxazine class has retained its standing, not only for well-established products such as C.I. Basic Blue 3 (1.33) but also as bright blue members of the direct and reactive dye ranges containing the dioxazine chromophore (1.35) [57]. Bluish-violet pigments of exceptionally high tinctorial strength have also been derived from the dioxazine chromophore.

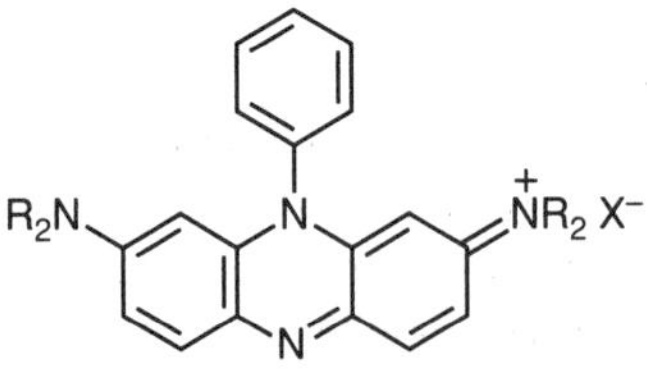

1.32

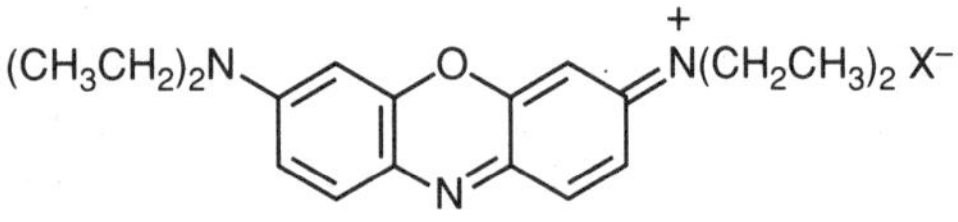

1.33

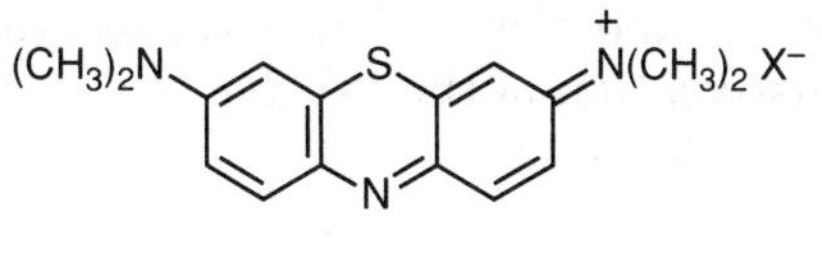

1.34

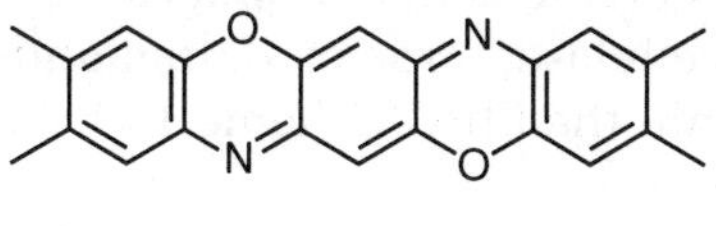

1.35

1.5.17 Indamine, indophenol and lactone dyes

These three chemical classes listed in the *Colour Index* are no longer of any practical significance, and are mentioned here only for their resemblance to quinonoid and ketone dyes already discussed. The chromophores are respectively quinonimine (1.36), with amino or hydroxy auxochromes in the indamines and indophenols, and a lactone ring (1.37) with a hydroxy auxochrome.

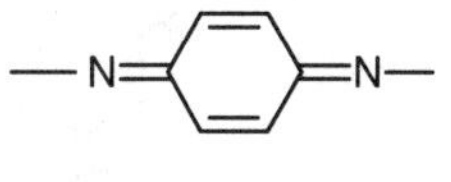

1.36

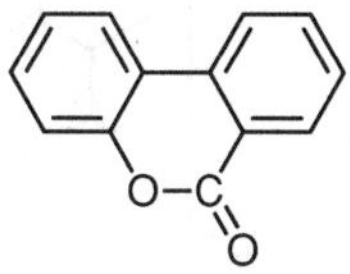

1.37

1.6 APPLICATION RANGES OF DYES AND PIGMENTS

As mentioned earlier, the nineteen application categories forming the respective series of C.I. generic names are not restricted to colorants. In the following discussion, attention is focussed mainly on the colorant ranges because this is where the major variations in chemical type are found. Thus it is necessary to depart from the alphabetical order of listing in the *Colour Index*. The application ranges of most interest for dyeing cellulosic fibres (vat, sulphur, reactive and direct dyes) are discussed first, followed by those for synthetic fibres and wool (disperse, basic, acid, mordant dyes) and finally the non-textile colorants (leather and food dyes, solvent dyes and pigments). The colorant precursors and uncoloured products listed form more homogeneous groups that are best discussed at length in the appropriate parts of Volumes 1 and 2. There is much variety of chemical structure among fluorescent brightening agents, but the characteristic classes represented are mostly different from those in colorants.

1.6.1 Vat dyes

A vat dye is a water-insoluble colorant containing two or more keto groups. It can thus be brought into aqueous solution by a reduction process (vatting), which converts the vat dye into its alkali-soluble enolic (leuco) form. As the soluble sodium enolate the leuco vat dye has substantivity for cellulose. The application of vat dyes to cellulosic fibres (virtually the only fibre type on which their outstanding fastness properties can be exploited) thus proceeds in four stages: (a) reduction and dissolution, (b) absorption by cellulose, (c) re-oxidation and (d) association of the vat dye molecules within the fibre (Scheme 1.1).

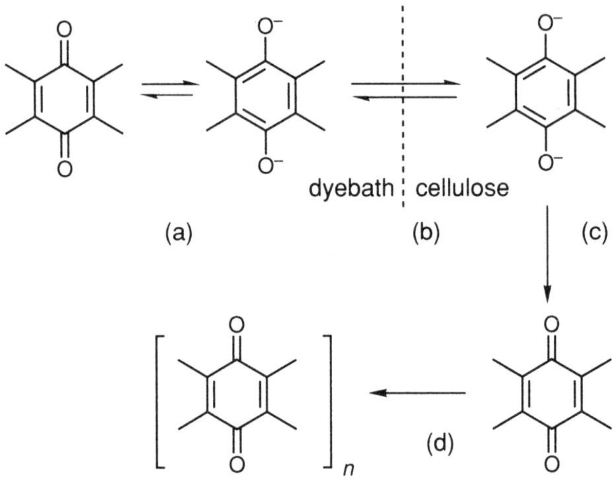

Scheme 1.1

Vat dyes are used to colour both fibre components in pale depths on polyester/cellulosic fibre blends but coloration of the polyester component in this case is more closely analogous to disperse dyeing (see section 1.6.5). Anthraquinone disperse dyes resemble the simpler anthraquinone vat dyes and in both instances it is exclusively the water-insoluble keto form that is absorbed by the polyester fibre. Further discussion of application methods and structural requirements for vat dyes follows in Chapter 3.

Approximately 80% of all vat dyes belong to the anthraquinonoid class (see Table 1.2) and this certainly includes the leading products in all sectors of the colour gamut. Indigo derivatives and the sulphurised vat types contribute mainly to the navy blue sector, whereas thioindigoids and indole-thianaphthenes occupy the red–brown region. In this and succeeding tables the figures again relate to all dyes of known chemical type listed and are given in percentage terms to permit easier comparison between the distributions in different hue sectors. The total number of dyes in each of the primary sectors (yellow, red or blue) is, of course, almost always larger than those in the non-primary sectors.

The vat dyes section of the *Colour Index* incorporates a subgroup called solubilised vat dyes. These are sodium salts of sulphuric acid esters of the parent leuco vat dyes. In contrast to the leuco compounds, the vat leuco esters dissolve readily in water at neutral pH. They have relatively low substantivity for cellulose and thus have been used mainly in continuous

TABLE 1.2

Percentage distribution of chemical classes in vat dye hue sectors

Chemical class	Distribution in hue sector/%								% of all vat dyes
	Yellow	Orange	Red	Violet	Blue	Green	Brown	Black	
Anthraquinonoid	94	92	65	67	62	90	83	90	79
Indigoid			11		17		2		5
Thioindigoid		4	15	14			4		4
Indole-thianaphthene			2	14	5		2	10	4
Aminoketone	3		5			5	9		3
Sulphurised vat	3				11	5			3
Miscellaneous		4	2	5	5				2

dyeing and printing. In the presence of an oxidant in mineral acid solution (sodium nitrite and sulphuric acid, for example) the leuco ester is rapidly decomposed and the insoluble vat dye regenerated. Thus application of a vat leuco ester represents a simpler (but more limited and costly) alternative to conventional dyeing methods via the alkaline leuco compound.

1.6.2 Sulphur dyes

These resemble the vat dyes in certain ways. Although indeterminate in constitution and usually mixtures of different chemical species (see section 6.4), they all contain the characteristic disulphide group in the insoluble form. A sulphur dye (D–S–S–D in Scheme1.2) can therefore be brought into aqueous solution by reduction to the alkali-soluble (leuco) form (D–S⁻). The soluble sodium thiolate form of the leuco sulphur dye has substantivity for cellulose. Thus the application of sulphur dyes to cellulosic fibres is a three-stage process broadly similar to that already outlined for vat dyes (Scheme 1.2).

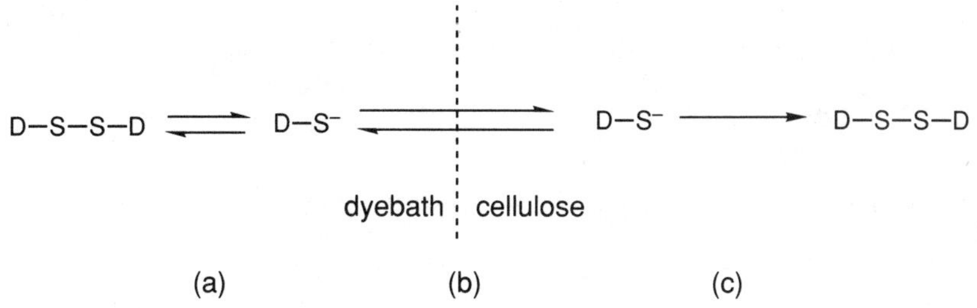

Scheme 1.2

In the *Colour Index* the sulphur dyes are subdivided into four subgroups. Conventional insoluble disulphide brands, taking the generic C.I. Sulphur dye designation, are converted to the leuco compound by adding sodium sulphide to a boiling aqueous dispersion of the parent dye. C.I. Leuco Sulphur dyes are marketed as concentrated solutions (or dispersions) of the pre-reduced dye with a small excess of reducing agent. They can be added directly to a dyebath prepared with additional reducing agent and are particularly convenient for continuous dyeing. C.I. Solubilised Sulphur dyes are sodium salts of thiosulphato derivatives of the parent dyes. Like the C.I. Solubilised Vat dyes they are highly soluble in water and low in substantivity for cellulose. Addition of alkali and reducing agent converts them to the leuco sulphur form. The *Colour Index* refers to a fourth subgroup, the C.I. Condense Sulphur dyes, but these are only of historical interest. They were water-soluble alkyl- or aryl-thiosulphato derivatives, mostly of phthalocyanine dyes. In the presence of an alkaline reducing agent (sodium sulphide or polysulphide) they undergo a polycondensation reaction to give an insoluble coloured polymer within the fibre, the chromophoric groupings being linked together via disulphide bridges as in conventional sulphur dyeings.

Apart from their extensive contribution to the dyeing of cellulosic fibres, sulphur dyes are of little further interest. They have been applied to a limited extent on leather capable of resisting alkali (oil-tanned leather, for instance) but only the water-soluble brands are suitable.

1.6.3 Reactive dyes

These dyes are capable of reacting chemically with a substrate under suitable application conditions to form a covalent dye–substrate bond. The characteristic structural feature is thus the possession of one or more reactive groupings of various kinds (see Chapter 7), the most important being vinyl sulphone and several halotriazine and halopyrimidine systems [22,23]. Unsaturated groups such as acrylamide or vinyl sulphone (1.38) react with cellulose, for example, by addition to the double bond, whereas the dyes with an activated halogen substituent (1.39; X = Cl or F) undergo a nucleophilic substitution reaction (Scheme 1.3).

$$\text{D–SO}_2\text{CH=CH}_2 + \text{HO–Cellulose} \longrightarrow \text{D–SO}_2\text{CH}_2\text{CH}_2\text{O–Cellulose}$$

1.38

$$\text{D–X} + {}^{-}\text{O–Cellulose} \longrightarrow \text{D–O–Cellulose} + \text{X}^{-}$$

1.39

Scheme 1.3

The introduction of this specific means of ensuring the fixation of the dye freed the colour chemist from many of the structural limitations implicit in the design of most other ranges of dyes, so that virtually any chemical class could be employed as a chromogen [58]. In practice, however, the reactive dye chemist has relied heavily on the azo chromophore (see Table 1.3), usually unmetallised but sometimes as metal complexes in the duller hues, such as navy, brown and black. Anthraquinone, dioxazine, formazan and phthalocyanine chromophores in the bright blue-green sector are the exceptions, but in every other hue sector azo dyes account for over 95% of reactive dye structures.

TABLE 1.3

Percentage distribution of chemical classes in reactive dye hue sectors

	Distribution in hue sector/%								% of all reactive
Chemical class	Yellow	Orange	Red	Violet	Blue	Green	Brown	Black	dyes
Unmetallised azo	97	90	90	63	20	16	57	42	66
Metal-complex azo	2	10	9	32	17	5	43	55	15
Anthraquinone				5	34	37		3	10
Phthalocyanine					27	42			8
Miscellaneous	1		1		2				1

By far the most important application segment for reactive dyes is the dyeing and printing of cellulosic fibres. Their impact in wool dyeing has been less dramatic but the exceptionally high wet fastness requirements of machine-washable knitwear, for example, do justify the use of reactive dyes, in spite of their higher cost relative to conventional dyes for wool. Reactive dyes are suitable for dyeing glove leather and similar materials where resistance to water and dry-cleaning solvents is essential. A limited range of disperse reactive dyes for nylon was marketed for many years but has become obsolete.

1.6.4 Direct dyes

These are defined as anionic dyes with substantivity for cellulosic fibres, normally applied from an aqueous dyebath containing an electrolyte. The forces that operate between a direct dye and cellulose include hydrogen bonding, dipolar forces and non-specific hydrophobic interaction, depending on the structural nature and polarity of the dye. Apparently multiple attachments are important, since linearity and planarity of molecular structure seem to be desirable features. The sorption process is reversible and numerous attempts have been made to minimise desorption by suitable aftertreatments (see Chapter 3). The two most significant non-textile outlets for direct dyes are the batchwise dyeing of leather and the continuous coloration of paper.

In every hue sector, at least 70% of direct dyes are unmetallised azo structures (see Table 1.4). In contrast to all other application ranges, the great majority of them are disazo or polyazo types, the former predominating in the brighter yellow to blue sectors and the latter in the duller greens, browns, greys and blacks. The copper-complex direct dyes are mostly duller violets, navies and blacks derived from the disazo and polyazo subclasses. Stilbene and thiazole dyes, the only non-azo contributors in the yellow to red sector, bear certain similarities to the disazo and polyazo structures (see sections 1.5.1 to 1.5.3). The dioxazine and phthalocyanine chromophores are represented in bright blue direct dyes of high light fastness.

TABLE 1.4

Percentage distribution of chemical classes in direct dye hue sectors

	Distribution in hue sector/%								% of all direct dyes
Chemical class	Yellow	Orange	Red	Violet	Blue	Green	Brown	Black	
Monoazo	4	7	14	4		8	3		5
Disazo	58	53	71	83	52	21	22	23	49
Polyazo	8	16	10	3	31	64	67	68	33
Copper-complex azo			3	9	12	4	1	8	5
Stilbene	17	23	2			1	7	1	5
Thiazole	13	1							1
Dioxazine				1	3				1
Phthalocyanine					2	2			1

1.6.5 Disperse dyes

These dyes have affinity for one or, usually, more types of hydrophobic fibre and they are normally applied from fine aqueous dispersion. Although pure disperse dyes have extremely low solubility in cold water, such dyes nevertheless do dissolve to a limited extent in aqueous surfactant solutions at typical dyeing temperatures. The fibre is believed to sorb dye from this dilute aqueous solution phase, which is continously replenished by progressive dissolution of particles in suspension. Alternatively, hydrophobic fibres can absorb disperse dyes from the vapour phase. This mechanism is the basis of many continuous dyeing and printing methods of application of these dyes. The requirements and limitations of disperse dyes on cellulose acetate, triacetate, polyester, nylon and other synthetic fibres will be discussed more fully in Chapter 3. Similar products have been employed in the surface coloration of certain thermoplastics, including cellulose acetate, poly(methyl methacrylate) and polystyrene.

Yellow disperse dyes are known from a wide range of chemical classes (see Table 1.5) but azo dyes derived from heterocyclic coupling components have gained in importance in recent years. The monoazo subclass dominates throughout the disperse dye range and there are no metal-complex disperse dyes for hydrophobic fibres, although they do exist as a subgroup of 'acid dyes' for wool. Anthraquinone derivatives have traditionally dominated the bright red to bright green series but here again cost considerations have favoured the introduction of new monoazo dyes, often with heterocyclic amines as diazo components, for the dyeing of polyester fibres [15]. Browns and blacks on polyester and acetate fibres have always been provided almost entirely by azo disperse dyes. The disperse range contains relatively few homogeneous green, brown or black dyes, however; many of the commercial navy, green, brown and black products are in fact mixtures of two or more components (see section 3.2.3).

TABLE 1.5

Percentage distribution of chemical classes in disperse dye hue sectors

	Distribution in hue sector/%								% of all disperse dyes
Chemical class	Yellow	Orange	Red	Violet	Blue	Green	Brown	Black	
Azo	48	92	73	47	27	30	100	100	59
Anthraquinone	6	2	25	53	72	65			32
Nitro	16	3							3
Aminoketone	8	2	1		1	5			2
Methine	14								2
Quinophthalone	4								1
Miscellaneous	4	1	1						1

1.6.6 Basic dyes

There has been recurring controversy regarding the apparent synonymity of the terms 'basic dye' and 'cationic dye' [59]. The Society of Dyers and Colourists defines a basic dye as 'characterised by its substantivity for the acidic types of acrylic fibres and for tannin-mordanted cotton', whereas a cationic dye is defined as one 'that dissociates in aqueous solution to give a positively charged coloured ion'. Since the second of these describes a chemical feature rather than an application property, the term 'basic dyes' will be adopted in this book for this application range, as in the *Colour Index*. This category includes those weakly basic dyes (1.40) that are essentially uncharged under neutral or alkaline conditions and only form a cation by protonation in acidic solution, such as chrysoidine (C.I. Basic Orange 2). The powerful dye–fibre forces responsible for the high wet fastness of basic dyes on acrylic fibres are believed to include electrostatic attraction to anionic sites in the fibre and non-specific hydrophobic interaction.

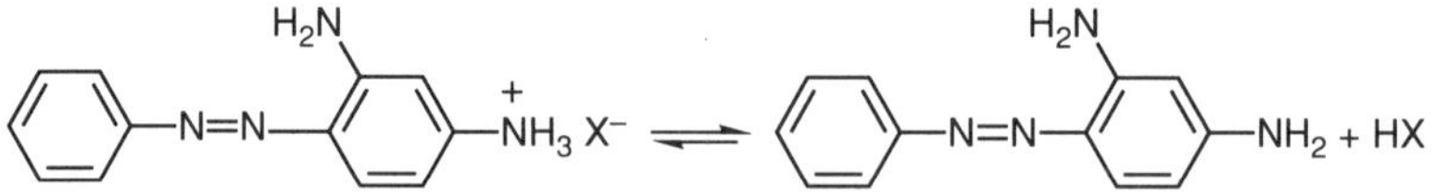

1.40

Many brilliantly coloured and tinctorially strong basic dyes for silk and tannin-mordanted cotton were developed in the early years of the synthetic dye industry. Most of these belonged to the acridine, azine, oxazine, triarylmethane, xanthene and related chemical classes; their molecules are usually characterised by a non-localised positive charge. Thus in crystal violet (1.29) the cationic charge is shared between the three equivalent methylated *p*-amino nitrogen atoms. A few of these 'traditional basic' dyes are still of some interest in the dyeing of acrylic fibres, notably as components of cheap mixture navies and blacks, but many new 'modified basic' dyes have been introduced since the 1950s for acrylic and modacrylic fibres, as well as for suitable variants of nylon and polyester.

Most of these newer products are azo or anthraquinone types, often with a localised quaternary ammonium group isolated from the chromophoric grouping by a saturated aliphatic chain, as in C.I. Basic Red 18 (1.41). Such products often exhibit better light fastness than the traditional delocalised types. Improved azomethine, methine and polymethine basic dyes of good light fastness have also been developed. In contrast to the more specialised traditional classes, the azo and methine dyes have contributed to the basic dye range across the entire spectrum of hues (see Table 1.6) and now account for a clear majority of all basic dyes listed in the *Colour Index*.

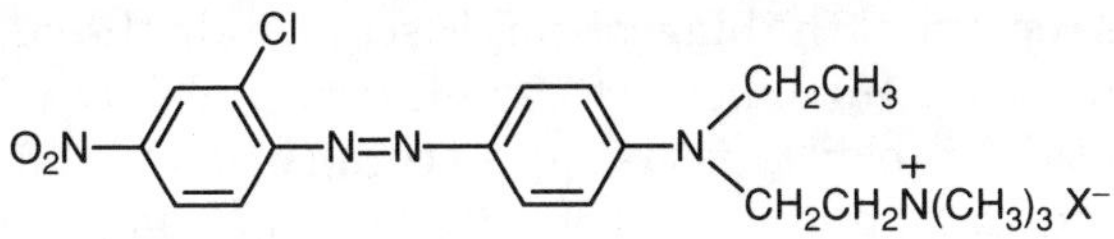

1.41

TABLE 1.6

Percentage distribution of chemical classes in basic dye hue sectors

	Distribution in hue sector/%								% of all basic
Chemical class	Yellow	Orange	Red	Violet	Blue	Green	Brown	Black	dyes
Azo	33	51	55	38	34	12	100	60	43
Methine	41	13	23	21	2	10			17
Triarylmethane			5	18	23	56			11
Acridine	12	36							7
Anthraquinone				5	15				5
Azine			6	13	5			20	5
Oxazine					11			20	3
Xanthene			11	5					3
Miscellaneous	14				10	22			6

Traditional basic dyes found a niche in the past as bright colours on certain types of leather, but as on cotton this usage is now insignificant. They are still of interest for colouring paper made from unbleached and mechanical pulps, however, where light fastness is of little consequence.

1.6.7 Acid dyes

These are defined as anionic dyes characterised by substantivity for protein fibres. Levelling acid dyes and 1:1 metal-complex dyes are normally applied to wool from strongly acidic solutions, but the 1:2 metal complexes and many milling acid dyes have considerable substantivity even from neutral dyebaths (see Chapter 3). Wool, silk and nylon contain basic groups and the uptake of levelling acid dyes by nylon at acidic pH can usually be related to the amine end-group content of the fibre. Under neutral dyeing conditions, however, non-specific hydrophobic interaction makes a considerable contribution, reinforcing the electrostatic bonding between such fibres and acid dyes of higher wet fastness.

Azo acid dyes represent the biggest single segment in the current *Colour Index*, contining about 1600 distinct chemical entities [44]. The share of monoazo 1:2 metal-complex types is approaching 600 and rising. Monoazo types are supreme in all hue sectors of the azo acid range except unmetallised browns and blacks, where disazo structures are more numerous. Anthraquinone and triarylmethane acid dyes have traditionally provided most of the brilliant violet to green members of this range but demand for them has declined, especially the triarylmethane types, as bright azo alternatives have appeared. The significance of nitro yellow-browns, xanthene red-violets and phthalocyanine blues has always been marginal in the acid dye range.

TABLE 1.7

Percentage distribution of chemical classes in acid dye hue sectors

Chemical class	Distribution in hue sector/%								% of all acid dyes
	Yellow	Orange	Red	Violet	Blue	Green	Brown	Black	
Unmetallised azo	65	55	64	22	17	15	79	47	48
Metal-complex azo	31	42	29	44	21	39	13	46	31
Anthraquinone			2	18	36	22	3	3	10
Triarylmethane				12	16	16			5
Xanthene	1	1	4	3					2
Azine		1	1	1	4			1	1
Phthalocyanine					5	1			1
Nitro	1	1					5		1
Miscellaneous	2				1	7		3	1

The first acid dye, Orange I (1.42; Acid Orange 20), was discovered in 1876. All but a handful of the acid dyes developed since then were evaluated initially with wool dyeing in mind. In terms of adaptability to the coloration of other substrates, however, acid dyes have proved pre-eminent. This is the main reason for their number and variety. As well as the dyeing and printing of nylon and protein fibres, acid dyes are important for the coloration of leather, paper, jute, wood and anodised aluminium. Most of the permitted dyes for food and cosmetics were originally marketed long ago for the dyeing of wool. Some solvent-soluble dyes are formed by co-precipitation of acid dyes and basic dyes. Metal toners are concentrated pigments prepared by precipitating anionic dyes, often of the unmetallised azo acid type, as their water-insoluble metal salts (see section 2.2.1).

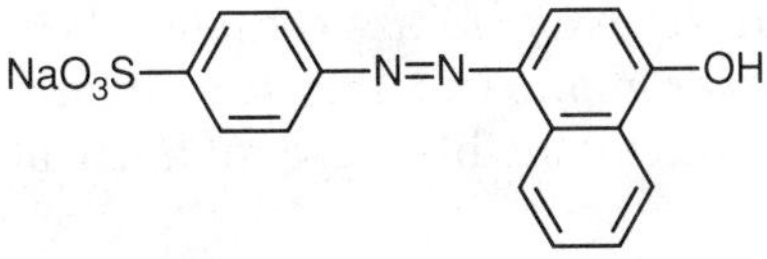

1.42

1.6.8 Mordant dyes

The somewhat ambiguous term 'mordant dye' is defined as a dye that is fixed with a 'mordant'. A mordant is itself defined as 'a substance, usually a metallic compound, applied to a substrate to form a complex with a dye which is retained by the substrate more firmly than the dye itself'. Unfortunately, the mordant dyes category in the *Colour Index* follows the more restricted convention of excluding both chelatable direct dyes capable of complexing with copper salts and also those basic dyes that were formerly applied to tannin-mordanted cotton, both of which subgroups conform to the above less than precise definition. A more logical generic name for those chelatable anionic dyes capable of being fixed to wool by aftertreatment with a dichromate would be the more popular term 'chrome dyes' or 'chrome mordant dyes', but the designation C.I. Mordant dye is well entrenched. Since the range is in slow commercial decline in favour of premetallised acid dyes, the ambiguity is of decreasing significance too.

In all hue sectors except violet and blue, where triarylmethane, anthraquinone and oxazine dyes make notable contributions, 80% or more of chrome dyes are of the monoazo type, with *o,o*′-dihydroxyazo substituents (1.43). Although most chrome dyes are only of interest for wool dyeing [60], a minority are applicable to most of those other substrates where acid dyes are found useful.

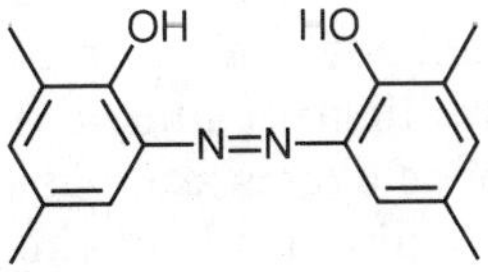

1.43

Natural dyes and pigments occupy a separate section of the *Colour Index* but most of them are mordant dyes in the broader sense defined above [61,62]. They have virtually no commercial significance in the coloration industries nowadays, except for indigo in its synthetic form.

1.6.9 Leather dyes

The application range designated by this generic name in the *Colour Index* incorporates those acid, direct and mordant dyes with substantivity for leather and satisfactory fastness on that substrate [28,63]. It is a commercially important sector, the number of products listed being exceeded only by the complete acid and direct ranges. As expected from the sources of this selection, about 85% of leather dyes are azo compounds (35% disazo, 30% monoazo, 20% metal-complex monoazo) and the remainder are mainly yellow-orange stilbenes and violet-blue-green anthraquinone or triarylmethane types.

1.6.10 Food dyes

Freedom from toxicity is clearly the primary criterion for dyes used in foods and cosmetics, followed by high solubility and chemical stability in the appropriate medium of incorporation. The possibility that they may affect consumers adversely is causing increasing concern [64–66]. Legislation over many years has increasingly restricted the usage of synthetic colorants to certain ‘permitted’ products that have shown no harmful effects when rigorously tested. There is as yet no agreed international list of permitted food colours, however. Only about forty long-established colorants of relatively simple constitution, mostly monoazo yellows and reds, appear on the permitted lists of the more legislatively critical countries. Triphenylmethanes and other chemical classes provide the few violets, blues and greens, whereas browns and blacks are disazo types. The natural carotenoid dyes are of outstanding importance for colouring edible fats and oils.

1.6.11 Solvent dyes

Solvent dyes are characterised by their solubility in one or more organic solvents. Numerous media are of interest in practice, the most significant being alcohols, ethers, esters, ketones, chlorinated solvents, hydrocarbons, oils, fats and waxes. The many technical applications of solvent-soluble dyes are discussed in section 2.9. More than half the solvent dyes listed in the *Colour Index* are unsulphonated azo compounds (27% monoazo, 18% monoazo metal-complex, 9% disazo) and a further 7% are complexes formed by co-precipitation of acid dyes and basic dyes, mostly yellows and reds. Unsulphonated anthraquinones lead the field in the bright violet-blue-green sectors (17%). Xanthene reds (5%) and triarylmethane blues (5%) are the other noteworthy segments represented, together with a miscellany (12%) drawn from almost all other chemical classes.

1.6.12 Pigments

In contrast to solvent dyes, pigments are substances in particulate form. They are essentially insoluble in the media into which they are incorporated, and are mechanically dispersed therein in order to modify the colour and/or light-scattering properties of such media. The complete range of pigments may be conveniently divided into five subgroups on the basis of their chemical constitution and mode of preparation for use:

(a) *Organic pigments* (approximately 60%): concentrated non-ionic colorants from various chemical classes, including about 40% monoazo and disazo compounds (yellow-orange-brown-red sectors) and 10% anthraquinones (mostly red-violet-blue).
(b) *Inorganic pigments* (approximately 20%): coloured insoluble materials from inorganic sources; they include coloured metal salts such as C.I. Pigment Yellow 34, complex salts such as C.I. Pigment Blue 27, metal oxides such as C.I. Pigment White 6 and elemental carbon such as C.I. Pigment Black 6.
(c) *Metal toners* (approximately 15%): concentrated anionic monoazo dyes (mainly yellows, oranges and especially reds) precipitated as their insoluble metal salts, typically those of barium, aluminium and calcium.
(d) *Pigment lakes* (approximately 5%): insoluble metal salts of acid or mordant dyes, or basic dyes rendered insoluble by complexing with copper(II) hexacyanoferrate, phosphomolybdic acid or phosphotungstic acid, precipitated in the presence of an inert substrate such as alumina, the latter being a necessary and integral component of the product.
(e) *Extended pigments*: concentrated pigments or toners diluted with an extender such as alumina or calcium carbonate, which does not form an integral component of the product.

Metal toners and pigment lakes are somewhat unstable under alkaline conditions owing to partial dissociation of the precipitated complex. Lakes and extended pigments are usually tinctorially weaker than concentrated pigments or toners because of the presence of the substrate components. These limitations tend to restrict the versatility of the lesser subgroups, especially lakes, which have steadily declined in importance. The chemistry and innumerable practical applications of pigments are discussed in detail in Chapter 2.

1.6.13 Azoic components and compositions

This is the most important of the precursor ranges of products classified in the *Colour Index*. Azoic dyes have a close formal relationship to azo pigments and some are chemically identical with them, although they are used in a totally different way. Azoic components are applied to produce insoluble azo dyes within the textile substrate (almost always a cellulosic fibre, and nowadays normally cotton). Corresponding azoic components for the dyeing of cellulose acetate, triacetate and polyester fibres were once commercially important [67], but are now obsolescent because of environmental hazards and the time-consuming application procedure.

In dyeing with azoic dyes, an insoluble azo compound is produced within the fibre by the coupling of an azoic diazo component (a diazotised arylamine) with an azoic coupling component (usually nowadays an arylide of BON acid or a related member of the Naphtol AS (HOE) series). Azoic dyes provide only relatively dull violets and navy blues, costly greens, and browns or blacks of limited versatility, but are tinctorially strong, bright and economical in the orange-scarlet-red sectors.

The two types of azoic component form separate generic series in the *Colour Index*. There is also a third series, designated C.I. Azoic compositions. Now obsolescent, these prepared mixtures of an azoic coupling component and a stabilised diazo component were intended mainly for textile printing. The stabilisation prevented coupling until the printed fabric was steamed under appropriate conditions to regenerate the active diazonium salt and promote formation of the azoic dye.

1.6.14 Developers

This short range of simple coupling components, comprising phenols, naphthols and 1-phenyl-3-methyl-5-pyrazolone, is seldom used nowadays. Developers were once used as aftertreatments for certain direct dyes for cellulosic fibres and disperse dyes for cellulose acetate, each containing a primary arylamine substituent capable of diazotisation, to enhance their normally low wet fastness properties. The application technique was prolonged, however, and the improvement in performance was disappointingly slight. The usage of these developers has now fallen to negligible levels.

1.6.15 Oxidation bases

Like azoic components, these are precursors of a distinct chemical class as well as an application category, but are too restricted in attainable hues to be described as a 'range'. This cheap but unpleasant method of continuous coloration, once popular for cotton printing in visually attractive illuminated resist styles under aniline black or paramine brown colours, has been superseded by environmentally more acceptable and controllable alternative processes. Oxidation dyeing was simply the *in situ* formation of a polymeric black or dark brown azine colorant of indeterminate constitution by acidic oxidation of a simple arylamine such as aniline or *p*-phenylenediamine on the fibre, using a chlorate or dichromate as oxidant and a copper salt catalyst. Although no longer significant in textile dyeing and printing, coloration techniques of a similar kind have remained the primary method of colouring human hair [68], feathers and natural furs.

1.6.16 Ingrain dyes

This is another somewhat ambiguous category in the context of the *Colour Index* classification. An 'ingrain' dyeing process is defined as the formation of a colorant '*in situ* in the substrate by the development and coupling of one or more intermediate compounds'. This imprecise description clearly embraces azoic dyeing and the application of oxidation bases, but the generic term 'C.I. Ingrain dye' is limited to a third small group of now obsolete colorant precursors, again too restricted in hue to be regarded as a 'range'. They were marketed for the textile printing of cellulosic fabrics under the trade names Alcian (ICI) and Phthalogen (BAY); both types resulted in the insolubilisation of copper phthalocyanine or related pigments within the fibre, although the respective techniques of synthesis and application differed considerably. Reactive phthalocyanine dyes have superseded these early approaches to the attainment of fast bright blues on cellulosic fibres [69].

1.6.17 Fluorescent brightening agents

These essentially colourless compounds have been defined as substances 'that, when added to an uncoloured or a coloured substrate, increase the reflectance of the substrate in the visible region by converting u.v. radiation into visible light and so increase the whiteness or brightness of the substrate'. The air of uncertainty in the final clause of this definition reflects the general indecision among practitioners in this field about whether to call them 'fluorescent brightening agents' (or FBAs), as in the *Colour Index*, or 'fluorescent whiteners' (FWs).

Since their commercial introduction during the 1940s as components of proprietary detergents and laundry preparations, these products have also found usage in the whitening of paper and textile materials. Disperse FBAs are available for whitening hydrophobic fibres, and solvent-soluble FBAs impart fluorescence to oils, paints, varnishes and waxes. Approximately 75% of commercially established FBAs are stilbene derivatives with inherent substantivity for paper and cellulosic textiles, but the remainder come from about twenty different chemical classes. These include aminocoumarins (6%), naphthalimides (3%), pyrazoles (2%), acenaphthenes, benzidine sulphones, pyrazolines, stilbene-naphthotriazoles, thiazoles and xanthenes (each about 1%). Further details are given in Volume 2.

1.6.18 Reducing agents

As noted earlier, the specific inclusion in the *Colour Index* of a group of textile auxiliaries under this generic name seems anomalous, if not aberrant. The justification claimed is that they are indispensable for the application of vat dyes and in discharge printing or stripping processes. A similar level of interdependency could be claimed for certain other classes of auxiliaries, as will be discussed in Volume 2. One has only to visualise textile printing without thickeners or emulsifying agents to realise this. A comprehensive coloration auxiliaries index, comparable in scope and presentation to the *Colour Index*, is long overdue. The way in which reducing agents are dealt with already in the latter publication would form a reasonable basis for a reference work of considerable potential value to commercial and technical activities within the coloration industries. The laudable efforts of the regulatory authorities to register and monitor the trade and use of chemical products in the interests of environmental and industrial safety would be assisted by the existence of a sister work to the *Colour Index* in the field of coloration auxiliaries or, better still, the broader category of speciality chemicals for textile treatment.

REFERENCES

1. F M Rowe, *The development of the chemistry of commercial synthetic dyes (1856–1938)* (London: Royal Inst. of Chem., 1938).
2. W H Cliffe, *J.S.D.C.*, **72** (1956) 563; **73** (1957) 312.
3. L E Morris, *Dyer*, **114** (1956) 770.
4. J Boulton, *J.S.D.C.*, **73** (1957) 81.
5. F M Rowe, *J.S.D.C.*, **46** (1930) 121.
6. J Bruce, *J.S.D.C.*, **57** (1941) 305.
7. K H Saunders, *J.S.D.C.*, **60** (1944) 81.
8. C M Whittaker, *J.S.D.C.*, **61** (1945) 201; **72** (1956) 557.
9. W Bradley, *Recent progress in the chemistry of dyes and pigments* (London: Royal Inst. of Chem., 1958).
10. R D Welham, *J.S.D.C.*, **79** (1963) 98, 146, 181, 229.
11. J Shore, *Textilveredlung*, **12** (1986) 207.
12. R A Peel, *J.S.D.C.*, **68** (1952) 496.
13. W H Cliffe, *J.S.D.C.*, **79** (1963) 353.
14. B Campbell, *J.S.D.C.*, **82** (1966) 303.
15. J F Dawson, *Rev. Prog. Coloration*, **3** (1972) 18; **9** (1978) 25; **14** (1984) 90; *J.S.D.C.*, **99** (1983) 183.
16. R K Fourness, *J.S.D.C.*, **72** (1956) 513; **90** (1974) 15; *Rev. Prog. Coloration*, **10** (1979) 61.
17. A N Derbyshire, *J.S.D.C.*, **90** (1974) 273.
18. U Mayer and E Siepmann, *Rev. Prog. Coloration*, **5** (1973) 65.
19. R Raue, *Rev. Prog. Coloration*, **14** (1984) 187.
20. P Ackroyd. *Rev. Prog. Coloration*, **5** (1973) 86.
21. M Ahmed, *Coloring of plastics: theory and practice* (New York: Van Nostrand-Reinhold, 1979).
22. F W Beech, *Fibre-reactive dyes* (London: Logos Press, 1970).
23. *The chemistry of synthetic dyes*, Vol. 6, Ed. K Venkataraman (New York and London: Academic Press, 1972).
24. P Rosenthal, *Rev. Prog. Coloration*, **7** (1976) 23.
25. D M Lewis, *Rev. Prog. Coloration*, **8** (1977) 10.
26. F R Alsberg, *Rev. Prog. Coloration*, **12** (1982) 66.
27. I D Rattee, *Rev. Prog. Coloration*, **14** (1984) 50.
28. K Eitel, E Rau and J Westphal, *Rev. Prog. Coloration*, **14** (1984) 119.
29. D A W Adams, *J.S.D.C.*, **75** (1959) 22.
30. A K Sarkar, *Fluorescent whitening agents* (Watford: Merrow Publishing, 1971).
31. D Barton and H Davidson, *Rev. Prog. Coloration*, **5** (1973) 3.
32. R Williamson, *Fluorescent brightening agents* (Amsterdam and New York: Elsevier, 1980).
33. P Gregory and P Gordon, *Organic chemistry in colour* (Berlin and New York: Springer-Verlag, 1983).
34. *Developments in the chemistry and technology of organic dyes*, Ed. J Griffiths (London: Society of Chemical Industry, 1984).

35. R W Horobin, *J.S.D.C.*, **91** (1975) 4; *Rev. Prog. Coloration*, **11** (1981) 101.
36. B C Burdett, *J.S.D.C.*, **98** (1982) 114.
37. J W Slater, *Manual of colours and dyewares* (London: 1870).
38. R Benedikt and E Knecht, *Chemistry of coal-tar products* (1886).
39. W R Richardson, *J.S.D.C.*, **1** (1885) 251.
40. G Schultz and P Julius, *Tabular survey of synthetic organic dyes*, 1st Edn (1888).
41. A Lehmann, *Schultz's Farbstofftabellen*, 7th Edn (1931–32).
42. F Beffa and E Steiner, *Rev. Prog. Coloration*, **4** (1972) 60.
43. F Beffa and G Back, *Rev. Prog. Coloration*, **14** (1984) 33.
44. N Ward, Private communication.
45 R Löwenfeld and K Opitz, *Melliand Textilber.*, **61** (1980) 160.
46. A Schmid, *Textilveredlung*, **11** (1985) 341.
47. E Coates, *J.S.D.C.*, **83** (1967) 95.
48. J Griffiths, *Colour and constitution of organic molecules* (London: Academic Press, 1976)
49. J Fabian and H Hartmann, *Light absorption of organic colorants* (Berlin and New York: Springer-Verlag, 1980).
50. J Griffiths, *Rev. Prog. Coloration*, **11** (1981) 37; *J.S.D.C.*, **14** (1984) 21.
51. H Kuhn, *Angew. Chem.*, **71** (1959) 93.
52. H Zollinger, *Azo and diazo chemistry* (London: Interscience, 1961).
53. K H Saunders and R L M Allen, *Aromatic diazo compounds*, 3rd Edn (London: Edward Arnold, 1985).
54. F H Moser and A L Thomas, *Phthalocyanine compounds* (New York: Reinhold, 1963).
55. F M Hamer, *Chemistry of heterocyclic compounds*, Vol. 18 (cyanine dyes and related compounds) (New York and London: Interscience, 1964).
56. A Albert, *The acridines*, 2nd Edn (London: Edward Arnold, 1967).
57. A H M Renfrew, *Rev. Prog. Coloration*, **15** (1985) 15.
58. I D Rattee, *J.S.D.C.*, **85** (1969) 23.
59. H Zollinger, *J.S.D.C.*, **88** (1972) 447.
60. F R Hartley, *J.S.D.C.*, **85** (1969) 66.
61. *Chemistry and biochemistry of plant pigments*, Ed. T W Goodwin (London: Academic Press, 1965).
62. R H Thomson, *Naturally occurring quinones* (London: Academic Press, 1971).
63. W R Dyson and A W Landmann, *Rev. Prog. Coloration*, **4** (1972) 51.
64. H E Nursten, *Rev. Prog. Coloration*, **4** (1972) 7.
65. D F Anstead, *Rev. Prog. Coloration*, **4** (1972) 13.
66. *Developments in food colours*, Ed. J Walford (London: Elsevier; Vol. 1, 1980; Vol. 2, 1984).
67 *The dyeing of synthetic-polymer and acetate fibres*, Ed. D M Nunn (Bradford: SDC, 1979).
68. J F Corbett, *Rev. Prog. Coloration*, **4** (1973) 3; **15** (1985) 52.
69. J Shore, *Dyer*, **167** (June 1982) 28.

CHAPTER 2

Organic and inorganic pigments; solvent dyes

David Patterson

2.1 ORGANIC PIGMENTS AS COLORANTS

2.1.1 General introduction

The term 'colorants' includes both dyes and pigments, and it is necessary at the beginning of any account of pigments to understand the difference between the two types of colorant. Pigments are insoluble in the materials that they are used to colour, whereas most dyes are soluble in them. Furthermore, pigments and solvent dyes are insoluble in water, while other dyes are to a greater or lesser extent water-soluble. Vat dyes are also water-insoluble and can be used as pigments. They are thought of as dyes only because they are first made water-soluble by reduction. In this form they diffuse into suitable fibres in which they are then rendered insoluble again by oxidation.

A further, and most important, difference between pigments and dyes is that pigments are used as colorants in the physical form in which they are manufactured. This is in marked contrast to dyes, which are first dissolved in water to form a dyebath, and which diffuse as single molecules from these solutions into textile fibres to colour them. Even vat dyes go through this process, as mentioned above. While the physical form of dyes is not without its importance, and becomes increasingly so for the less soluble dye classes, it does not dominate the required properties as does that of pigments.

Because of the requirement of insolubility in water, solvents and the medium that the pigment is being used to colour, coloration with pigments is entirely different from the application of dyes. It is a process of dispersion instead of dyeing, with all that this implies. Both organic and inorganic compounds can be used as pigments. But the media that they are used to colour are (in decreasing order of importance) paints, plastics and printing inks [1], all of which are organic. Since 'like dissolves in like' is a wide general rule, in certain instances the solubility of organic (covalent) pigments in organic (covalent) media gives rise to considerable technical problems.

The chemical nature of many organic pigments is closely similar to that of dyes; except for the phthalocyanines, every chemical class of pigments found its first application as dyes. Since there are no inorganic dyes in the strict sense of the word, the chemistry of inorganic pigments has no parallel in the field of dyes.

Finally, passing mention must be made of the two most important organic pigments in the world: chlorophyll and haemoglobin. They are vital in the literal sense of that much over-used word but have no use as colorants for manufactured products, except for the relatively minor use of chlorophyll to colour food.

2.1.2 Pigment history

The discoveries of archaeologists indicate that the first pigments used by humans were the earth pigments. These coloured minerals are widely distributed throughout the world and are very resistant to decomposition by all weathering processes and light; indeed, without these properties they would have disappeared long ago. When found in the form of lumps, earth pigments were simply ground to a powder and then mixed into fatty materials to form rather crude paints. Examples of such materials are the bright red pigment vermilion (mercury(II) sulphide, HgS), yellow orpiment (arsenic trisulphide, As_2S_3), green malachite (basic copper(II) carbonate, $CuCO_3.Cu(OH)_2$), blue lapis lazuli (natural ultramarine), white chalk (calcium carbonate, $CaCO_3$) and black charcoal (carbon from many sources). The *Colour Index* gives some more examples, but the best sources of information about these materials are artists' books such as that by Gettens and Stout [2].

For many centuries, pigments have been derived from the colouring matters found occurring naturally in many plants and even in some animals. Examples are the red pigment madder and the blue indigo, both extracted from plants, cochineal and lac lake both from insects, the much-prized Tyrian purple derived from certain shellfish, logwood extracts ranging in colour from red to brown and black depending on the precipitant, and finally sepia obtained from cuttle fish [3].

The methods used for making pigments from these and other natural dyes were more like recipes than scientific procedures and were probably derived from the work of alchemists and herbalists. The former spent their lives trying to prepare gold by dissolving all kinds of cheaper materials in acids and then re-precipitating them (hence their discovery of many precipitants), while the latter sought to extract compounds of medicinal value from plants and some of their extracts must have included natural dyes.

Among the precipitants employed were tannic acid, tartar emetic, rosin soaps, fatty acid (stearic and oleic) soaps, sulphonated oils (Turkey red oil), 'earth lakes' (mixed natural silicates), phosphates, casein and arsenious acid. The fastness properties of these pigments were relatively poor, both to water and to light. Furthermore, the colour of the pigments made by these methods largely depended on the particular precipitant or combination of precipitants used, a fact also exploited by mediaeval dyers to extend the range of colours that they could produce from the very limited number of natural dyes available to them. Many of the methods for making these pigments [4] are of historical interest only, except a few which are mentioned later.

2.2 DYES CONVERTED INTO PIGMENTS

2.2.1 Dyes made insoluble by precipitation on substrates

With the introduction of synthetic dyes, attempts were made to prepare pigments from them by methods based on those that had been used with the natural dyes. Very many azo dyes can be made insoluble by precipitating them as the salts of heavy metals in the presence of so-called alumina hydrate. This method will be treated in detail later and here only the making of pigments from acid dyes and basic dyes will be mentioned. The use of alumina hydrate as a basic substrate for making pigments from dyes is now, however, of diminishing commercial importance.

Alumina hydrate is made by slowly adding a 10% solution of sodium carbonate to a well-stirred 10% solution of aluminium sulphate, thus producing a white gelatinous precipitate of aluminium hydroxide. It is a gross over-simplification to regard alumina hydrate simply as $Al(OH)_3$, however. It is colloidal in nature and its composition is dependent on such factors as the concentrations and proportions of solutions used, the temperature and the speed of agitation during the process. Even prolonged washing of the precipitate does not remove all the sulphate ions from alumina hydrate and it is necessary to subject it to ignition to remove the two or three water molecules loosely bound in each alumina hydrate micelle. By careful control of these variables, however, a consistent product can be prepared with the required absorption power for precipitated dyes. The physical chemistry was established many years ago [5].

Although there is a current decline in the demand for pigments of this kind, mainly for the printing ink industry, twenty years ago large amounts (400 tonnes annually in USA alone) of methyl violet (C.I. Basic Violet 1) were precipitated with tannic acid and antimony salts as C.I. Pigment Violet 3, for use in copying paper and typewriter ribbons. Erioglaucine (2.1; C.I. Acid Blue 9) was precipitated as its barium salt on alumina hydrate to make Peacock Blue Lake (C.I. Pigment Blue 24) for use as a transparent trichromatic blue in printing inks. Eosine (2.2; C.I. Acid Red 87) was similarly precipitated as its barium or aluminium salt on alumina hydrate

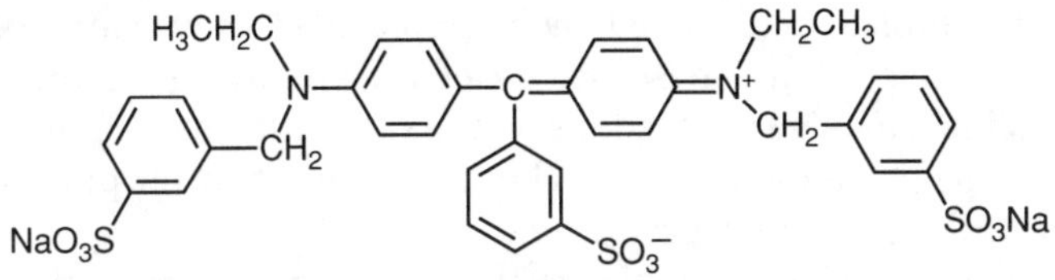

2.1

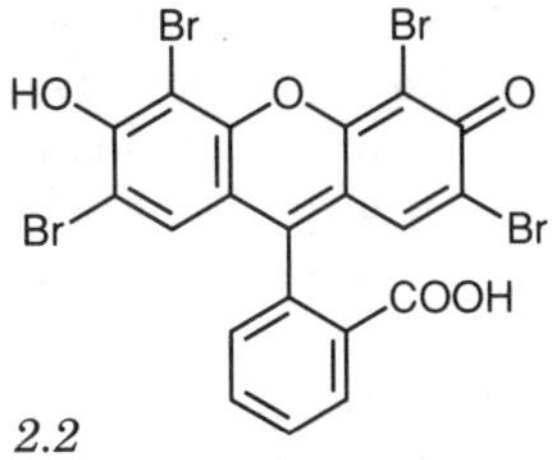

2.2

to give Bronze Red (C.I. Pigment Red 90), which as its name implies has an attractive yellowish bronzing effect in printing inks.

This process of making dyes into pigments was called 'laking' and the pigments so made were called 'lakes'. Owing to changes in both the printing industry and the textile industry the use of these dyes has greatly diminished. Printers no longer make their own inks, and the demand from dyers for the particular acid dyes and basic dyes used in making lakes has almost disappeared.

2.2.2 Dyes made water-insoluble by precipitation with complex acids

In the early years of this century efforts to improve the properties of pigments prepared from basic dyes took a fresh direction: precipitation of the pigments using the mordants that were employed when the same substances were being used as dyes. The light fastness of the resulting pigments remained poor, however.

Another approach was much more successful. It had been noted in biochemical work that some amines could be precipitated by treatment with acid in the presence of sodium phosphate and sodium tungstate or molybdate. When these precipitants were tried on basic dyes, pigments were produced with a higher light fastness than that of the parent dyes. This important discovery was patented by the Bayer company in Germany just before the First World War, but the shortage of tungsten and molybdenum resulting from the demand for these metals for use in armoured steel during the war delayed the exploitation of this discovery until peace returned.

Between the two world wars it was found that there was no need to isolate the complex acids used as precipitants: they could be used in solution. The use of mixed solutions of phosphotungstic and phosphomolybdic acids was also studied, as well as the effects of changing the pH of the solutions used and of various heat treatments after precipitation on the colour of the pigments produced.

The Second World War again resulted in a shortage of tungsten and molybdenum and in an effort to overcome this the so-called 'copper toners' were made by co-precipitating basic dyes and copper(II) hexacyanoferrate. The resulting products were less brilliant in colour than their predecessors, and of lower strength and light fastness. Despite being cheaper, they found only limited use when supplies of tungsten and molybdenum resumed after the war had ended. About twenty-five years ago several manufacturers produced ranges of about fifteen pigments of this chemical type, comprising blues, greens, yellows and reds. Some were made by precipitating mixtures of dyes: for example, blue pigments of increasingly green hue could be prepared by adding minor but increasing amounts of a yellow dye to a blue dye. Commercially significant differences in technical properties, as well as in price, could also be achieved by changing the ratio of tungsten to molybdenum in the complex acids.

The chemistry of these complex phosphotungstic (PTA) and phosphomolybdic (PMA) and mixed (PTMA) acids was not fully understood until X-ray diffraction methods were used to determine their structures in the early 1950s.

The Miolati–Rosenheim formulae such as can be found in the BIOS reports [6] represented the complex anions of phospho-12-tungstic acid as $\{H_3[P(W_2O_7)_6]\}^{4-}$, on the basis that $W_2O_7^-$ ions are coordinated to the phosphorus atom within the overall complex ion. X-ray analysis has now shown that the structure of the compounds is built up from WO_6 octahedra. By sharing two oxygen atoms a W_2O_{10} unit (which does not exist independently) can be formed. In the phospho-12-tungstic acid the sharing process is continued so that there are four W_3O_9 (hypothetical) units clustered round the central PO_4 group in the form of a tetrahedron.

The overall structure of the anion is shown in structure 2.3. Since the phosphorus atom has five positive charges, the overall negative charge on the ion is three, i.e. $[P(W_3O_{10})_4]^{3-}$. The hydrated free acid is thus phospho-12-tungstic acid, i.e. $H_3[P(W_3O_{10})_4].14H_2O$. There are other complex phosphotungstic acids known with different phosphorus:tungsten ratios, but only the 1:12 series has found use in pigment making. Since tungsten and molybdenum are isomorphous in these compounds, the use of any ratio of these two metals is theoretically possible in preparing the complex acids. In practice, ratios of 85:15 or 60:40 seem to have been found satisfactory for most purposes. Molybdenum is less expensive than tungsten and thus it might seem advantageous to use lower ratios, but the pigments thus produced tend to darken on exposure to light, a rather undesirable

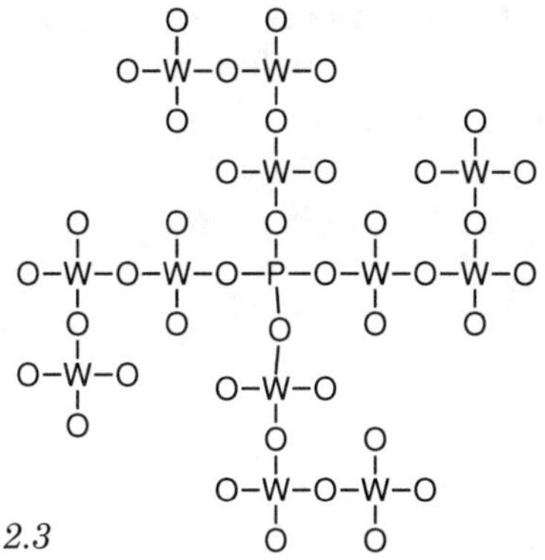

2.3

property. The ratio of dye cations to complex anions in the pigments is 4:1.

In view of their declining importance these pigments will not be treated in more detail. They were generically known as Fanal pigments, and the most important were C.I. Pigment Blues 1, 3 and 8, Pigment Greens 1, 2 and 3, Pigment Violets 1, 2 and 3 and Pigment Red 81.

2.2.3 Vat dyes converted into pigments

Although vat dyes can be regarded as pigments in view of their water insolubility, they were first used as dyes. When prepared for use as dyes, their particle size is an important technical factor that governs to a large extent their ease of reduction. Particle size was found to be equally important when attempts were made to use vat dyes as pigments. Commercial success was only achieved when methods for further decreasing and controlling the particle size were found, so that tinctorially strong pigments of bright hues resulted.

The impetus for these developments came mainly from the need for pigments of high light fastness in car finishes, particularly when a fashion for light-coloured cars arose. The necessary white pigment, titanium dioxide, was available, but to make the desired off-white paints small amounts of coloured pigments had to be mixed with the titanium dioxide. In such formulations many pigments, particularly azo pigments, show much diminished light fastness, from ISO ratings of 7 to ratings of 3 or 4.

The light fastness of vat dyes is generally good, so they were tried out for this purpose. In the physical form available at the time the low brilliance and poor tinctorial strength shown by many vat colours was unacceptable and therefore work was put in hand to alter the purity, particle size distribution and in some instances the crystal modification to make them suitable for pigmentary use. Vesce [7] of Harmon Colors and Gaertner [8] of Ciba–Geigy have given accounts of these developments.

The chemical classes of vat colours from which suitable pigments have been developed are thioindigos, anthraquinones, perinones and perylenes.

Of the large numbers of products tested, only about a dozen met all the stringent property standards demanded. Examples are compounds 2.4 to 2.11 (Table 2.1).

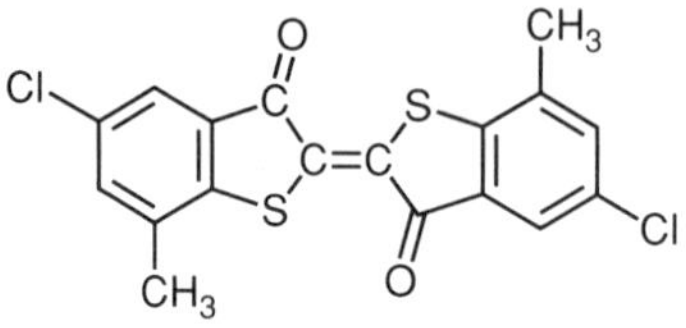

2.4

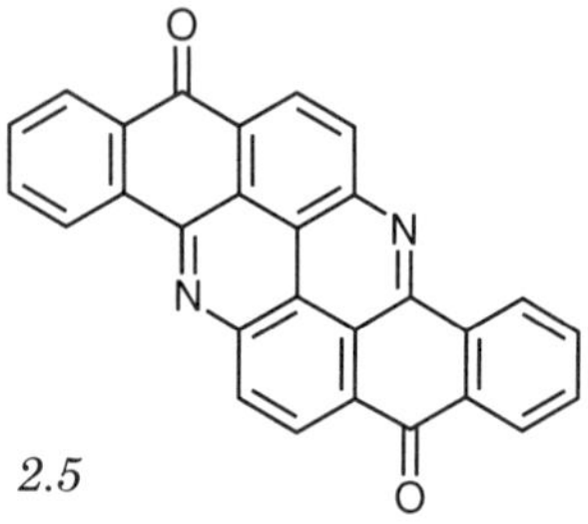

2.5

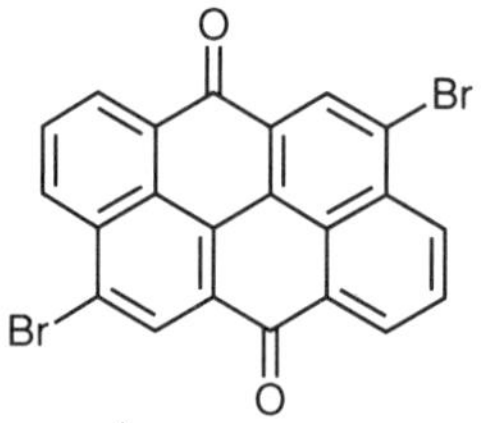

2.6

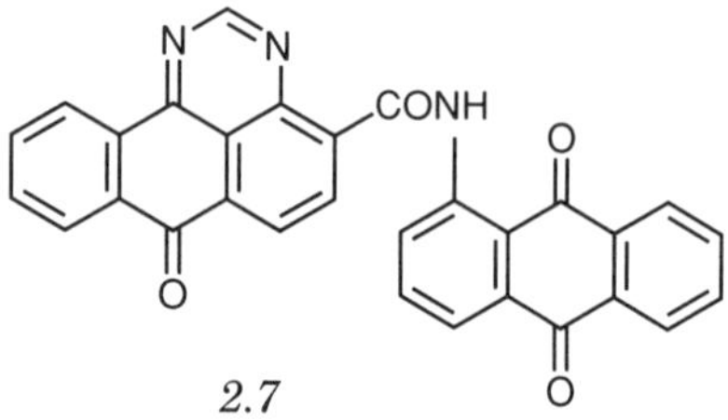

2.7

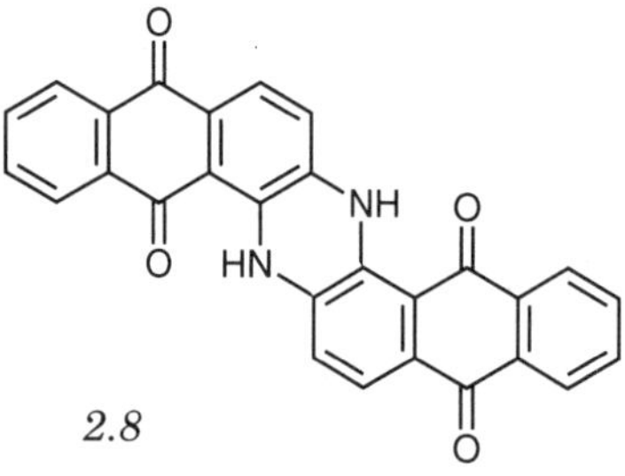

2.8

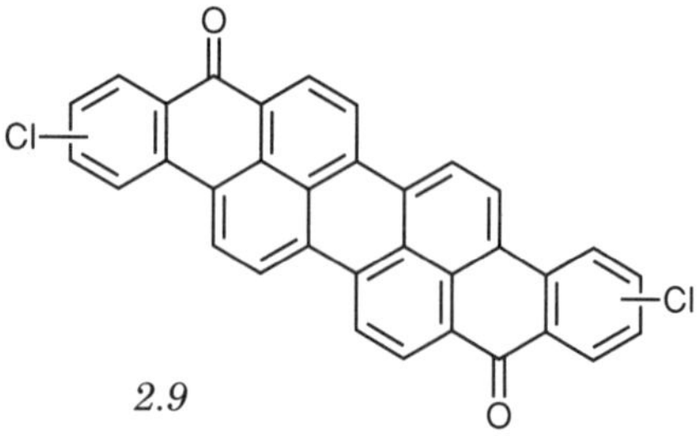

2.9

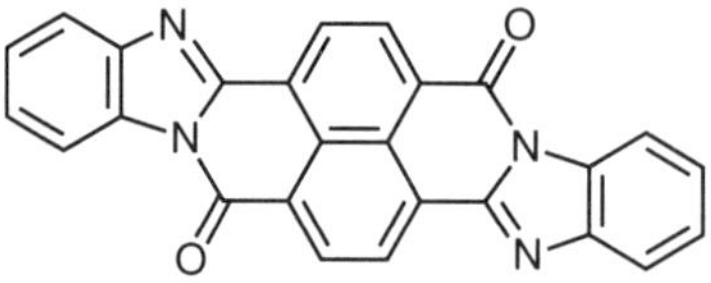

2.10

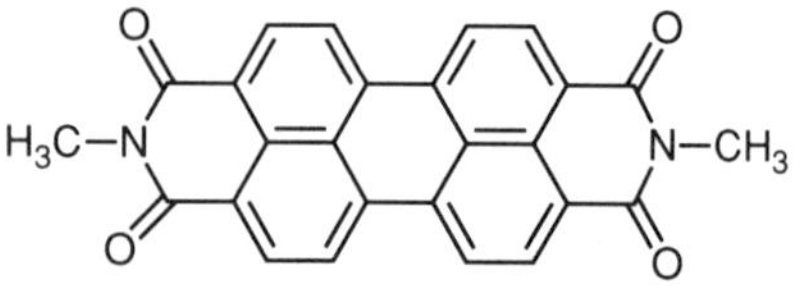

2.11

TABLE 2.1

Typical pigments converted from vat dyes

Structure no.	Commercial name	C.I. Vat	C.I. Pigment
2.4	Thioindigo Red Violet	Violet 2	Violet 36
2.5	Flavanthrone Yellow	Yellow 1	Yellow 24
2.6	Anthanthrone Orange	Orange 3	Red 168
2.7	Anthrapyrimidine Yellow	Yellow 20	Yellow 108
2.8	Indanthrone Blue	Blue 4	Blue 60[a]
2.9	Isoviolanthrone Violet	Violet 1	Violet 31
2.10	Perinone Orange	Orange 7	Orange 43[b]
2.11	Perylene Maroon	Red 23	Red 179

[a] α-form only
[b] *trans* form only

The methods used to convert these vat dyes into a suitable physical form in terms of size distribution (and, in some cases, crystal structure) for use as pigments have been carefully guarded industrial secrets, revealed only in patents. The general principles are clear, however. The original particles are disintegrated by bringing them into solution, where they are generally present as single molecules. These are then re-precipitated by reducing their solubility in this solution, under controlled conditions of nucleation and growth such that the resulting precipitate has the required structure and size distribution. Mean particle sizes of less than 1 μm (1000 nm) are generally needed.

The first method for carrying out the disintegration and re-precipitation processes is the reduction of the dye followed by its re-oxidation under controlled conditions. If this does not give the required result then the second method is used: this is a fine grinding operation, of which there are two types. The first comprises dry grinding in the presence of an inert substance such as salt, which can be removed by aqueous washing when the particle size of the dye has been sufficiently decreased. Alternatively, grinding in the presence of an organic solvent can be used. It is possible that this operates because local heating under the intense shearing conditions in the mill causes the dye particles to pass temporarily into solution. On release from the local high-temperature zone the solubility is exceeded and the dye comes out of solution at a smaller particle size and in some cases with a different crystal structure. Sometimes hybrid processes of grinding in the presence of both salt and an organic solvent have been used. More has been published about the methods used in the case of phthalocyanine and quinacridone pigments, which will be mentioned later.

Because of the empirical nature of these size-reduction processes, their successful industrial operation has depended on the monitoring of progress by means of advanced methods of measuring particle size distributions. Amongst these have been centrifugal, X-ray, electron microscope, surface area and more recently laser light-scattering methods [9]. These have been necessary because the size range involved is below the resolution limit of the ordinary microscope using visible light.

2.3 AZO PIGMENTS

Azo pigments comprise by far the largest chemical class of compounds from which pigments are made, and reflect the wide range of aromatic amines that can be diazotised and the large number of compounds to which the resulting diazo compounds can be coupled, not all of course yielding pigments. The hues of the azo pigments range from red, yellow and orange to blue, in descending order of numbers of each hue.

The first commercial use of the diazotisation and coupling reaction in the field of colorants was in the making of the so-called 'ice colours' by the firm of Read, Holliday and Sons in 1880. It is perhaps appropriate that by diazotising *p*-nitroaniline and then coupling the product to 2-naphthol a dye results if the diazotised *p*-nitroaniline has diffused into a fibre before coupling, but a pigment results if the diazo component is added to the coupler in the absence of any fibre. Para red (2.12) is both a dye and a pigment.

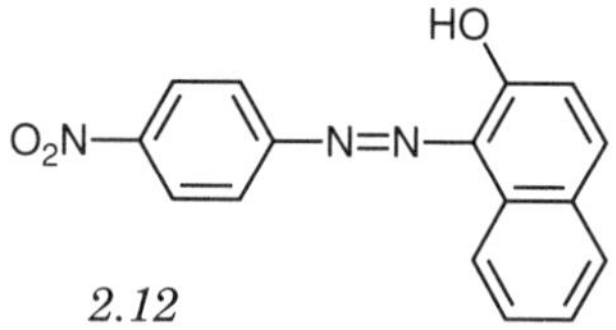

2.12

Since the chemistry of the reactions involved in making both azo dyes and azo pigments is the same, discussion of its details has been left until Chapter 4. Only the structures of azo pigments and the special features of their preparation, largely concerned with preparing the pigments in their optimum physical form appropriate to their end use, are considered here. To do this, a classification of azo pigments is needed, as shown in Figure 2.1.

The primary division separates azo pigments into those that are metal-free and those that contain a metal atom in their structure. This means that those in the former group are entirely covalent in nature while those in the latter are not. An entirely covalent molecule possesses one property that is highly desirable in a pigment: insolubility in water. But at the same time covalent molecules are to some extent soluble in the organic media

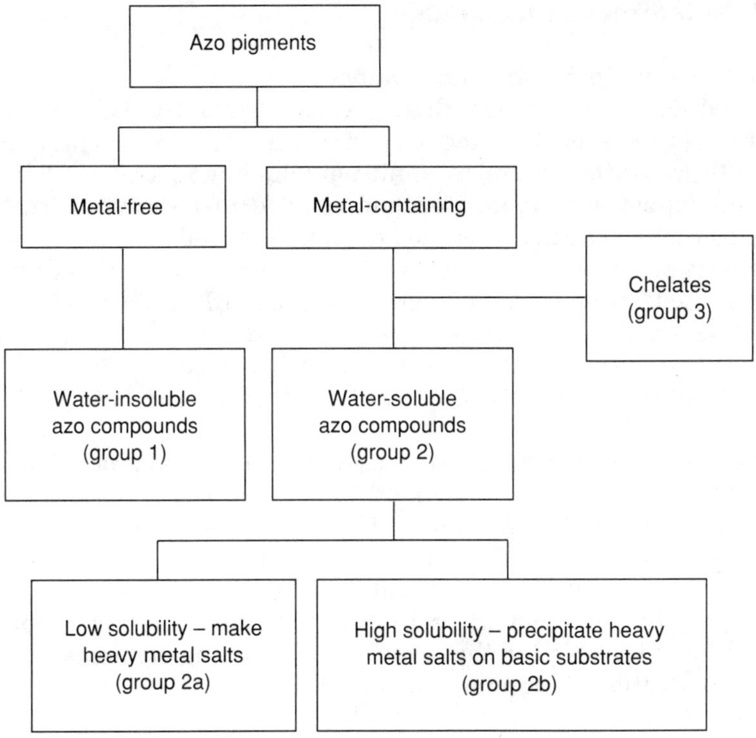

Figure 2.1 – Classification of azo pigments

and solvents in which these pigments are generally used – paints, printing inks and plastics. The result is that the pigments have a tendency to diffuse in these media, giving rise to colour changes and poor fastness properties.

The need to make azo pigments that are metal-containing arises from the requirement to reduce the aqueous solubility of some azo colorants to a sufficiently low level to make them usable as pigments. The problem arises when either the amine or the coupling component from which the colorant is prepared (or both) contain substituents such as hydroxy or acidic groups. Depending on their number and their positions in the molecule, these can confer some degree of aqueous solubility on the resulting azo compound. The means adopted to reduce this solubility to an acceptable level depends on its degree and causes the pigment to fall into either group 2a or group 2b of the classification shown in Figure 2.1. All the pigments in group 2b are made from azo dyes. Those in group 3, of which there are only very few, are metal chelates.

2.3.1 Metal-free azo pigments

Pigments in group 1 made from 2-naphthol

Para red (2.12), which has already been mentioned, was the earliest pigment of this type. Toluidine red (C.I. Pigment Red 3) was first made in 1905. To make it, the amine *m*-nitro-*p*-toluidine (called MNPT in the pigment industry) is diazotised and coupled to 2-naphthol, the same coupling component that was used to make para red.

There is probably more published information about the making of C.I. Pigment Red 3 than any other pigment, since in BIOS report 1661 [10] no less than six detailed works processes are described. While all these processes produced pigments with the same chemical structure, the hues of the resulting products differed sufficiently for them to be sold under different trade names.

Since one of the recurrent problems of the industrial production of azo pigments is to keep the coloration properties of the various batches constant within very close limits, it is useful to see the effects of changes in the process variables. They are set out in Table 2.2 [11]. In order to see which of these many variables had the greatest effect on the pigment properties, Hildreth and Patterson [12] investigated the effects of changing the following variables on the colour of the resulting pigments:
(a) concentration of coupling component
(b) pH of coupling
(c) temperature of coupling
(d) effect of the use of dispersing agent in the coupling bath.

The effects of the changes in the preparation methods were assessed by measuring the colour of the pigments in draw-downs and the particle size distributions of the powders. It was found that changes in all the listed variables could produce changes of several NBS units of colour difference (one unit being about the commercial limit of acceptability between pigment batches). The strongest pigments were those with the largest proportion of particles below 500 nm in size and the largest surface area. The need for the strictest possible control of process variables to reduce batch-to-batch variability was clearly shown, but the variables needing the closest control were not the same for all the pigments investigated.

Both para red and C.I. Pigment Red 3 are prepared using 2-naphthol as the coupling component. Other pigments made from the same coupling component, all discovered before 1910, are compounds 2.13 to 2.15 (Table 2.3,). All these pigments, except C.I. Pigment Orange 5, are made by the direct diazotisation of the appropriate amine, that is, using sodium nitrite and hydrochloric acid at temperatures around 0°C and then adding the diazotised amine (the 'diazo') to an alkaline solution of the coupling component. There are problems in carrying out these diazotisations be-

TABLE 2.2

Batch process for preparation of toluidine red

Conditions at various stages	Helio Red RBL	Lithol Fast Scarlet GRN	Lithol Fast Scarlet RB	Lithol Fast Scarlet RN	Hansa Red B
Diazotisation					
m-Nitro-*p*-toluidine suspn/mol l^{-1}	1	1.25	3.3	1.25	1.25
Temperature/°C	2–5	0	0	0	0
Diazo concn/mol l^{-1}	0.59	0.71	1.1	0.71	0.71
Carbon addition	No	No	Yes	No	No
Coupling component					
Concn of soln of 2-naphthol in caustic soda	0.36	0.25	0.35	0.25	–
2-Naphthol, whether precipitated	Yes (HCl)	No	Yes (HCl)	No	Yes (HCl)
Concn 2-naphthol when coupled /mol l^{-1}	0.21 suspn	0.061 soln	0.245 suspn	0.082 soln	Suspn
pH of coupling soln	5–8	10	8	10	–
Coupling conditions					
Temperature/°C	40	10	25	25	36
Method of adding diazo component	Below surface	Above surface	Above surface	Above surface	Below surface
Time of addition/h	2	2.5	1	2.5	3
Subsequent treatment					
Heating after coupling	50°C	None	Heat to 90°C	95°C (30 min)	Heat to 70°C over 2.5 h
Washing method	Water (2 h)	Water	None	Water	Water
Drying temp./°C	50–55	50–55	Made into paste	45	50–55

cause of the hydrophobic nature of many aromatic amines. Two methods of overcoming this are frequently used. Either the hydrochloride of the amine is formed, by stirring it with concentrated hydrochloric acid, or the particle size of the amine is reduced, by dissolving it in concentrated sulphuric acid and then re-precipitating. Both methods aim to increase the

TABLE 2.3

Typical group 1 azo pigments (from 2-naphthol)

Structure no.	Commercial name	C.I. Pigment
2.13	Fire Red	Red 4
2.14	Red Toner	Red 6
2.15	Permatone Orange	Orange 5

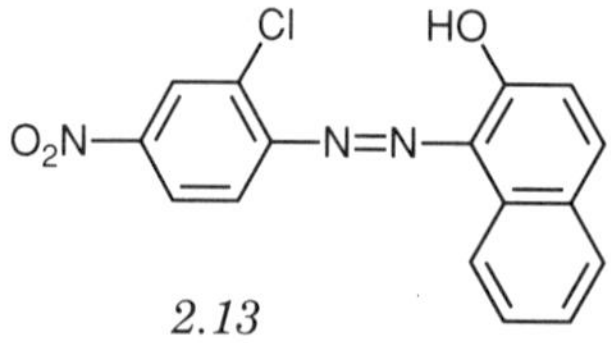

2.13

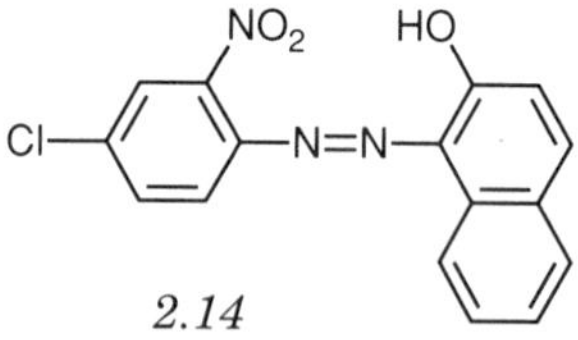

2.14

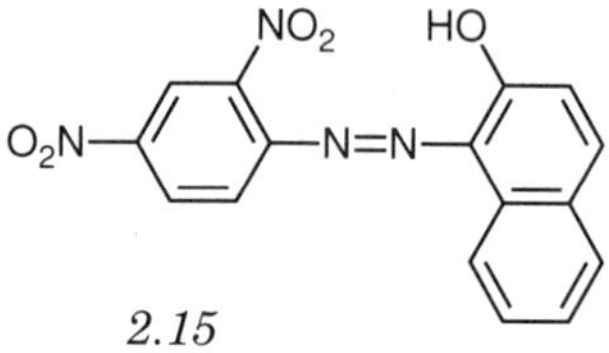

2.15

rate of diazotisation, which tends to be slow on account of the low temperature and the small surface area for chemical attack, particularly if the amine particles have aggregated because of unfavourable storage conditions or other reasons.

Another technical problem in diazotising these aromatic amines is how to decide when the diazotisation is complete, especially when the diazo component is sparingly soluble and therefore appears in the form of a suspension. (It is essential that the amine should be completely diazotised, because any unreacted amine will remain in the precipitated pigment product with adverse effects on the colour and fastness properties.) A widely used method is to ensure that free nitrous acid is still present at the end of the time which experience has indicated to be adequate. A further test is then carried out some time later to check that a small excess still remains. The presence of nitrous acid produces an immediate and intense dark blue colour on starch–iodide paper when the diazo component is spotted on to it. If this colour is not obtained the cause may be lack of acid or of nitrite. The amount of acid is adjusted if a test with pH paper shows insufficient to be present. Retesting with starch–iodide paper will then indicate whether more sodium nitrite is needed or not.

The amine that has be be diazotised to make C.I. Pigment Orange 5, 2,4-dinitroaniline, cannot be diazotised by the usual direct method. Nitrosylsulphuric acid, $HSO_3.NO$, is needed, and this is so corrosive that lead-lined vessels must be used for handling it. Even though this pigment has an unusually high light fastness in reduction (5–6 at 1/25 International Standard Depth, which is two points above most of the others in this group), the high costs of its production have led to a decline in its use.

Pigments in group 1 made from acetoacetanilides

Early this century several yellow pigments were made by coupling substituted nitroanilines to acetoacetanilides. They are commonly known as the Hansa Yellows; those that are greener in hue are identified by the increasing number before the 'G' in the suffix to their names [2.16–2.19 (Table 2.4)]. The G (C.I. Pigment Yellow 1) and 10G (C.I. Pigment Yellow 3) pigments are the most widely used, the latter being much greener than the former, but having only about one-third the tinctorial strength. The light fastness of all the pigments is 7 in full strength, but falls to 4 in 1/25 International Standard Depth, with the exception of C.I. Pigment Yellow 3, which is rated 6+.

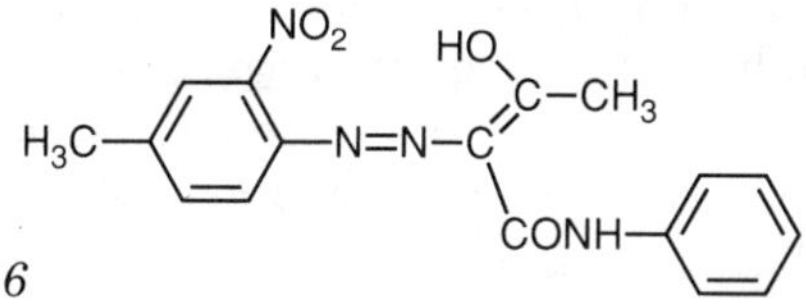

2.16

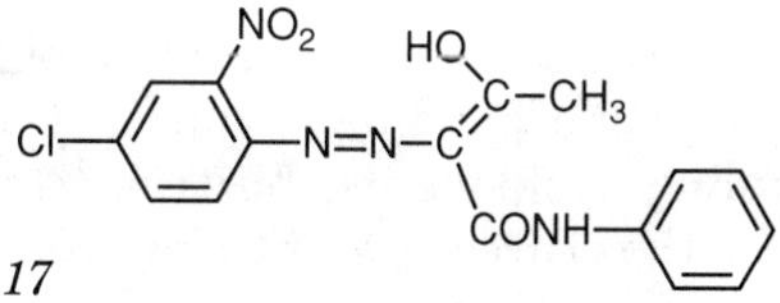

2.17

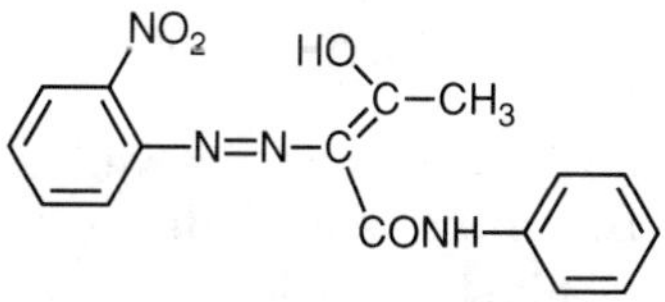

2.18

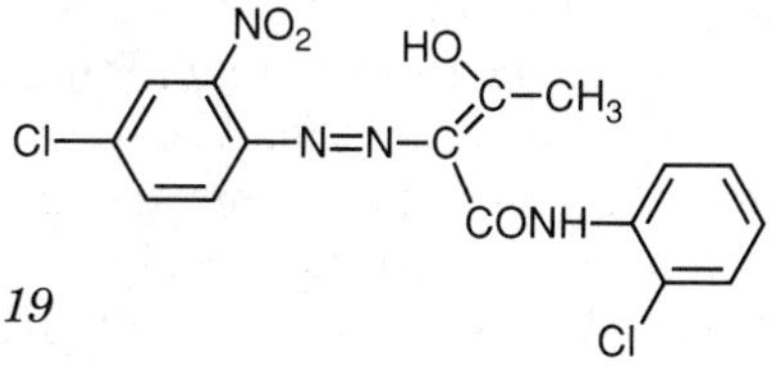

2.19

TABLE 2.4

Typical group 1 azo pigments (from acetanilides)

Structure no.	Commercial name	C.I. Pigment
2.16	Hansa Yellow G	Yellow 1
2.17	Hansa Yellow 3G	Yellow 6
2.18	Hansa Yellow 5G	Yellow 5
2.19	Hansa Yellow 10G	Yellow 3

The manufacturing factors already mentioned control the exact hue of these pigments too. In particular, the transparency of the pigment is sensitive to the pH of coupling, and several manufacturers make both opaque and transparent versions of C.I. Pigment Yellow 3.

All the above yellow pigments are monoazo compounds, but disazo pigments can also be prepared from derivatives of the aromatic diamine benzidine, which can be tetrazotised by the direct method to produce both dyes and pigments. Benzidine itself is a carcinogen, a fact that some chemical companies discreditably suppressed. However, all the disazo pigments (as opposed to dyes) that have appeared on the market have been made from substituted benzidines, generally 3,3′-dichloro or 3,3′-dimethoxy derivatives, and the evidence for the carcinogenic properties of these compounds is less convincing. They are generally classified as 'cancer suspect' agents.

The hues of these diaryl yellow pigments are generally redder than those of the corresponding monoazo yellow pigments made from the same coupling components, but their main advantage is that the tinctorial strength is three or four times greater. This is particularly valuable when making printing inks, an application in which the flow properties of the inks (which depend mainly on the amount of pigment in the formulation) must be very carefully controlled.

Diaryl yellows have another advantage over monoazo yellows in that they show less blooming in plastics. Blooming is a tendency of pigment molecules to migrate to the surface of the plastic articles that they are used to colour. It can occur in moulding processes and also more slowly when in use. In either case it leads to colour loss and sometimes to contamination of articles coming into contact with the surface. The reason for the better performance of the diaryl pigments is that their doubled molecular size leads to lower diffusion coefficients in the plastics, but the matter is complicated by the presence of organic plasticisers and other additives in some types of plastics. On the other hand, the light fastness of the disazo yellow pigments is usually less than that of the monoazo types.

The most widely used pigment of this type is C.I. Pigment Yellow 12 (2.20), made by the direct tetrazotisation of 3,3′-dichlorobenzidine, starting with the hydrochloride (during the reaction the green solution turns yellowish-brown), followed by the coupling of one mole of the product to two moles of acetoacetanilide.

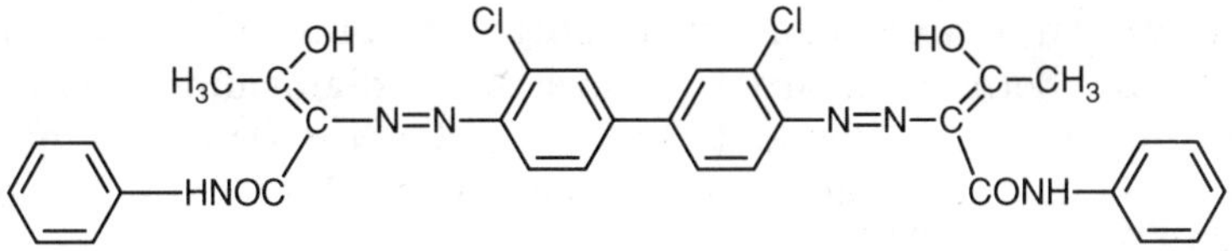

2.20

The possibilities of preparing pigments by introducing substituents into both the diazo and coupling components have been explored. Substituting methoxy groups instead of chlorine in benzidine yields orange rather than yellow pigments, while substituents in the 2- and 2′-positions in the benzidine rings makes the molecule non-planar with loss of colour strength and a hypsochromic shift in hue. C.I. Pigment Yellow 13 is made by the use of acetoacet-*m*-xylidide as the coupling component.

Pigments in group 1 made from derivatives of BON acid

In about 1910 problems were encountered when pigments made from the coupling component BON acid were reduced with alumina hydrate in the making of printing inks. The attractive bluish-red hue of the pigments turned brown, according to Zitscher [13]. The problem was solved by using anilides of BON acid as coupling components. Six pigments were produced, all red or maroon in colour. They are often called the Permanent Reds. Permanent Red FRR (2.21; C.I. Pigment Red 2) is derived from Naphtol AS (C.I. Azoic Coupling Component 2) and C.I. Pigment Red 18 (2.22) from Naphtol AS-BS (C.I. Azoic Coupling Component 17).

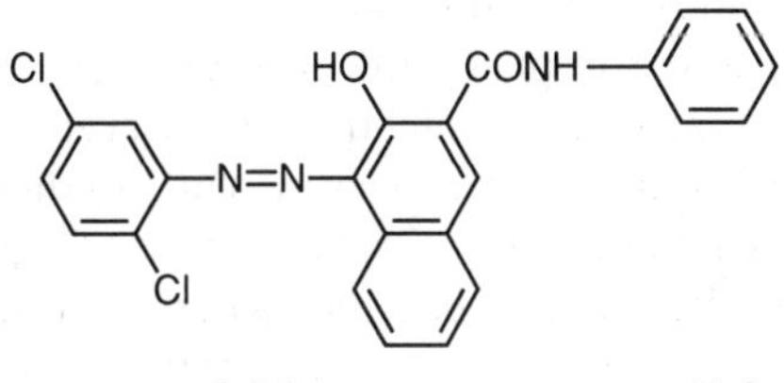

2.21

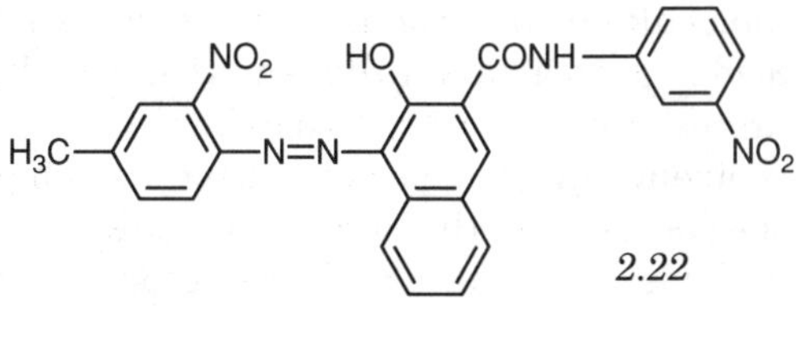

2.22

In the 1920s the hue range of pigments with this type of chemical structure was further extended by using other Naphtols. The pigments are maroon and have a light fastness of about 7 in full strength. Being rather complex hydrophobic structures they tend to have poor solvent fastness properties; this is a considerable disadvantage in the formulation of paints for vehicles, which are applied by spraying methods and are thinned with solvents.

Azo pigments are almost all in the colour range from yellow to orange and red, but research led to an extension of this range to brown (2.23; C.I. Pigment Brown 1) and, more surprisingly, to blue. The two blue azo pigments are both made from tetrazotised *o*-dianisidine; coupling to Naphtol AS gives C.I. Pigment Blue 25 (2.24; X = H) and coupling to Naphtol AS-OL (C.I. Azoic Coupling Component 20) gives C.I. Pigment Blue 26 (2.24; X = OCH_3).

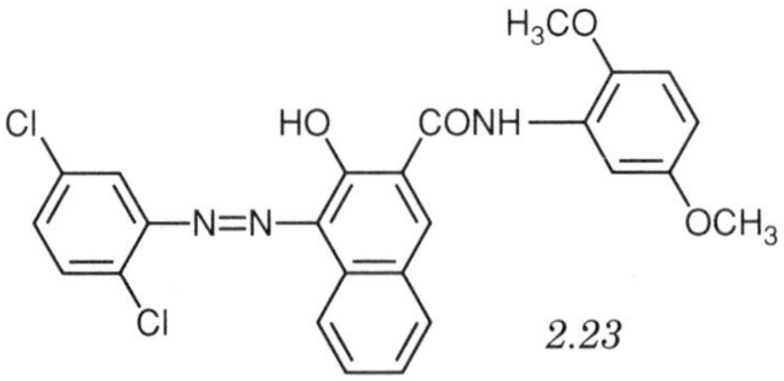

2.23

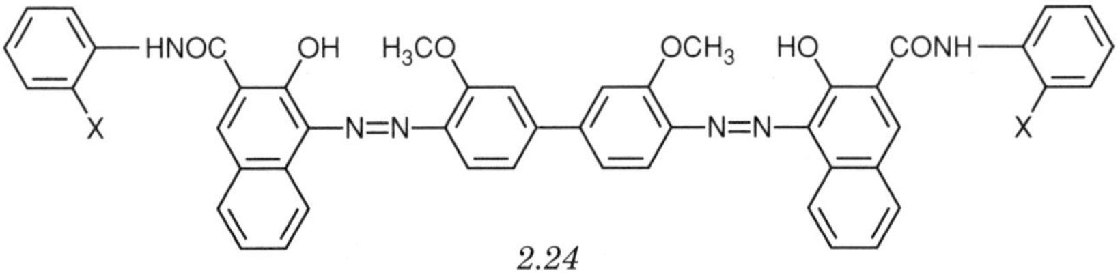

2.24

Azo condensation pigments

The general experience is that while metal-free azo pigments have good chemical resistance their completely covalent nature leads to a tendency to bleed in organic solvents and to migration and blooming in plastic formulations. It was noted, however, that the problems of migration and blooming decreased when the overall size of the molecule was increased. Attention turned to the possibility of doubling the size of the pigment molecules by first condensing one mole of benzidine with two moles of BON acid to form a dinaphthol and then coupling this to two moles of diazotised aromatic amine, as shown in Scheme 2.1. Difficulties in the synthesis were soon encountered, because after one mole of diazotised amine has coupled the solubility of the product in water is so low that it comes out of solution and does not couple with the second mole of diazotised amine.

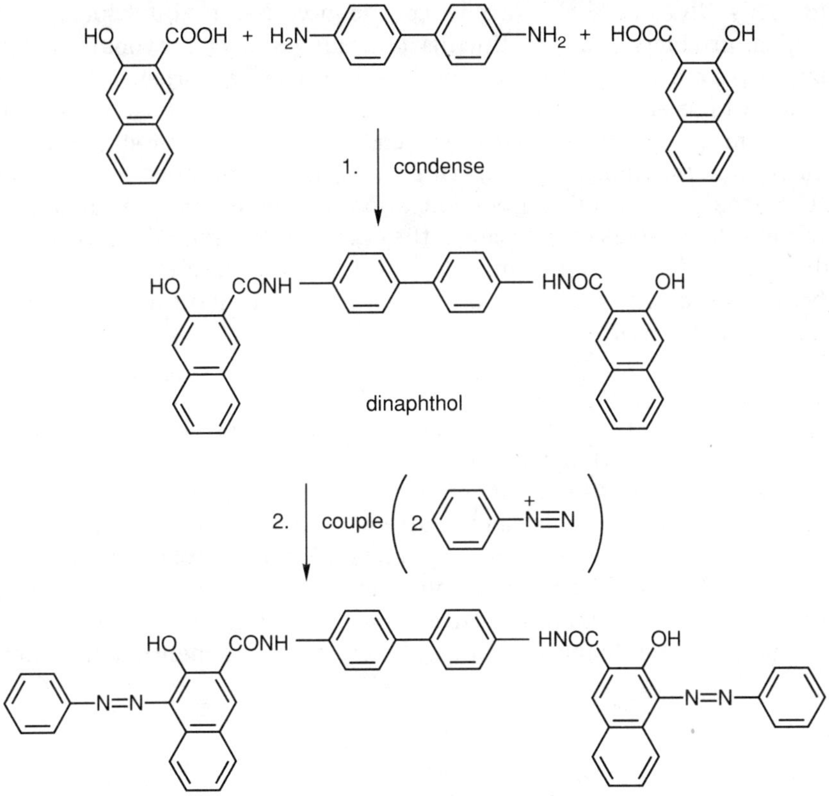

Scheme 2.1

The solution to this difficulty, discovered by Schmid of Ciba in the early 1950s, is to reverse the order of the two steps. The diazotised amine is first coupled with BON acid and then two moles of this intermediate are condensed with one mole of benzidine. The benzidine is in the form of its hydrochloride and the condensation is carried out in a high-boiling solvent such as nitrobenzene, when yields of 95% can be obtained. These pigments have been called 'azo condensation pigments' in view of their method of synthesis.

Gaertner has given some interesting insights into the problems of developing new products in a field of organic synthesis which has been the subject of well-funded and intensive research for nearly a hundred years [8]. He stated that the Ciba company had used their new method to prepare well over 10 000 new pigments. They were given the commercial name of Cromophtals and initially eleven new pigments were marketed. Not all pigments now termed Cromophtals are of this chemical type, however, and allowing for this there are now about ten azo condensation pigments being sold. This illustrates the large expense involved in introducing new technical products and the difficulty of finding a place for them in a very competitive market, particularly as their initial selling price must be fixed at a level high enough to recoup their development costs. New pigments made from intermediates that are already in large-scale production seem to have the best chance of competing with existing products.

The small range of azo condensation pigments have been able to establish themselves only because they combine outstanding fastness to light, heat, solvents and chemicals, and have consequently found acceptance for use in paints, printing inks and plastics and in the mass coloration of some synthetic fibres.

Benzimidazolone pigments

A further solution to the problem of reducing the reactivity of the carboxyl group in BON acid when this compound is used as a coupling component in the preparation of azo pigments was found in 1964 when the Hoechst company introduced their benzimidazolone pigments. A series of new Naphtol AS derivatives was made by condensing 5-aminobenzimidazolone (2.25) with BON acid. Other new coupling components have been made by condensing this compound with acetoacetanilides, which are used as coupling components for the Hansa Yellow pigments mentioned earlier.

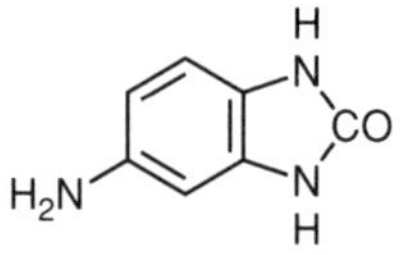

2.25

Two examples of benzimidazolone pigments are C.I. Pigment Yellow 120 (2.26) made by diazotising dimethyl 5-aminoisophthalate and coupling with *N*-(2-oxo-5-benzimidazolyl)acetoacetamide and C.I. Pigment Brown 25 (2.27) made from diazotised 2,5-dichloroaniline and 2-hydroxy-3-*N*-(2′-oxo-5′-benzimidazolyl)naphthalamide. Benzimidazolone pigments have improved migration fastness and lower bleed in paint and PVC formulations, possibly because of hydrogen bonding between the carbonamide groups in adjacent molecules in the crystal lattices of these compounds.

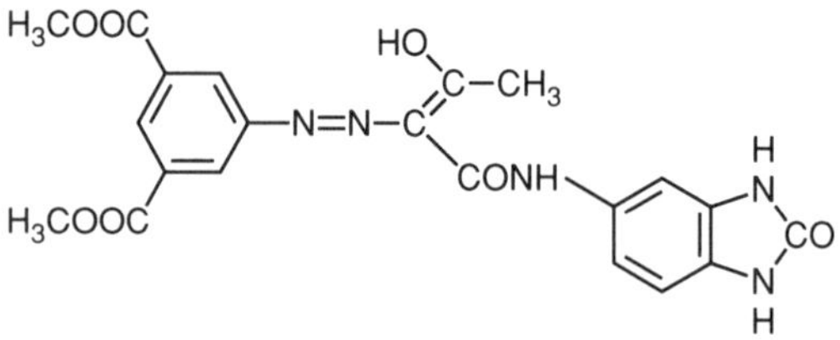

2.26

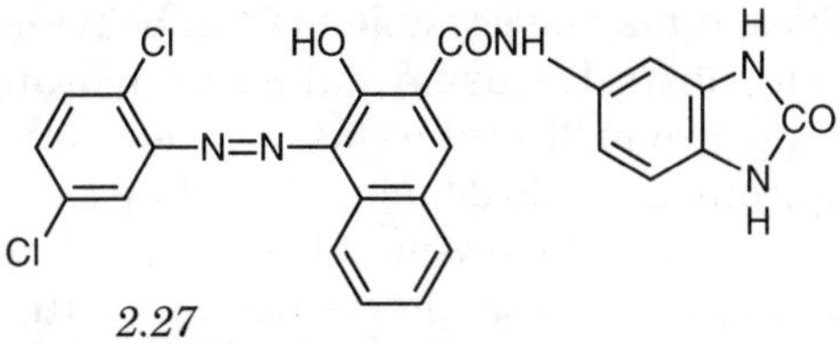

2.27

Pigments in group 1 made from pyrazolones
The last group of metal-free azo pigments are the pyrazolone compounds, the most commonly used examples being made from the coupling component 1-phenyl-3-methyl-5-pyrazolone, also used in making azo dyes. The two most important pigments are the reddish-yellow monoazo Hansa Yellow R (2.28; C.I. Pigment Yellow 10), in which the amine diazotised is 2,5-dichloroaniline, and the disazo pigment C.I. Pigment Orange 13 (2.29), made from 3,3′-dichlorobenzidine.

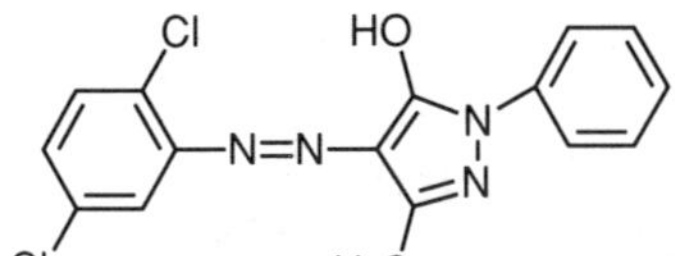

2.28

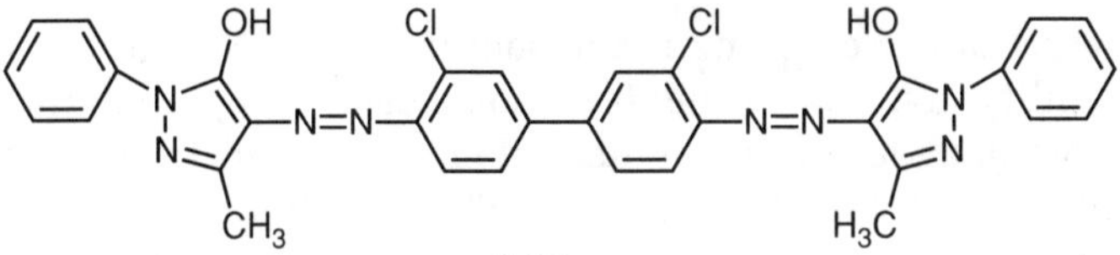

2.29

In making both of these pigments the coupling component must be dissolved in concentrated caustic soda solution before diluting it prior to the coupling stage itself.

2.3.2 Metal-containing azo pigments

As mentioned at the beginning of this section, certain azo colorants have a solubility in water which is low, but not sufficiently low for them to be usable as pigments. They are azo colorants prepared from amines and coupling components containing hydroxyl or carboxyl groups in their sodium salt form as a result of alkaline coupling conditions.

In many instances the aqueous solubility can be reduced to acceptably low levels by replacing sodium with a heavier metal, most commonly calcium, barium or manganese. These colorants fall into group 2a of the classification of azo pigments shown in Figure 2.1. When the solubility of the sodium-containing azo colorants is higher, so that they can be used and regarded as azo dyes, it may be possible to form heavy-metal salts only in the presence of basic substrates, usually alumina hydrate. These pigments consequently fall into group 2b of the classification. Whether a particular pigment falls into group 2a or 2b depends therefore on the method used to make it in the form of these heavy-metal salts. The hues of the products formed using various heavy-metal precipitants are often so different that they are sold as different pigments.

Pigments in group 2a

The first pigments in this group appeared in about 1909 and were termed the Lithols. Lithol Red (2.30; C.I. Pigment Red 49) is made by diazotising Tobias acid (2-naphthylamine-1-sulphonic acid) and coupling it to 2-naphthol. Diazotised Tobias acid is not very soluble and forms a dispersion, not a solution. This gives rise to difficulties in deciding when the diazotisation reaction is completed, which can be overcome by the method described in section 2.3.1.

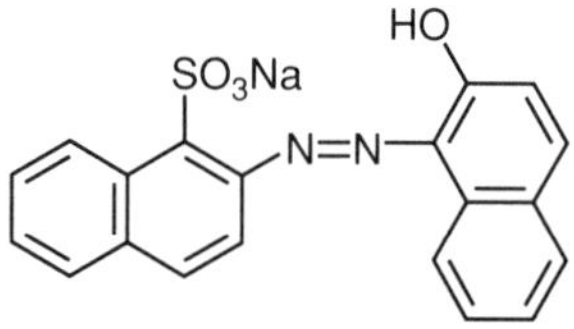

2.30

The product of the coupling is the sodium salt, which is yellowish-red with a solubility just low enough for it to be used as a pigment in printing inks. The pigment found wider use when converted to its barium and calcium salts, which are of a bluer hue. The light fastness in full strength is only 3, falling to 1 at 1/3 ISD, which is too low for modern requirements.

Lake Red C (2.31; C.I. Pigment Red 53), made by diazotising CLT acid (5-amino-2-chlorotoluene-4-sulphonic acid) and coupling to 2-naphthol, is a pigment that has also found considerable use in printing inks. The sodium salt gives prints with a strong bronze surface finish, which seems to depend on the presence of three moles of water of crystallisation to each mole of pigment. If these are driven off by strong heating the bronzing effect is lost and cannot be regained. The barium salt found wider pigment applications.

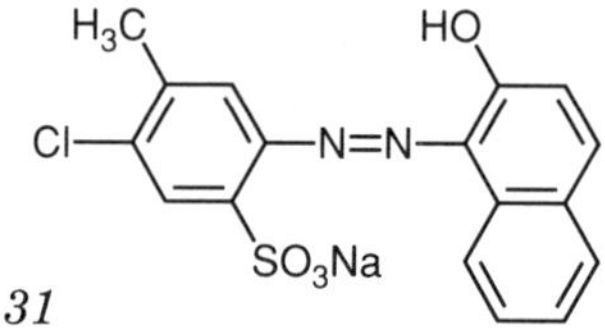

2.31

Tobias acid diazotised and then coupled to BON acid has been used as the manganese salt (C.I. Pigment Red 63) to give a bordeaux shade, but two other pigments (2.32; C.I. Pigment Red 48, and 2.33; C.I. Pigment Red 57) made from the same coupling component are of much greater current importance. The first is made from 2B acid (4-amino-2-chlorotoluene-5-sulphonic acid) and the second from 4B acid (4-aminotoluene-3-sulphonic acid). They are called 2B and 4B toners as might be expected, whilst compound 2.33 is also named Lithol Rubine. The 2B toner is converted into strontium, barium, calcium and manganese salts, whilst the 4B toner is used as the calcium salt. The principal use for both pigments is in printing inks, particularly oil-based types, but both are also used in paints and plastics. Their light fastness in full strength is 5 or 6, falling to about 4 in 1/3 ISD. The effect of light in full strength is to cause a darkening rather than a loss of colour. Fastness to heat is also quite good, as is solvent fastness. This combination of properties has led to their wide availability and use.

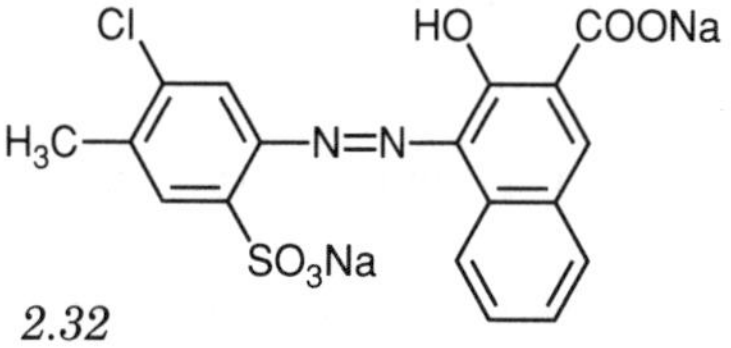

2.32

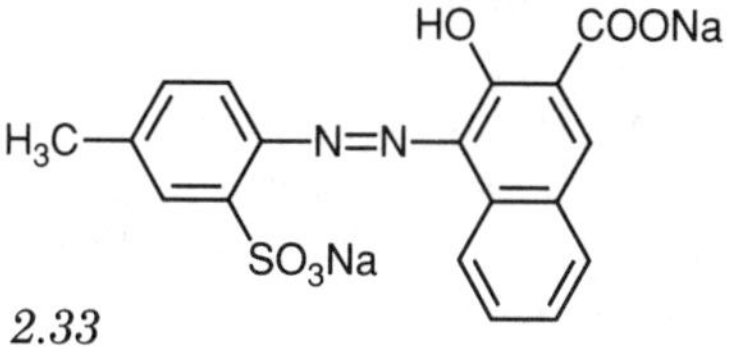

2.33

Rosination

In the preparation of all these pigments in the form of their heavy-metal salts, the method of metal exchange is to heat the sodium salt with a solution of the chloride of the required heavy metal. Very great technical advantages result if this metal exchange is carried out in the presence of rosin soap. The resulting pigments are more easily dispersed, of greater tinctorial strength and higher brilliance, and give a glossier appearance to printing inks. Rosination is therefore a most important part of pigment technology.

Rosin is a natural product derived, along with turpentine, from the resin collected from certain American pine trees. It is graded by colour and the grade WW (water white) is used in the pigment industry. The acidic components originally contain about 40% L- and D-pimaric acid but during processing these are converted to abietic acid (2.34). Neoabietic, dehydroabietic and dihydroabietic acids are also present, according to Harris [14]. Rosin soap is made by dissolving the rosin in a solution of sodium hydroxide at a temperature which is as low as possible, although high enough to keep the soap in solution – it is almost insoluble in cold water. The hot solution, containing also the dissolved heavy-metal chloride, is added to a dispersion of the sodium salt of the pigment. The whole is brought to the boil and, at a temperature between 80 and 90°C, a marked colour change indicates that metal exchange has occurred. The amount of rosin soap used comprises some 15–30% of the mass of the final pigment.

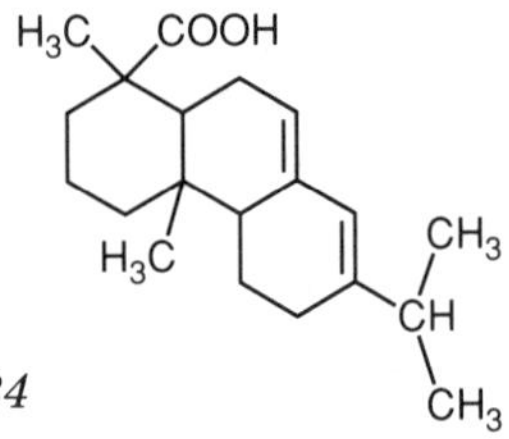

2.34

It acts as a non-diluting substrate. Morris [15] showed that the rosin soap has the effect of increasing the proportion of particles in the pigment with a size below 250 nm, but could find no evidence for chemical combination.

Attempts to fractionate rosin to identify any specially active constituent have been unsuccessful, as have attempts to find better general dispersing agents, though various manufacturers have produced ranges of easily dispersible pigments containing other additives.

Pigments in group 2b prepared from azo dyes

Many azo dyes prepared as sodium salts can be made into pigments by precipitating them as the salts of heavy metals, but only in the presence of basic substrates; the most commonly used substrate is alumina hydrate, as mentioned in section 2.2.1. The pigment is prepared by running solutions of the dye and a salt of the precipitating metal, generally barium or aluminium, into a suspension of alumina hydrate. An intimate mixture of the precipitated heavy metal salt and alumina hydrate is formed and filtered off. The colloid chemistry of such systems, along with the technical and colour properties of the resulting pigments, is complex and depends on the dye:alumina hydrate ratio and the pH of the mixture. Weiser and Porter [16] studied the preparation of pigments from the dye Orange II (2.35; C.I. Acid Orange 7).

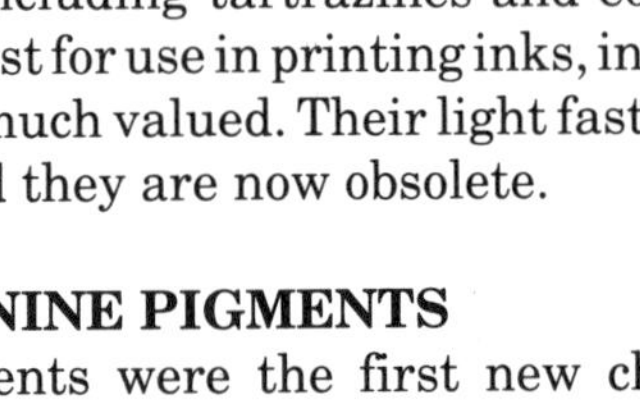

2.35

Many other dyes, including tartrazines and eosines, have been made into pigments in the past for use in printing inks, in which their soft texture and bright hues were much valued. Their light fastness was generally only 1 on the ISO scale and they are now obsolete.

2.4 PHTHALOCYANINE PIGMENTS

Phthalocyanine pigments were the first new chromophore type to be introduced into the field of the colour chemistry of organic compounds concerned with pigments. Before their production, all the known organic pigments had been developed either by making dyes insoluble or by synthesising new insoluble azo compounds. Their development is therefore interesting both technically and scientifically.

It began in 1928 when a blue impurity was found in phthalimide during its manufacture at Scottish Dyes (later part of ICI). The method of preparation being used was to pass ammonia through molten phthalic anhydride in an iron vessel [17]. It is now known that the blue impurity was iron phthalocyanine. It was found to be a very stable material and it was decided to determine its structure. (It is of interest that another important blue pigment – ultramarine – was also first synthesised as the result of the investigation of a blue impurity in an industry not directly related to colour making, in this case sodium carbonate manufacture at St Gobain in France; see section 2.7.2.)

With hindsight it can be guessed that other ways of solving the impurity problem in making phthalimide, such as the use of a ceramic vessel, might well have solved the initial difficulty but would have resulted in missing the discovery of what has turned out to be a most important chromophore. (Two other preparations of phthalocyanine had already been reported in the chemical literature but did not lead to the recognition of its usefulness, the first in 1907 [18] and the second in 1927 [19].) Linstead and his co-workers investigated the chemical structure of the blue impurity, and its novel structure was finally established by Robertson [20] using X-ray diffraction.

Copper phthalocyanine (2.36; C.I. Pigment Blue 15) first appeared commercially in 1935 under the ICI trade name of Monastral Blue B.

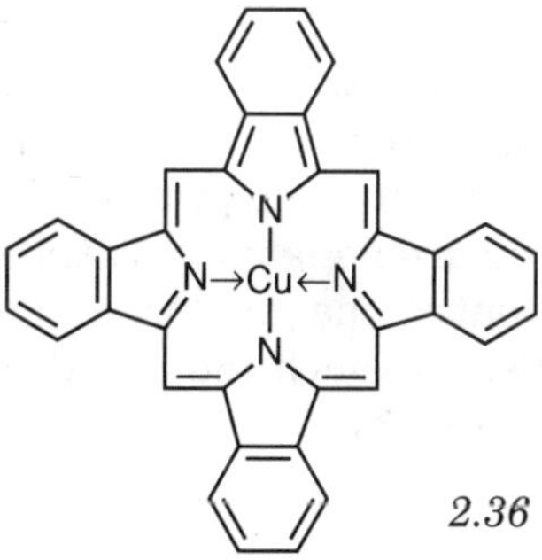

2.36

2.4.1 Manufacture of copper phthalocyanine [21,22]

Two processes have been used for making copper phthalocyanine, but neither yields a form suitable for immediate use as a pigment since the physical form and size distribution of the resulting particles is far from the optimum. The product is generally referred to as 'crude blue' and requires purification and a sometimes complex series of finishing processes before it is in a form ready for use as a pigment.

The solvent process

The reactants are phthalic anhydride, urea and copper(II) chloride, which are heated in a high-boiling aromatic solvent such as 1,2,4-trichlorobenzene, nitrobenzene or *m*-dinitrobenzene in the presence of a catalyst, usually ammonium molybdate. The solvent also acts as a heat-transfer agent. On heating to 120°C an exothermic reaction begins and this temperature is maintained for about an hour. The temperature is then raised to 160–180°C and kept constant for 6–12 hours. During this time ammonia and carbon dioxide are evolved, together with some solvent; the reaction is complete when ammonia evolution ceases.

The remaining solvent is then removed by either steam or vacuum distillation. The yield is 90–95%.

Dry bake methods

As the name implies, no solvents are used in this process and the expensive solvent recovery stage is not needed. Instead of phthalic anhydride, the more expensive phthalonitrile is used a reactant. Urea is not necessary as a source of nitrogen, but seems to have been used in some manufacturing processes. The copper needed comes from copper(I) chloride. Attempts at running continuous processes on moving heated copper belts have been made, but mechanical difficulties in handling the reactants and products after hard solid masses had formed were only partly solved by adding neutral inorganic diluents such as sodium sulphate or common salt to improve the fluidity. The overall result was that until a few years ago the solvent process was in almost exclusive use.

Carcinogenic by-products

Interest in the dry bake process was renewed when polychlorobiphenyls (PCBs) were recognised as carcinogens and found to be an impurity in crude blues produced by the solvent process. The mechanism of their formation, when trichlorobenzene is used as the solvent, is by hydrogen abstraction from this solvent by the urea present leading to the formation of trichlorophenyl radicals. Two such radicals can then combine (Scheme 2.2) to form a PCB molecule (2.37).

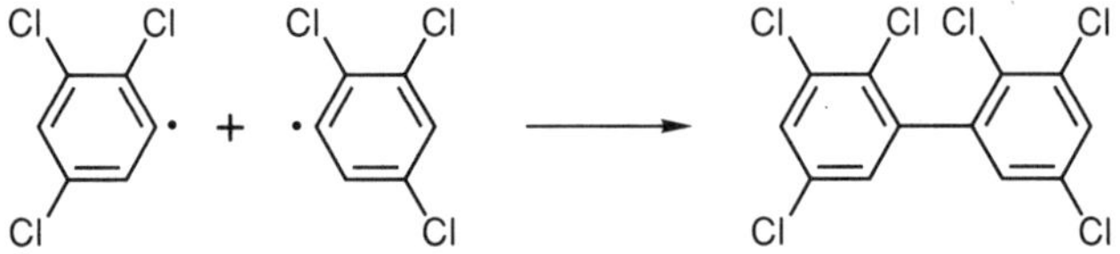

2.37

Scheme 2.2

This reaction can be suppressed by adding antioxidants such as hydroquinone or sodium phosphite, which yield hydrogen radicals (H·), which

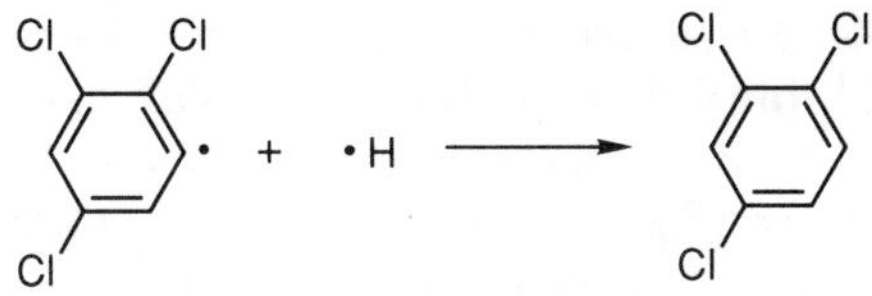
2.38

Scheme 2.3

can convert the trichlorophenyl radicals back to the parent substance (2.38) (Scheme 2.3).

Other high-boiling solvents such as nitrobenzene and 2-nitrotoluene can also be used in the solvent process, but 4-nitrobiphenyl, a listed carcinogen, is produced when nitrobenzene is employed. One way to avoid the formation of substituted biphenyls is to employ a polysubstituted toluene as solvent, the substituents being chlorine or alkyl groups, mainly propyl. These are difficult to prepare but they have been patented by Toyo Inks.

Another approach is to use high-boiling aliphatic solvents with minimal aromatic content; boiling points are in the ranges 150–250°C. Much higher ratios of solvent to reactants are needed than when aromatic solvents are used, owing to the high viscosity of the system.

Yet another method of avoiding PCB formation in the solvent process is to dispense with the need for urea as a reactant by using the more expensive phthalonitrile instead of phthalic anhydride.

Isolation of crude blue

No method of synthesis produces crude blue in pigmentary form. The solvent (if any) and the impurities must first be removed. Vacuum or steam distillation are used to remove the solvent, the former being generally cheaper. Impurities can then be removed by thorough washing for about two hours at 90–95°C, first with 10% aqueous sodium hydroxide solution and then with 10% aqueous hydrochloric acid, followed by filtration, pressing, water washing and drying. From the purified crude blue, copper phthalocyanine blue pigments in either the α- or β-crystal forms can be isolated. By introducing various amounts of halogen atoms into the molecule a series of green pigments can be derived (see section 2.4.2).

Conversion of crude blue into blue copper phthalocyanine pigments

Purified crude blue exists in the most stable crystal form of copper phthalocyanine, the β-form, but the particles are much too large to make a good pigment. The β-form was that prepared by Robertson [20] from which he proved the molecular formula, as mentioned earlier. Three other crystal structures of copper phthalocyanine have been reported but only one of them, the α-form, is of interest in the pigment industry, though mixtures of the two have caused various technical problems. These have arisen because although both forms are blue, the α-form has a reddish-blue hue, while the β-form is a greenish-blue. The α-form has a tendency to revert to the more stable β-form in the presence of the organic solvents that are constituents of many paint formulations, with a consequent shade change and sometimes a loss of colour strength. Hence there is a need to prepare solvent-stable α-form pigments. Since β-form crystals also tend to grow and lose strength in the presence of organic solvents, there is a need to render them solvent-stable too.

Methods for making both forms solvent-stable were the subject of many patents and closedly guarded industrial secrets, but much of the mystery was cleared up by two papers by Gerstner [23] and Smith and Easton [24] published in 1966, by which time X-ray diffraction, electron microscopy and disc centrifuge particle sizing methods had been brought to bear on the problem.

Converting the β-form to the α-form

The β-form of crude blue can be converted to the α-form by complete disintegration of the crystals. This can be readily achieved by dissolving them in concentrated sulphuric acid, thus forming a solution of single molecules, and then re-precipitating as α-form crystals by running the solution into a large volume of water. Careful control of the flow rate, coupled with stirring and temperature control (since heat is evolved in the process) and perhaps the presence of surfactants, can give tinctorially strong α-form pigments. Other methods, termed 'acid slurrying' or 'permutoid swelling', involve the use of insufficient sulphuric acid to bring about complete solution. Copper phthalocyanine sulphate crystals are formed; these are then drowned in water as before, forming α-form crystals. Yet a third method of conversion is by prolonged salt grinding in a ball mill in the absence of any solvent. The α-form so produced, however, is not solvent-stable.

Solvent-stable forms

Smith and Easton [24] give the scientific basis of many of the patent claims. Salt grinding of either α- or β-form and treatment with solvent either during or after grinding gives the β-form. In the absence of solvent the product is the α-form, irrespective of the starting material. However, mixtures of α- and β-forms can result from salt grinding in the presence of solvent, depending on the grinding efficiency and the time of grinding in the mill. Two stages, first grinding and then solvent treatment, are often easier to control. For example, if the grinding process in the absence of solvent is stopped when about one-third of the material remains as the β-form, exposure to solvent will then cause a rapid reversion of the α-form back to the β-form, giving solvent-stable β-form. Solvent-stable α-form pigment can be made by incorporating some copper monochlorophthalocyanine in the pigment or by carrying out the salt grinding in the presence of compounds that can be adsorbed on the surface of the growing α-form crystals, hindering further growth and stabilising them against reversion to the β-form.

2.4.2 Other phthalocyanine derivatives

The discovery of the phthalocyanine chromophore gave rise to much research, particularly in view of its high fastness properties. Many other metallic derivatives have been prepared and many substituents introduced into the phthalocyanine nucleus. Sulphonation has given some soluble dyes, which have even been converted back into pigments by making them insoluble again as barium salts. The only widely produced colorants are a range of green pigments made by halogenating copper phthalocyanine using chlorine either alone or mixed with bromine.

There are 16 hydrogen atoms in the phthalocyanine molecule that can in theory be replaced by halogens. Of these, eight can be replaced relatively easily, while more severe conditions are needed to remove the last eight. The products of different manufacturers vary in the chlorine content, 14 or 15 atoms per molecule being usual. As the number of chlorine atoms increases, the green pigments become increasingly yellow in shade. Halogenation can be carried still further by using mixtures of chlorine and bromine containing increasing amounts of bromine. The yellowest pigments have twelve bromine and two chlorine atoms per molecule.

Direct chlorination of copper phthalocyanine can be carried out by passing the gas into a molten eutectic of aluminium chloride and sodium chloride at a temperature of 180–200°C, the copper phthalocyanine content being about 20% of the eutectic by mass. Iron(III) chloride is present as a catalyst [25]. Other methods include (a) chlorination in trichlorobenzene at 200°C using 10% antimony trisulphide as a catalyst, (b) chlorination under a pressure of 5–10 atmospheres at a lower temperature of 160–170°C, a process in which the release of hydrogen chloride gas is used to control the pressure, and (c) dry chlorination in a fluidised bed. Yet other processes have used sulphuryl chloride (SO_2Cl_2), thionyl chloride ($SOCl_2$) or sulphur dichloride (SCl_2) as chlorinating agents.

The crude polychlorinated copper phthalocyanine requires a finishing process to develop the full potential colour strength, but the problems are less than those experienced with copper phthalocyanine itself, as no problems of polymorphism occur. Finishing is carried out by acid pasting of the crude green in oleum or chlorosulphonic acid. The crystallinity as shown by the sharpness of the X-ray diffraction pattern can be improved by solvent treatment as there is no tendency for this to cause crystal growth. The light fastness rating (ISO 8) and solvent resistance (generally 5) are both as high as those of phthalocyanine blues, making them very high-quality pigments.

A great deal of research has been carried out to try to find organic pigments of red colour to match phthalocyanines in these two fastness requirements. Red azo pigments, of which there are very large numbers, show considerable falls in light fastness, particularly in reductions with titanium dioxide. Pigments made from vat dyes, mentioned earlier, have met the need to some extent, but there is still a gap in the organic colour pigment range for very bright reds of the highest fastness properties.

2.5 QUINACRIDONE PIGMENTS

The synthesis of linear *trans*-quinacridone was reported in 1935 by Liebermann [26] and was cursorily looked at as a red vat colorant but not developed commercially. It was more than twenty years later that the DuPont company introduced these compounds as pigments under the trade name of Cinquasia. Their chemical structures are based on C.I Pigment Violet 19 (2.39). As with the phthalocyanines, this compound can exist in several polymorphic forms; in this case there are three, termed α-, β- and γ-forms, only the last two being useful as pigments. The first three pigments were called:

– Cinquasia Red B	γ-form, size above 1000 nm
– Cinquasia Red Y	γ-form, size below 1000 nm
– Cinquasia Violet R	β-form.

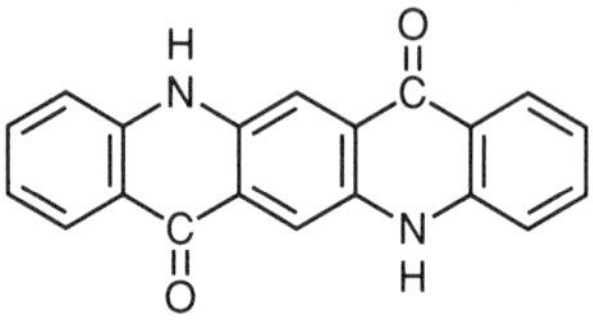

2.39

Quite apart from the problems of chemical synthesis, it is clear that the application of physico-chemical methods to pigment manufacture, first needed in the development of the phthalocyanines, were essential to the successful introduction of these new pigments. X-ray diffraction methods are needed to recognise the crystal forms. The β-form was made by milling the crude product as synthesised with a large amount of salt in the presence of xylene, while the preparation of the γ-form required wet milling with dimethylformamide, followed by removal of the solvent and the salt with aqueous acid. Since the Red B and Red Y pigments differed only in their particle size, particle sizing methods capable of measuring particles less than 1000 nm in diameter were obviously necessary. Such methods have only been available in the last thirty years or so.

The chemical synthesis of linear *trans*-quinacridones and the substituted derivatives that have subsequently appeared on the pigment market is a complicated multi-stage process, making the pigments very expensive and sustainable only where high-fastness red pigments are essential, as in the car industry. There are four synthesis routes, details of which have been given by Pollack [27].

The DuPont company used the succinic acid ester process, in which diethyl succinate is first cyclised to diethyl succinylsuccinate (2.40) in a sodium/alcohol mixture. One mole of the product is next condensed with two moles of aniline under oxidising conditions, forming diethyl 2,5-dianilino-3,6-dihydroterephthalate (2.41). Ring closure at 250°C gives dihydroquinacridone, from which linear quinacridone is obtained by oxidising away the two extra hydrogen atoms.

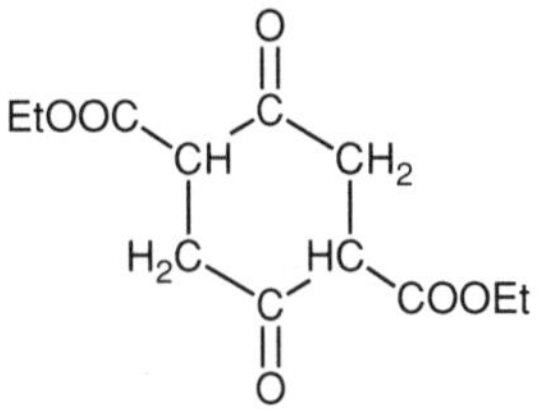

2.40

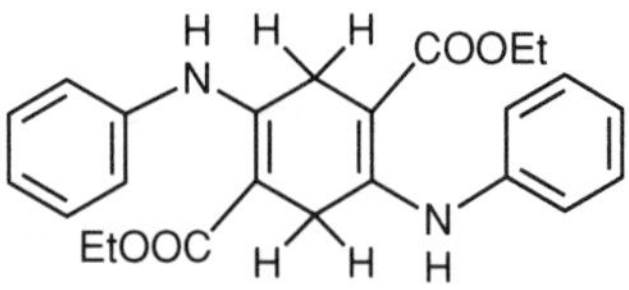

2.41

The Sandoz company used the dibromoterephthalic acid method. This acid was made from *p*-xylene by brominating it to form 2,5-dibromo-*p*-xylene and then oxidising this to 2,5-dibromoterephthalic acid. Reaction of one mole of this with two moles of an arylamine in the presence of copper(II) acetate gives a 2,5-bis(arylamino)terephthalic acid, which can be ring-closed to a linear quinacridone. Unsymmetrical substitution is possible.

The BASF route started from hydroquinone, which was converted to 2,5-dihydroterephthalic acid by a Kolbe–Schmitt type reaction. One mole of this acid was treated with two moles of aniline, both components being in the form of a suspension in aqueous methanol when run into a small amount of a solution of vanadium(III) chloride and sodium chlorate. Gentle heating gave a 95% yield of 2,5-bis(arylamino)benzo-1,4-quinone-3,6-dicarboxylic acid. Next, ring closure to the *trans*-quinacridonequinone took place in the presence of concentrated sulphuric acid at 60–80°C. This was then reduced to the required crude pigment by zinc or aluminium powder in caustic soda under pressure in an aluminium chloride/urea melt or by the use of a sulphuric acid/polyphosphoric acid mixture.

The fourth synthesis, developed by the Swiss Lonza Company and known as the diketene process, is similar to the succinic acid ester process of DuPont in that the central ring of the quinacridone molecule is totally synthesised. Acetone vapour is passed over an electrically heated wire to form first ketene, $H_2C=C=O$, which then dimerises to form diketene (2.42). Chlorination in the presence of alcohol yields ethyl 4-chloroacetoacetate, $ClCH_2COCH_2COOEt$. This reaction proceeds at low temperature but is exothermic to a considerable extent and can only be successfully carried out on a large scale in special reactors having a high surface area. Diketene can

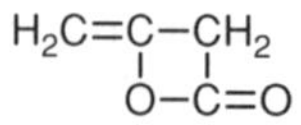

2.42

also react with aromatic amines to make the acetoacetarylides, which are the coupling components needed to prepare Hansa Yellows.

In the manufacture of quinacridone pigments only the first and last of the four routes outlined have been operated commercially. They are followed by the milling processes necessary to give products with the crystal structure and particle size required for their use as pigments.

2.6 ISOINDOLINONE PIGMENTS

Another new chemical class of pigments, called Irgazins, was produced in 1964 by the Geigy company [28]. They are made by condensing two moles of 4,5,6,7-tetrachloroisoindolin-1-one with one mole of an aromatic diamine such as *p*-phenylenediamine to give the yellow pigment Irgazin Yellow 3RLT (2.43; CGY, C.I. Pigment Yellow 110) of high light fastness and heat resistance.

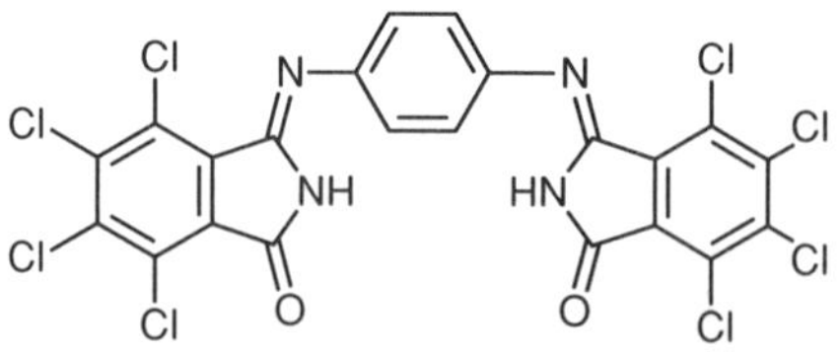

2.43

2.7 INORGANIC PIGMENTS

Inorganic pigments have a special place in pigment chemistry: firstly, because the art galleries of the world are filled with oil paintings painted almost entirely with inorganic pigments and, secondly, because white inorganic pigments (and there are no white organic pigments) play a unique role in the artistic and technological use of all pigments. This is because white pigments are used to confer the property of opacity on the paints and inks prepared from coloured pigments, both organic and inorganic. Paints and inks are surface coatings applied to many surfaces such as wood, metal, canvas, textiles and paper for decorative or protective purposes. Pigments used in colouring today's immense range of plastics may also make them opaque. But there is nothing comparable in the field of dye usage to the place of white pigments in pigment technology, since natural fibres are already opaque and synthetic fibres are generally made opaque on manufacture, often by incorporating white pigment in them.

Many coloured inorganic pigments have special uses for which their coloured organic counterparts are unsuitable, such as the colouring of glass, which requires the ability to withstand very high temperatures. Carbon black in one of its many forms is, of course, the pigment most widely used in printing inks.

2.7.1 White inorganic pigments

The primary requirement for a compound to be of a white appearance is that it should not absorb any of the radiation in the visible part of the electromagnetic spectrum, that is, of wavelengths between 400 and 650 nm; in practice this is seldom attainable and slightly less than 100% transmission can be tolerated, but there must be no marked absorption bands. The additional requirement for a substance to be usable as a white pigment is that it should have as high a refractive index throughout the visible spectrum as possible. A third requirement is that it should be in the form of particles fairly uniform in size, with diameters of the order of 1000 nm.

Because pigments are used as dispersions of particles in media of a refractive index of about 1.5, this is the lower limit for their practical use. There are several inorganic white compounds with refractive indices between 1.5 and 1.63, which are called 'transparent whites' or 'extenders'. They include barytes (barium sulphate), gypsum ($CaSO_4.2H_2O$), whiting ($CaCO_3$), china clay, talc and silica. They confer very little opacity and are mainly used to thicken paints in which they are incorporated.

White pigments, some of rather variable composition, that have been used in the past include white lead (refractive index 1.9), zinc oxide (1.9), lithopone ($ZnS–BaSO_4$, 1.84), zinc sulphide (2.3) and antimony oxide (2.2). They have all been eclipsed by titanium dioxide.

Titanium dioxide

As a pigment this is used in two crystalline forms, rutile (refractive index 2.71) and anatase (2.51). It is these high refractive indices which, combined with modern manufacturing methods, have ensured its pre-eminence.

The ore used for making commercial titanium dioxide is the black ilmenite, nominally $FeO.TiO_2$ but containing small amounts of many impurities, depending on the source. The development of manufacturing methods to prepare pure titanium dioxide in the form of either rutile or anatase took many years. Anatase pigments appeared in the mid-1930s and rutile pigments in the 1940s. One of the problems was to remove the last traces of iron and other metallic impurities which distorted the crystalline lattice, leading to off-white colours.

There are now two chemical processes for making titanium dioxide pigments in widespread use. In the sulphate process finely ground ilmenite is digested in sulphuric acid and the iron is reduced and separated as iron(II) sulphate. The titanium(IV) sulphate is hydrolysed by steam to a hydrous oxide, which is thoroughly washed to remove soluble impurities and finally calcined at temperatures about 1000°C to give the anatase form of titanium dioxide.

In the chloride process, developed in about 1960, the ore is converted to titanium(IV) chloride by heating it to 800°C with chlorine in the presence of carbon, which combines with the released oxygen. The pure chloride is then oxidised to titanium dioxide at 1000°C and the chlorine formed is recycled. Technical problems arise because the oxidation of titanium(IV) chloride is not sufficiently exothermic to make the reaction self-sustaining. These problems can be overcome by pre-heating the reactants and by burning carbon monoxide in the reactor to raise the temperature. By careful control of the conditions, it is possible to produce pure rutile particles of a mean size of 200 nm.

Titanium dioxide is non-toxic, in marked contrast to pigments such as white lead, and is widely used as a pigment in utensils and on surfaces coming into contact with food in industry, as well as in almost all other uses calling for white pigment. It is not the easiest pigment to disperse, and after a long struggle to produce it in pure form most manufacturers now produce a range of quite heavily coated forms, each prepared to ease dispersion and its maintenance in particular media.

Both crystal forms of titanium dioxide are photochemically active, anatase more so than rutile. Under ultra-violet light the pigment releases active oxygen, which degrades paint films causing ‘chalking’ and tenders synthetic fibres. In the dark, atmospheric oxygen regenerates the titanium dioxide and the cycle recurs. In applications where this is a major problem, rutile-form pigments containing additives such as antimony(III) sulphide (Sb_2S_3) are therefore used.

2.7.2 Coloured inorganic pigments

Many kinds of coloured inorganic pigments were formerly widely made [4,28], including yellow lead chromates and iron blues, but the recognition of the dangers from lead poisoning and other technical changes have led to a rapid decline in their use.

The coloured inorganic pigments in current production are generally those that have particular advantages which organic pigments cannot match.

Ultramarine

This is the synthetic form of the blue mineral lapis lazuli, which was imported from China and India in the Middle Ages. It is characterised by high light fastness (ISO 7–8), an attractive blue colour with excellent resistance to alkalis and all organic solvents, easy dispersibility and non-toxicity, but it has poor resistance to acids. It was first made by Guimet in France in 1830 after having been noticed as an impurity in the glass furnaces at St Gobain. It is now made by calcining a mixture of the ingredients shown in Table 2.5, which are first finely ground and formed into bricks. The reaction takes 4–5 days in a sealed muffle furnace at 800°C, during which time the sodium carbonate reacts with the alumina and silica to form a zeolite, a hydrated silicate of calcium and aluminium. This reacts with the sodium polysulphides formed at the same time to give

TABLE 2.5

Ingredients used in manufacture of ultramarine

Ingredient	Amount/%
China clay	30
Sodium carbonate	32
Sulphur	30
Silica	4
Rosin	4

primary green ultramarine. As the furnace cools, air diffuses into it and the primary green oxidises to form raw blue ultramarine. The oxidation process can be accelerated by blowing in sulphur dioxide. The product at this stage contains 20–25% sodium sulphate, which is leached out with water. The remainder is then ball-milled in the wet state until the size of the particles is in the range 500–6000 nm. The particles are next separated into size fractions by sedimentation or centrifugal methods. The coarsest fraction is used for laundry blues, the intermediate fraction in plastics coloration and the finest in high-grade printing inks.

The formula as deduced from chemical and X-ray structural analysis is $Na_6Al_6Si_6O_{24}S_4$. There is an open framework of SiO_4 tetrahedra with shared corners. Some silicon ions are replaced by aluminium ions, however, and to preserve electrical neutrality some sodium cations are needed. Generally there are too many of these and the overall balance is maintained by having S^{2-}, S_2^{2-} and S_3^{2-} ions present. Consequently, chemical formulae containing fractional numbers of atoms are possible.

The colour of inorganic compounds is associated with charge transfer within the structure. This can be brought about by specific light absorption, so that the variability of ultramarine composition that can be achieved opens up many possibilities for colour changes. Pure starting materials, particularly the absence of iron impurities, are necessary in making consistent pigments.

Cadmium pigments

A range of mixtures containing cadmium sulphide and varying amounts of zinc sulphide and cadmium selenide can be made, the resulting colours being as shown in Table 2.6. The outstanding brightness of shade of these pigments cannot be matched by any other inorganic pigment. Equally bright organic pigments have lower heat stability than have cadmium

TABLE 2.6

Colours obtainable with cadmium pigments

Colour	Components
Primrose	CdS, ZnS
Yellow	CdS
Orange	CdS, 0.2 CdSe
Red	CdS, 0.4 CdSe
Maroon	CdS, 0.7 CdSe

pigments, which can withstand temperatures of 400°C. Other advantages of the cadmium pigments are better solvent fastness and high migration fastness, particularly in plastics, including ABS, polyolefins and extended PVC.

The starting material for making these pigments is cadmium sulphate, which must be free from iron, nickel and copper impurities. Cadmium sulphide is precipitated from the sulphate solution by adding an alkaline solution of pure sodium sulphide under controlled conditions of pH, temperature and addition rate. The product is in the cubic crystal form, which is converted to the required hexagonal form by calcination at 500–600°C in the absence of air.

Pigment containing zinc sulphide is made by adding zinc sulphate to the cadmium sulphate starting material, and the zinc sulphide appears in solid solution in the final product. The pigments consisting of mixtures of cadmium sulphide and selenide are made rather differently. An alkaline slurry of cadmium carbonate is made by mixing solutions of cadmium sulphate and sodium carbonate. At the same time selenium is added to a solution of sodium sulphide to form a solution of sodium sulphoselenide as a source of S–Se polyanions. This must be kept alkaline to prevent the precipitation of selenium. Mixing of this solution in the correct proportions with the cadmium carbonate slurry results in the precipitation of cadmium sulphoselenide in the desired proportions. After washing, the precipitate is calcined to give the pigment.

Both cadmium and selenium are poisonous and there are regulations about the limits of free cadmium in the pigment (100 p.p.m.) and on the disposal of waste liquors from manufacturing plants.

As well as the uses in plastics already mentioned, these pigments are used in car paints and stoving enamels, and also in leather finishes and rubber.

Iron oxide pigments

Naturally occurring iron oxide pigments are widely distributed geographically and can be found in a wide range of colours from black to reds and yellows, depending on composition and crystal structure. Ochres of many kinds and from different sources were widely used in oil paintings by the Old Masters. The use of magnetic iron oxides in recording tapes has stimulated their investigation at advanced technological level, but in the field of pigments their use has declined to that of cheap fillers for controlling paint flow properties and of facing colours for building bricks.

Iron(II) sulphate is a by-product in many industrial processes, such as the manufacture of titanium dioxide, the pickling of steel sheet before galvanising and the reduction of aromatic nitro compounds to amines using iron catalysts. Conversion of waste iron(II) salts to usable iron oxide pigments where the quality requirements are not too stringent is therefore a useful proposition, since it uses up chemicals that would otherwise be regarded as waste products.

Yellow iron oxides can be prepared by precipitating iron(II) hydroxide from a solution of iron(II) sulphate by adding sodium hydroxide and then bubbling air through the heated suspension to oxidise it to α-iron(III) oxide hydrate (α-FeOOH or $Fe_2O_3.H_2O$). An alternative method uses a single step of passing a mixture of air and ammonia into the iron(II) sulphate solution, followed by a heating stage. Seeding of the precipitated dispersion with a previously prepared pigment of the required crystal form helps to control the quality of the final product.

Red, brown and black iron oxides are prepared by first heating green iron(II) sulphate crystals to remove six of the molecules of water of crystallisation (leaving $FeSO_4.H_2O$) and then calcining the product to iron(III) oxide with the evolution of sulphur dioxide and trioxide.

The crystallography of iron oxides is complex because there are two oxides, FeO and Fe_2O_3, as well as α FeOOH, which can occur in mixed crystals. Kittel [29] has given a full account of the chemistry and Feitknecht [30] has discussed the theory of the colours of these pigments.

2.8 HOW PIGMENTS ACT AS COLORANTS

This chapter has stressed the importance of preparing pigments in the correct physical form, that is, a form from which a dispersion of very small particles can be made. In the case of polymorphic pigments there is also the need to ensure that the particles are in one crystal form only. To understand the overriding importance of the physical form of pigments it is necessary to look in some detail at the optical behaviour of tiny particles dispersed in transparent media. Figure 2.2 shows a single pigment particle embedded in a solid transparent organic medium. These have a refractive index of about 1.5.

The refractive index of a medium is defined as the ratio of the velocity

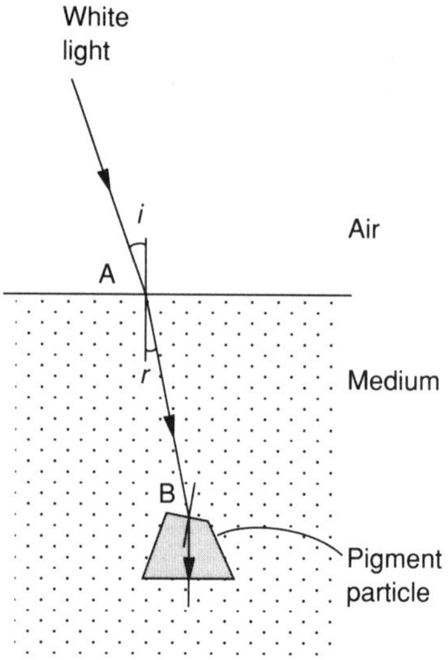

Figure 2.2 – Optics of a single pigment particle in a solid transparent organic medium

at which light travels *in vacuo* (in air is near enough for the present purpose) to the velocity at which it travels in the medium. The ray of white light entering the medium at A slows down, with the result that its path is bent towards the normal at this point. Snell's law of refraction (for that is what is happening here) is given by Eqn 2.1:

$$n = \frac{\sin i}{\sin r} \tag{2.1}$$

where n is the refractive index. The ray then travels on and strikes the pigment particle at the point B. Here it is again refracted and, because the pigment particle has a higher refractive index than the medium, its direction of travel is again deflected as shown. If the angle of incidence at the back face of the crystal is too great, total internal reflection may occur.

So far the nature of the pigment particle has not been defined. If it is a particle of a red pigment as shown in Figure 2.3, for example, some wavelengths of the white light will be absorbed as the ray passes on its way through the particle. The longer the path, the more light will be absorbed. When the ray emerges from the particle at C it will be again refracted, this time away from the normal and it will be red in colour. If the particle is too large or the absorption too high, all the light can be absorbed and the colouring power of the particle wasted.

If, on the other hand, the particle had been of a white pigment, it would have been much smaller in size, say 200 nm, below that of the wavelength

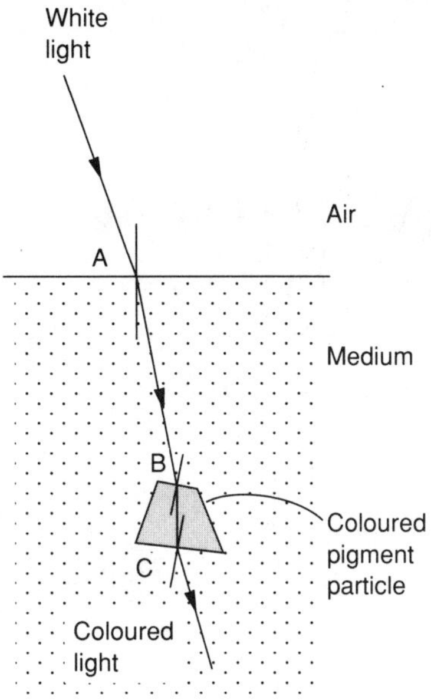

Figure 2.3 – Effect on light scattering of a single coloured particle

of visible light (400–650 nm). Then, as shown in Figure 2.4, the light will be scattered in many directions, some of it passing back out of the medium to the eye of an observer.

The combination of a coloured pigment particle and a white pigment particle shown in Figure 2.5 gives an idealised combination of the two effects, the coloured particle producing only light absorption and the white pigment producing only scattering. In practice most pigment particles produce both scattering and absorption in some degree.

This analysis itself oversimplifies the optical effects. Every refraction is accompanied by some reflection, and effects due to dispersion (the variation of refractive index with the wavelength of light refracted) have been ignored. Scattering itself is wavelength-dependent (inversely proportional to the fourth power). Absorption coefficients depend on the plane of polarisation of light with respect to crystal planes (coloured crystalline pigments are birefringent and also exhibit dichroism).

In a concentrated dispersion of white and coloured pigments, each of a more or less extended size range, a detailed analysis of ray directions becomes impossible. It is possible to deduce some general principles, however. These are that white pigments should be of a narrow size distribution, peaking in the region of 200 nm, and that coloured pigments should be of larger sizes, peaking around 500 nm, depending on the value of their absorption coefficients.

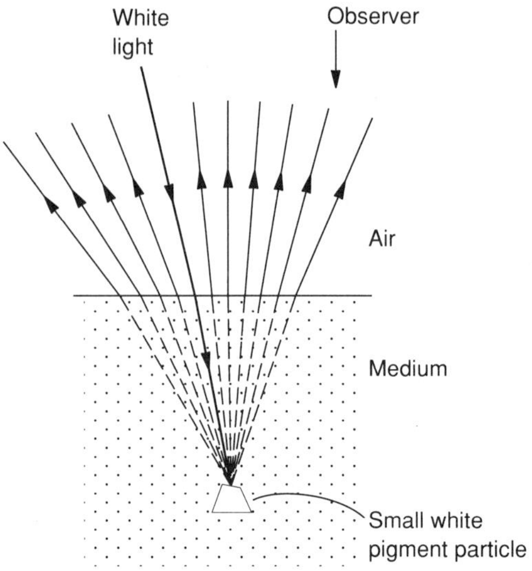

Figure 2.4 – Scattering produced by single white pigment particle of size below the wavelength of light

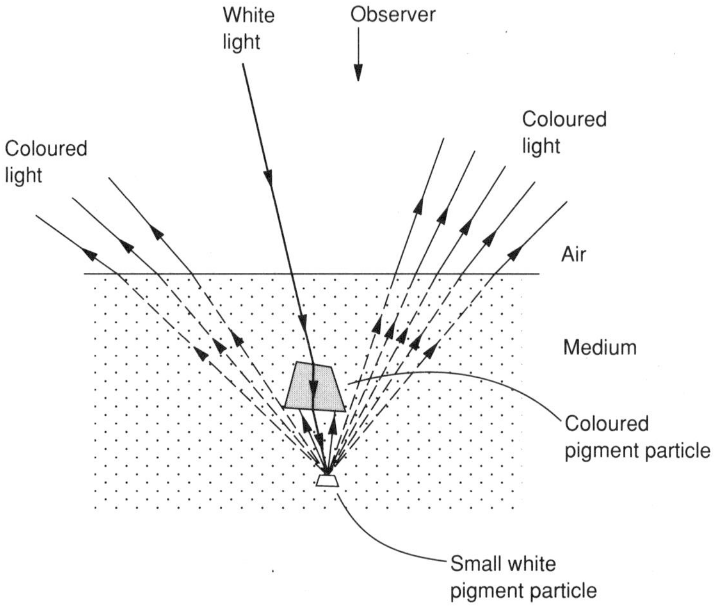

Figure 2.5 – Effects on light scattering of a combination of coloured pigment particle and white pigment particle

This analytical approach is difficult to apply to individual pigments because physical data relating to refractive index, dispersion curves and the absorption curves in the solid state are not available. A colligative approach, based on the Kubelka–Munk analysis which characterises pigments by only two constants, an absorption and a scattering coefficient, has been applied with considerable success to the computation of the proportions of pigments in mixtures needed to match a given colour. Much of the book *Colour physics for industry* is devoted to this topic [31].

There are also problems arising from the electrical interactions between different pigments and between pigments and the media in which they are dispersed, particularly when these are liquid paints and inks. They are discussed in the textbooks by Parfitt and Apps (amongst others), given in the bibliography at the end of this chapter.

2.9 SOLVENT DYES

Most textile dyes are insoluble in organic solvents and are not desorbed in the dry-cleaning processes to which garments and fabrics are customarily subjected. Pigments are required to be insoluble both in water and in organic solvents. That leaves still another class of colorants, which are water-insoluble but soluble in organic solvents, the so-called solvent dyes.

Solvent dyes have specialised uses as markers for the identification of different hydrocarbon fractions produced in oil refineries. Although the dyes are used at low concentrations, the vast volumes involved can make the amounts of dye required quite large. These dyes are also used in coloured transparent varnishes and lacquers for the furniture and leather industries, where the surface texture of the manufactured articles must be protected but not obscured. Similar needs for transparent colouring finishes arise in the packaging and printing industries. For example, attractive coloured effects can be imparted to aluminium foil forming parts of disposable packages, the durability of printed labels on bottles can be improved and printing can be carried out directly on transparent wrappers made of cellophane, PVC or polythene. Metallic pigments in the form of very fine flakes can be transformed to 'gold' by coating with suitably coloured lacquers.

There have probably been few, if any, research programmes directed to the synthesis of solvent dyes. Rather, the need has been met by a selection or modification of known dyes to give the solubility in the solvents required in special applications.

The *Colour Index* lists a surprisingly large total of 800 solvent dyes, although in only a few instances are the chemical structures revealed. Of these, most of the yellow and orange examples are monoazo or disazo dyes, while among the blues there are anthraquinones. Among the monoazo dyes, chromium complexes are sometimes formed to mask any water solubility in the parent dye. In others, insolubility in water but high solubility in solvents are produced by the reaction of sulphonic acid groups with long-chain alkylamines.

2.10 CONCLUSION

The diversity of chemical structures involved in pigment chemistry is much less that that in dyes, but extends from organic to inorganic types. The technology involved in the use of pigments is one of dispersion and not that of dyeing. As a result, even a compound that is chemically suitable for pigment use may not be acceptable in practice unless it can be prepared in a physical form that allows its potentialities to be fully exploited.

The range of pigments commercially available is diminishing. There are several reasons for this. One is a rationalisation in the range of intermediates available from the chemical industry for their preparation; another is that some have been recognised as toxic. Other pigments have disappeared because of demands from consumers for higher standards of fastness, particularly to light. Yet another factor is the growth of instrumental colour match prediction, which, in theory at any rate, provides the possibility of matching wide colour ranges by mixing only three or four pigments. Other technical problems such as dispersion stability will doubtless prevent such a simple solution in practice, however.

REFERENCES

1. L M Valentine, *J. Oil Col. Chem. Assocn*, **67** (1984) 157.
2. R J Gettens and G L Stout, *Painting materials* (New York: Dover Publications, 1966).
3. A G Perkin and A E Everest, *The natural organic colouring matters* (London: Longmans, 1918).
4. A W C Harrison, *The manufacture of lakes and precipitated pigments* (London: Leonard Hill, 1930).
5. L S Pratt, *The chemistry and physics of organic pigments* (New York: Wiley, 1947)
6. BIOS Report 1661 (London: HMSO, 1947) 16–48.
7. V C Vesce, *Official Digest*, **28** (1956) 1; **31** (1934) 530.
8. H Gaertner, *J. Oil Col. Chem. Assocn*, **46** (1963) 13.
9. H J Flath, *Textiltechnik*, **33** (2) (1983) 111.
10. BIOS Report 1661 (London: HMSO, 1947) 49, 83, 98, 99, 100.
11. D Patterson, *Bunsengesell. Phys. Chem.*, **71** (1967) 271.
12. J D Hildreth and D Patterson, *J.S.D.C.*, **79** (1962) 519.
13. A Zitscher, *J.S.D.C.*, **70** (1954) 530.
14. J C Harris, *J. Amer. Chem. Soc.*, **70** (1948) 3671.
15. A Morris, MSc thesis, University of Leeds (1964).
16. H B Weiser and E E Porter, *J. Phys. Chem.*, **31** (1927) 1704.
17. N H Haddock, *J.S.D.C.*, **61** (1945) 68.
18. W Braun and O Tcherniac, *Ber.*, **40** (1907) 2709.
19. H de Diesbach and E van der Weid, *Helv. Chim. Acta*, **10** (1927) 886.
20. J Robertson, *J.C.S.*, (1935) 615.
21. BIOS Report 960 (London: HMSO, 1947) 47.
22. FIAT Report 1313, Vol. 3, 273.
23. W Gerstner, *J. Oil Col. Chem. Assocn*, **49** (1966) 954.
24. F M Smith and D Easton, *J. Oil Col. Chem. Assocn*, **49** (1966) 614.

25. FIAT Report 1313, Vol. 3, 285.
26. H Liebermann, *Ann.*, **518** (1935) 245.
27. P Pollack, *Chimia*, **30** (1976) 357.
28. Gy, US P 2 973 358 (1961).
29. H Kittel, *Pigments* (Stuttgart: Wissenschaft Verlag, 1960).
30. W Feitknecht in *Pigments* (London: Elsevier, 1967).
31. *Colour physics for industry*, Ed. R McDonald (Bradford: SDC, 1987).

BIBLIOGRAPHY

As well as the abstracted volume from the *Colour Index*, entitled 'Pigments and solvent dyes', many pigment manufacturers issue pattern cards giving technical information about their products. In addition to the books mentioned in the list of references above, the following books contain considerable chapters dealing with pigments and their technology.

1. E A Apps, *Printing ink technology* (London, Leonard Hill, 1958).
2. K McLaren, *The colour science of dyes and pigments*, 2nd Edn (Bristol: Adam Hilger, 1986).
3. W M Morgans, *Pigments for paints and inks* (Manchester: Selection and Industrial Training Ltd, 1979).
4. F H Moser and A L Thomas, *Phthalocyanine compounds* (New York: Reinhold, 1963).
5. Y L Meltzer, *Phthalocyanine technology* (New Jersey: Noyes Data, 1970).
6. G W Parfitt and K S W Sing, *Characterisation of powder surfaces* (London: Academic Press, 1976).
7. G W Parfitt, *Dispersion of powders in liquids, with special reference to pigments*, 3rd Edn (London: Applied Science Publishers, 1981).
8. D Patterson, *Pigments, an introduction to their physical chemistry* (London: Elsevier, 1967).
9. T C Patton, *Pigment handbook* (3 vols) (New York: Wiley, 1973).

CHAPTER 3

Dye structure and application properties

John Shore

3.1 DYE CHARACTERISTICS AND CHEMICAL STRUCTURE

Virtually all commercial dyeing and printing processes take place by the application of a solution or a dispersion of the dye to the textile material followed by some type of fixation process. The dye solution or dispersion is almost always an aqueous medium, and a major objective of the fixation step is often to ensure that the coloured textile exhibits satisfactory fastness to subsequent treatment in aqueous wash liquors. In view of the overriding importance of water as a transfer medium in dyeing and printing it seems reasonable to begin this section with a discussion of the properties of dye solutions and dispersions.

3.1.1 The solubility of dyes in water

In a pure ice crystal each oxygen atom is surrounded by four hydrogen atoms disposed in a tetrahedral configuration. Two of them are covalently bonded to the oxygen atom, conveniently forming an angle of 120°. The other hydrogen atoms from two other molecules are thus located equidistantly on opposite sides of the plane of the H–O–H structure. In steam, on the other hand, each H–O–H molecule is rapidly and freely in motion, frequently colliding with other molecules but not capable of forming stable hydrogen bonds with them at such an elevated temperature. The association between water molecules in the liquid state represents an intermediate condition. Molecular motion is slower than in steam and the formation, motion, collision and re-formation of clusters of two or more molecules is a characteristic feature of the system [1].

This special nature of the hydrogen atom permits the two electronegative oxygen atoms in neighbouring water molecules to approach one another more closely than they could have done in the absence of the intervening hydrogen atom. The hydrogen atom normally consists of a positively charged nucleus (a proton) and a much smaller orbiting electron. The proton is thus virtually unscreened, and when interposed between two much larger electronegative oxygen atoms it minimises the force of repulsion between them. This reduction in repulsion is known as a hydrogen bond [2].

The idealised two-dimensional representation of a seven-molecule cluster shown here (3.1) illustrates that some molecules are more firmly held than others. In reality, of course, each cluster is three-dimensional in shape and the individual molecules adopt a more random orientation than indicated here. As the temperature of the liquid solvent increases and the cluster becomes more mobile, the loose network of hydrogen bonds (indicated by broken lines) holding the molecules together is weakened. In a collision the cluster is more likely to lose one or more fringe members (a) depending on a single hydrogen bond than those held in place by two (b), three (c) or four (d) links. The average cluster size thus decreases as the temperature rises, and a higher proportion of free H–O–H molecules is formed as the boiling point is approached, heralding the incipient evaporation as steam.

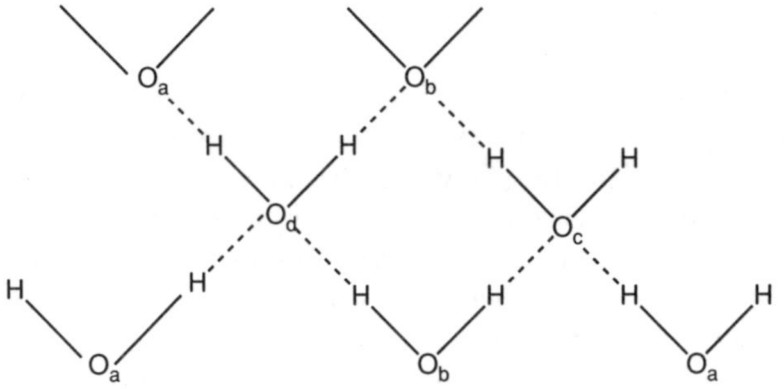

3.1

The dissolution of an inert non-polar solute molecule in water has been described in terms of the 'iceberg' concept, by which transient local clusters of water molecules are stabilised by cohesive association around the non-polar molecule [3]. The enveloping cluster surrounding the latter is believed to contain a higher proportion of multi-linked members (b–d) than are present in the typical freely mobile clusters characteristic of pure water.

Hydrophobic interaction between inert non-polar solute molecules arises as a direct consequence of the 'iceberg' effect. There is always a tendency for pairs of such stabilised cohesive clusters to merge and redistribute, forming a larger and more stable envelope surrounding two or more solute molecules. These molecules then behave as if a binding force exists between them, that is, a hydrophobic interaction [4]. An analogous process of clustering within the aqueous envelope to favour preferential interaction between the hydrophobic portions of dissolved solute ions is similarly evident in concentrated solutions of dyes and surfactants. A dissolved ionic solute of this kind will always have a more stable envelope of water molecules surrounding the hydrophobic portion of the ion. This asymmetric distribution will favour coalescence of solute molecules in concentrated solution to form a solute micelle or a dye aggregate. Once formed, however, such an associated moiety is further stabilised by the inner envelope of charged ionic groupings oriented outwards towards the aqueous phase.

Many anionic dyes (see Chapter 1) depend on the possession of one or more sulphonic acid groups for their solubility in water. Dye sulphonic acids normally have pK values within the range of pH 1–2 and are thus fully ionised under dyeing conditions as either the free acid or the sodium salt [5,6]. The mutual electrostatic repulsion between dye sulphonate anions favours their uniform separation and distribution in dilute aqueous solution. At higher concentrations, however, this repulsion is counterbalanced by mutually attractive forces of various kinds operating at shorter range [7]:

(a) hydrogen bonding
(b) van der Waals forces
(c) hydrophobic or non-polar bonding.

The most important types of hydrogen bond, in order of relative strength, are the hydroxy–ether, hydroxy–amine and amino–amine linkages (Figure 3.1). Thus one of the most significant forces of attraction between dye molecules containing alcoholic or phenolic hydroxy groups is precisely that which governs the mutual interaction in clusters of water molecules. Many water-soluble dyes contain primary, secondary or tertiary amine groupings (see Chapter 1), all of which are capable of participating via hydrogen bonding in dye–dye association or dye–water solubilisation mechanisms. The van der Waals forces include dipole–dipole interaction between atoms of unlike polarity in neighbouring molecules and less specific π-bonding or dispersion forces between unsaturated groupings and aromatic nuclei. Hydrophobic bonding between non-polar regions of dye ions is particularly significant for those acid dyes of higher wet fastness solubilised by a single sulphonate group and for most basic dyes, which normally contain only one quaternary nitrogen atom (see section 1.6.6).

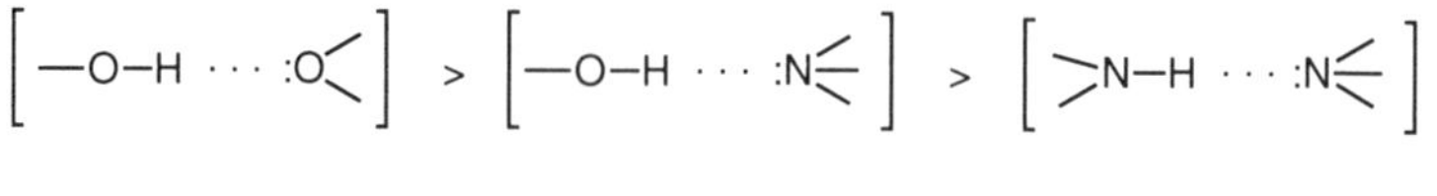

Figure 3.1 – Types of hydrogen bond in order of relative strength (from the left: hydroxy–ether, hydroxy–amine and amino–amine)

The solubility of ionic dyes in water normally increases with temperature, since their enhanced mobility at higher temperature favours electrostatic repulsion between ions, rather than closer approach to form aggregates by means of the short-range attractive forces discussed above and ultimately precipitation from solution at the saturation limit. Addition of an inorganic electrolyte, on the other hand, normally lowers this solubility limit at a given temperature, although anomalous behaviour has been observed in some instances [8]. Such additions enhance the ionic character of the aqueous phase and help to stabilise the structure of dye aggregates by forming an electrical double layer within the envelope of clustered water molecules surrounding them.

In contrast, the addition of a water-miscible non-ionic solute such as urea or an alcohol to an aqueous solution containing aggregates of dye ions has a disaggregating action at a given temperature and electrolyte concentration. Urea has a strong tendency to inhibit the clustering of water molecules arising from the competitive influence of urea–water hydrogen bonding. Consequently the stabilising 'iceberg' cluster surrounding each dye aggregate tends to disintegrate, and dye–urea hydrogen bonding or dipole–dipole attractive forces further disrupt the hydrophobic interactions between pairs of dye ions.

Non-ionic solutes with a directional polar and non-polar structural arrangement, such as long-chain alkyl polyoxyethylene adducts, destabilise the loose structure of dye aggregates by a somewhat different mechanism. These additives readily form micelles in which the hydrophobic alkyl groups form the core and the oxyethylene chains are hydrogen-bonded within an extensive shell of water molecules clustered around them. The formation of these large clusters denudes the less firmly held envelopes surrounding the dye aggregates, which become smaller and more unstable. The lower dielectric constant of the surfactant solution enhances the electronegative forces of repulsion between monomeric dye anions. Hydrogen bonding or dipolar interaction between the oxyethylene chains and individual dye anions further minimises the probability of hydrophobic attraction between pairs of dye anions.

All aqueous solutions of anionic or cationic dyes have lower surface tension than pure water. Solutions of the free sulphonic acids exhibit lower values than solutions of the sodium salts of anionic dyes. The decrease in surface tension is more dependent on the degree of sulphonation of the dye than on its relative molecular mass, although alkyl substituents can exert a further lowering effect [9]. Thus a relatively hydrophilic tetrasulphonated disazo dye such as C.I. Direct Blue 1 (3.2) has comparatively little influence on surface tension, whereas a monosulphonated acid milling dye such as C.I. Acid Yellow 72 (3.3) can exert a mild detergency effect [10].

Measurements of the surface tension of aqueous solutions of various sulphonated and unsulphonated phenylazonaphthol dyes showed that the degree of surface activity (that is, the lowering of surface tension) tends to increase progressively with the degree of the alkyl substitution in the series of dyes [11]. The surface-active characteristics of such dye anions

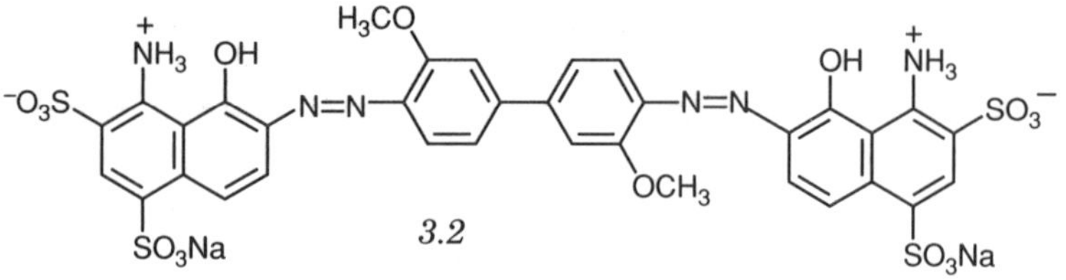

3.2

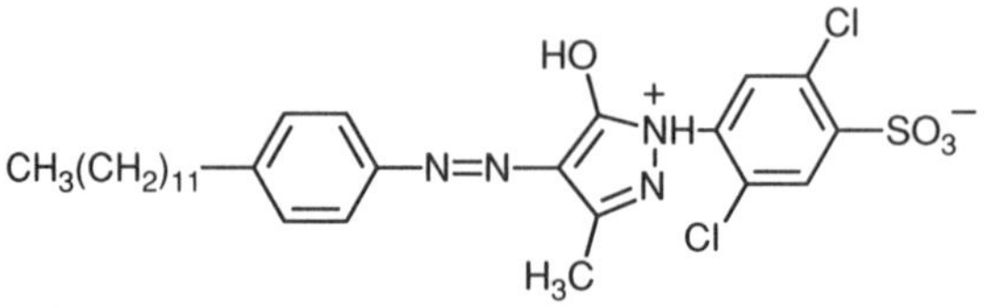

3.3

ensure that they tend to become more concentrated at the interface between the dyebath and the fibre surface, just as they do at the air–water interface of the dye solution. Foaming of dye solutions can be a serious practical problem with relatively hydrophobic structures dissolved by means of a single solubilising group.

3.1.2 Aggregation of dyes in aqueous solution

Dyes often exist in aqueous solution as aggregates of several molecules or ions rather than individual moieties, as discussed above. It is not necessary to assume that all component molecules in a dye aggregate are in the same state of ionisation. Unlike surfactants, dye molecules in general do not possess structures with directional polar and non-polar segments, and thus do not associate as micelles of ordered structure with a well-defined hydrophobic core and a charged hydrophilic sheath. Dye aggregates should be envisaged as relatively amorphous in composition with zones of more or less polar character distributed within them, although there will still be a tendency for the individual molecules to become oriented with their ionised groupings towards the aqueous phase.

Aggregation is promoted by a high ratio of relative molecular mass to ionic group content, that is, by a high value for the equivalent mass per sulphonic acid substituent in the case of sulphonated anionic dyes. Multisulphonate dye anions may be stabilised as dimers in which the two components adopt a planar arrangement with the sulpho groups located if possible at opposite ends of the dimer. More hydrophobic structures of higher equivalent mass, especially monosulphonates, can readily associate together to form aggregates of larger sizes. Dimers are formed first, growing further by accretion of more dye anions to form lamellar micelles in which the dye moieties are stacked like cards in a pack [7]. Planar chromophores, especially the sulphonated phthalocyanines, are particularly prone to such stacking.

Equilibrium constants for dimerisation (K_D) have been determined [7] for typical dyes from several different classes (Table 3.1). They are all within the same order of magnitude (log K_D = 3–4 at 25°C). Since the molecules of typical *o*-hydroxyazo dyes normally form intramolecular hydrogen bonds, the contribution of hydrogen bonds of the intermolecular type may be relatively unimportant for such dyes. Thus the heats of dimerisation[1] of Orange II (–10.5 kcal/mol) and Solochrome Violet R (–9.2 kcal/mol) are similar, even though the latter has twice as many phenolic hydroxy groups per molecule [7]. It seems likely that with most dyes the forces governing aggregation are mainly of the van der Waals type, including dipole–dipole and dispersion forces; these are able to operate through many layers of stacked dye molecules. An attempt to obtain equilibrium constants for the dimerisation of four *p*-substituted phenylazo-1-naphthol-6-sulphonic acids failed because of the superimposed effect of acid–base, tautomeric (hydroxyazo–quinonehydrazone) and oligomerisation equilibria [12].

TABLE 3.1

Equilibrium constants for dimerisation of dyes [7]

Dye	Structure	Log K_D
Acridine Orange (C.I. Basic Orange 14)	3.4	4.02
Magdala Red (C.I. Basic Red 6)	3.5	3.78
Methylene blue (C.I. Basic Blue 9)	3.6	3.77
Rhodamine 6G (C.I. Basic Red 1)	3.7	3.36
Solochrome Violet R (C.I. Mordant Violet 5)	3.8	3.29
Orange II (C.I. Acid Orange 7)	3.9	3.12

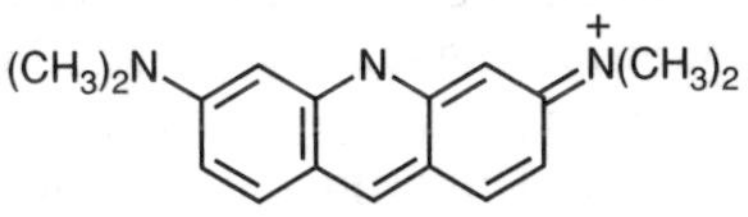

3.4

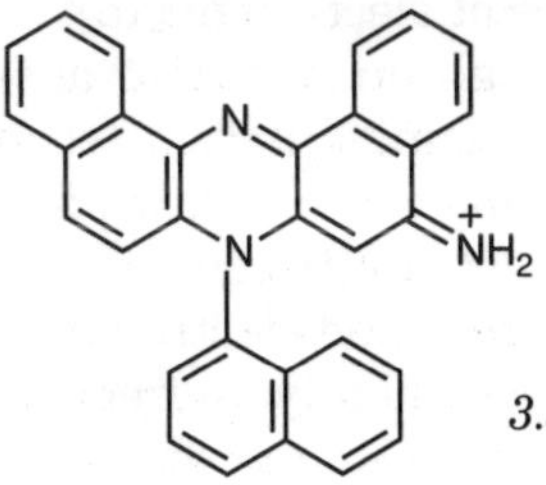

3.5

[1] In the tables and text of this chapter there are frequent references to the thermodynamic data expressed in kcal/mol, as quoted by the original authors of the work cited. If necessary, these values can be converted to the SI units of kJ mol^{-1} by means of the conversion factor: 1 kcal = 4.18 kJ.

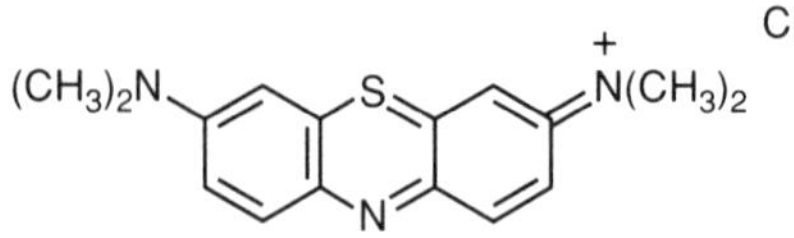

3.6

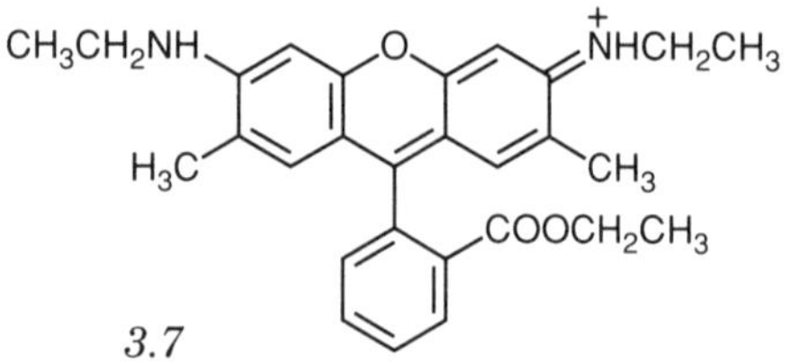

3.7

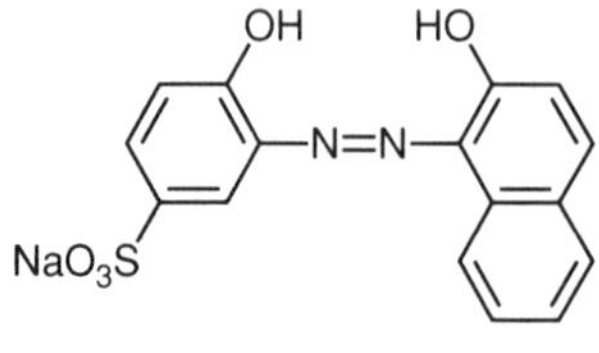

3.8

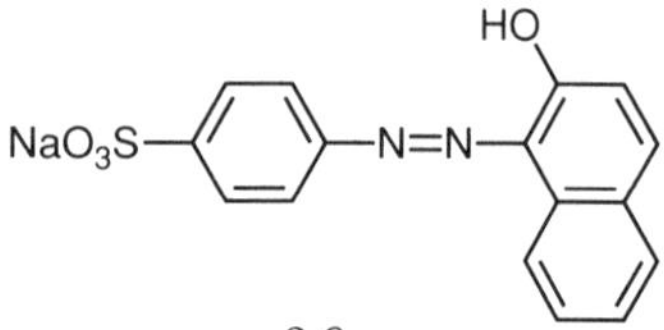

3.9

Most measurements of aggregation number (the average number of molecules/ions per aggregate) have been carried out on solutions of direct dyes, since these dyes aggregate more readily than most of the other classes of water-soluble dyes. Regrettably, aggregation numbers determined for the same dye under the same conditions using different methods of measurement are seldom consistent. Most investigators agree, however, that such solutions contain a mixture of aggregates of various sizes in dynamic equilibrium, somewhat analogous to the mixed clusters of molecules in pure water. As individual ions, dimers or trimers are removed by adsorption on to the fibre surface during dyeing, larger aggregates break down to maintain a similar overall distribution of aggregate sizes [13,14]. It is even more certain that the extent of aggregation of a direct dye decreases with increasing temperature and increases with increasing concentration of added electrolyte or of the dye itself. Although many direct dyes are highly aggregated at room temperature, the degree of aggregation is often negligible under normal dyeing conditions at the boil even in the presence of electrolyte [15].

Although quantitative agreement between the aggregation numbers obtained using different techniques of measurement is poor, meaningful conclusions can be drawn from them regarding the relative aggregative tendencies of different direct dye structures. Under given conditions of temperature and electrolyte concentration, C.I. Direct Orange 26 (3.10) is much more highly aggregated than Chrysophenine G (3.11; C.I. Direct Yellow 12). Both are disulphonated disazo dyes, but presumably the two phenolic groups and especially the urea residue in the former dye provide much greater scope for intermolecular hydrogen bonding than the etherified phenolic groups in the latter.

A detailed comparison of C.I. Direct Red 81 (3.12; X = H), a dye that has good migration properties but low affinity and poor fastness to washing,

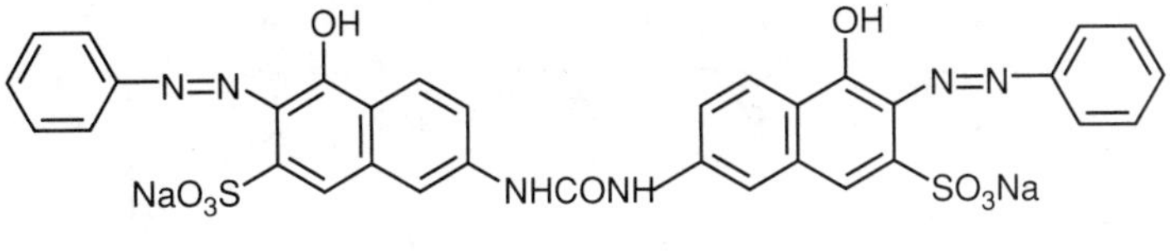

3.10

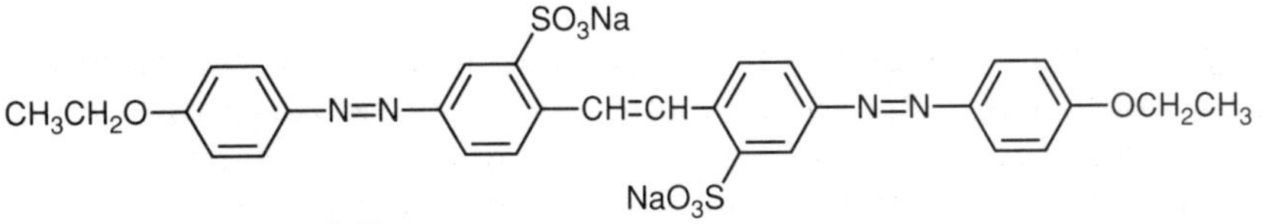

3.11

with its *p*-benzoylamino analogue (3.12; X = NHCOPh) revealed that the higher affinity and wet fastness but inferior migration of the latter could be correlated with an increased tendency to self-associate, presumably due to the greater scope for hydrogen bonding via the amide groupings.

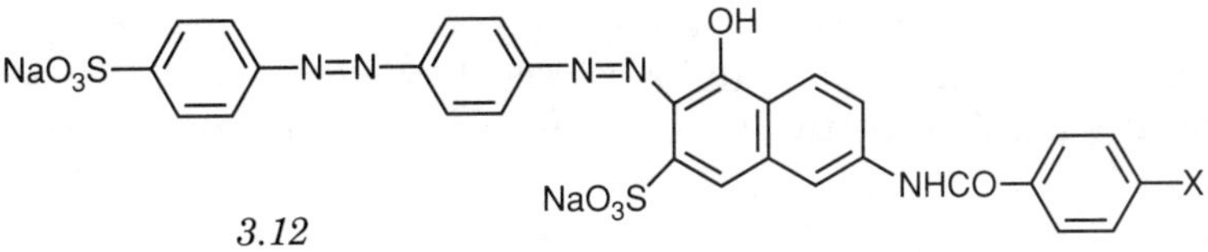

3.12

Three disazo analogues derived from the zwitterionic naphthionic acid – Congo red (C.I. Direct Red 28) (3.13; X = Y = H), Benzopurpurin 4B (C.I. Direct Red 2) (3.13; X = CH_3, Y = H) and its *m*-tolidine isomer (3.13; X = H, Y = CH_3) – are all highly aggregated [16,17], no doubt because of electrostatic intermolecular attraction between the matching pairs of protonated primary amino groups and sulphonate anions of the naphthionic acid residues. The two tolidine derivatives aggregate more readily than does Congo red, although the non-planar *m*-tolidine dye is not so highly aggregated as the coplanar *o*-tolidine structure in Benzopurpurin 4B, which is more sensitive to flocculation by salt and several times more substantive to cellulose than its *m*-tolidine isomer. C.I. Direct Blue 1 (3.2) is considerably less aggregated than these naphthionic acid derivatives [18] because the charges on the protonated amino groups are satisfied by intramolecular association with their neighbouring sulphonate groups, and the dye anion thus has a net negative charge of two.

Fewer results are available to define the aggregation characteristics of solutions of acid dyes. An increase in the size of the planar aromatic nuclei in monoazo acid dyes (for example, replacement of a phenyl by a naphthyl residue) increases the tendency to aggregate. Conversely, increasing the

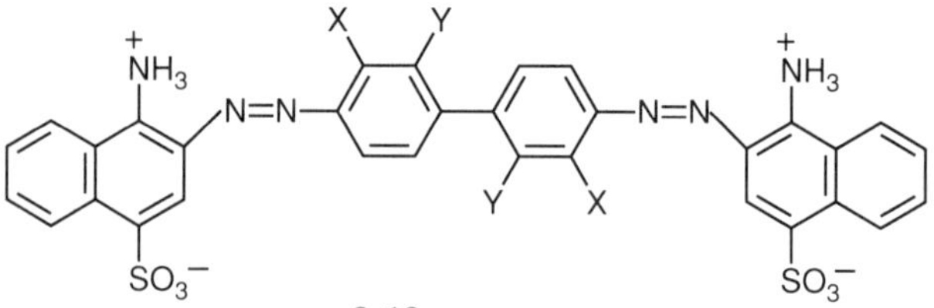

3.13

number of sulphonic acid groups minimises aggregation, especially if they are widely separated from one another and from the azo group [19]. Light-scattering studies demonstrated freedom from aggregation with the monoazo monosulphonate Orange II (3.9) and many simple monoazo di- or tri-sulphonates, but typical disulphonated milling dyes such as C.I. Acid Red 111 (3.14) showed salt-sensitive aggregation. With the disazopyrazolone disulphonate C.I. Acid Orange 63 (3.15) the aggregation appeared to be virtually independent of the concentrations of dye and electrolyte at low temperatures. Aggregation greatly decreased above about 60°C, in good agreement with the anomalous dyeing properties of this dye [20]. It has been suggested that it may form stable micelles analogous to those of surfactants, since the dye molecules may associate at low temperatures through the zwitterionic sulphanilic acid residues, and this effect is apparently highly sensitive to temperature in the 50–70°C region [21].

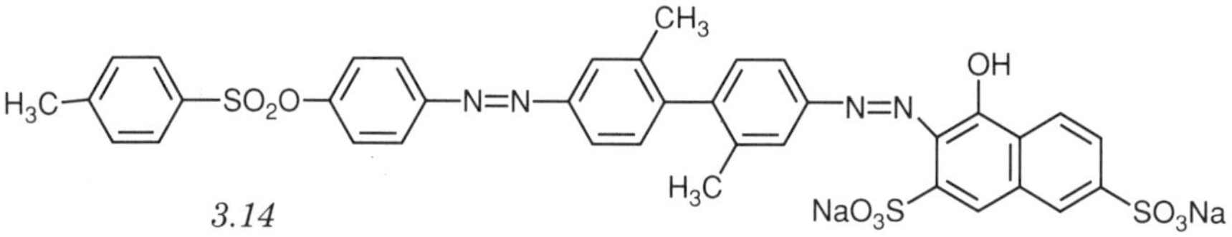

3.14

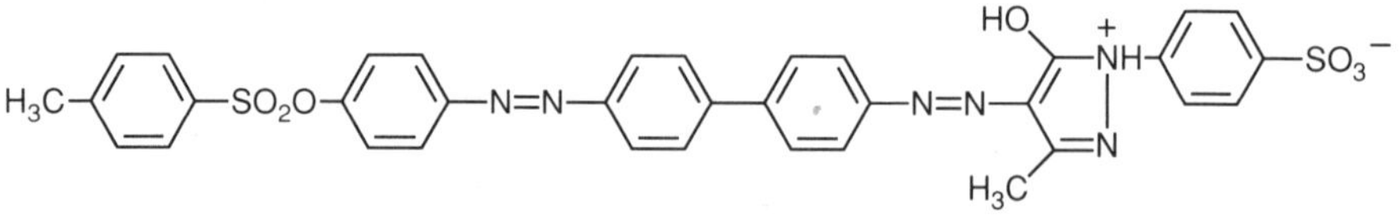

3.15

Owing to experimental difficulties, knowledge of aggregation behaviour in solutions of leuco vat dyes is even scantier [22,23]. Investigations based on absorption spectra have shown that, depending on concentration and temperature, planar molecules such as the violanthrone derivatives C.I. Vat Blues 19, 20 and 22 and the perylene tetracarboxydiimide derivatives C.I. Vat Reds 23 and 32 are mainly present as monomers or dimers in leuco vat solutions. Violanthrones that do not have a coplanar structure because of the presence of alkoxy groups in the 16,17 positions, such as C.I. Vat Blue 16 and C.I. Vat Greens 1, 2 and 4, are always present as single molecules under vat dyeing conditions.

3.1.3 Decomposition of water-soluble dyes during dyeing

The primary objective of an exhaust dyeing process is to transfer as much as possible of the dye in the system from the dyebath into the fibre, provided this can be achieved with the dye uniformly distributed throughout the material at the end of the process. If dyeing is unduly prolonged, either to improve the levelness or to attain an exact colour match to standard, decomposition of a minor proportion of the dye present may occur. The rate of decomposition is normally much more rapid in the dyebath phase than on the fibre and the onset of degradation can sometimes be observed as a change in colour of the exhaust liquor. This effect usually occurs as a hypsochromic shift, such as reddening of a blue or yellowing of a red solution.

Many direct dyes show instability of this nature if applied at temperatures above the boil – when dyeing polyester/cellulosic blends, for example. A frequent cause of the problem is reduction of the azo linkages, and the decomposition may be accelerated in the presence of viscose, which has a reducing action under alkaline dyeing conditions [24]. Direct dyes with exposed azo groups free from *o*-substituents – for example, the two azo links in Chrysophenine G (3.11) – are particularly prone to reductive decomposition. However, some dyes with protonated *o*-aminoazo groups, such as Congo red (3.13; X = Y = H) and Benzopurpurin 4B (3.13; X = CH_3, Y = H) are also reduction-sensitive [25]. In the relatively unstable trisazo structure of C.I. Direct Green 33 (3.16), the most vulnerable of the three azo groups is the unprotected central linkage, so that the likely initial products of reduction are two monoazo dyes (3.17 and 3.18), resulting in a hypsochromic reddening of the blue-green solution.

Dyes containing azo groups protected by *o,o'*-substituents, especially *o*-sulphonate groups, are generally relatively resistant to reductive breakdown during dyeing. Azothiazole dyes such as C.I. Direct Yellow 29 and especially copper-complex structures such as C.I. Direct Blue 98, in which both azo groups are fully protected as the metal complex (3.19), show excellent stability in high-temperature dyeing conditions [26]. Buffering with ammonium sulphate confers a stabilising influence when dyes of lower stability have to be used on viscose. Mild oxidants such as sodium *m*-nitrobenzenesulphonate are occasionally useful and more powerful inorganic oxidants, such as potassium chlorate or dichromate, can be most effective if used with care. Excessive use of these additives can result in oxidative decomposition of azo dyes, however.

Direct dyes of the diphenylurea type, such as C.I. Direct Orange 26 (3.10) and C.I. Direct Red 79 (3.20), are particularly prone to decomposition in

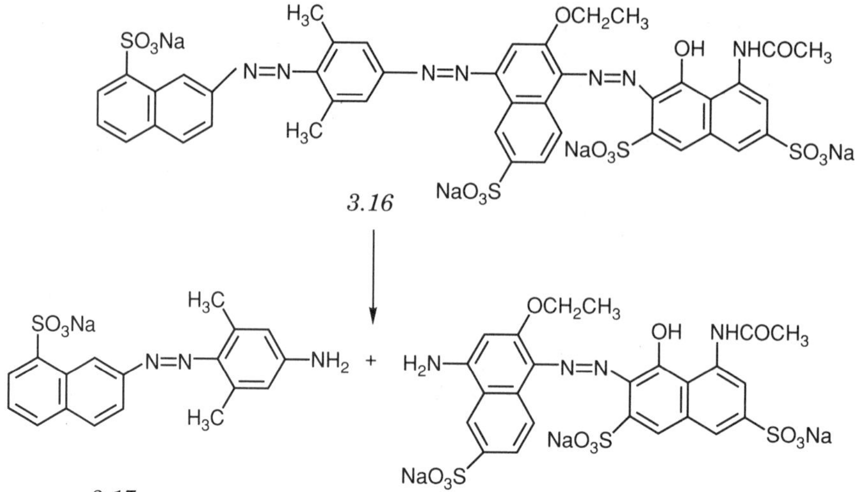

3.16

3.17

3.18

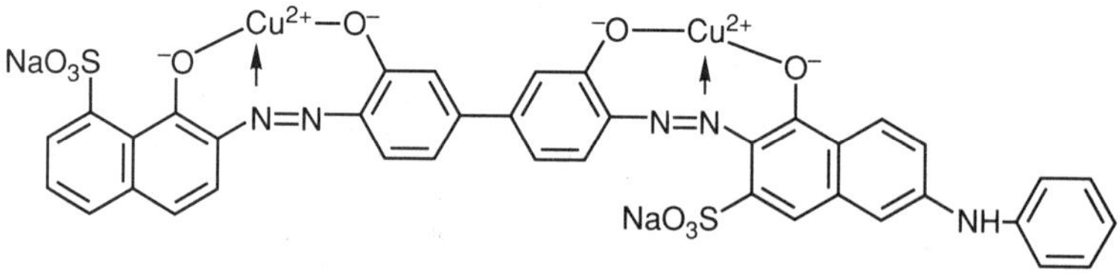

3.19

high-temperature dyeing because molecular breakdown can occur by hydrolysis [27] as well as by the reductive mechanism under alkaline conditions. Thus the ureido linkage may be broken by hydrolysis to give monoazo dye fragments (3.21) or the azo groups can be reduced and naphthylamine breakdown products formed.

The mechanism of degradation of water-soluble azo dyes by dilute solutions of sodium hypochlorite has been investigated in detail. A study of a series of 2-arylazo-1-naphthol-3,6-disulphonic acids (3.22) demonstrated that degradation involves initial attack at the nitrogen atom of the dye in its hydrazone form by the electrophilic chloronium ion (Cl^+). Sulphonic and carboxylic acid groups located *ortho* to the azo linkage are particularly effective here also in retarding the rate of azo decomposition. The arylazonaphthol dyes exist almost exclusively as the hydrazone tautomer in aqueous solution (see section 4.5.2) and this is more susceptible to attack by Cl^+ ion than is the keto form [28].

With some leuco vat dyes, chemical changes may take place in alkaline reducing media at high temperatures [29]. Many vat dyes contain benzoylamino substituents to confer higher affinity for cellulosic fibres. Alkaline hydrolysis of such groups may take place with the loss of benzoic acid

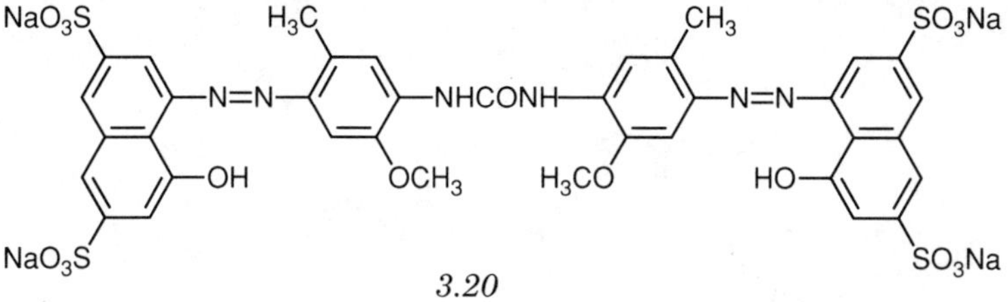

3.20

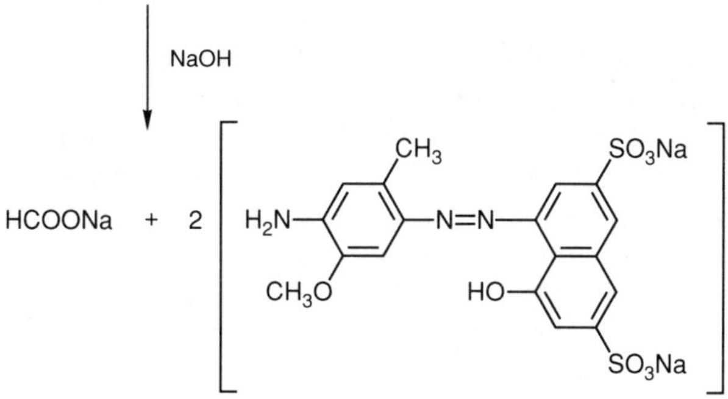

3.21

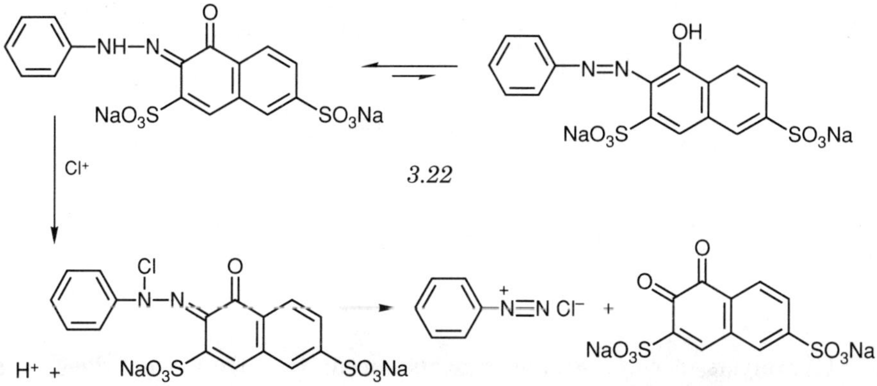

3.22

and a gradual lowering of substantivity, leading to unexpectedly low exhaustion of the dyebath. Certain halogenated indanthrone derivatives [for example C.I. Vat Blue 6 (3.23) and its monochloro analogue C.I. Vat Blue 14] suffer dehalogenation under excessively hot alkaline conditions. Furthermore, indanthrone (C.I. Vat Blue 4) and its derivatives can lose oxygen atoms as a result of over-reduction at temperatures above 60°C. Reoxidation gives a different product and it is no longer possible to regenerate the indanthrone starting material. The tendency for over-reduction increases with time and temperature of reduction and with the concentrations of alkali and hydrosulphite. Additions of glucose or sodium nitrite have a stabilising influence on these leuco solutions.

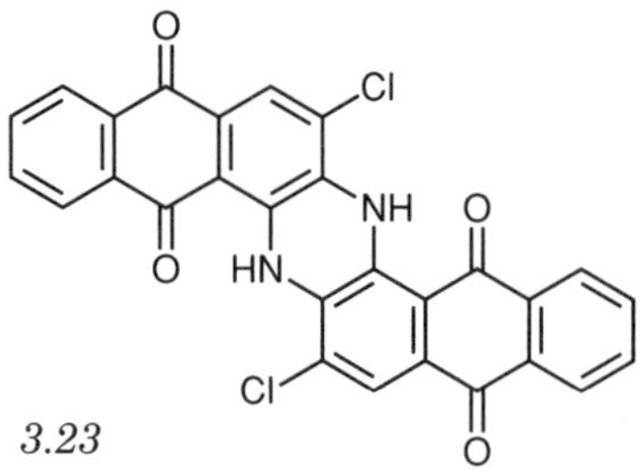

3.23

The azo linkage in some acid dyes is sensitive to the mild reducing action in hot alkaline media of the cystine and cysteine groups of wool. C.I. Acid Blue 113 (3.24) is cheap and shows good all-round wet fastness on wool and nylon, but it tends to turn brown during dyeing if exposed to alkaline reducing conditions at high temperatures. The degradation involves reduction to metanilic acid and a brown monoazo dye, the rate increasing rapidly above pH 7 [30].

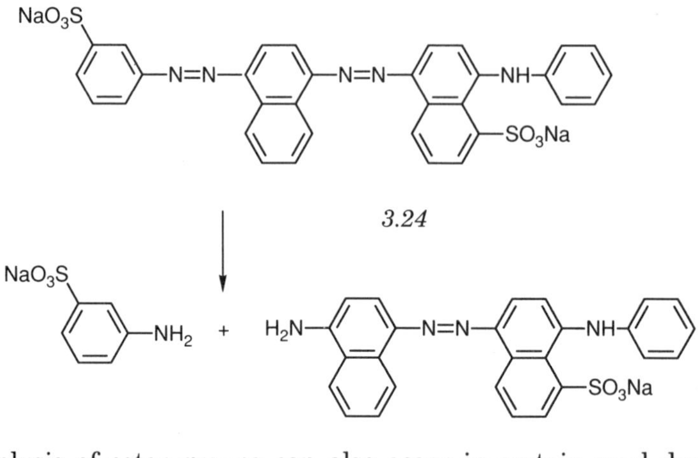

3.24

Hydrolysis of ester groups can also occur in certain wool dyes. For example, C.I. Acid Red 1 (3.25) loses its acetyl group on prolonged dyeing in an acid dyebath, the hue becoming bluer and duller and the dyeing much less fast to light. The anthraquinone derivative C.I. Acid Blue 45 (3.26) is sensitive to oxidation (by sodium nitrite for example), especially in the presence of trace metal ions such as those of iron(III) or copper(II), to form a dull reddish-blue tetrahydroxydisulphonate, which exhibits inferior levelling properties.

Mordant dyes are notoriously troublesome from the viewpoint of colour matching because the hue of the chromium complex usually differs greatly from that of the unmetallised parent dye (see Chapter 5). If other metal ions are present in the treatment bath or on the fibre during chroming, the colour obtained is likely to differ from that of the chromium complex.

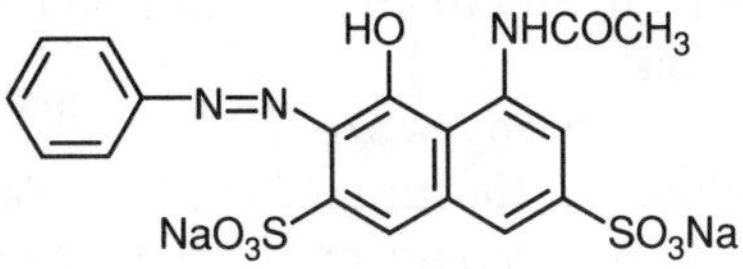

3.25

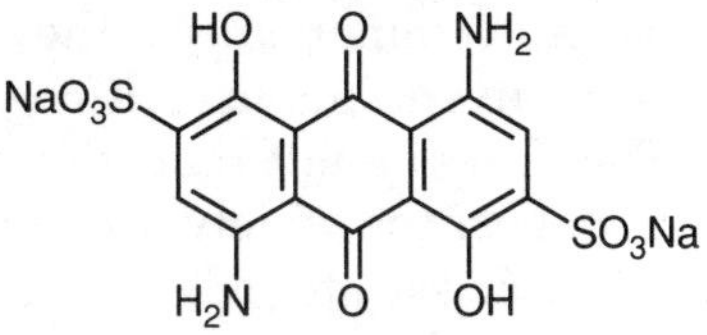

3.26

Certain important mordant dyes, including C.I. Mordant Blacks 11 and 17 (3.27 and 3.28), are particularly sensitive to traces of iron and copper. The hue of the black dyeings obtained is redder in the presence of copper and browner with iron contamination. The fastness to light and wet treatments may also prove inferior under these conditions. Even certain 1:2 metal-complex acid dyes show similar effects in the presence of these impurities, indicating that the iron and copper complexes are more stable than the desired chromium complex. Sequestering agents of the sodium polyphosphate type can help to minimise these problems.

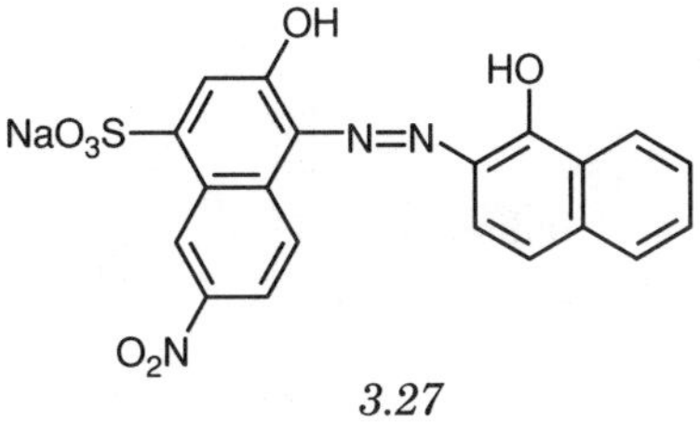

3.27

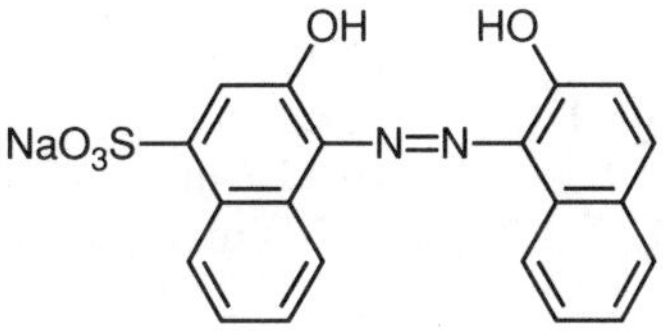

3.28

3.1.4 Rates of reduction and redox potentials of vat dyes

The rate of reduction of a vat dye depends partly on the intrinsic chemical properties of the dye and partly on the size and physical form of the dispersed particles undergoing this reaction. The physical factors are much less important than the chemical aspects [31]. The vatting process entails conversion of the insoluble quinone form into the soluble sodium enolate (see section 1.6.1). The reaction appears to take place in two stages at room temperature. Extremely rapid reduction to the hydroquinone is followed by slower dissolution of the latter in the alkaline solution. At higher temperatures, however, the dissolution rate approximates more closely to the rate of reduction [32]. Temperature and hydrosulphite concentration are the more important variables; the reduction rate is much less dependent on dye or alkali concentration.

The ease of reduction of a vat dye depends on its chemical structure in relation to the reducing power of the agent selected. In practice this question is seldom critical because the reduction potentials given by the commoner reducing agents are low enough to permit the reduction of all vat dyes of practical interest. Indeed, such dyes would never have been exploited commercially if they had not responded effectively to reduction under conventional conditions of application.

The times of half-reduction of a few typical vat dyes are listed in Table 3.2. Most of the widely used dyes give values in the range 25–500 s. Few dyes are as slow to reduce as C.I. Vat Red 1 and even fewer are reduced as quickly as flavanthrone. In fact when these times were measured the average particle size of the sample of flavanthrone under test was greater than that of the C.I. Vat Green 1 sample, which was reduced at least ten times more slowly [31].

TABLE 3.2

Times of half-reduction of typical vat dyes [31]

C.I. Vat	Chemical name	Time/s
Yellow 1	Flavanthrone	<5
Blue 17	16,17-Dichloroviolanthrone	31
Orange 9	Pyranthrone	36
Green 1	16,17-Dimethoxyviolanthrone	50
Red 1	4,4′-Dimethyl-6,6′-dichlorothioindigo	2880

Vat dye reduction is a reversible reaction leading to the formation of an equilibrium mixture of the oxidised and reduced forms. The relative rates of reduction of vat dyes bear no obvious relation to their intrinsic reduction potentials, which determine the position of the equlibrium reached when the vat dye quinone is dissolved under a given set of reducing conditions. The relative stabilities of the reduced and oxidised forms can be measured by the reduction potential developed at a bright platinum electrode (Table 3.3). An unstable leuco compound tends to revert to the oxidised form by transfer of electrons to the electrode, developing a high negative reduction potential. On the other hand, a dye with a small negative or positive reduction potential is easy to reduce since the leuco form is relatively stable.

Few vat dyes are as difficult to reduce as the dibenzoylaminoanthraquinones, in which the amide NH groups are hydrogen-bonded to the keto oxygen atoms in the quinone form, or as easy to reduce as tetrabrominated indigo. Most dyes of commercial interest have reduction potentials within the range –200 to –300 mV. It is interesting to note from Tables 3.2 and 3.3 that C.I. Vat Yellow 1 is inherently less stable in the reduced form than is C.I. Vat Red 1, even though the yellow quinone is reduced approximately

TABLE 3.3

Standard reduction potentials of vat dyes in pyridine [31]

C.I. Vat	Chemical name	Potential/mV
	1,5-Dibenzoylaminoanthraquinone	–490
	1,4-Dibenzoylaminoanthraquinone	–400
Orange 9	Pyranthrone	–270
Green 1	16,17-Dimethoxyviolanthrone	–250
Yellow 4	3,4,8,9-Dibenzopyrenequinone	–240
Yellow 1	Flavanthrone	–230
Red 6	4-Methyl-6-chloro-6′-methoxythioindigo	–220
Blue 1	Indigo	–220
Blue 4	Indanthrone	–220
Blue 20	Violanthrone	–220
Blue 17	16,17-Dichloroviolanthrone	–210
Red 1	4,4′-Dimethyl-6,6′-dichlorothioindigo	–190
Blue 5	5,5′,7,7′-Tetrabromoindigo	–70

a thousand times faster than is the red indigoid structure. Although standard reduction potentials have to be measured in anhydrous solvent solution for practical reasons, the results appear to be in general agreement with the qualitative behaviour of vat dyes in aqueous media. Thus indigoid dyes are more stable in the leuco form than are the anthraquinone derivatives in general, but they tend to be reduced more slowly and need a higher temperature for complete reduction.

3.1.5 Dispersion properties and aqueous solubility of disperse dyes

An aqueous dispersion of a disperse dye contains an equilibrium distribution of solid dye particles of various sizes. Only the smallest oligomers present, possibly only single molecules or dimers in true solution, are capable of becoming adsorbed by cellulose acetate or polyester fibres. Dyeing therefore takes place from a saturated solution, which is maintained in this state by the presence of undissolved particles of dye. As dyeing proceeds the smallest insoluble particles dissolve at a rate appropriate to maintain the saturated solution.

Surface-active agents are invariably incorporated in the formulation of commercial disperse dye powders, grains or liquid brands. Further additions of surfactant to the dyebath may be made during batchwise or continuous application. The initial purpose of the agents included by the manufacturer is to prevent the finely divided particles of dye from agglomeration during milling or spray drying. They also help to maintain concentrated dispersions in a stable condition during transportation and handling, that is, to minimise agglomeration and settling during storage of liquid brands. In the dyeing process, where the dispersion is much less concentrated, the function of the surfactant addition is to control a process of dynamic change rather than to preserve a static equilibrium for as long as possible. Here the agents play an active part in accelerating the disintegration and dissolution of dispersed particles and thus in increasing the concentration of dissolved dye present in the dyebath. This has the secondary effect of increasing the rate at which dye can be adsorbed on to the fibre surface. During cooling of the dyebath surrounding the dyed material, the presence of the surfactant helps to retard agglomeration or crystallisation of dye particles from the saturated solution as it cools and to minimise redeposition of insoluble dye on to the surface of the fibres.

Numerous disperse dyes are marketed in a metastable crystalline form that gives significantly higher absorption than the corresponding thermodynamically stable modification. Certain dyes can be isolated in several different morphological forms. For example, an azopyrazole yellow disperse dye (3.29) was prepared in five different crystal forms and applied to cellulose acetate fibres. Each form exhibited a characteristic fibre saturation level, the thermodynamically less stable modifications giving the higher values [33]. In a later investigation three different crystal modifications of each of three monoazo dyes, C.I. Disperse Yellows 3 and 42 and C.I. Disperse Red 54, were characterised in terms of melting point, X-ray diffraction pattern and dyeing behaviour on polyester fibres. Different fibre saturation values were found for the various forms of Yellow 42 and Red 54, but the three modifications of Yellow 3 (3.30) did not differ in this respect [34]. Dispersing agents may influence the aqueous solubilities of crystallographic modifications of the same dye to differing extents, so that the rate of change from a metastable to a more stable form on heating may be modified in the presence of an agent. On the other hand, a metastable form of C.I. Disperse Yellow 3 showed no tendency to revert to the stable form in the absence of a dispersing agent [35].

The saturation solubility of a disperse dye in water at a given temperature may be defined as the maximum amount of pure dye that will give a

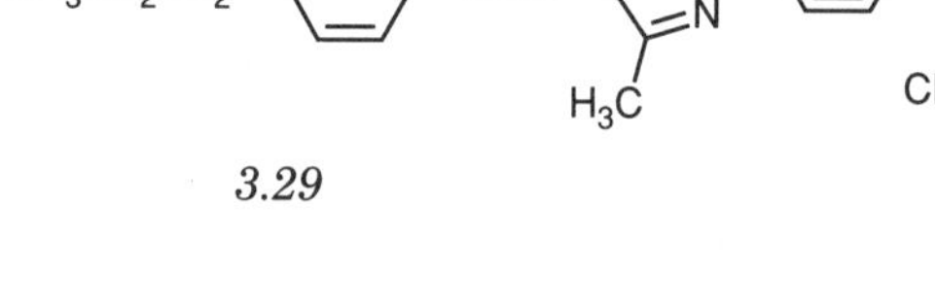

3.29

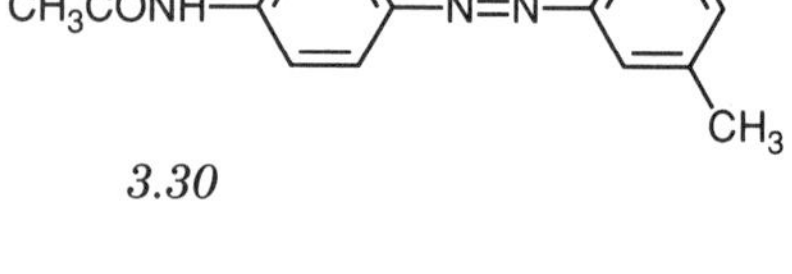

3.30

single homogenous phase when dissolved in a given volume of pure water. The qualification of the term 'saturation solubility' implies that any further addition of dye would remain undissolved as a separate solid phase. Solubilisation of a disperse dye refers to the marked increase in saturation solubility that occurs when a water-soluble additive, usually a surfactant or water-miscible solvent, is added.

The saturation solubility of pure disperse dyes in water is often extremely low at ambient temperatures. A survey of about forty typical dyes for cellulose acetate yielded values within the range 0.1 to 32 mg/l [36]. Raising the temperature to 80°C results in a considerable increase in saturation solubility, but the solubilising power of surfactants is essential for the satisfactory dyeing of cellulose acetate with the less soluble dyes. Dyes for polyester tend to be more hydrophobic in structural characteristics. These require considerable solubilisation by agents even at 130°C in order to give satisfactory dyeing behaviour and freedom from agglomeration or redeposition problems.

The hydrocarbon groupings in disperse dye molecules represent the most hydrophobic portions and these facilitate the agglomeration of such moieties in an aqueous environment. Substituents containing electronegative atoms (such as nitrogen, oxygen and sulphur), especially when bearing hydrogen atoms (as in =N–NH–, $-NH_2$, –CONH–, –OH, –SH groups), are able to interact strongly with water to form hydrogen bonds and thus to contribute to solubility in water [2].

When a water-miscible co-solvent such as acetone or ethanol is added to an aqueous suspension of a disperse dye there is a gradual increase in the amount of dissolved dye that is proportional to the quantity of co-solvent added. A disperse dye suspension responds somewhat differently when a surface-active solubilising agent is added. At low concentrations of this type of additive the saturation solubility of the dye remains relatively little changed, but when the critical micelle concentration of the surfactant is reached, a sudden and marked increase in dye solubility is observed [37]. When micelles of the surfactant begin to form in the solution, these provide a more amenable environment into which the dye molecules can transfer. Above the critical micelle concentration the increase in solubility of the dye is directly proportional to the concentration of surfactant present [2,35].

Much of the available published data on the aqueous solubility of disperse dyes was accumulated many years ago in the course of studies of the mechanism of dyeing cellulose acetate with disperse dyes. Most of the dyes examined were Class A or Class B types for dyeing cellulose acetate; particular attention was given to monoazo dyes derived from aniline or *p*-nitroaniline coupled to *NN*-disubstituted anilines with ethyl or 2-hydroxyethyl substituents and to 1,4-disubstituted anthraquinones containing potential hydrogen-bonding groups (–OH, $-NH_2$, $-NHCH_3$ or $-NHCH_2CH_2OH$).

In the azo series, as Table 3.4 indicates, unsubstituted aniline as a diazo component gave more soluble dyes than did *p*-phenylenediamine or *p*-nitroaniline, in that order. Introduction of an additional aryl ring resulted in a marked decrease in solubility, as shown by comparing C.I. Disperse Orange 1 with Orange 3, or phenylazo-*NN*-bis-(2-hydroxyethyl)aniline (3.32; X = Ph) with its 1-naphthylazo analogue. Hydroxyethyl groups are readily solvated in water by hydrogen bonding and therefore confer much higher solubility than alkyl substituents in analogous dyes. Examples are given in Table 3.4 from both the monoazo (3.33) and the anthraquinone (3.34) series. The solubility is approximately doubled if two hydroxyethyl groups are introduced instead of one. Aminoanthraquinone dyes are markedly more soluble than their hydroxy analogues, presumably because a primary amino group can form hydrogen bonds with oxygen atoms in two water molecules, but a phenolic group can bond with only one molecule in this way.

TABLE 3.4

Aqueous solubility of disperse dyes of related structure [38–42]

		Solubility/mg l^{-1}				
	Temp./°C:	60	80	80	90	100
C.I. Disperse	Structure Lit. ref.:	38	39	40	41	42
	3.31: X = H, Y = H	141				
	3.31: X = NH_2, Y = H	74				
Orange 3	3.31: X = NO_2, Y = H	3			16.6	
Orange 1	3.31: X = NO_2, Y = Ph				1.1	
	3.32: X = Ph	214				
	3.32: X = 1-naphthyl	7.8				
Red 19	3.33: X = Y = CH_2CH_2OH		26.4			
Red 1	3.33: X = CH_2CH_3, Y = CH_2CH_2OH		7.7			
	3.33: X = Y = CH_2CH_3		*ca* 0.2			
Blue 23	3.34: X = Y = $NHCH_2CH_2OH$			61		
Blue 3	3.34: X = $NHCH_3$, Y = $NHCH_2CH_2OH$			30		
Blue 14	3.34: X = Y = $NHCH_3$			2		
Violet 1	3.34: X = Y = NH_2			21.7		
Red 15	3.34: X = OH, Y = NH_2			11.7		38
	3.34: X = Y = OH					17

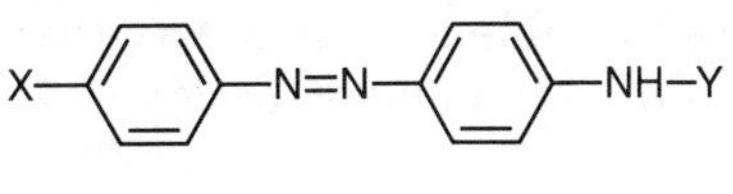

3.31

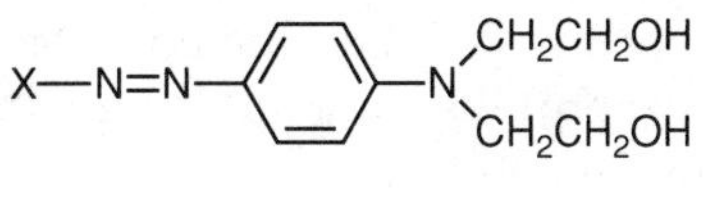

3.32

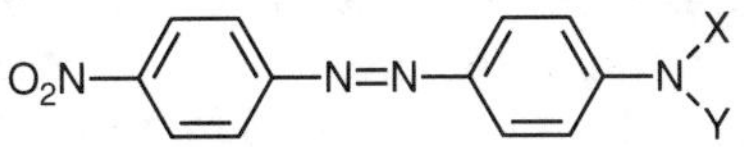

3.33

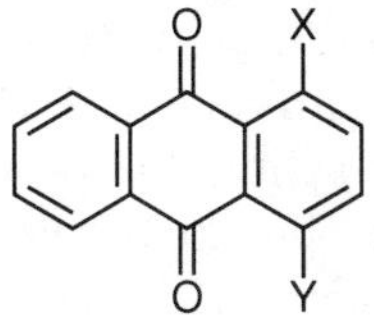

3.34

3.2 DYEABILITY OF FIBRES IN RELATION TO DYE STRUCTURE

The chemical forces that operate between dye molecules and fibre-forming polymers are essentially the same as those already mentioned in earlier discussions of clustering of water molecules, dye solubilisation in aqueous media and the aggregation or agglomeration of dyes. In dyeing systems depending on an essentially ionic mechanism, such as the dyeing of acrylic fibres with basic dyes or the application of anionic dyes to nylon, silk or wool, a decisive role is played by electrostatic attraction between the charged dye ions and sites of opposite charge in the substrate. Shorter-range forces make a notable contribution to the affinity of the dye ion but their function is essentially supportive. In all other dyeing systems, however, the charge (if any) on the dye molecule or ion does not enhance affinity. In fact the fibrous polymer often carries a charge of the same sign as that of the approaching dye moiety and measures must be taken to minimise any forces of repulsion arising from this similarity. Under these circumstances the less specific shorter-range forces (hydrogen bonding, van der Waals or dispersion forces, hydrophobic interaction) make a much more substantial contribution to dye–fibre bonding.

3.2.1 Dyeing of cellulosic fibres with anionic dyes

The decisive influence of the degree of sulphonation for the dissolution of direct dyes in water has been discussed earlier. In general, the lower the equivalent mass (that is, the ratio of relative molecular mass to ionic group content), the higher will be the aqueous solubility of the dye. Exceptions from this simple rule can arise, however, if a direct dye structure contains protonated amino groups (as in 3.2 or 3.13, for example). As already noted, the converse of this general trend is that dye molecules of higher equivalent mass tend to aggregate together more readily because of the greater opportunity for hydrophobic interaction. Such dyes in the presence of a cellulosic substrate will tend to escape more readily from the aqueous dyebath by adsorption on to the fibre surface. It is reasonable to assume that the dye molecules will become adsorbed as far as possible in such a way that the hydrophilic sulphonate ionic groups are directed outwards towards the aqueous phase.

Just as addition of an electrolyte to a solution of direct dye tends to lower the electrostatic repulsion between the negatively charged dye anions and promote aggregation, the effect of the presence of an inorganic salt when dyeing cellulose is to overcome the long-range forces of repulsion between the dye anions and the negatively charged fibre surface. The closer approach then allows hydrogen bonding and other short-range attractive forces to operate between the dye molecules and the glucoside units of the fibrous polymer. When dyeing cellulose, this favourable influence of electrolyte addition is as useful with anionic leuco forms of vat and sulphur dyes as it is with sulphonated direct or reactive dyes.

Hydrogen bonding between the hydroxy groups of cellulose and centres of electronegativity (nitrogen, oxygen and sulphur atoms) in the dye molecule, especially those substituted with hydrogen atoms (as in =N–NH–, $-NH_2$, –CONH–, –OH and –SH), is widely acknowledged to contribute to adsorption and retention of the dye molecules. Stereochemical studies have demonstrated that hydrogen bonding is possible for almost any cellulosic hydroxy group in the amorphous regions of the fibre. Measurements of the heat of dyeing confirm that a typical adsorption process entails the formation of at least two hydrogen bonds per molecule of dye adsorbed [43,44] and it is obvious that retention of the adsorbed molecule will be favoured the higher the number of potential dye–fibre bonding points and the more widely they are spaced along the length of the dye molecule. The more hydrogen bonds that a dye can form with the glucosidic polymer, the more readily it can compete with and rupture the fibre–fibre hydrogen bonds in order to penetrate more deeply into the amorphous structure of the polymer.

The recognition that the existence of conjugation in all organic dye molecules was essential to provide intense colour (see section 1.4) and that an increase in the length of the conjugated chain tended to result in a bathochromic shift of the main waveband in the absorption spectrum led to the adoption of related ideas in the context of the theory of substantivity. Many direct dyes with high affinity for cellulose are disazo or trisazo structures in which the azo groups are located in *para* positions relative to one another, so that the longest conjugated chain through the azo-linked aryl nuclei tends to be as near linear as practicable. The presence of an amide group conjugated via a 1,4-phenylene or 2,6-naphthylene residue to the nearest azo group, as in C.I. Direct Red 81(3.12; X = H), enhances the substantivity of the molecule by lengthening the conjugated chain, a modification that also causes a bathochromic shift. A benzoylamino substituent in this position tautomerises mainly to the enol form, because *N*-acylated derivatives of C.I. Direct Red 81 are much less substantive.

A methine group oriented *para* to an azo group through an intervening aryl nucleus boosts substantivity even more than does an amide group in a similar position, as was demonstrated for a series of azostilbene and related dyes (3.35; X = CH=CH, CONH or CO) with their thiophene (3.36) analogues (Table 3.5). Affinity increased in the order of increasing electron mobility (CH=CH > CONH > CO) and was greater for the heterocyclic derivatives than for the corresponding phenyl-substituted dyes [45]. A series of analogues of the most substantive of these dyes (containing both the thiophene ring and the methine linkage) proved to be highly sensitive to the nature of the coupling component at the opposite end of the conjugated chain, however (Table 3.6). The presence of a protonated amino group favours substantivity (naphthionic acid > NW acid, H acid > chromotropic acid) but the extra sulpho group in the three disulphonated couplers affects it adversely.

TABLE 3.5

Substantivity and structure of *p*-substituted monoazo naphthionic acid derivatives [45]

Structural types	X	Affinity/kcal mol^{-1}
Thiophenyl	CH=CH	3.78
Phenyl	CH=CH	3.41
Thiophenyl	CONH	3.13
Phenyl	CONH	2.87
Thiophenyl	CO	2.10
Phenyl	CO	1.68

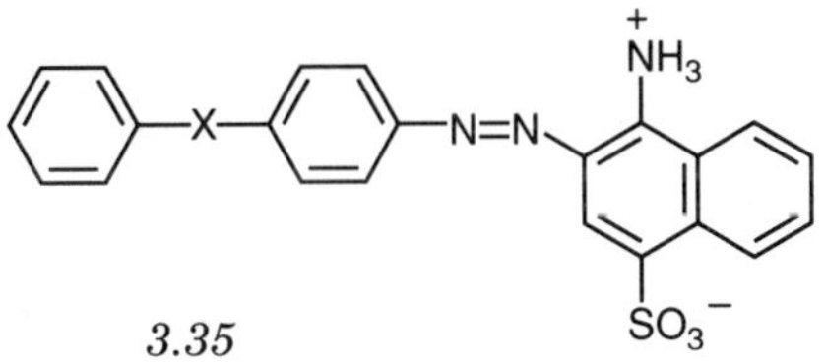

3.35

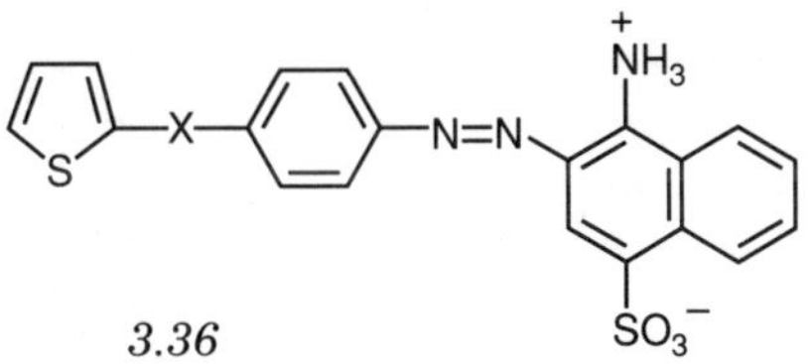

3.36

The significance of conjugation as a contributor to the substantivity of dyes for cellulose is not always easy to distinguish from the effect of the degree of linearity of the molecule. The molecules of almost all direct dyes and most of the highly substantive dyes in other classes used for the dyeing of cellulosic fibres possess flexible chains of aryl nuclei linked by azo or

TABLE 3.6

Substantivity of thiophenyl monoazo dyes with various coupling components [45]

Coupling component	Structure	Affinity/kcal mol^{-1}
Naphthionic acid	3.37: Y = Y1	3.78
NW acid	3.37: Y = Y2	2.31
H acid	3.37: Y = Y3	2.26
Chromotropic acid	3.37: Y = Y4	1.69
R acid	3.37: Y = Y5	1.50

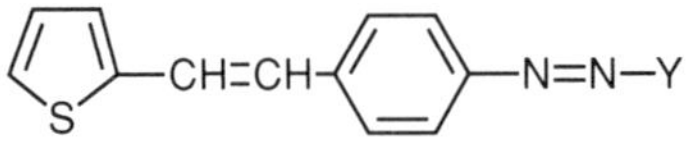

3.37

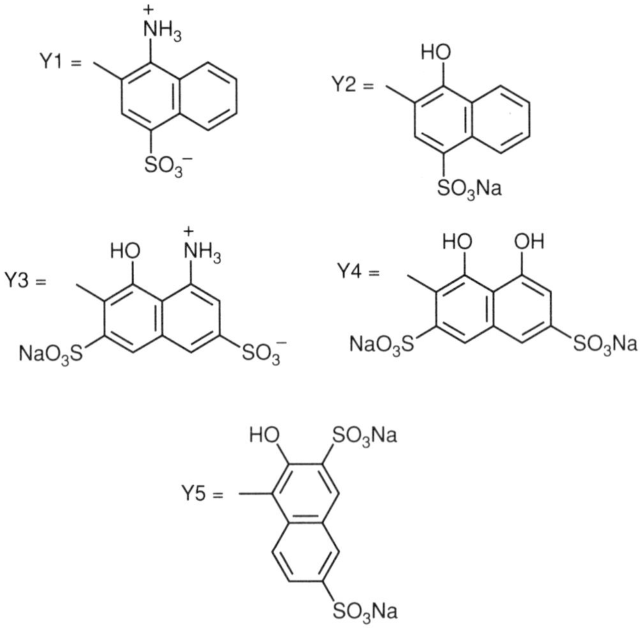

other groups (as in structure 3.37). Such structures can readily adopt a near-linear spatial conformation, as far as the angular constraints of the participating bonds will allow. Azo groups attached to central naphthylene residues are frequently oriented in 1,4- or 2,6-positions relative to one another, and typical central diamines that give linear dyes on tetrazotisation include *p*-phenylenediamine, benzidine, *p,p'*-diaminostilbene, *p,p'*-diaminodiphenylamine and *p,p'*-diaminodiphenylurea. For similar reasons, 1,4-disubstituted arylaminoanthraquinone vat dyes are markedly more substantive than their 1,5-disubstituted analogues [46].

The ability to adopt an extended configuration has been recognised for many years [47] to be a desirable feature of substantive dyes. It helps to explain why J acid is such a popular choice as a central component in unsymmetrical disazo dyes. For example, it is much easier for aniline→J acid→H acid (3.38) with the 2,6-naphthylene substitution pattern to adopt a linear conformation than for the similar disazo dyes aniline→γ acid→H acid (3.39) and aniline→H acid→J acid (3.40) with 2,7- and 2,8-substitution patterns respectively.

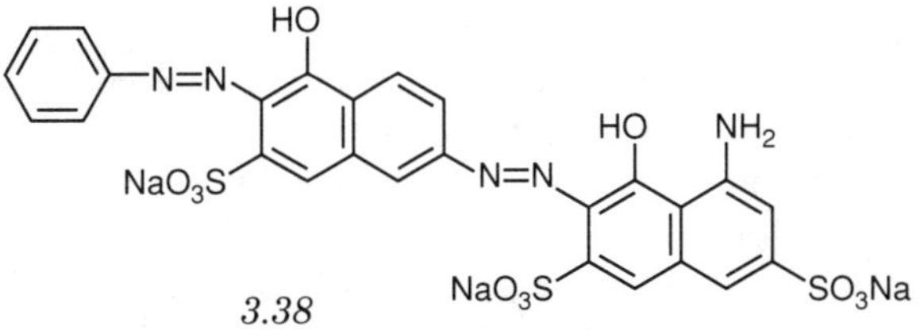

3.38

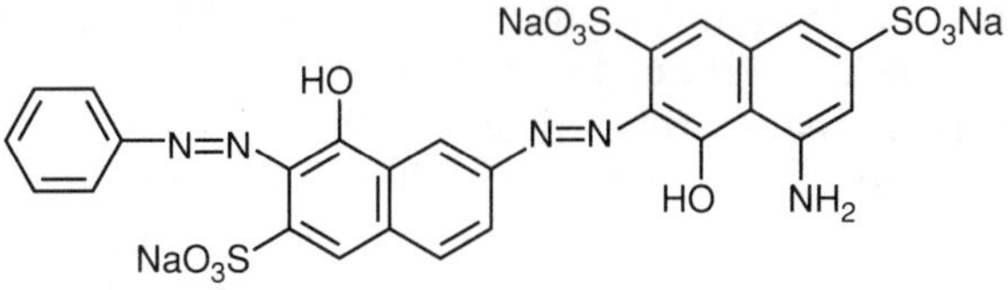

3.39

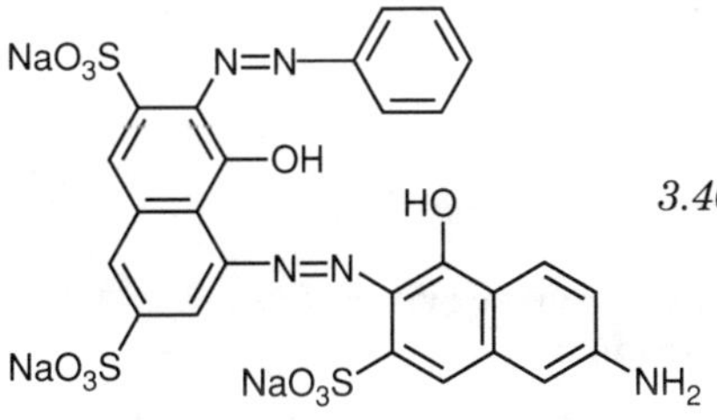

3.40

A further refinement of the concept that linear conjugated aryl–azo–aryl–azo–aryl sequences and analogous systems represented the preferred design for highly substantive molecules was the stipulation [48] that the aryl nuclei must be able to adopt a coplanar conformation. This favours the adsorption of one dye molecule on to another during the formation of laminar aggregates, as well as the multi-point adsorption of a dye molecule on to two or more glucoside units of a polymer segment via hydrogen-bonding sites. There are unequivocal examples that clearly demonstrate the validity of this interpretation. Thus Benzopurpurin 4B (3.13; X = CH_3, Y = H) and other 3,3′-disubstituted analogues of Congo red (3.13; X = H, Y = H) are markedly more substantive than their '*meta*' analogues with 2,2′-disubstitution (such as 3.13; X = H, Y = CH_3). The substituents *ortho* to the central bond in these symmetrical disazo structures inhibit free rotation of the phenylene nuclei and thus prevent the adoption of a coplanar conformation [49].

The scope for intramolecular hydrogen bonding between azo group and coupling component in symmetrical disazo structures of this kind also contributes significantly to the maintenance of coplanarity and hence to the affinity for cellulose. The number of commercial direct dyes in which coupling has been directed *para* to the amino or hydroxy substituent in the coupling component is much smaller than those manufactured by *ortho* coupling. Within the series of *o*-tolidine symmetrical disazo dyes in which the naphthionic acid of Benzopurpurin 4B (3.41; X = SO_3^-, Y = H) is replaced in turn by Laurent's acid, Cleve's 6- and 7-acids, and finally Peri acid (Table 3.7), there is a marked lowering of affinity for the 6-, 7- and 8-sulpho derivatives compared with the first two members of the series [50]. Naphthylamine-4- and 5-sulphonic acids can only couple *ortho* to the amino substituent and the derived dyes preferentially form an extended coplanar arrangement (3.41) by intramolecular hydrogen bonding of the protonated amino group to the more distant nitrogen atom of the neighbouring azo group. In the remaining members of the series, however, coupling readily takes place in the *para* position (3.42) and the protonated amino groups are no longer able to form intramolecular hydrogen bonds. Free rotation is possible on either side of the azo linkage and the preferred coplanar arrangement is no longer preferentially stabilised.

Vat dye molecules are relatively hydrophobic and are usually capable of assuming a coplanar configuration that favours aggregation in a laminar

TABLE 3.7

Substantivity of symmetrical disazo dyes derived from naphthylamine monosulphonic acids [50]

Coupling component	Structure	Affinity/kcal mol^{-1}
Naphthionic acid	3.41: X = SO_3^-, Y = H	6.4
Laurent's acid	3.41: X = H, Y = SO_3^-	6.7
Cleve's 6-acid	3.42: X = SO_3^-, Y = Z = H	5.4
Cleve's 7-acid	3.42: Y = SO_3^-, X = Z = H	5.3
Peri acid	3.42: Z = SO_3^-, X = Y = H	5.6

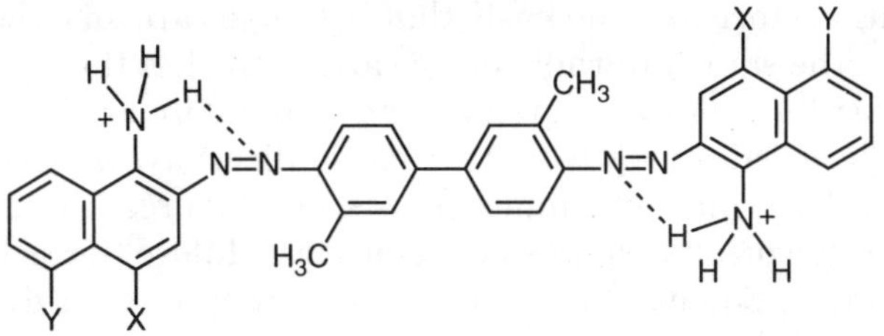

3.41

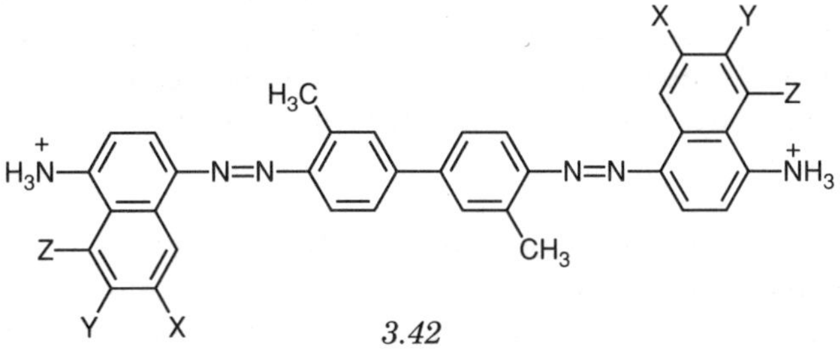

3.42

arrangement as well as the formation of multiple attachments to a segment of the cellulose macromolecule. Hydrogen bonding can be invoked to account for the favourable influence of arylamino or benzoylamino substituents on the affinity of the parent anthraquinonoid nuclei in vat dye structures. Purely carbocyclic quinones such as pyranthrone, dibenzanthrone and dibenzopyrenequinone must depend mainly on coplanarity, hydrophobic interaction and hydrogen bonding between dye keto groups and cellulose hydroxy groups. In any closely related series of vat dyes the affinity does seem to increase with molecular size, but the differences between members of different classes emphasise the specific influence of characteristic structural features, especially nitrogen-containing groupings, on the forces of affinity.

A discussion of the relationships between direct or vat dye structure and substantivity for cellulose cannot ignore the contribution of multi-layer adsorption of dye molecules within the pore structure of the fibre [51]. The wide difference in substantivity between Congo red (3.13; X = Y = H) and the monoazo acid dye aniline→naphthionic acid, essentially the 'monomer' analogue of the 'dimeric' Congo red, may be interpreted in terms of their relative tendency to form multi-layers within the fibre pores as a result of dye–dye aggregation. Saturation adsorption values of these two dyes on viscose fibres at pH 9 and 50°C corresponded to monolayer coverage levels of approximately 90 and 11 m^2/g of internal surface respectively [52]. In view of the smaller molecular area and greater mobility of the acid dye 'monomer', higher uptake than the direct dye would be anticipated if there were only a limited area of surface available for true monolayer adsorption. The much higher adsorption of Congo red at saturation equilibrium can be explained satisfactorily on the basis that both dyes are adsorbed initially on essentially the same internal surface area, but that the direct dye has a much greater tendency to aggregate and so build up readily into multilayers [51]. The greater length of the 'dimeric' direct dye molecule relative to the acid dye 'monomer' of similar cross-sectional area favours penetration into the intermicellar spaces between the cellulose fibrils followed by multi-point adsorption on to a polymer segment or another dye molecule already adsorbed. Inorganic electrolytes assist powerfully in promoting these interactive forces leading to multi-layer adsorption of direct dyes, since they are present in the intermicellar pore spaces as well as the external phase. Acid dyes aggregate much less readily in solution (see section 3.1.2) and are thus less likely to form multi-layered aggegates in the fibre pores.

3.2.2 Dyeing of amide fibres with anionic dyes

Wool, silk and nylon contain both basic and acidic groups, amongst which by far the most important are amino and carboxyl groups respectively. Just like the parent amino acids from which protein fibres are derived, all these polymeric amides show zwitterionic characteristics at pH values close to the isoelectric point (that is, the pH at which the fibre contains equal numbers of protonated basic and ionised acidic groups). As the pH decreases below this point, the carboxylate anions are progressively neutralised by the adsorption of protons and the fibre acquires a net positive charge (Scheme 3.1).

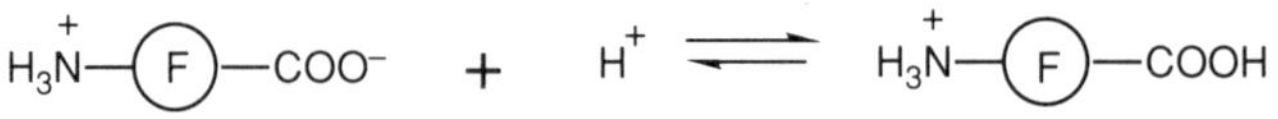

Scheme 3.1

Conversely, as the pH rises above the isoelectric point, the fibre becomes negatively charged as a result of deprotonation of the amino groups by adsorption of hydroxide ions or other simple anions (Scheme 3.2).

Scheme 3.2

The capacity of nylon and protein fibres to adsorb simple organic or inorganic acids is closely equivalent to their respective content of accessible amino groups. More complex dye anions, however, differ in their affinity for these fibres owing to non-polar bonding between the hydrophobic portions of the dye molecule (alkyl substituents and unsubstituted aryl nuclei) and non-polar side-chain or main-chain segments in the fibrous polymer. The contributions to dye anion affinity of electrostatic attraction to protonated amino groups in the fibre and hydrophobic interaction with non-polar segments tend to mask any further influence of hydrogen bonding forces, although this form of bonding seems likely to reinforce the effects of the other factors. According to adsorption studies on wool, silk and nylon [53,54], the tendency of proton donors to form hydrogen bonds, probably with the enolic form [–C(OH)=N–] of the characteristic amide grouping in the fibre, is in the order ArOH > $ArNHCOCH_3$ > $ArNH_2$, ArNHR, AlkOH.

As a general rule in the dyeing of the amide fibres, the lower the dyebath pH – and therefore the greater the number of protonated dyeing sites in the substrate phase – the more rapid is the rate of initial adsorption and the subsequent approach towards equilibrium exhaustion. Consequently the attainment of a safely controllable rate of dyeing becomes easier the higher the pH at which dyeing begins. Eventually, however, a pH is reached at which the dyebath is no longer exhausted in a reasonable time. Dyeing conditions are usually selected to achieve a compromise of a moderate initial rate of dyeing for optimum levelling, followed by a gradual lowering of pH by acid release to achieve as much exhaustion as possible at the end of the process. The selection of the preferred initial pH and rate of acidification is dependent on the affinity of the dyes at neutral pH values. These characteristics are dependent on the structural features and hydrophilic–lipophilic balance of the dye molecule.

The degree of sulphonation of an acid dye (see section 1.6.7) has a major influence on the profile of exhaustion values reached at different pH values on a specific substrate. At low pH in the absence of electrolytes all acid dyes show high exhaustion, irrespective of their sulpho group content. In neutral solution, on the other hand, polysulphonated dyes generally show much lower exhaustion than dye monosulphonates [55]. Above the isoelectric point the fibre is negatively charged and the forces of electrostatic repulsion increase with the degree of sulphonation [56]. Hydrophobic interaction permits adsorption of a dye monosulphonate on to the fibre surface in such a way that the charged group is directed outwards towards the dyebath phase.

The rate of dyeing of acid dyes of relatively small molecular size decreases with increasing degree of sulphonation. Simple monoazo monosulphonates such as Orange II (3.9) are rapidly absorbed but readily desorbed again. Thus they migrate easily but show extremely poor fastness to wet treatments. There are two main types of levelling acid dye:

(a) disulphonated dyes of relative molecular mass about 400–600 that are somewhat sensitive to dyeability differences in the substrate and show the lowest wet fastness of all dye classes used on wool
(b) monosulphonated dyes of lower r.m.m. (300–500) and slightly higher wet fastness that migrate more readily and cover dyeing irregularities (such as carbonising faults) more effectively.

In an early study of the dyeing of mixtures of normal and chlorinated wool with a series of naphthionic acid→2-naphtholsulphonate dyes [57], the trisulphonated C.I. Acid Red 18 (3.43; X = Y = SO_3Na) showed a most marked contrast. Penetration of the hydrophobic epicuticle of normal wool by this hydrophilic dye was negligible in the initial stage of dyeing but chlorinated fibres were rapidly and deeply dyed. The monosulphonated C.I. Acid Red 88 (3.43; X = Y = H), however, revealed relatively little contrast in dyeability between the two types of wool, whereas C.I. Acid Red 13 (3.43: X = SO_3Na, Y = H) gave a degree of contrast intermediate between those of the other two dyes. Apparently the relatively hydrophobic and potentially hydrogen-bond-forming 2-naphthol residue in C.I. Acid Red 88 is almost as effective a means of attachment to the epicuticle of normal wool as the negatively charged naphthionic acid residue is towards the hydrophilic surface of chlorinated wool.

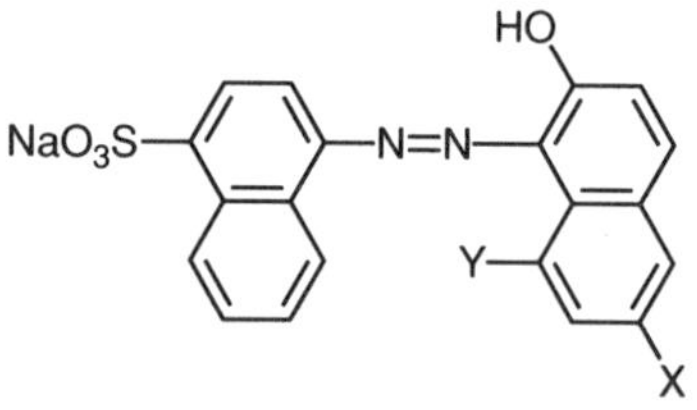

3.43

There are three main subclasses of milling acid dyes:

(a) monosulphonated dyes of relative molecular mass 500–600 that are somewhat more hydrophobic than the monosulphonated levelling acid dyes; like them they migrate and cover well but they are a little inferior to disulphonated milling dyes in wet fastness and thus are sometimes described as 'half-milling' dyes
(b) disulphonated dyes of high r.m.m. (600–900) that diffuse much more slowly than typical levelling acid dyes; they exhibit correspondingly higher wet fastness but their migration and coverage properties are inferior
(c) certain disulphonated milling acid dyes containing higher alkyl substituents (butyl, octyl, dodecyl, for example) to confer higher neutral dyeing affinity, better coverage of dyeability irregularities and exceptionally good wet fastness; they are sometimes described as 'super-milling' dyes to distinguish them from category (b), the largest of the three subclasses.

Direct evidence has been provided to demonstrate the contribution of hydrophobic interaction to the substantivity of milling dyes for wool [58]. In a series of zwitterionic monosulphonated phenylazopyrazolone dyes (Table 3.8), substantivity for wool increased with chain length in the alkyl

TABLE 3.8

Exhaustion of substituted phenylazopyrazolone dyes on wool [58]

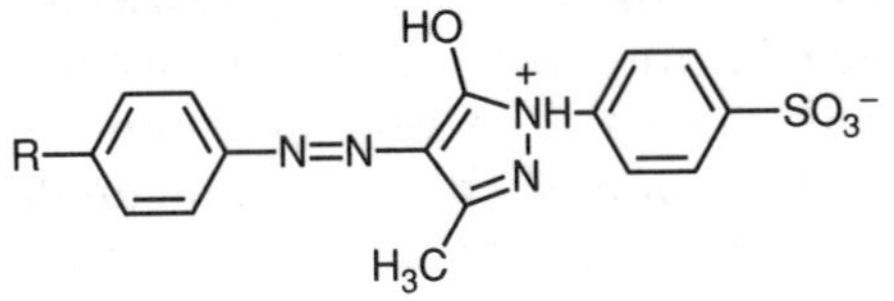

3.44

Dye	Substituent (R)	Exhaustion/%
A	Cyclohexyl	85
B	n-Hexyl	65
C	Phenyl	59
D	Ethyl	17
E	Hexatrienyl	10
F	Hydrogen	8
G	2-Hydroxyethyl	3

series (i.e. B >> D > F) but the increased polarisability of alkenyl groups almost completely negated this effect (B >> E). Cyclisation of the saturated alkyl group produced maximum substantivity (A > B) and formation of an unsaturated aryl ring had a similar effect (C >> E). Finally, a hydroxy substituent almost completely inhibited the hydrophobic interaction attributable to the ethyl group (D > G). It was evident from these results that coulombic interaction and hydrogen bonding played insignificant parts in these equilibria.

Chrome dyes were once the most important range of fast dyes for wool. Various mordanting processes were used but eventually aftertreatment of the dyed wool with potassium dichromate became by far the most significant. Dichromate (Cr^{VI}) anions are rapidly reduced to the trivalent state by the cysteine residues in wool and the chromic (Cr^{III}) ions then react with the *o,o′*-dihydroxyazo ligand group in a typical mordant dye. Chromic salts must be used as mordants on silk fibres, which lack the reducing properties necessary in the dichromate method.

The practical advantages of chrome dyes include economic build-up to full depths, good level dyeing properties (relative molecular mass 300–600) and excellent wet fastness after chromium complex formation. Unfortunately a marked change in colour of the dyeing during aftertreatment complicates the colour matching stage. Further drawbacks include the prolonged dyeing procedure, oxidative damage to the wool and the environmental hazards associated with chromium-containing effluent.

The 1:1 metal-complex dyes (see section 5.4.2) are almost all monosulphonates of r.m.m. 400–500. This gives them dyeing characteristics not unlike those of the monosulphonated levelling acid dyes, with good migration and coverage of damaged wool. In fabric dyeing it is usual to apply the 1:1 metal-complex dyes at pH 2, sacrificing optimum physical properties of the wool in favour of better migration and coverage [59,60].

The 1:2 metal-complex dyes that were originally developed following the successful adoption of the 1:1 metal-complex types are symmetrical structures free from ionic solubilising groups and are applied from aqueous dispersion. They show high neutral-dyeing affinity and very good fastness to light, although they are significantly inferior to chrome dyes for fastness to the most severe wet treatments. Such hydrophobic properties promote coverage of dye-uptake variations in wool but their high affinity can cause rapid initial strike and they diffuse and migrate slowly (relative molecular mass 700–900). Most dyes of this type now used on wool or nylon are solubilised by one non-ionised but polar group, such as sulphonamide ($-SO_2NH_2$), methylsulphonamide ($-SO_2NHCH_3$) or methyl sulphone ($-SO_2CH_3$), on each of the dye ligand groups. These are more expensive to manufacture than the traditional non-polar structures, however, and share most of their drawbacks.

Over the last thirty years advances have been made in developing dyeing methods and auxiliary products to improve the level-dyeing and coverage properties of sulphonated 1:2 metal-complex dyes, leading to the widespread adoption of the more economical monosulphonates in uses where unsulphonated dyes were previously considered essential. Sulphonated metal-complex dyes may be divided into two subclasses:

(a) unsymmetrical monosulphonated dyes of r.m.m. 700–900; the presence of the solubilising group reduces manufacturing costs and minimises the staining of adjacents in severe wet tests, but these dyes are more sensitive to dye-affinity differences than are the unsulphonated 1:2 metal-complex types

(b) disulphonated dyes of r.m.m. 800–1000, many of them symmetrical in structure; although often cheaper to manufacture than the unsymmetrical monosulphonates and the unsulphonated types, these slow-dyeing complexes have intrinsically poor levelling and migration properties.

If the wool cuticle is damaged by localised chemical attack, abrasion or exposure to light, more rapid strike of dye occurs on the exposed cortical cells. At low dyeing temperatures all dyes are taken up preferentially by damaged fibres and fibre tips [61], but at higher temperatures the more polar the dye the stronger is the preferential absorption by damaged fibres [62,63]. Dyes of low r.m.m., particularly milling acid and sulphonated 1:2 metal-complex dyes, cover such differences well by subsequent migration away from the tip towards the undamaged root and good coverage can be achieved using cationic auxiliaries. These interact with anionic dyes to form hydrophobic complexes that are less sensitive to differences in fibre dyeability. Chrome dyes reveal a marked contrast between root and tip, since both the mordant and the dye are preferentially absorbed by the damaged tips and the reducing action of the cystine breakdown products resulting from wool damage accelerate the conversion of dichromate (Cr^{VI}) to chromic (Cr^{III}) ions.

Normal nylon 6.6 [poly(hexamethylene adipamide)] fibres contain only about 35–45 milliequiv./kg of amine end-groups, and this limits the build-up attainable with relatively hydrophilic levelling acid dyes on this substrate. Differential acid-dyeable nylon variant yarns can be made by modifying the amine end-group content – for example, by 'capping' about half of the terminal amine groups using an appropriate excess of adipic acid in manufacture to give a pale-dyeing yarn (15–20 milliequiv./kg), or by adding excess diamine and roughly doubling the amine end-group content to give deep-dyeing nylon (60–70 milliequiv./kg). Ultra-deep-dyeing yarns (> 80 milliequiv./kg) are made by adding basic compounds such as *NN'*-bis(2-aminoethyl)piperazine during manufacture. Basic dye uptake can be enhanced by adding excess adipic acid or by introducing sulpho groups, perhaps by copolymerisation with 5-sulphoisophthalic acid. The use of acid, basic and disperse dyes to achieve multi-coloured effects on differential-dyeing nylon materials has been described in detail elsewhere [64].

Similar effects can be achieved by pretreating normal nylon with acid-dye resist agents such as syntans or with fibre-reactive chemicals. More varied design possibilities are attainable by space treating of yarn or by fabric printing with a paste containing the modifying agent. Durable acid-dye resist treatments include acylation with an acid chloride in the presence of a swelling agent and a basic catalyst, or application of a colourless anionic reactant, such as 4-(dichlorotriazinylamino)benzenesulphonic acid, that reacts with amine end-groups at 60–100°C, imparting excellent durable resistance to acid dye uptake. Basic-dye resists are more difficult to find but an aminosiloxane derivative appears to be quite effective [65].

3.2.3 Dyeing of ester fibres with disperse dyes

Since disperse dyes and ester fibres (cellulose esters and polyester) do not contain any ionic groups, dye–fibre attraction clearly originates in hydrogen bonding, dipole–dipole interaction and dispersion forces. Simple un-ionised proton donors, such as phenol, are readily adsorbed by cellulose acetate and this has been taken as evidence of hydrogen bonding between the phenolic hydroxy groups and keto groups in the polymer [66]. Most disperse dye structures contain proton-donating groups such as hydroxy and amino, as well as proton acceptors such as keto, methoxy, azo and dimethylamino groups. Refractometric studies have demonstrated that the α-keto hydrogen atoms in the acetate grouping $-OCOCH_3$ are sufficiently polarisable to take part in hydrogen bonding with proton acceptors [67]. Thus all the polar groups in a disperse dye molecule can contribute to its affinity for an ester fibre and not merely those with proton-donating character.

Monolayer experiments have indicated that face-to-face complexing occurs between azo disperse dye molecules and the hexa-acetylcellobiose units of cellulose triacetate, with the molecular segments lying parallel [68,69]. The relatively hydrophobic disperse dye molecule leaves the aqueous dyebath and penetrates between the polymer chains, entering a non-polar environment where water molecules are unable to follow. New sites for adsorption become accessible as the dye molecule diffuses further into the amorphous regions of the fibres. The more numerous the substituents on the dye molecule that can interact with the polymer segments, the more able it is to break the interchain bonds of the fibre structure. Thus the greater the bonding capacity of a disperse dye molecule, the higher will be its saturation limit on the fibre.

It is possible to account for the absorption of all disperse dyes by cellulose acetate, triacetate and polyester fibres in terms of the combined action of powerful hydrogen bonding and weak dispersion forces. The latter forces (π-bonding) between aryl rings in the dye and terephthalate residues make a more important contribution on the polyester substrate, but on polypropylene only weak aliphatic dispersion forces can operate [70]. Provided the dye molecule is planar and can lie flat against the cyclic nuclei in the ester polymer segments, all polar substituents on the fringes of the rings have the potential to contribute to the overall strength of the dye–fibre multiple bonding. This accounts for the much higher affinity of planar disperse dye molecules compared with non-planar analogues (Table 3.9).

TABLE 3.9

Partition coefficient and saturation value of disperse dyes on secondary cellulose acetate [71]

Dye structure	Substituent (R)	Partition coeff.[a]	Satn value/mol kg^{-1}
3.45	$COOCH_2CH_3$	998	0.218
3.46	$COOCH_2CH_3$	39	0.0319
3.45	$CONHCH_2CH_2OH$	194	0.135
3.46	$CONHCH_2CH_2OH$	2.6	0.0031
3.45	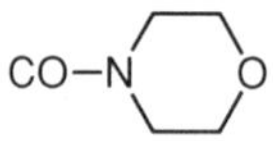	216	0.212
3.46	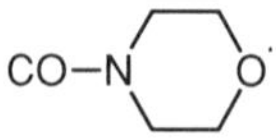	17	0.018

[a] Partition coefficient = {[Dye] (mol/kg) in acetate}/{[Dye] (mol/l) in water}

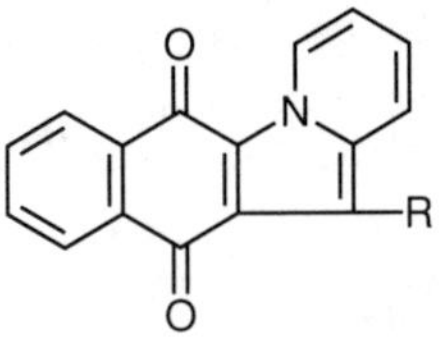

3.45 (planar)

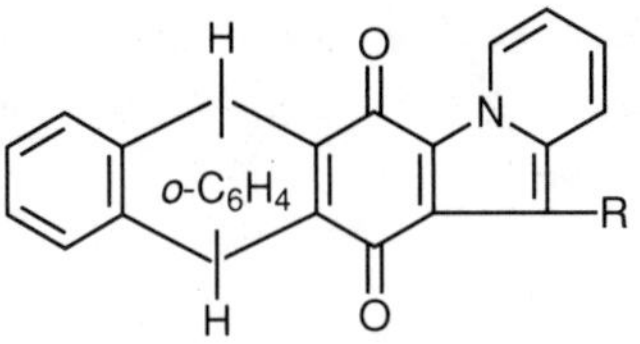

3.46 (non-planar)

Measurements of aqueous solubility and partition coefficient between secondary cellulose acetate and water were compared for 30 disperse dyes, and an approximate inverse relationship was postulated [39]. This can only be valid to a limited extent, however, because these two parameters are related by the saturation solubility of the dye in cellulose acetate; this property varies from dye to dye and is not directly related to aqueous solubility. The solubilities of four dyes in a range of solvents were compared with their saturation values on cellulose acetate. Solubilities in benzene showed no significant correlation. With the other solvents the degree of correlation increased in the order: ethanol < ethyl acetate < 20% aqueous diethylene glycol diacetate. The last-named compound was suggested as a model with groups similar to those in cellulose acetate [72]. Dyes containing hydrophilic *N*-2-hydroxyethyl groups combine with free hydroxy groups in cellulose acetate by hydrogen bonding, whereas dyes with more hydrophobic *N*-alkyl substituents depend much more on non-polar bonding. Hence dyes of the latter type have more affinity for cellulose triacetate or polyester than those of the former type, which are more suitable for acetate dyeing. Unsymmetrical disubstituted dyes often give higher absorption on secondary acetate than their symmetrical analogues (Table 3.10), presumably because they can participate in both types of dye–fibre bonding mechanism.

TABLE 3.10

Solubility of disubstituted dyes in secondary cellulose acetate [72]

Dye structure	X	Y	Solubility at 80°C/g kg^{-1}
3.47	H	H	*ca* 2.4
3.47	H	OH	18
3.47	OH	OH	14
3.48	H	H	*ca* 4.6
3.48	H	CH_2OH	22
3.48	CH_2OH	CH_2OH	6.6

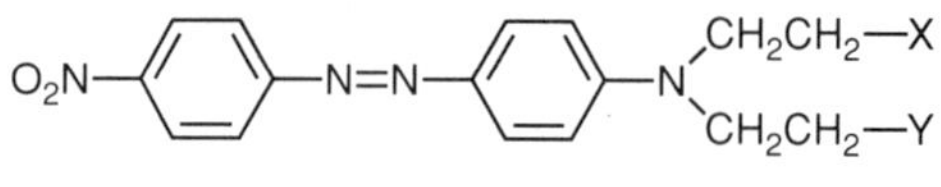

3.47

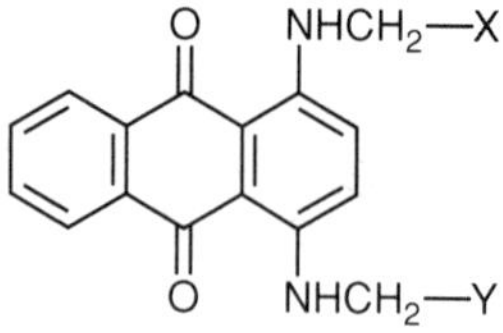

3.48

In order to achieve efficient build-up to heavy depths when dyeing cellulose acetate at 80°C it is customary, particularly for navy blues, to use a mixture of two or more components of similar hue. If these behave independently, each will give its saturation solubility in the fibre. In practice, certain mixtures of dyes of closely related structure are 20–50% less soluble in cellulose acetate than the sum of their individual solubilities predicts [73]. Dyes of this kind form mixed crystals in which the components are able to replace one another in the crystal lattice; the melting point depends on composition, varying gradually between those of the components, and the mixed crystals exhibit lower solubility than the sum of solubilities of the component dyes [74]. Dyes of dissimilar molecular shape do not form mixed crystals, the melting-point curves of the mixtures show a eutectic point and they behave additively in mixtures with respect to solubility in water and in the fibre.

Derivatives of diaminoanthrarufin (3.49; X = Y = H) and its 1,8-dihydroxy-4,5-diamino isomer (diaminochrysazin) have been among the most useful of anthraquinone dyes for ester fibres [75]. For example, methylation of diaminoanthrarufin gives C.I. Disperse Blue 26, a mixture of several components. Study of the pure *N*-alkylated derivatives from this base confirmed that monosubstitution (3.49; X = H, Y = alkyl) gives mid-blue dyes with excellent dyeing properties and acceptable fastness on polyester, but the bisalkyl dyes (3.49; X = Y = alkyl) are greener and inferior in application properties. Mixtures of the unsubstituted base with alkylated components, as obtained industrially, were especially advantageous for build-up to heavy depths, however [76].

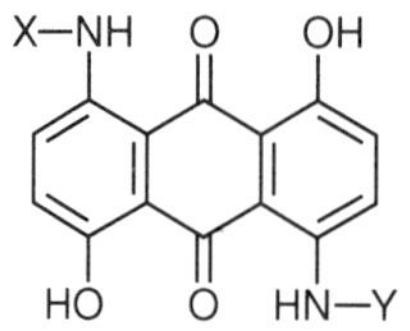

3.49

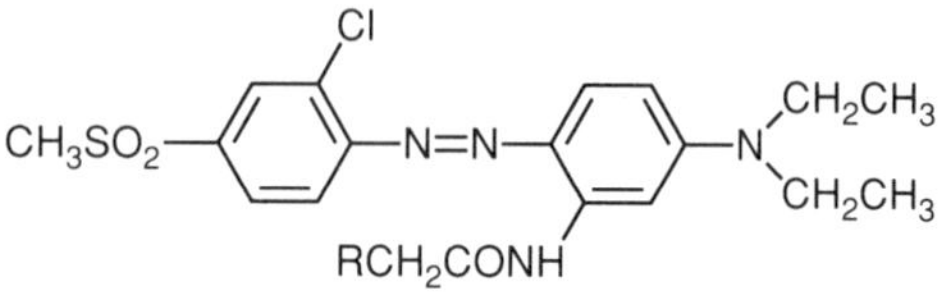

3.50

One of the criteria that determine the substantivity of disperse dyes for ester fibres is that the solubility parameter of the dye must be within one unit [$(cal/cm^3)^{1/2}$] of that of the fibre [77]. For example, the acetylamino dye

(3.50; R = H) with a solubility parameter (s) of 11.3 has much greater substantivity to polyester ($s = 10.8$) than has the propionylamino analogue (3.50; R = CH_3, $s = 13.0$). The effects of dyeing auxiliaries on the exhaust dyeing process can also be interpreted by reference to the solubility parameters of fibre, dye and auxiliary [78].

3.2.4 Dyeing of acrylic fibres with basic dyes

Most acrylic copolymer fibres contain sulphonate and sulphate end-groups on the polymer chains, derived from the inorganic redox catalysts employed to initiate the polymerisation reaction (potassium persulphate and sodium bisulphite). The adsorption of basic dyes at low pH occurs mainly on these strongly acidic groups [79–82]. Carboxylic acid groups are sometimes present, however, resulting in dyeing behaviour that is considerably more dependent on dyebath pH [83]. The concentration of strongly acidic dye sites in an acrylic fibre, as determined by sulphur analysis or by non-aqueous titration with a suitable base, agrees extremely well with saturation values derived from Langmuir isotherms defining the uptake of basic dyes [80,84]. This correspondence provides excellent confirmation that the dyeing mechanism is essentially a process of ion exchange.

The rate of dyeing of basic dyes on these fibres is highly sensitive to temperature. This originates in the response of the segmental mobility of the polymer chains to changes in temperature. At or near the glass transition temperature (T_g), the rate of diffusion of dye into the polymer structure increases markedly owing to the onset of enhanced mobility of the chain segments. Water acts as a plasticising agent for this process, by lowering the T_g of the acrylic polymer by 30–35°C [85]. This acceleration of the rate of dyeing as soon as T_g is exceeded, together with the virtual irreversibility of the electrostatic interaction between the dye cation and the anionic dyeing site in an otherwise non-polar local environment, means that careful control of application conditions is essential if level dyeing is to be achieved.

Difficulties of incompatibility can arise with mixtures of basic dyes on acrylic fibres because of competition for the limited number of dyeing sites available and the differences between dyes in terms of affinity and rate of diffusion. The rate of uptake of each dye when applied in the presence of another is invariably lower than when the dye is applied alone at the same concentration. Competition effects of this kind can lead to serious practical problems unless the dyes are carefully designed and selected to have similar dyeing characteristics [80,81,86,87]. Dyes with exceptionally low affinity and rapid rates of diffusion have been developed, offering improved migration properties on acrylic fibres [88,89]. These dyes have a migration behaviour not unlike that of monosulphonated acid dyes on nylon.

With the majority of basic dyes for acrylic fibres, however, levelling by migration is extremely limited because of their high affinity. A low rate of increase of dyeing temperature (0.5 degC/min or less) over the range in which the diffusion rate is increasing rapidly gives an improved degree of control. Colourless cationic agents, described as retarders, provide temporary competition for the dye sites in the early stages of adsorption of the dye. The amount of retarder applied is adjusted according to the proportion of unoccupied sites remaining when the dyeing equilibrium is attained, in order to minimise the risk of undesirable restraining effects. Since dye uptake is related to the number of unoccupied sites remaining, the temporary blocking action of the retarder effectively reduces the rate of dyeing [90].

3.3 APPLICATION PROPERTIES AND CHEMICAL STRUCTURE

The choice of dyes for a required colour on a given substrate almost always involves one or more compromises between conflicting requirements. Similarly there is seldom an 'ideal' dye structure of a desired hue that embodies all possible attractive features with regard to application and fastness properties. Many bright azo dyes have only moderate fastness to light, and the incorporation of a metal atom to form a complex (see Chapter 5) often entails a significant loss in brightness. The gain in aqueous solubility provided by an extra sulpho group in a direct or reactive dye often has to be paid for by a decrease in affinity for cellulose. An increase in the hydrophobic character of an acid dye for wool or nylon may improve the neutral-dyeing affinity at the expense of impaired level-dyeing properties. Likewise, the migration behaviour of a disperse dye is often affected adversely by an increase in molecular size intended to confer better wet fastness on cellulose acetate or better sublimation properties on polyester. Further examples of such conflicting requirements will become evident in the remaining sections of this chapter.

3.3.1 Dyeing kinetics and dye structure

The rates of uptake of direct dyes by cellulose vary extremely widely, as indicated by the data in Table 3.11. The time of 'half-dyeing' (the time to reach an exhaustion level half of that attained at equilibrium) on viscose for the disazo J acid derivative C.I. Direct Red 23 is approximately 400 times that for the rapidly diffusing stilbene dye C.I. Direct Yellow 12 [91]. Coefficients of diffusion for some of these dyes in cellophane film have been measured [24] and these values do account for some of the differences in dyeing rate, although the inverse relationship between diffusion coefficient and time of half-dyeing is only approximate. There is a slight tendency for the relatively smaller molecules to diffuse more quickly, in that structures 3.51 to 3.53 contain only four or five aryl nuclei and one or two sulpho groups, whereas structures 3.60 to 3.62 have six to eight aryl rings and two to four solubilising groups.

TABLE 3.11

Diffusion coefficients in cellophane film and times of half-dyeing on viscose for direct dyes at 90°C [24,91]

C.I. Direct	Structure	Diffusion coeff. in cellophane film $\times 10^{14}/m^2\ s^{-1}$	Time of half-dyeing on viscose yarn/min
Yellow 12	3.51	150	0.26
Red 81	3.52	50	0.84
Yellow 59	3.53		1.0
Yellow 50	3.54		1.5
Red 31	3.55	25.3	1.7
Blue 67	3.56		5.7
Red 2	3.57	7.7	8.9
Blue 1	3.58	13.3	15.9
Violet 53	3.59		25.3
Red 26	3.60		43.8
Black 22	3.61		43.8
Red 23	3.62		100

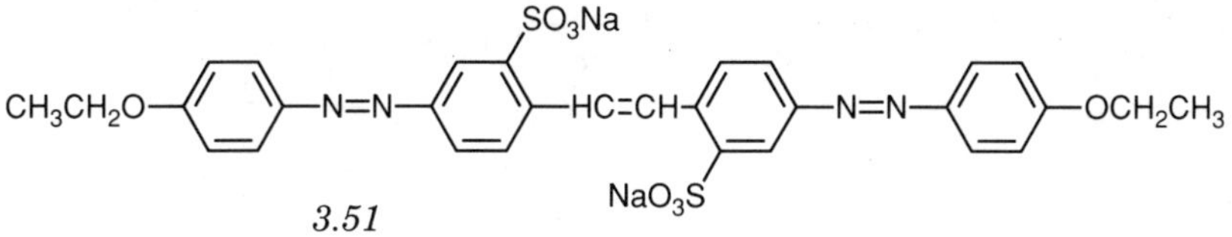

3.51

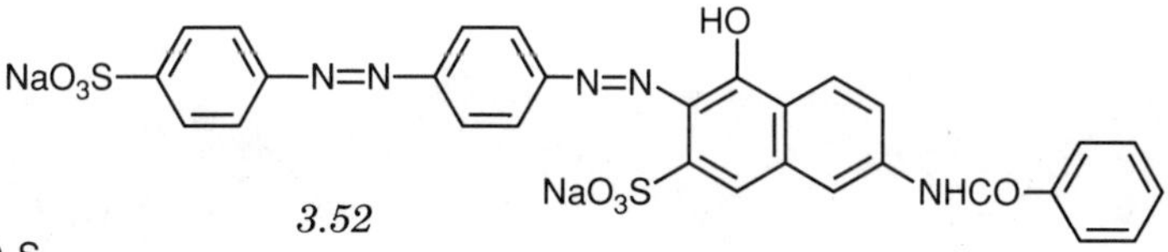

3.52

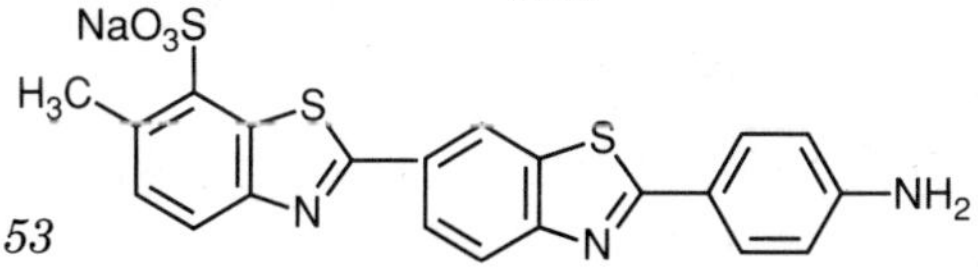

3.53

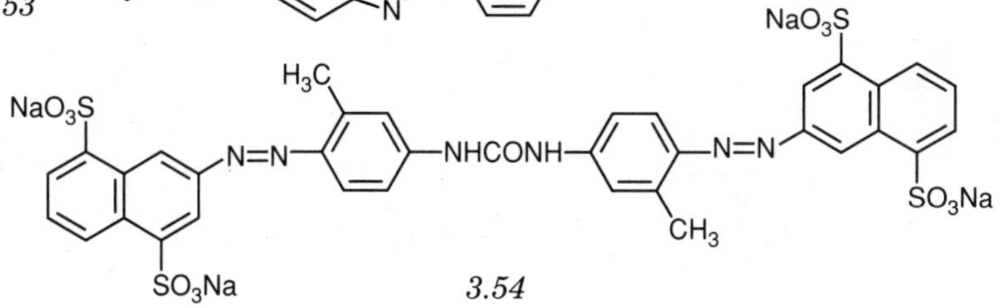

3.54

3.55

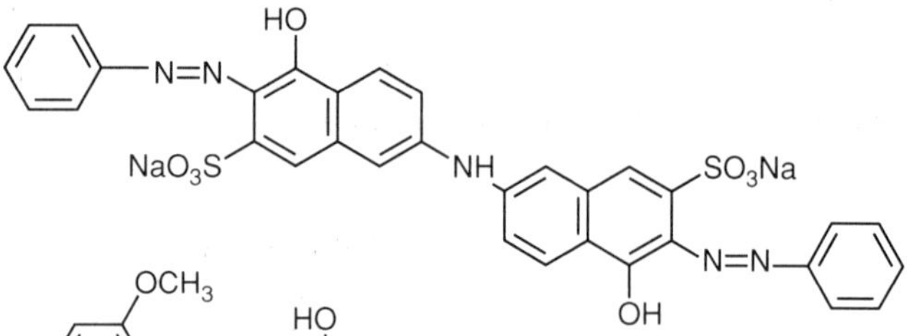

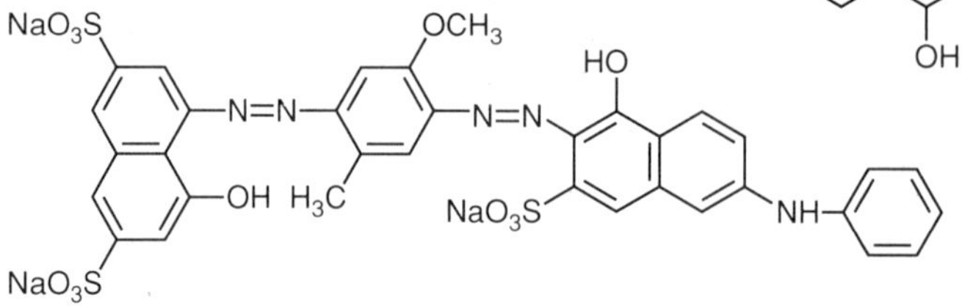

3.56

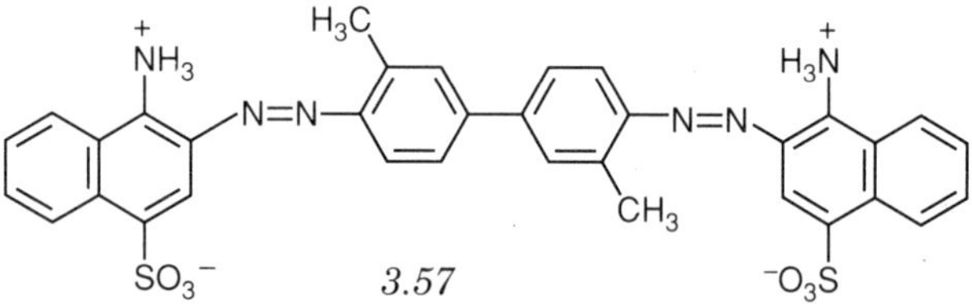

3.57

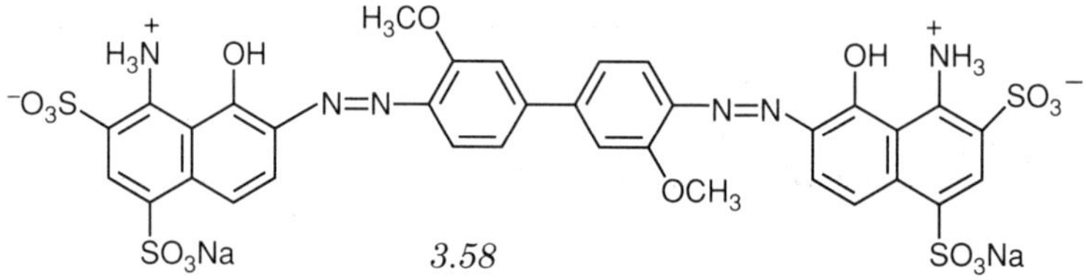

3.58

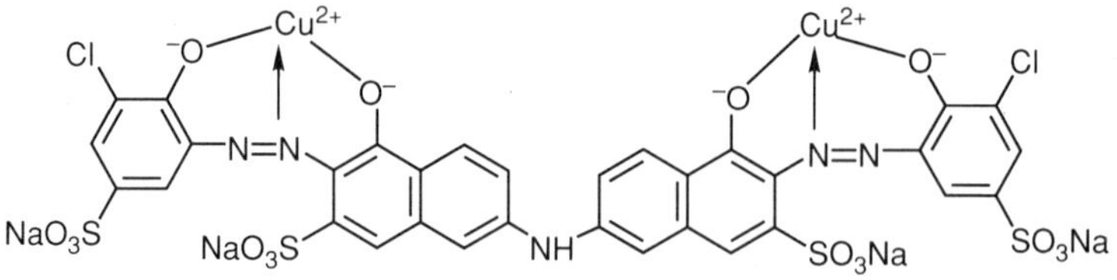

3.59

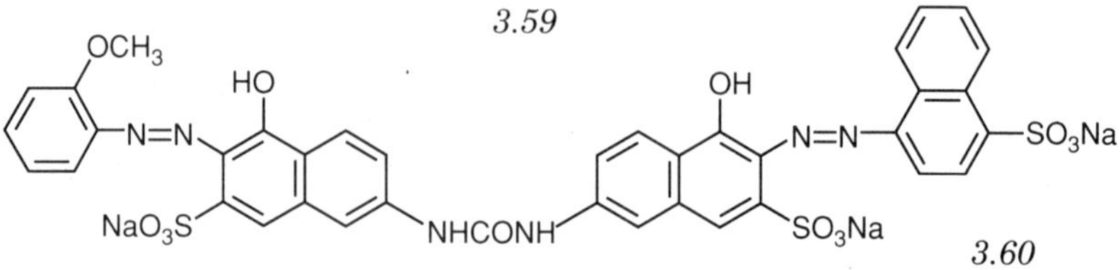

3.60

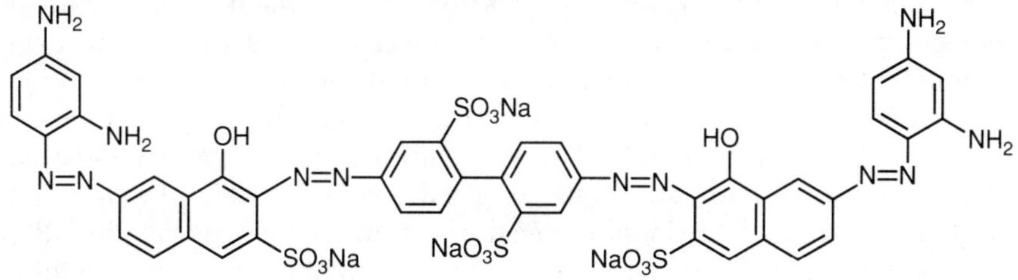

3.61

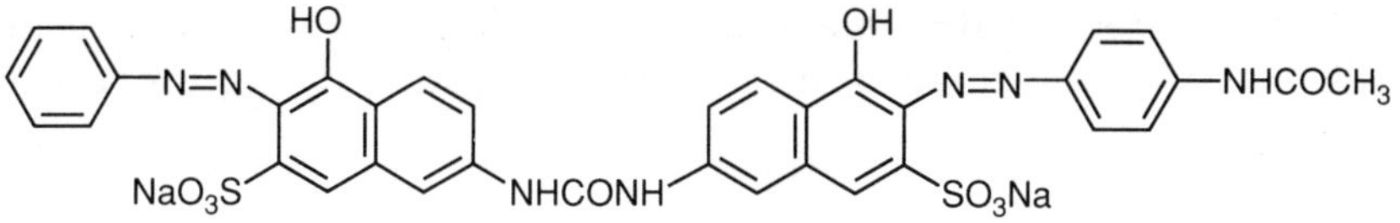

3.62

A much more decisive factor in determining the rates of dyeing of many direct dyes from aqueous salt solutions is their tendency to aggregate (section 3.1.2), since this greatly retards their mobility into the water-swollen voids of the substrate. Ureido-linked J acid residues (C.I. Direct Reds 23 and 26), especially if there is an acetanilide residue elsewhere in the molecule (as in C.I. Direct Red 23), provide opportunities for hydrogen-bond stabilisation of the weaker π-bonding attraction between aryl rings in neighbouring component molecules of an aggregate. Electrostatic bond stabilisation can occur if the individual dye molecules have some degree of zwitterionic character due to the presence in them of primary amino and sulpho groups (C.I. Direct Yellow 59, Red 2, Blue 1 and Black 22), or of copper(II) ions and ionised phenolic groups (as in C.I. Direct Violet 53).

The kinetic behaviour of anionic dyes on the amide fibres tends to be much more closely related to molecular size and hydrophilic–lipophilic balance. Thus within a related series of dye structures it is possible to discern more specific relationships. For example, in the series of substituted aniline→R acid dyes (3.63; R = H, methyl, n-butyl or n-dodecyl) the logarithm of the rate of dyeing is inversely proportional to the molecular volume [92].

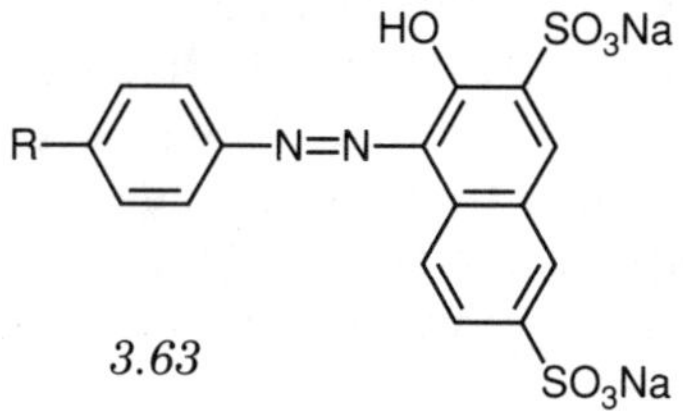

3.63

A somewhat similar inverse relationship between dyeing rate and molecular size was invoked to interpret the results of a study of the degree of penetration of three anthraquinone blue dyes (3.64 to 3.66) into nylon fibres after dyeing under various conditions [93]. After only 5–10 minutes at 90°C, C.I. Acid Blue 25 (3.64) had penetrated the nylon almost completely, with no evidence of ring dyeing. C.I. Acid Blue 138, a disulphonated analogue with an additional dodecyl group in the anilino residue (3.65), showed a marked difference in degree of penetration after dyeing for one hour at neutral pH (40% at 90°C, compared with 100% at 100°C). The extremely slowly diffusing disulphonated dye C.I. Acid Blue 175 (3.66), with seven interlinked rings able to form a coplanar arrangement, attained only 40–50% penetration of the substrate even after dyeing for two hours at 100°C.

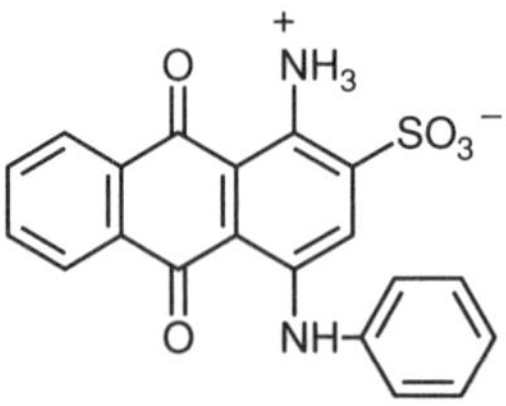

3.64

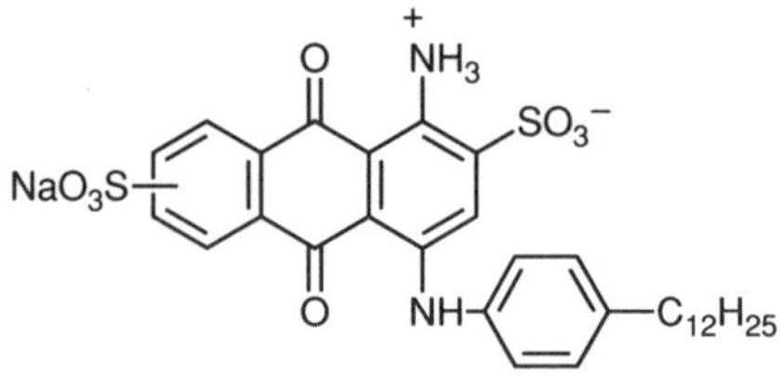

3.65

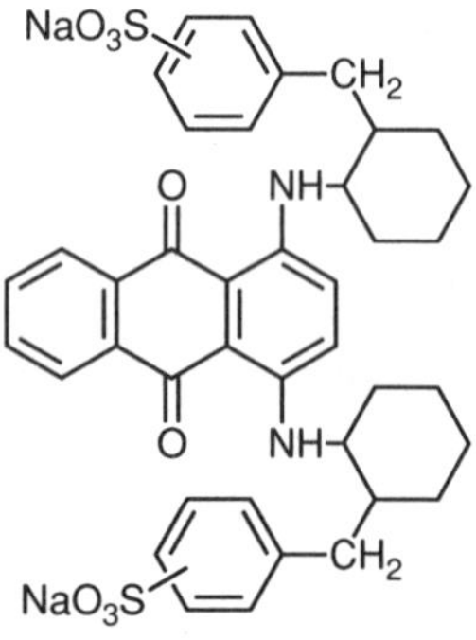

3.66

The relative rates of diffusion of disperse dyes in hydrophobic fibres tend to exhibit an approximately inverse relationship to molecular size, although here again the overall trend is somewhat obscured by specific forces of interaction between hydrogen-bonding substituents in the dye and polymer chain segments. C.I. Disperse Orange 3 (relative molecular mass 242) diffuses about five times faster than C.I. Disperse Blue 24 (r.m.m. 328) in nylon and polyester but only about 1.5 times faster in cellulose acetate [94]. These differences suggest that the orange azo dye interacts more strongly with acetyl groups in the regenerated fibre than does the blue anthraquinone dye, possibly by hydrogen bonding through the terminal primary amino group.

The relative rates of diffusion of C.I. Disperse Orange 3 on polyester, acetate and nylon indicated in Table 3.12 are reasonably consistent with those measured independently in the early days of polyester dyeing [95], as shown in Table 3.13. These figures, for three disperse dyes of almost equal relative molecular mass, also emphasise a marked increase in

TABLE 3.12

Diffusion coefficients of disperse dyes in hydrophobic fibres [94]

		Diffusion coefficient at 100°C $\times 10^{15}/m^2 s^{-1}$		
C.I. Disperse	Structure	Nylon	Acetate	Polyester
Orange 3	3.67: X = Y = H	23.1	15.1	1.3
Blue 24	3.68: X = $NHCH_3$, Y = NHPh	4.2	10.6	0.3

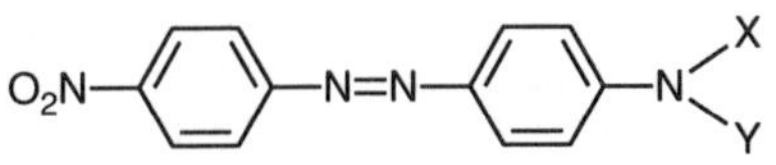

3.67

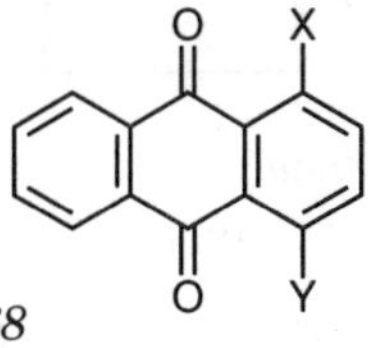

3.68

TABLE 3.13

Relative diffusion coefficients of disperse dyes [95,96]

			Relative diffusion coefficients			
					Polyester	
C.I. Disperse	Structure	R.m.m.	Nylon 85°C	Acetate 85°C	100°C	85°C
Red 15	3.68: X = NH_2, Y = OH	239	1000	286	34	1
Orange 3	3.67: X = Y = H	242	680	460	48	1
Violet 1	3.68: X = Y = NH_2	238	450	452	31	1

diffusion rate on polyester between 85 and 100°C, which is much less evident on nylon or cellulose acetate. Disperse dyes tend to diffuse more rapidly in polypropylene than in other hydrophobic fibres but to give relatively low saturation values indicative of poor affinity for this fibre. The results in Table 3.14 illustrate these differences for 1-anilinoanthraquinone (3.68; X = NHPh, Y = H) on cellulose acetate and polypropylene [70].

There is a roughly inverse relationship between the times of half-dyeing of a series of structurally related dyes and their saturation solubility in the appropriate substrate, as illustrated for four *p*-alkylamino derivatives of 1-anilinoanthraquinone on cellulose acetate (Table 3.15). It is interesting that methylamino and 2-hydroxyethylamino substituents confer good solubility in this substrate, but ethylamino groups are even less effective than isobutylamino groups in this respect [97].

TABLE 3.14

Saturation values and diffusion coefficients of 1-anilinoanthraquinone [70]

Dyed at 80°C	Saturation value/g kg^{-1}	Diffusion coeff. × $10^{14}/m^2\ s^{-1}$
Polypropylene	3.2	6.74
Cellulose acetate	6.0	1.12

TABLE 3.15

Times of half-dyeing and solubilities of disperse dyes in cellulose acetate at 85°C [97]

	Structure	Substituent (R)	Time of half-dyeing /min	Solubility /g kg^{-1}
3.69	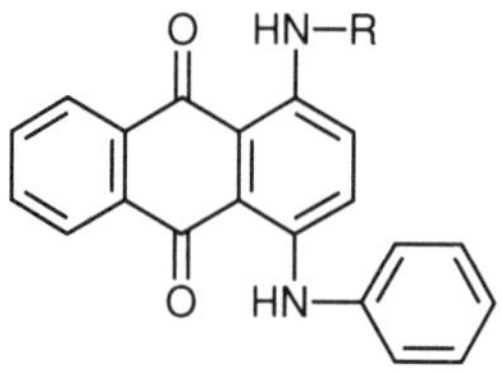	Methyl	30	8.8
		2-hydroxyethyl	35	4.2
		Isobutyl	–	4.1
		Ethyl	37	1.5

TABLE 3.16

Times of half-dyeing of disperse dyes on cellulose acetate at 85°C

C.I. Disperse	Structure	Time of half-dyeing/min
Yellow 1	3.70	0.75
Orange 3	3.67: X = Y = H	0.8
Violet 4	3.68: X = NH_2, Y = $NHCH_3$	1.6
Red 15	3.68: X = NH_2, Y = OH	2.4
Blue 3	3.68: X = $NHCH_3$, Y = $NHCH_2CH_2OH$	2.5
Violet 1	3.68: X = Y = NH_2	3.0
Red 19	3.67: X = Y = CH_2CH_2OH	3.3
Red 1	3.67: X = CH_2CH_3, Y = CH_2CH_2OH	4.0
Red 11	3.71	4.8
Blue 26	3.72: X = H	6.5
Black 1	3.73	6.6
Orange 5	3.74	11.5
Violet 8	3.75	12.0
Orange 1	3.67: X = H, Y = Ph	14.0
Red 13	3.76	16.5
Blue 14	3.68: X = Y = $NHCH_3$	16.5
Orange 13	3.77	19.0
Blue 26:1	3.72: X = CH_3	28.5

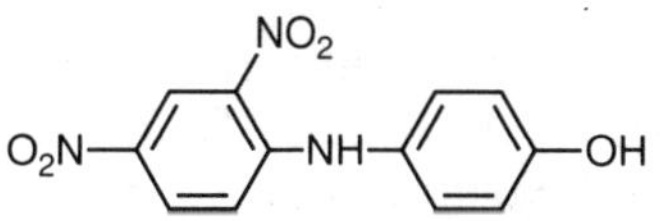

3.70

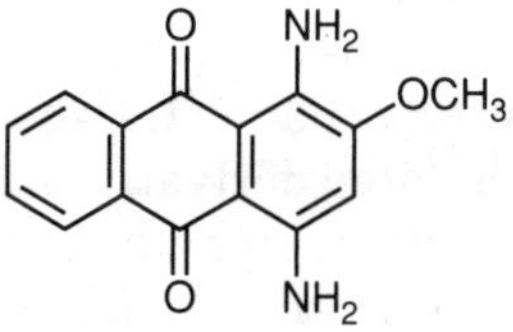

3.71

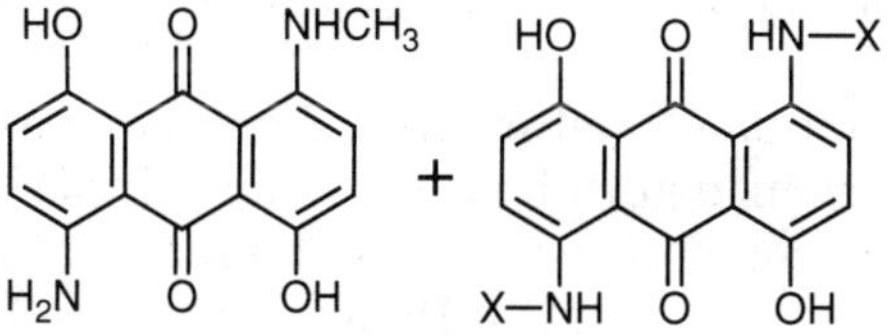

3.72

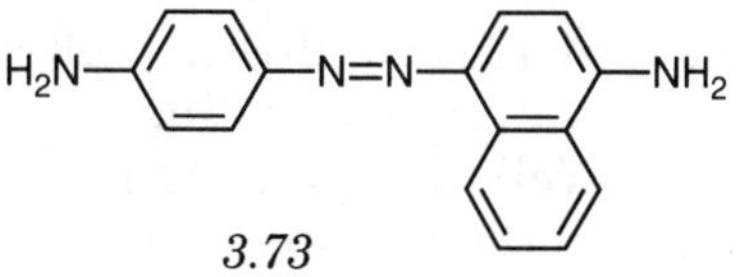

3.73

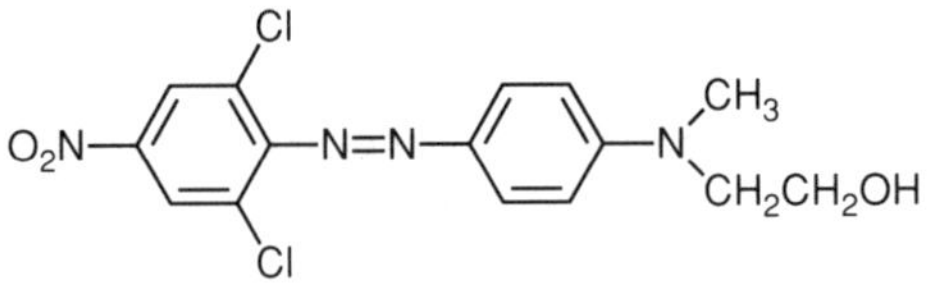

3.74

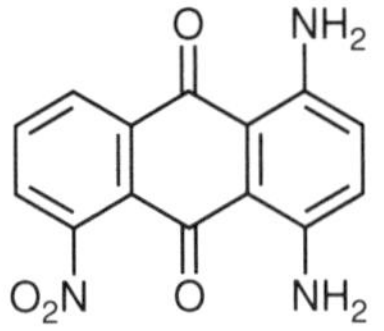

3.75

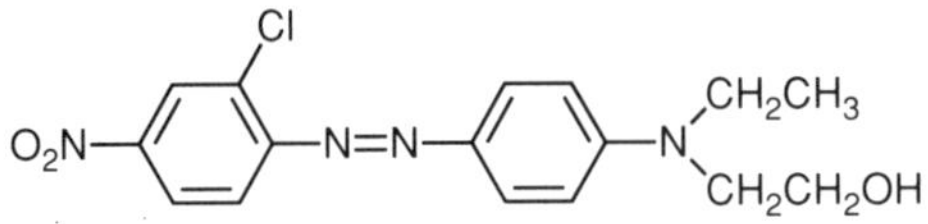

3.76

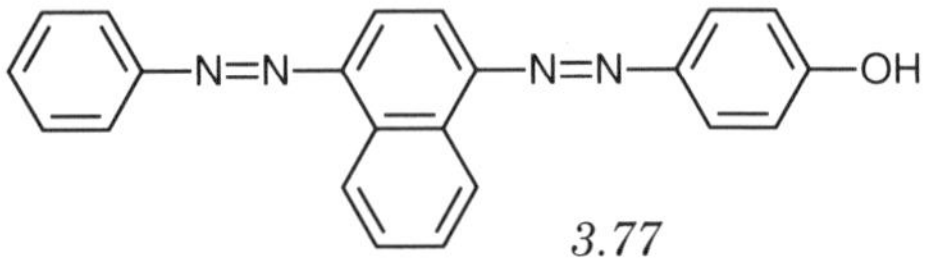

3.77

The list of times of half-dyeing on cellulose acetate [15] reproduced in Table 3.16 gives some significant pointers to the relationship between disperse dye structure and dyeing rate on this fibre. The most mobile molecular structure shown is the dinitrodiphenylamine dye C.I. Disperse Yellow 1, closely followed by the unsubstituted *p*-nitroaniline→aniline C.I. Disperse Orange 3. The introduction of one *N*-phenyl group into the latter structure (to form C.I. Disperse Orange 1) has a much more pronounced retarding effect than two ethyl or 2-hydroxyethyl groups (C.I. Disperse Red 1 or Red 19). In the series of 1,4-diaminoanthraquinone (3.68) derivatives, the unsymmetrically substituted types (C.I. Disperse Red 15, C.I. Disperse Violet 4, C.I. Disperse Blue 3) are absorbed much more rapidly than is the symmetrical analogue with two methylamino groups (C.I. Disperse Blue 14). This dye is greatly retarded in comparison with the unmethylated dye (C.I. Disperse Violet 1). In a similar way, C.I. Disperse Blue 26:1 containing mono- and di-methylated diaminoanthrarufin is absorbed several times more slowly than C.I. Disperse Blue 26, a mixture of the monomethyl and unmethylated derivatives. A 5-nitro substituent (in C.I. Disperse Violet 8) has a much greater effect than a 2-methoxy group (in C.I. Disperse Red 11) in retarding the dyeing rate of 1,4-diaminoanthraquinone itself (C.I. Disperse Violet 1).

Dyeing rate constants [98] and diffusion coefficients for many of the dyes listed in Table 3.16 have also been determined on nylon. Table 3.17 confirms that many of the trends noted above are also applicable to the nylon dyeing system. C.I. Disperse Yellow 1, C.I. Disperse Orange 3 and C.I. Disperse Violet 4 are again the three most rapidly adsorbed dyes listed, in that order. The diffusion coefficients place the first two dyes in reverse order, however, suggesting that the terminal phenolic group in C.I. Disperse Yellow 1 interacts more effectively with proton-acceptor sites in nylon than the primary amino terminal group in C.I. Disperse Orange 3.

TABLE 3.17

Diffusion coefficients and dyeing rate constants for disperse dyes on nylon at 80°C [15,98]

C.I. Disperse	Structure	Diffusion coeff. $\times 10^{15}/m^2 s^{-1}$	Rate constant
Yellow 1	3.70	11	1820
Orange 3	3.67: X = Y = H	23	940
Violet 4	3.68: X = NH_2, Y = $NHCH_3$		840
Blue 3	3.68: X = $NHCH_3$, Y = $NHCH_2CH_2OH$		570
Red 19	3.67: X = Y = CH_2CH_2OH		450
Violet 1	3.68: X = Y = NH_2	11	350
Red 15	3.68: X = NH_2, Y = OH		280
Red 1	3.67: X = CH_2CH_3, Y = CH_2CH_2OH	3	200
Orange 13	3.77	1	160
Red 13	3.76		130
Blue 14	3.68: X = Y = $NHCH_3$	11	120
Red 11	3.71	15	80
Blue 26	3.72: X = H	5	80
Blue 26:1	3.72: X = CH_3	17	68
Orange 1	3.67: X = H, Y = Ph	3	56

As before, an *N*-phenyl substituent (C.I. Disperse Orange 1) exerts a markedly greater retarding influence than *N*-ethyl-*N*-2-hydroxyethyl (C.I. Disperse Red 1) or *NN*-bis(2-hydroxyethyl) (C.I. Disperse Red 19) substitution of the parent structure (C.I. Disperse Orange 1). Once again in the 1,4-disubstituted anthraquinone series, the rate constants for the unsymmetrical structures (C.I. Disperse Red 15, Violet 4, Blue 3) are much higher than that for the symmetrical bis-methylamino analogue (C.I. Disperse Blue 14). The rates for the two mixed diaminoanthrarufin derivatives, however, are much closer on nylon than on cellulose acetate. The diffusion coefficients of these dyes do differ markedly, perhaps because the unmethylated component of C.I. Disperse Blue 26 interacts more strongly with proton-acceptor sites in nylon than does the more fully methylated mixture C.I. Disperse Blue 26:1.

The study of the degree of penetration of blue acid dyes in nylon mentioned earlier also included a further series of measurements for two blue disperse anthraquinone derivatives on polyester [93]. A fully penetrated dyeing with C.I. Disperse Blue 56 (3.78; X = Br) was achieved after 90 minutes at 120°C, but with C.I. Disperse Blue 83 (3.78; X = $PhOCOCH_3$ and PhOCOPh) complete penetration within this time of dyeing could only

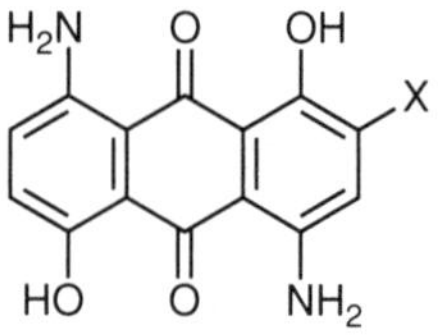

3.78

be attained at 140°C. In practice, the dyeing of polyester fibres is often restricted to 30–60 minutes at 125–130°C in order to minimise the release of oligomer into the dyebath or to preserve desirable fabric properties. When applying Class D dyes of the C.I. Disperse Blue 83 type these conditions can produce dyeings that are only 30–40% penetrated.

Diffusion coefficients for five disperse dyes on polyester are compared in Table 3.18 [40]. Some of these structures have also been included in the corresponding data for the nylon and cellulose acetate systems. As before, the unsymmetrical 1,4-diaminoanthraquinone derivative (C.I. Disperse Blue 3) diffuses significantly more rapidly than the symmetrical bis-dimethylamino analogue (C.I. Disperse Blue 14). It is interesting that C.I. Disperse Orange 13, which is one of the most slowly diffusing dyes on acetate (Table 3.16) and nylon (Table 3.17), shows a higher diffusion coefficient on polyester than the two monoazo dyes with a terminal nitro group (C.I. Disperse Reds 7 and 13).

TABLE 3.18

Diffusion coefficients of disperse dyes on unset polyester at 100°C [40]

C.I. Disperse	Structure	Diffusion coeff. $\times 10^{14}/m^2\ s^{-1}$
Blue 3	3.68: X = $NHCH_3$, Y = $NHCH_2$, CH_2OH	1.6
Orange 13	3.77	1.4
Red 7	3.79	1.25
Blue 14	3.68: X = Y = $NHCH_3$	1.0
Red 13	3.76	0.9

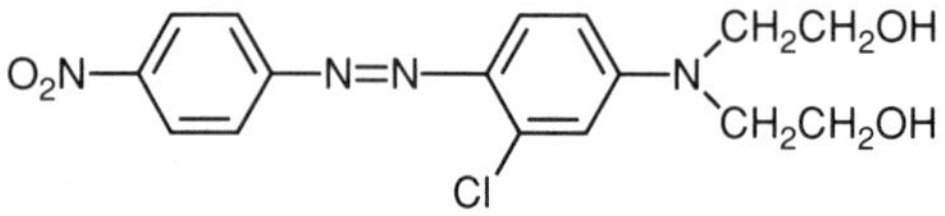

3.79

As in the case of disperse dyes on polyester, the diffusion coefficients of basic dyes on acrylic fibres tend to decrease as molecular size increases. Most basic dyes have a single cationic centre surrounded by a relatively compact hydrophobic system containing three or four aryl rings. Five typical structures are compared in Table 3.19, which presents their relative diffusion coefficients on acrylic fibre and on a basic-dyeable polyester variant in the presence and absence of 2 g/l of a diphenyl-type carrier [99]. Individual substituents exert only minor effects on the overall rate of diffusion. Thus the *N*-ethyl-*N*-2-hydroxyethyl substitution pattern in C.I. Disperse Blue 41 produces only slightly lower diffusion coefficients than the dimethylamino analogue (C.I. Disperse Blue 54) and even the dimethyltriazole diazo component in C.I. Disperse Yellow 25 has a minor retarding effect relative to the monomethylthiazole grouping in C.I. Disperse Red 29, even on basic-dyeable polyester.

TABLE 3.19

Relative diffusion coefficients of basic dyes on basic-dyeable polyester and acrylic fibres at 100°C [99]

C.I. Basic	Structure	Basic-dyeable polyester: No carrier	Basic-dyeable polyester: Carrier	Acrylic fibre
Red 29	3.80	1.1	1.9	3.6
Blue 54	3.81: X = CH_3, Y = CH_3	1.1	1.8	3.4
Blue 41	3.81: X = CH_2CH_3, Y = CH_2CH_2OH	1.0	1.8	3.1
Yellow 25	3.82	1.0	1.7	2.8
Yellow 11	3.83	0.7	1.6	2.6

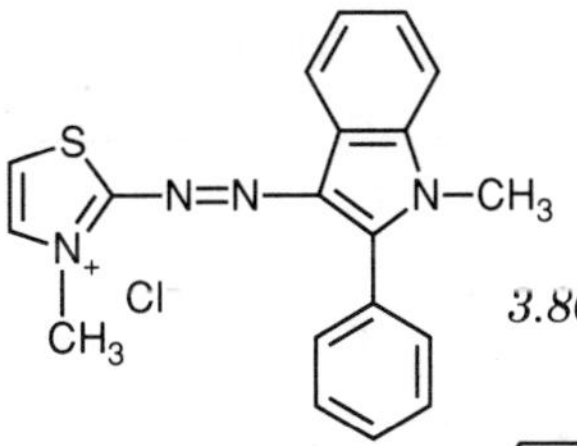
3.80

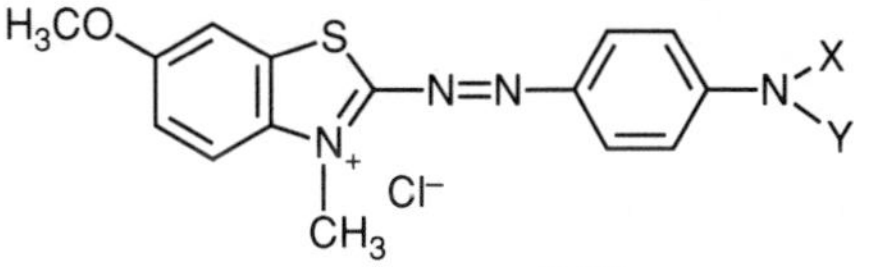
3.81

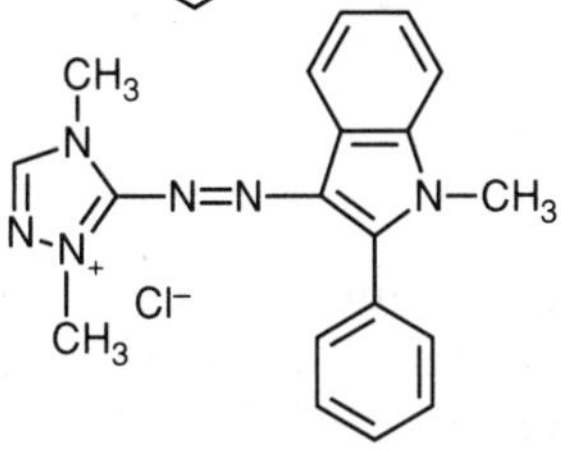
3.82

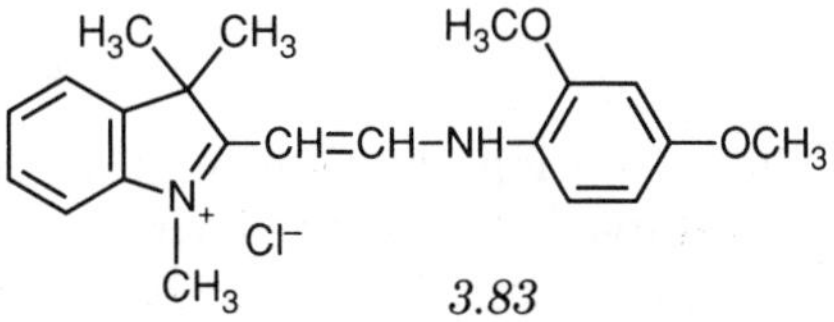
3.83

3.3.2 Dye structure and affinity for the fibre

As already noted in relation to kinetic behaviour, minor structural changes in direct dyes can have marked effects on dyeing properties. Within the closely related series of J acid disazo dyes illustrated in Table 3.20, it is seen that the insertion of an acetylamino group in C.I. Direct Orange 26 to form C.I. Direct Red 23 gives a significant boost to the affinity as well as a bathochromic influence on the colour of the dye. Comparing the other two members of the series, an additional aryl ring (C.I. Direct Red 26) has a much greater effect on the overall affinity of the molecule than a simple methyl group (C.I. Direct Red 24).

TABLE 3.20

Affinity of direct dyes for viscose at 60°C [44]

C.I. Direct	Structure	Affinity/kcal mol^{-1}
Red 26	3.84: X = 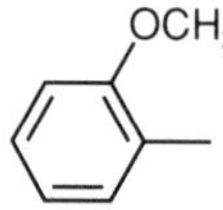 Y = 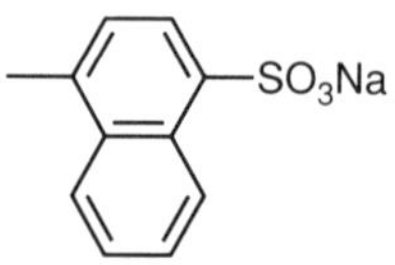	5.7
Red 23	3.84: X = 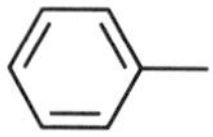 Y = 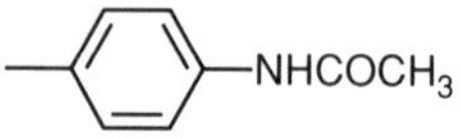	5.1
Orange 26	3.84: X = 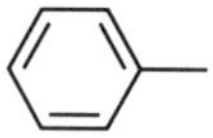 Y = 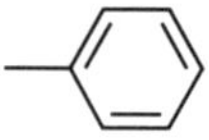	4.8
Red 24	3.84: X = 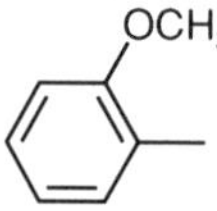 Y = 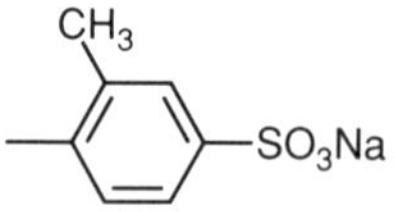	4.4

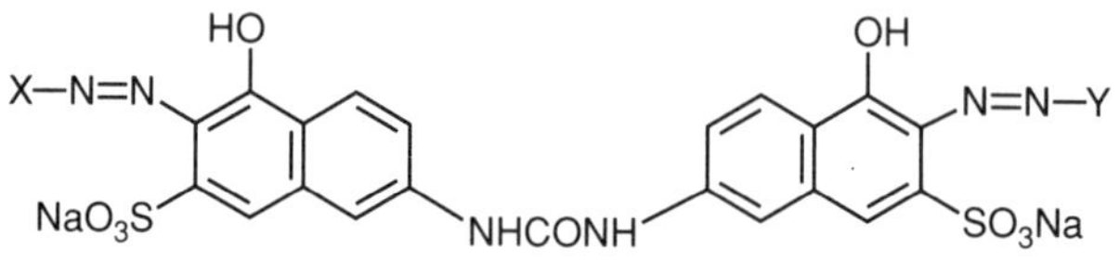

3.84

The attraction between cellulose and the leuco form of a simple polycyclic vat dye is believed to be mainly attributable to dispersion forces. Not surprisingly, therefore, affinity tends to increase with relative molecular mass in the series shown in Table 3.21. This is not the only factor, however, because the eight-ring pyranthrone (3.85) has higher affinity than the nine-ring violanthrone isomers (3.86 and 3.87). The configuration of the rings and the location of the keto groups are also significant, since violanthrone has greater substantivity than isoviolanthrone.

TABLE 3.21

Affinity of polycyclic vat dyes for cotton at 40°C [15,46]

Polycyclic vat dye	Structure	R.m.m.	Affinity/kcal mol^{-1}
Pyranthrone	3.85	406	5.56
Violanthrone	3.86	456	5.46
Isoviolanthrone	3.87	456	4.77
Dibenzopyrenequinone	3.88	332	3.76
Anthranthrone	3.89	306	2.86
Benzanthraquinone	3.90	258	1.18

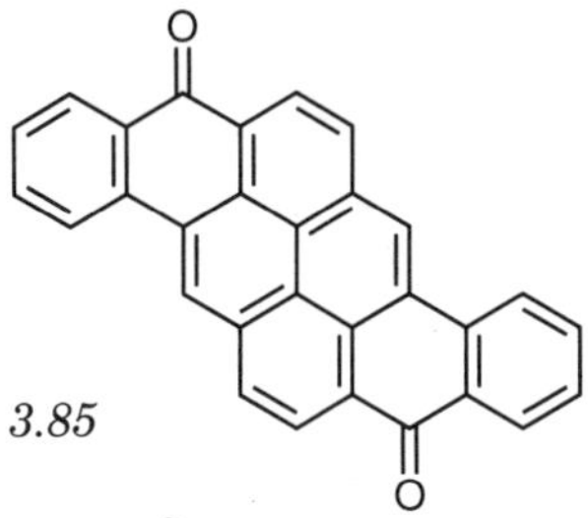

3.85

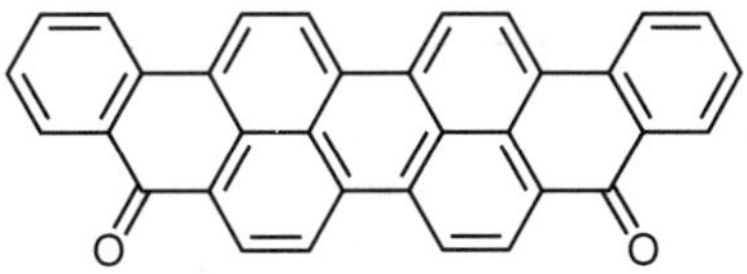

3.86

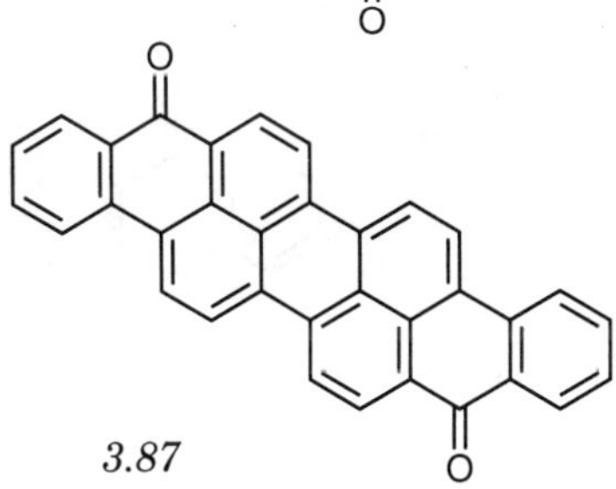

3.87

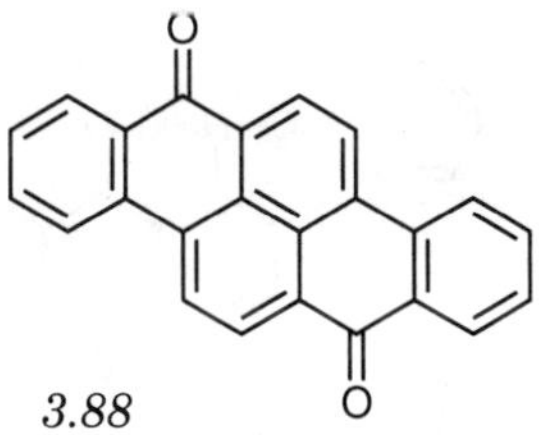

3.88

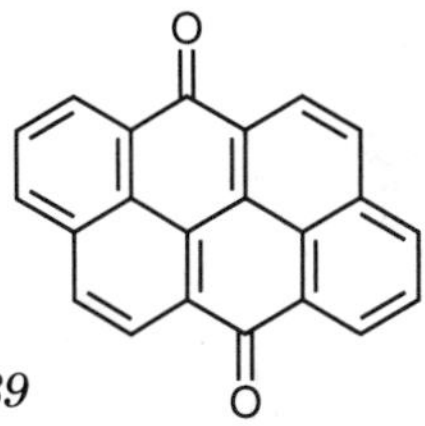

3.89

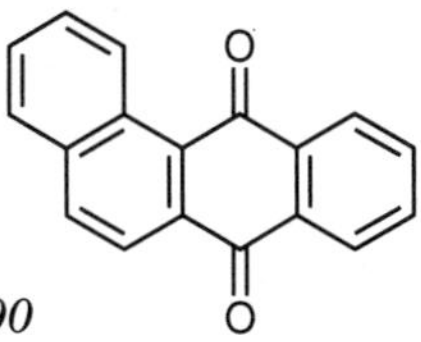

3.90

Hydrogen bonding is believed to have a notable influence on the substantivity of benzoylaminoanthraquinone vat dyes, as detailed studies have shown (Table 3.22). Certain other substituents are also important, since the non-amide 1-benzylaminoanthraquinone has virtually no affinity. Introduction of a second benzoylamino group into 1-benzoylaminoanthraquinone (3.91; X = H) approximately doubles the affinity, 1,4-substitution having a greater effect than 1,5-substitution. The enhanced affinity conferred by *p*-X-substituted benzoylamino groups increases with the dipole moment of the C–X bond (C–CH_3 0.41 < C–OCH_3 1.16 < C–Cl 1.56). Such substituents are almost equally effective in either *m*- or *p*-positions, but *o*-X-substituted analogues are much less substantive because coplanarity of

TABLE 3.22

Affinity of benzoylaminoanthraquinone vat dyes at 40°C on cotton [15]

Substituent (X)	Affinity/kcal mol^{-1} 3.91	3.92	3.93
Cl	2.30	3.73	4.69
CH_3O	2.18	3.46	3.82
CH_3	1.96	3.06	3.98
H	1.64	2.88	3.67

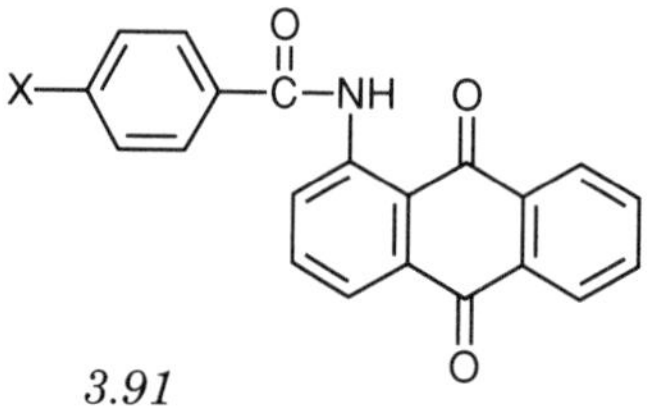

3.91

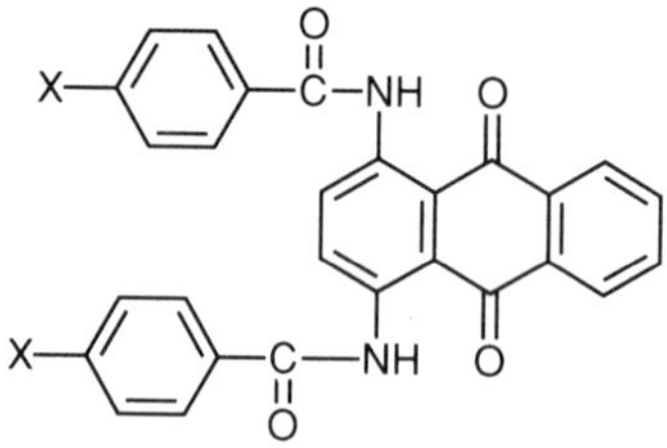

3.93

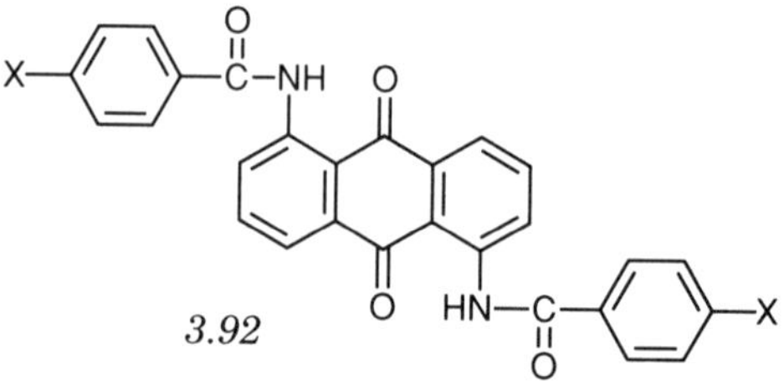

3.92

the aryl rings with the plane of the anthraquinone nucleus is inhibited [14].

Early studies of the affinity of acid dyes for wool revealed noteworthy correlations with dye structure. For example, in four pairs of monoazo dyes differing only by replacement of a benzene by a naphthalene nucleus, the affinity increase in each pair was consistently 1.1 to 1.5 kcal/mol. In three related pairs of dyes, an additional sulpho group reduced the affinity by about 1 kcal/mol [100,101]. In the series of alkylsulphuric acids [ethyl (–5.8) < octyl (–9.45) < dodecyl (–10.92) (units in kcal/mol)], the affinity increased by about –0.4 kcal per methylene group [102]. These results were supported by later measurements on two series of monoazo dyes containing alkyl chains of increasing length, the increment per methylene group again amounting to about –0.4 kcal in both series. The same investigator also revealed a close correlation between affinity and relative molecular mass for a series of substituted aniline→NW acid dyes [15].

Two further interesting series of systematic changes in structure have shed more light on the relative effects of alkyl and aryl groupings in conferring enhanced affinity for wool. The phenylazo member of the series shown in Table 3.23 (3.94: X = Ph) has an affinity of 4 kcal/mol and and the introduction of a 4-n-butyl substituent has almost as great an effect as the additional aryl ring formed by replacing the phenylazo by a naphthylazo grouping. Likewise, in the three anthraquinone derivatives illustrated in Table 3.24, replacing the 4-methylamino group by an n-butylamino group boosts the affinity by about 10%, but the effect of an anilino residue represents almost a 15% increase.

TABLE 3.23

Affinity of monoazo acid dyes for wool at 70°C [103]

	Substituent (X)	Affinity/kcal mol^{-1}
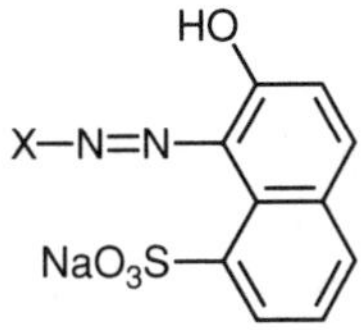 3.94	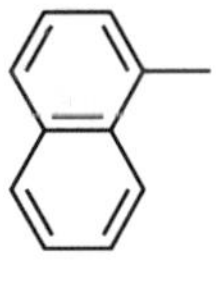	5.4
	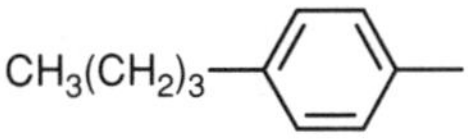	5.3
	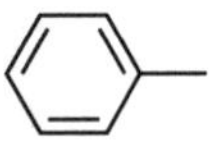	4.0

TABLE 3.24

Affinity of anthraquinone acid dyes for wool at 50°C and pH 4.6 [104]

	Substituent (X)	Affinity/kcal mol^{-1}
3.95 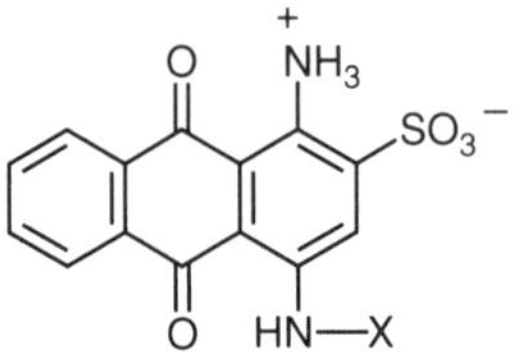	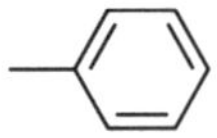	6.0
	$(CH_2)_3CH_3$	5.8
	CH_3	5.3

Careful measurements have been made of the affinity of many levelling and milling acid dyes for nylon using various experimental techniques, and these have been used to interpret the well-known problem of mutual blocking effects in mixture recipes on this substrate. Typical values for some monoazo acid dyes are given in Table 3.25 [105]. C.I. Acid Orange 7 (3.96; X = sulphanilic acid residue, Y = H) is the *ortho* isomer of C.I. Acid Orange 20 (3.99), which has a significantly higher affinity. An *o*-hydroxy group is capable of intramolecular hydrogen bonding to the azo linkage but a *p*-hydroxy is not and is free to form a hydrogen bond with, for example, a main-chain amide group in nylon. C.I. Acid Oranges 7 and 10 have the same phenylazo-2-naphthol chromophoric system but the additional sulpho group in the structure of the latter lowers the affinity by about –0.8 kcal/mol. Similarly the two extra sulpho groups in C.I. Acid Red 18 decrease the affinity by about –1.8 kcal/mol compared with the analogous C.I. Acid Red 88, the most substantive of the dyes listed in Table 3.25. C.I. Acid Orange 8 differs from C.I. Acid Orange 7 only in having an additional methyl substituent *ortho* to the azo group. This only raises the affinity by about –0.2 kcal/mol, but the change from phenylazo to naphthylazo in going from C.I. Acid Orange 7 to C.I. Acid Red 88 gives an increase of approximately –1.8 kcal/mol, virtually the same magnitude as the lowering effect of two sulpho groups.

In the series of three aniline→*N*-acyl H acid (3.97) dyes, the additional *p*-acetylamino grouping in C.I. Acid Violet 7 enhances the affinity of C.I. Acid Red 1 by about –0.25 kcal/mol, as well as producing a marked bathochromic shift in hue. Replacing the acetylamino group in the H acid residue of C.I. Acid Violet 7 by a *p*-tosylamino group has less effect on hue but raises the affinity by about 1.25 kcal/mol. Somewhat surprisingly, replacing the carboxyl group in the azopyrazolone structure C.I. Acid Yellow 23 by a methyl group and introducing two chloro substituents into the *N*-phenyl residue only increases the overall affinity by about –0.3 kcal/mol.

TABLE 3.25

Affinity of acid dyes for nylon 6.6 by chloride desorption at 75°C [105]

C.I. Acid	Structure			Affinity/kcal mol^{-1}
Red 88	3.96:	X = 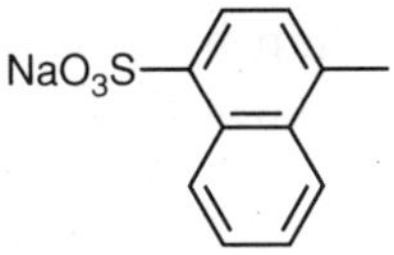	Y = H	7.66
Violet 5	3.97:	X = CH_3CONH	Y = 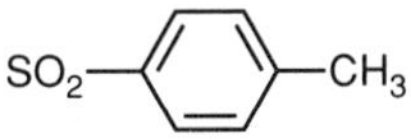	6.62
Orange 20	3.99			6.17
Orange 8	3.96:	X = 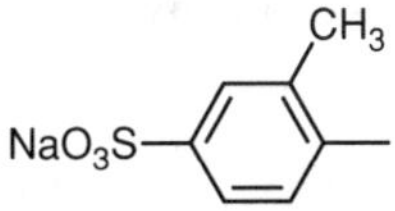	Y = H	6.03
Red 18	3.96:	X = 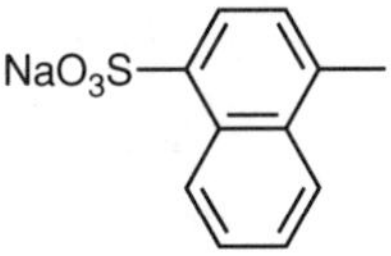	Y = SO_3Na	5.88
Orange 7	3.96:	X = 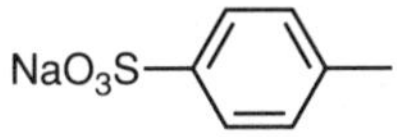	Y = H	5.82
Violet 7	3.97:	X = CH_3CONH	Y = $COCH_3$	5.37
Yellow 17	3.98:	X = CH_3	Y = Cl	5.31
Red 1	3.97:	X = H	Y = $COCH_3$	5.11
Yellow 23	3.98:	X = COOH	Y = H	5.01
Orange 10	3.96:	X = 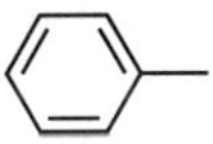	Y = SO_3Na	5.00

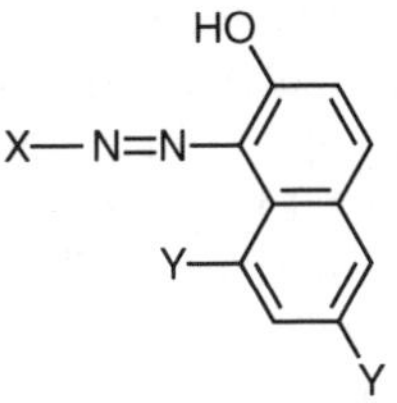

3.96

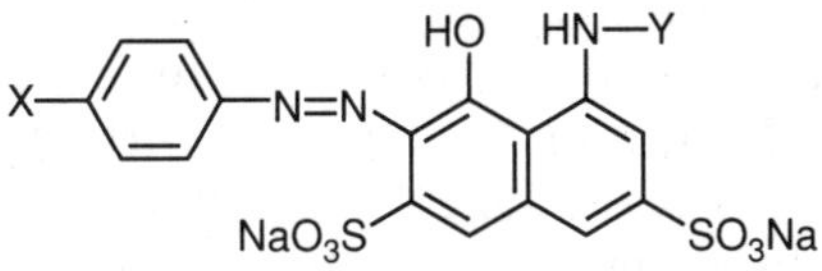

3.97

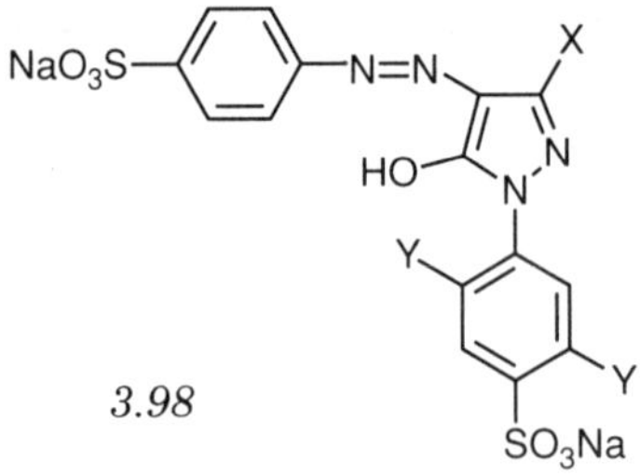

3.98

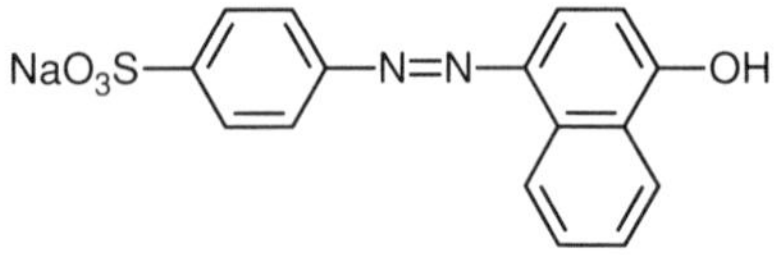

3.99

The yield of disperse dyes on polyester is limited by their slow rate of diffusion into the fibre rather than by inherently low substantivity. If dyeing times are sufficiently long, saturation values on polyester approach those on cellulose acetate and often exceed those on nylon (Table 3.26; for structures of these dyes see Tables 3.12 and 3.16).

TABLE 3.26

Saturation values of disperse dyes on hydrophobic fibres at 85°C [95]

		Saturation value/% o.w.f.		
C.I. Disperse	Structure	Acetate	Nylon	Polyester
Yellow 1	3.70	16.0	5.0	7.1
Red 15	3.68: X = NH_2, Y = OH	11.2	4.4	12.0
Violet 1	3.68: X = Y = NH_2	8.3	4.9	4.4
Orange 3	3.67: X = Y = H	5.1	2.0	4.1

Some early measurements of the partition of various aminoanthraquinone derivatives between cellulose acetate and ethanol (Table 3.27) dramatically illustrate the value of primary amino groups in conferring affinity for cellulose acetate. Indeed, the introduction of a *p*-hydroxy group into 1-aminoanthraquinone actually lowers the affinity for the fibre slightly, perhaps by favouring intramolecular hydrogen bonding and thus greater aggregation in the aqueous phase. Methylation of the 1-amino group promotes affinity, but introduction of a 2-methyl substituent into 1-aminoanthraquinone has a greater effect. The most impressive increases observed, however, are caused by further amino substitution. The partition coefficients of the 1,4-, the 1,8- and the 1,5-isomers are respectively about twice, three times and four times that of the monoamino analogue. Further amino substitution has a less marked effect, the 1,4,5,8-tetra-substituted derivative having a partition coefficient about five times that of 1-amino-anthraquinone.

TABLE 3.27

Partition coefficients of aminoanthraquinone derivatives between cellulose acetate and ethanol at 60°C [106]

Anthraquinone derivative	Partition coefficient
1,4,5,8-Tetra-amino	8.77
1,5-Diamino	6.64
1,8-Diamino	5.73
1,4-Diamino	2.80
1-Amino-2-methyl	2.67
1-Methylamino	2.38
1-Amino	1.79
1-Amino-4-hydroxy	1.59

3.3.3 Effect of dye structure on migration and levelling

Early investigators of the skittery dyeing of wool concluded that the differences in dyeability between individual fibres are attributable to the variations in extent of degradation that they have undergone in the fleece and during purification treatments [107,108]. The responses of levelling and milling acid dyes to such differences were studied in detail [109]. With typical levelling acid dyes such as C.I. Acid Red 1 (3.100; X = H) the proportion of ring-dyed fibres present gradually increases with dyeing temperature up to a maximum at about 70°C. As dyeing continues above this temperature the content of ring-dyed fibres gradually falls again and uniformly dyed fibres begin to predominate. Eventually the migration process ensures that a level and fully penetrated dyeing is achieved.

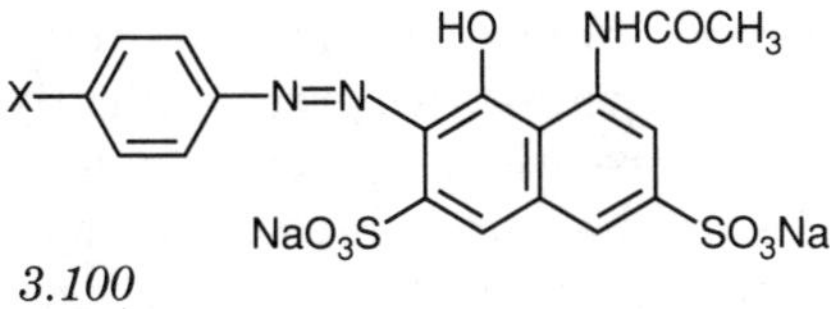

3.100

Milling acid dyes behave in a more individual way. The disazo monosulphonate C.I. Acid Red 151 (3.101) is adsorbed extremely rapidly at low temperatures, all fibres becoming ring-dyed. Little further adsorption then takes place until the temperature approaches 90°C. Diffusion into the fibre then accelerates greatly, so that the proportion of ring-dyed fibres rapidly falls to zero and soon all the fibres present become deeply and uniformly dyed. The disazo disulphonate C.I. Acid Red 85 (3.102), however, shows virtually no uptake at all below 70°C and ring dyeing is at a maximum in the range 80–90°C. Only after boiling for about 30 minutes does the proportion of fully penetrated fibres exceed that of ring-dyed ones. A level dyeing of satisfactory fastness requires at least an hour at the boil. More hydrophobic 'super-milling' dyes such as C.I. Acid Red 138 (3.100; X = dodecyl), the *p*-alkylated analogue of C.I. Acid Red 1, give ring dyeings at 70°C and above, but even after prolonged boiling it is difficult to attain more than a small proportion of fully penetrated fibres. Such dyes tend to be especially sensitive to skitteriness and to root–tip differences of dyeability in weathered wool.

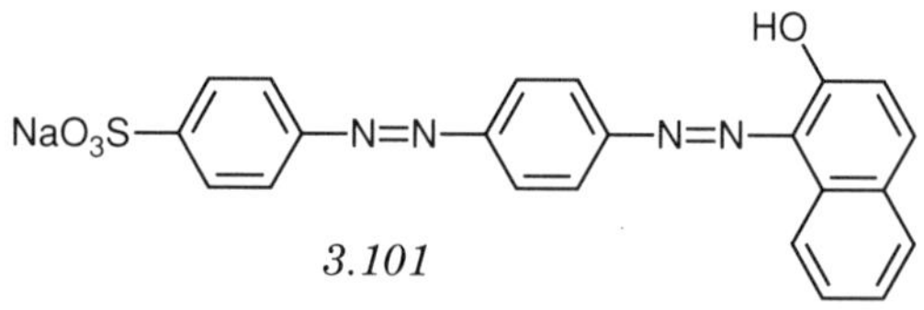

3.101

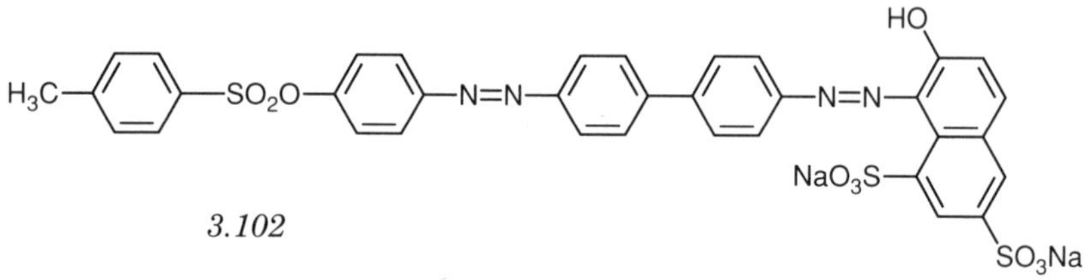

3.102

As already indicated in section 3.3.2, the introduction of a further sulpho group into an acid dye structure tends to lower the affinity by about 1 kcal/mol. The results in Table 3.28 for two series of phenylazonaphthol dyes with varying degrees of sulphonation clearly demonstrate that the increase in hydrophilic character brought about in this way significantly impairs the migration behaviour of the dye. Thus the monosulphonated levelling acid dye sulphanilic acid→1-naphthol has excellent migration properties, but the pentasulphonated analogue aniline-2,5-disulphonic acid→oxy-Koch acid only migrates with great difficulty.

Systematic studies of the relationship between disperse dye structure and levelling properties on the ester fibres have shown that in general levelling tends to decrease as molecular size increases. Thus an inverse linear correlation was found between the molar volume of a series of disperse dyes and their barriness rating on polyester [77]. As molecular size increased, migration became less effective in covering dye affinity

TABLE 3.28

Migration properties of phenylazonaphthol acid dyes on wool [14]

Structure (3.103: unspecified X and Y = H)	Basicity	Migration
$Y_4 = SO_3Na$	1	5
$X_4 = Y_4 = SO_3Na$	2	4
$X_2 = X_5 = Y_4 = SO_3Na$	3	3–4
$X_4 = SO_3Na$	1	5
$X_4 = Y_4 = SO_3Na$	2	4
$X_4 = Y_6 = Y_8 = SO_3Na$	3	3–4
$X_4 = Y_3 = Y_6 = Y_8 = SO_3Na$	4	3
$X_2 = X_5 = Y_3 = Y_6 = Y_8 = SO_3Na$	5	2

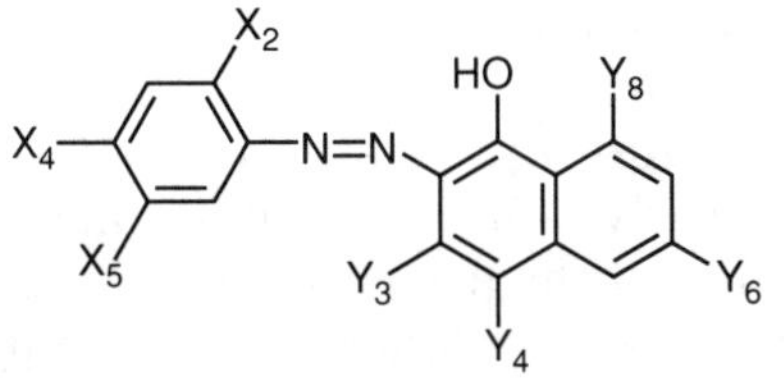

3.103

differences in a textured polyester fabric prone to show barriness. Molecular size is not the only relevant factor, however, because disperse dyes with significant aqueous solubility show better migration properties than less soluble dyes of similar molecular size.

An interesting comparison of the migration behaviour of a related series of alkylaminothiazine basic dyes on acrylic fibres is outlined in Table 3.29.

TABLE 3.29

Migration properties of thiazine basic dyes on acrylic fibre [110]

Structure (3.104)	Migration
$R = X = Y = Z = H$	5
$R = Y = CH_3$, $X = Z = H$	4
$R = X_1 = CH_3$, $X_2 = Y = Z = H$	2
$R = X = CH_3$, $Y = Z = H$	2–3
$R = X = CH_3$, $Y = H$, $Z = NO_2$	2
$R = X = CH_2CH_3$, $Y = Z = H$	1

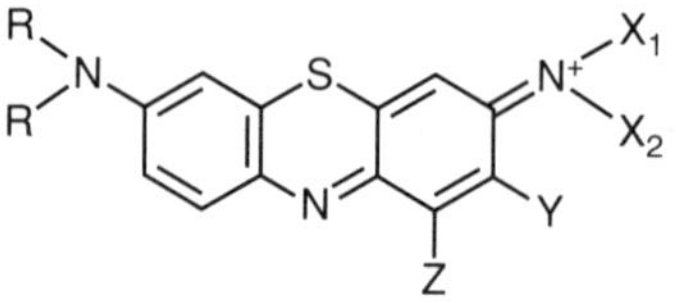

3.104

Methylene blue (C.I. Basic Blue 9), the best known of these, is the tetramethyl derivative with a migration rating of 2–3. Insertion of a nitro group *meta* to one of the dimethylamino groups interferes slightly with the migration property. As might be expected, the completely unalkylated diaminothiazine structure shows the best migration behaviour, whereas the tetraethyl analogue of methylene blue migrates only with great difficulty. Removal of one of the methyl groups from the methylene blue structure surprisingly reduces the migration slightly, but the isomeric trimethylated derivative with one of the methyl groups *ortho* to the unmethylated amino group has a migration rating almost as good as the parent unmethylated diaminothiazine.

3.3.4 Dye structure and light fastness

The fastness to light of a dyed fibre depends on many factors. These include the inherent resistance of the dye chromophore to photochemical attack and the way in which this interaction is modified by the chemical nature of the polymer substrate. Numerous general reviews of the mechanisms of fading reactions and the influence of dye structure and aggregation behaviour on these processes have been published [111–119]. Kinetic evidence suggests that the light fastness of direct dyes on cotton is markedly dependent on the formation of aggregates within the fibre. This can sometimes lead to apparently anomalous differences in fastness rating for the same dye applied and aftertreated in various ways. Hence the light fastness of direct dyeings is determined more by their aggregation state than by structural features of the dye chromophore [120–122].

The fading behaviour of azoic dyeings on cotton is determined by the chemical characteristics of the diazo and coupling components used in the application process. Electron-withdrawing substituents in the phenylazo component, especially trifluoromethyl and nitro groups, improve light fastness while electron-donating groups; such as *o*-methyl or *o*-methoxy, usually lower it. Substituents in the anilide moiety of Naphtol AS combinations, however, influence light fastness in the opposite way. Thus an electron-donating methoxy group enhances fastness but strongly electron-withdrawing groups such as nitro have an adverse effect; for example, in the *o*-nitroaniline→Naphtol AS series (3.105) the light fastness is highest when R = *p*-methoxy and lowest when R = *m*-nitro [123].

The nature of the substrate has little effect on the light fastness of anthraquinone acid dyes on amide fibres. Most substituents hardly affect

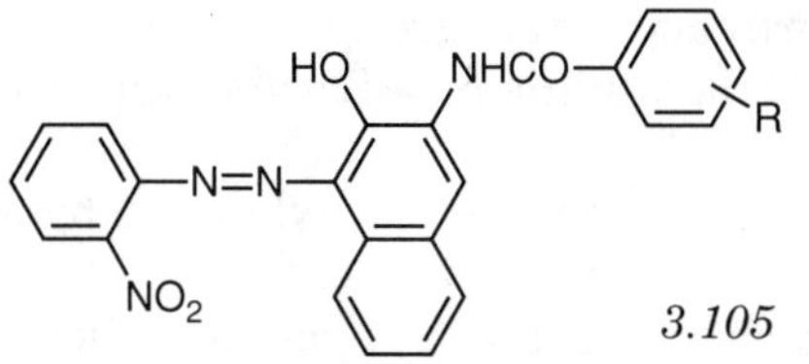

3.105

the low rate of fading of these structures, but photodealkylation has been shown to occur when alkylaminoanthraquinones are exposed to light. The fastness can be enhanced by incorporating arylamino in place of alkylamino groups [124,125]. Measurements of the fading of a series of monoazo dyes of the substituted aniline→R acid type on wool demonstrated that this was a reductive process influenced particularly by the nature of the *ortho* substituent in the diazo component. Anionic groups (sulpho, carboxyl) conferred higher light fastness but strongly electron-withdrawing unionised groups (nitro, chloro) had adverse effects [126]. A similar study of *p*-substituted aniline→γ acid dyes revealed that electron-donating groups such as ethoxy, amino and anilino markedly lowered the light fastness.

Aggregates of dye molecules normally fade less rapidly than do single molecules distributed uniformly through the substrate. When alkyl groups are introduced into acid dye molecules to give higher affinity and better wet fastness on wool, relatively short chains (C_4–C_6) can have a favourable influence on light fastness by promoting aggregation in the fibre [122]. If the chain length is increased (C_{12}–C_{16}) the fastness to light begins to fall markedly owing to the higher surface activity of the dye, tending to destabilise aggregates in favour of monomolecular distribution [127]. For example, the light fastness on wool of the phenylazopyrazolone dye (3.106; X = H) is 6 but that of the long-chain alkyl derivative (3.106; X = dodecyl) is only 4–5.

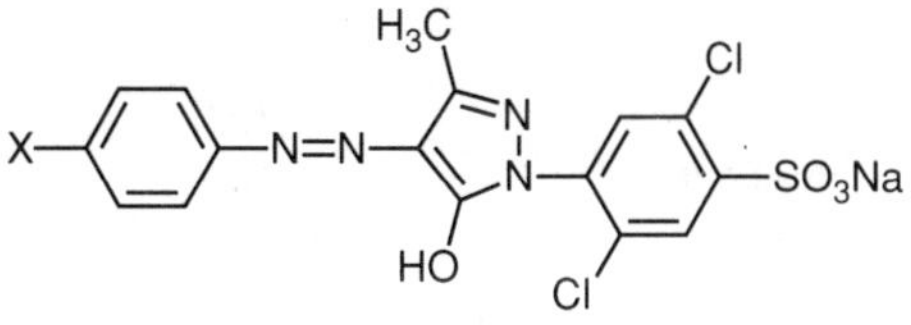

3.106

For the same disperse dyes on various hydrophobic fibres the highest light fastness is found on polyester, normally followed by cellulose acetate and triacetate, with nylon giving the lowest ratings [128]. Anthraquinone disperse dyes normally contain one or two primary or secondary amino groups in the 1- or 4-position. Tetra-amino derivatives have lower fastness than mono- or di-amino analogues. Light fastness on cellulose acetate or polyester may be enhanced by

(a) a hydroxy group *para* to an amino group

(b) acetylation of a primary amino group
(c) replacement of an alkylamino by an arylamino group.

Electron-donating groups (amino, methylamino, hydroxy, methoxy) in the 2-position, on the other hand, are extremely undesirable because, unlike similar substituents in the 1- or 4-position, they are unable to form intramolecular hydrogen bonds with the keto groups of anthraquinone and hence are highly susceptible to photo-oxidation [129].

The fading rates of substituted 4-phenylazo-1-naphthylamines on cellulose acetate were consistent with a photooxidative mechanism [130,131], as for substituted aminoazobenzene structures on polyester, but studies of the fading of phenylazo-2-naphthol dyes on polypropylene were indicative of photoreduction [132,133]. In simple azobenzenes electron-donating *o*-hydroxy groups markedly lower the light fastness, especially when *o*-nitro and *m*-methyl substituents are present. Electron-withdrawing *p*-nitro and especially *m*-nitro groups enhance the fastness, however, this effect being boosted by *m*-methyl and *o*- or *m*-methoxy in the opposite ring (Table 3.30) [134].

TABLE 3.30

Light fastness of substituted azobenzenes on polyester [134]

Azobenzene substituents	Light fastness
3-Nitro-3′-methoxy	7–8
3-Nitro-2′-methoxy-5′-methyl	7–8
4-Nitro-2′-methoxy-5′-methyl	7–8
3-Nitro	7
4-Nitro-3′-methoxy	7
4-Nitro	6–7
3-Nitro-2′-hydroxy-5′-methyl	6–7
2-Nitro	6
None	6
2-Nitro-3′-methoxy	6
4-Nitro-2′-hydroxy-5′-methyl	5
2-Nitro-2′-methoxy-5′-methyl	4–5
2-Nitro-2′-hydroxy-5′-methyl	3

Many of the most important dyes belong to the substituted *p*-phenylazoaniline type (Table 3.31). On polyester, electron-withdrawing groups in the phenylazo residue (*p*-nitro > *o*-cyano > *o*-chloro) improve light fastness, but the 2,4-dinitrophenylazo arrangement reduces it. *N*-Cyanoethyl- and *N*-acetoxyethyl-substituted coupling components confer much higher light fastness than do their *N*-ethyl or *N*-2-hydroxyethyl analogues [135]. On nylon, light fastness in the *p*-nitrophenyl series is enhanced by electron donation (methyl, methoxy) but not by electron withdrawal (cyano, nitro) at the *ortho* position [136].

TABLE 3.31

Light fastness of substituted aminoazobenzenes on nylon and polyester [135,136]

	Light fastness	
Structure	Nylon	Polyester
3.107: X = Y = CN		7
3.107: X = $OCOCH_3$, Y = CN		7
3.108: X = NO_2, Y = CN, Z = H	1–2	6–7
3.107: X = H, Y = CN		6
3.108: X = NO_2, Y = Cl, Z = H	3–4	6
3.108: X = NO_2, Y = Z = H	4	5
3.108: X = NO_2, Y = H, Z = CH_3		5
3.108: X = NO_2, Y = CH_3, Z = H	4–5	4–5
3.107: X = OH, Y = CN		4
3.108: X = $COCH_3$, Y = H, Z = CH_3		4
3.108: X = SO_2CH_3, Y = H, Z = CH_3		4
3.107: X = Y = H		3–4
3.108: X = NO_2, Y = OCH_3, Z = H	4	3–4
3.107: X = OH, Y = H		3
3.108: X = $SO_2N(CH_2CH_3)_2$, Y = H, Z = CH_3		3
3.108: X = Y = NO_2, Z = H	1	2
3.108: X = Y = H, Z = CH_3		2
3.108: X = OCH_3, Y = H, Z = CH_3		2

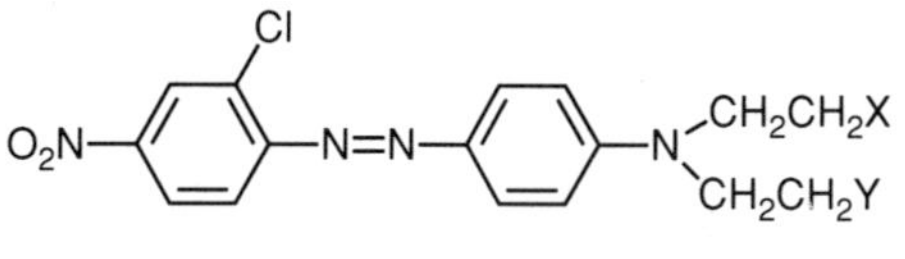

3.107

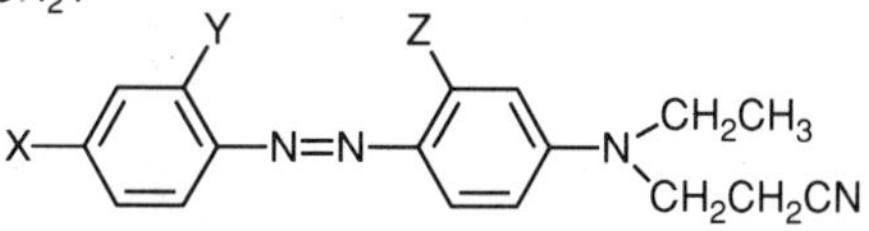

3.108

3.3.5 Dye structure and wet fastness

The fastness of a dye to wet treatments, such as washing or perspiration, is a function of kinetic (diffusion) and thermodynamic (affinity) effects. Certain general trends can be discerned, but conclusions derived from simple structural changes in closely related structures have only limited applicability [137]. Relationships between the wet fastness and molecular structure of direct dyes on cellulosic fibres are somewhat tenuous because of the major obscuring influence of hydrogen bonding and other forces of association between dye molecules leading to aggregation within the substrate (see sections 3.1.2 and 3.3.4). In general the wet fastness of disperse dyes on hydrophobic fibres tends to increase with the molecular size of the dye and with the increasingly hydrophobic character of the substrate (polyester > triacetate > nylon > acetate). Intramolecular hydrogen bonding has a significant influence on the diffusion behaviour of disperse dyes in polyester [138].

Acid dyes are applied to wool at acid pH but are washed under neutral or mildly alkaline conditions. Affinity forces between the dye and the substrate tend to retard desorption. The higher the degree of sulphonation in the series of naphthylazo-2-naphthol dyes illustrated in Table 3.32, the higher is the rate of desorption into sodium borate solution at 40°C. With a series of monosulphonated acid dyes the desorption rate under these conditions was inversely related to the affinity of the dye for wool at pH 3 and 60°C (Table 3.33) [56].

TABLE 3.32

Basicity and desorption rate of monoazo acid dyes on wool [56]

Substituents (3.109)	Basicity	Desorption rate/% min^{-1}
R = X = Y = Z = SO_3Na	4	16.0
R = Y = Z = SO_3Na, X = H	3	12.1
R = X = H, Y = Z = SO_3Na	2	5.2
R = SO_3Na, X = Y = Z = H	1	1.6

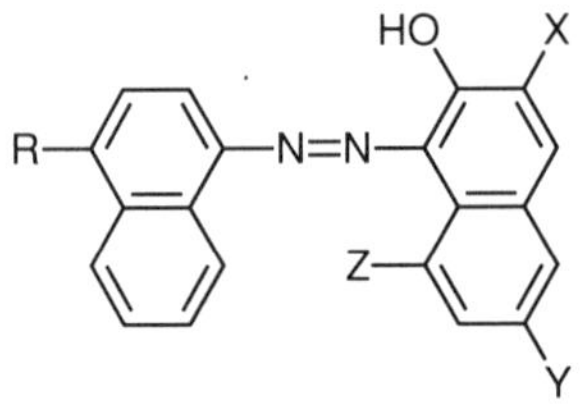

3.109

TABLE 3.33

Affinity and desorption rate of monosulphonated acid dyes on wool [56]

C.I. Acid	Structure	Affinity /kcal mol^{-1}	Desorption rate/% min^{-1} 40°C	60°C
Orange 7	3.110	4.4	6.4	18.5
Yellow 36	3.111	5.2	5.4	16.6
Violet 43	3.112	6.5	3.0	9.1
Blue 25	3.113	7.1	2.0	6.0

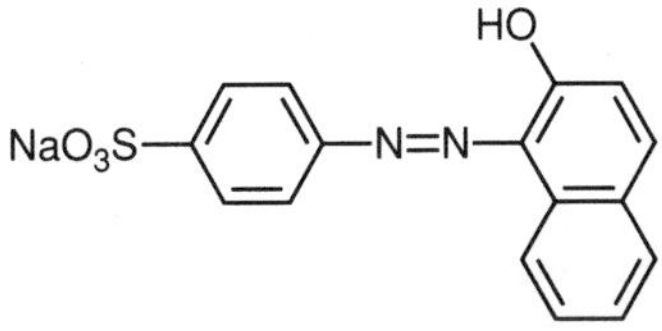

3.110

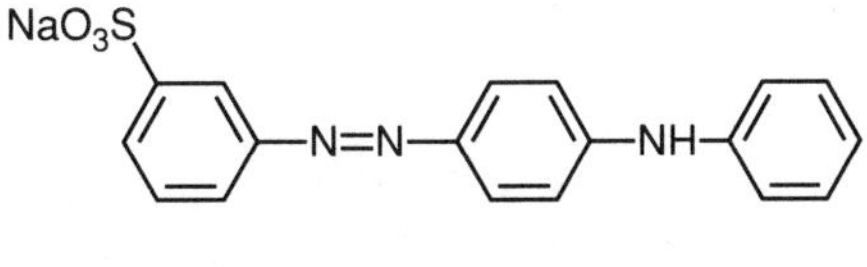

3.111

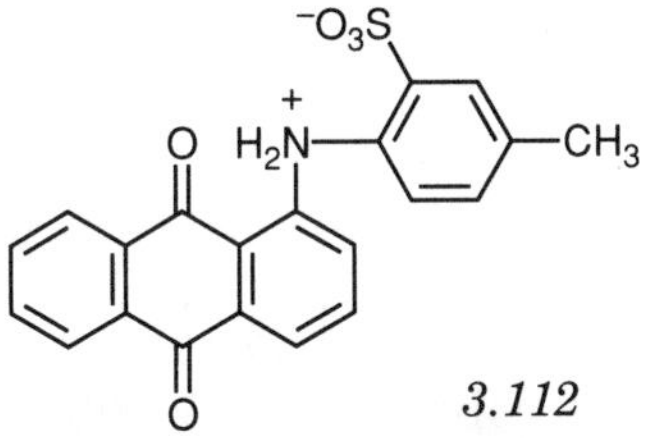

3.112

3.113

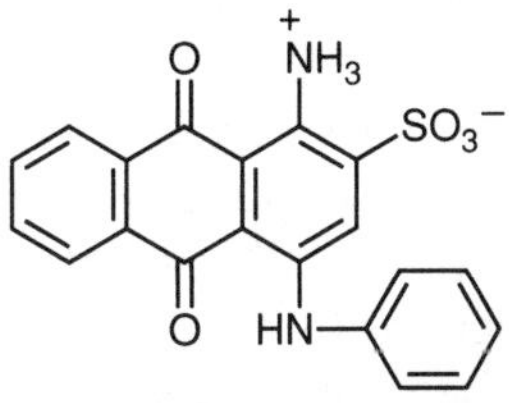

The increase in affinity for wool imparted to acid dye structures by the inclusion of long-chain alkyl groups has a noteworthy effect on wet fastness. Thus the incorporation of such a group into the levelling dye C.I. Acid Red 1 (3.100; X = H) to form the 'super-milling' dye C.I. Acid Red 138 (3.100; X = dodecyl) raises the fastness to ISO 2 washing (effect on pattern) from 2–3 to 4–5. Similar effects for a series of alkylated phenylazo-2-naphthol-6-sulphonate dyes are recorded in Table 3.34.

A similar linear relationship exists between the standard affinity for nylon and the number of carbon atoms in the substituent R of a structural series of anthraquinone acid dyes (3.115; R = C_1–C_4 alkyl) [139]. The fastness to ISO 2 washing of these dyeings increased consistently with the relative molecular mass of the dye.

TABLE 3.34

Structure and wash fastness of monoazo dyes on wool [137]

Substituents (3.114)	Wash fastness (effect on pattern)
X = $CH_3(CH_2)_{11}$, Y = H	4
X = $CH_3(CH_2)_3$, Y = H	3
X = CH_3CH_2, Y = H	2
X = H, Y = CH_3CH_2	2
X = CH_3, Y = H	1–2
X = Y = H	1

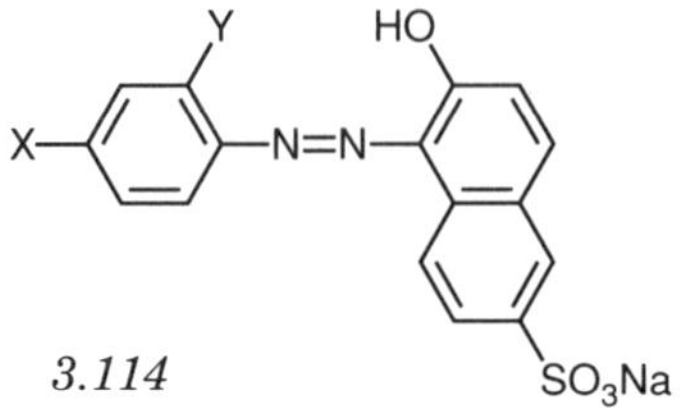

3.114

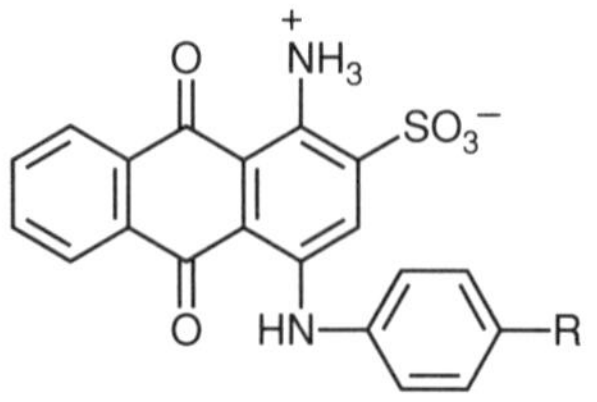

3.115

3.3.6 Dye structure and heat fastness

This fastness requirement is only of practical significance for disperse dyes on hydrophobic fibres. Dyes of low fastness are sublimed from the surface of the heated fibre at a rate dependent on the temperature of treatment. The diffusion coefficient of the dye in the dry polymer controls the rate at which the dye lost from the surface by volatilisation is replenished from the interior of the fibre. Dye decomposition during the heat fastness test can have a major practical significance in determining the overall rating attainable with a given dye structure [140].

Inverse relationships between the vapour pressure of disperse dyes and their molecular size and polarity have been established [141]. It is evident from Table 3.35 that the 12% increase in relative molecular mass of the symmetrical disubstituted structure C.I. Disperse Blue 14 compared with its unsymmetrical 1-methylamino analogue C.I. Disperse Red 9 has profound effects, retarding the diffusion rate approximately sevenfold and lowering the vapour pressure almost 700-fold [142]. Both factors have a beneficial influence on heat fastness. *p*-Aminoacetanilide→*p*-cresol (C.I. Disperse Yellow 3) is similar to C.I. Disperse Blue 14 in molecular size but diffuses twice as rapidly and vaporises much more readily, possibly because it has only two aryl rings and a more linear conformation.

TABLE 3.35

Vapour pressure, diffusion coefficient and structure of disperse dyes for polyester [142]

C.I. Disperse	Structure	R.m.m.	Vapour pressure at 160°C × 10^5/cmHg	Diffusion coeff. × 10^{13}/m^2 s^{-1}
Red 9	3.116: X = H	237	692	6.87
Yellow 3	3.117	269	13.5	2.39
Blue 14	3.116: X = $NHCH_3$	266	1.01	(1.2)

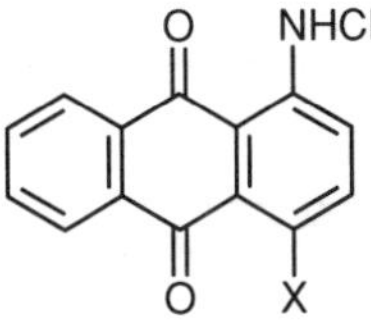

3.116

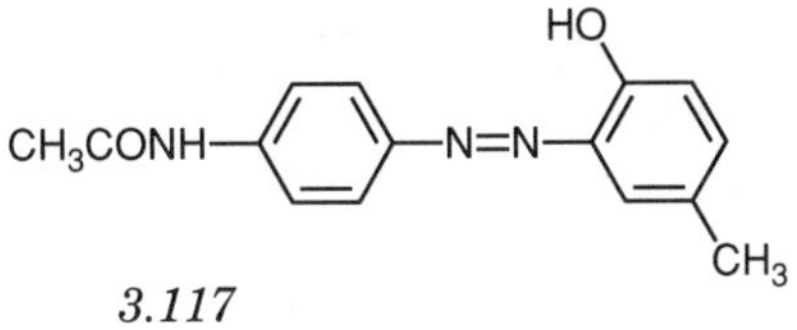

3.117

The requirements of disperse dyes for the transfer printing of polyester fabrics are almost exactly opposite to those of the dyer. Adequate sublimation during the transfer step can normally be achieved only with dyes of low relative molecular mass (< 350) that are substantially free from highly polar substituents [143]. For steric reasons, however, there are some exceptions to this general rule. Thus 2,4,6-trisubstituted phenylazo groups, which often form the basis for useful brown and navy blue members of the aminoazobenzene class, can interfere with the coplanarity of such structures. This may hinder dye–dye and dye–fibre bonding to such an extent that dyes of r.m.m. exceeding 400 can sometimes be used for transfer printing.

Dyes of high r.m.m. with numerous and/or bulky substituent groups, however, almost invariably show higher fastness to sublimation than simpler aminoazobenzene or aminoanthraquinone analogues [137,144]. The influence of the number and orientation of bromo substituents on heat fastness in two systematic series of multibrominated derivatives of 1,5-diaminoanthrarufin (3.118) and 1,8-diaminochrysazin (3.119) has been investigated [145].

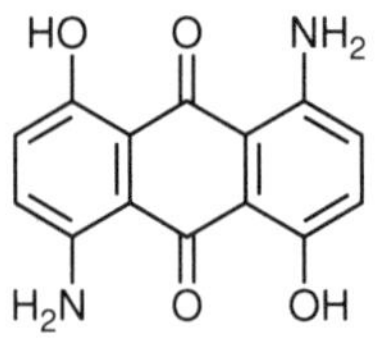

3.118

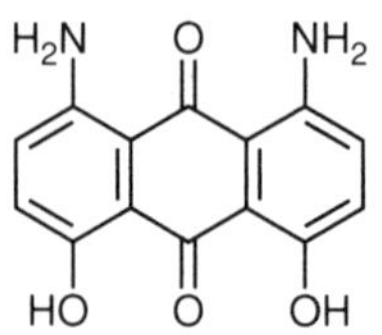

3.119

The relationships between dye structure and heat fastness for various phenylazopyrazolone dyes (Table 3.36) demonstrate the significance of changes in molecular size and polar substituents for this property [136]. In the 3.120 series the tetramethylene sulphone ring enhances heat fastness far more than does a conventional *N*-phenyl group, even though the changes in molecular size are relatively modest. A 3-carbethoxy substituent in the pyrazolone ring raises heat fastness about one point higher than does the 3-methyl analogue, a difference roughly equivalent to the insertion of an *N*-phenyl ring into the 3-methylpyrazolone parent dye. In the

TABLE 3.36

Structure and heat fastness of azopyrazolone disperse dyes on polyester [136]

Structure		Heat fastness
3.120:	$X = CH_3$, Y = 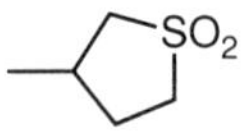	5
3.122		4–5
3.120:	$X = COOCH_2CH_3$, Y = Ph	3–4
3.121:	$X = NH_2$	3–4
3.120:	$X = CH_3$, Y = Ph	2–3
3.121:	X = OH	2–3
3.120:	$X = CH_3$, Y = H	1–2

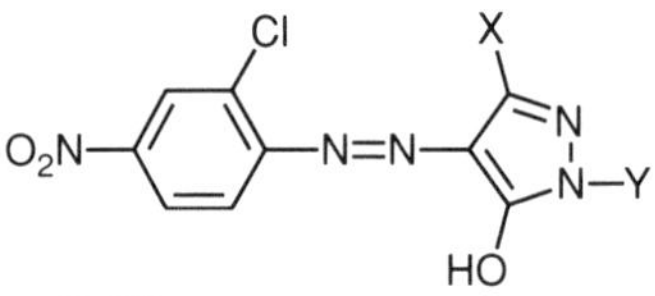

3.120

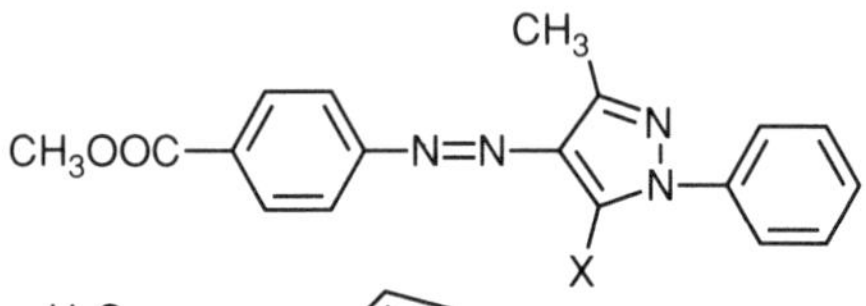

3.121

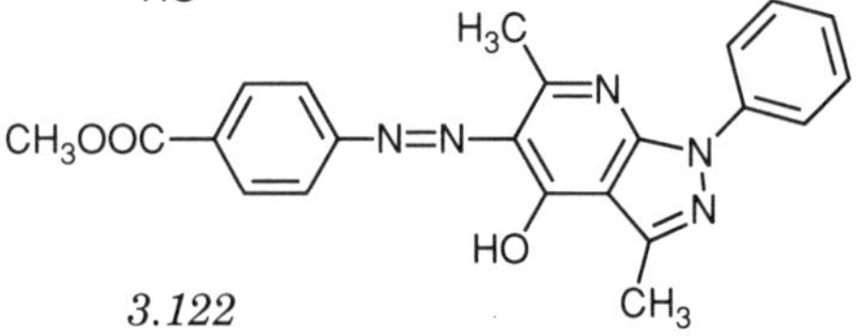

3.122

TABLE 3.37

Structure and heat fastness of aminoanthraquinone disperse dyes on polyester [136]

Structure			Heat fastness
3.124:	R =	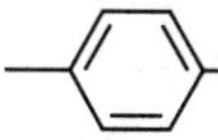 $-SO_2NH(CH_2)_3OCH_2CH_3$	4–5
3.123:	X =	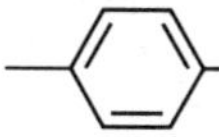 $-NHCOCH_3$	4–5
3.124:	R =	$-(CH_2)_6OH$	3–4
3.124:	R =	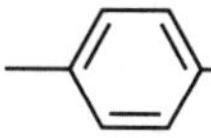 $-Cl$	3–4
3.123:	X =	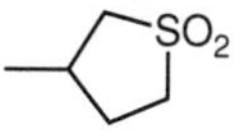	3–4
3.124:	R =	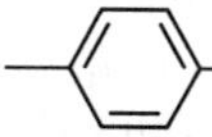 $-SCH_3$	2–3
3.124:	R =	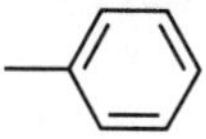	2–3
3.123:	X =	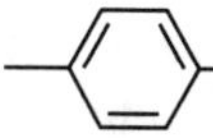 $-OCH_3$	2–3
3.123:	X =	$-CH_2CH_2COOCH_2CH_3$	2–3
3.124:	R =	$-CH_3$	1–2
3.123:	X =	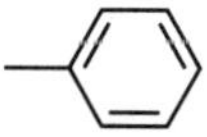	1–2
3.123:	X =	H	1–2

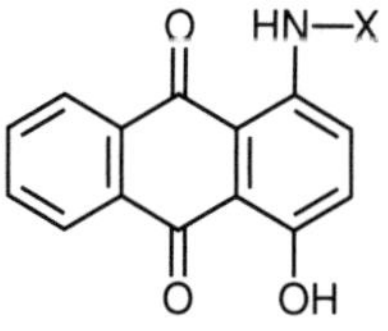

3.123

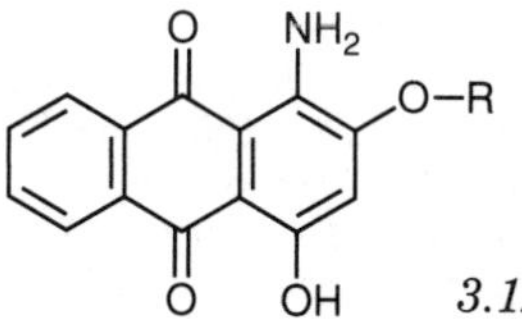

3.124

3.121 structure an improvement of this magnitude can be achieved simply by using the pyrazol-5-imine instead of the pyrazol-5-one as the coupling component. Surprisingly, however, a coupling component as bulky as 1-phenyl-3,6-dimethylpyrazolo[2,3*b*]pyrid-4-one in structure 3.122 still does not give a rating as high as 5, and this complex structure would be expected to exhibit poor level-dyeing properties.

Variations in the dialkylamino terminal substitution pattern of typical alkylaminoazobenzene derivatives have a greater effect on heat fastness than have substituents in the aryl nuclei of the diazo and coupling components [116]. The fastness can be significantly improved by incorporating polar groups capable of hydrogen bonding or dipole–dipole interaction with polymer segments in the fibre (acetylamino, acetoxy, cyano). In an analogous way, proton-donor groups (amino, hydroxy) in the 2-position of 1,4-disubstituted anthraquinone dyes enhance the fastness to heat because they are readily able to form intramolecular hydrogen bonds with the substrate or other dye molecules. Similar groups in the 1,4-positions are normally hydrogen-bonded intramolecularly to the keto groups of anthraquinone. The major drawback of electron-donating groups in the β-positions of anthraquinone dyes, however, is their usually unfavourable effect on light fastness (see section 3.3.4).

Trends indicating the capability of polar and non-polar aryl substituents to enhance the rather low heat fastness of 1-amino-4-hydroxyanthraquinone (3.123; X = H) and its 2-hydroxy derivative (3.124; R = H) are outlined in Table 3.37. In the 3.123 series of *N*-substituted derivatives a phenyl group offers little or no improvement. An ethyl propionate residue or a *p*-methoxyphenyl group produces a modest increase in fastness but more polar moieties such as tetramethylene sulphone or acetanilide residues are necessary to achieve high ratings. Methylation of the 2-hydroxy group in structure 3.124 improves the light fastness but not the fastness to sublimation. A phenyl or *p*-methylthiophenyl group raises heat fastness somewhat, but more marked improvements follow with *p*-chlorophenyl, 6-hydroxyhexyl and the complex alkoxyalkylsulphonamide grouping shown in Table 3.37.

REFERENCES

1. H S Frank and N Y Wen, *Disc. Faraday Soc.*, No. 24 (1957) 133.
2. F Jones, *J.S.D.C.*, **100** (1984) 66.
3. H S Frank and M W Evans, *J. Chem. Phys.*, **13** (1945) 507.
4. I D Rattee, *J.S.D.C.*, **69** (1953) 288; **90** (1974) 367.
5. W Ender and A Müller, *Melliand Textilber.*, **18** (1937) 906.
6. E Elöd and H G Fröhlich, *Melliand Textilber.*, **32** (1951) 622.
7. E Coates, *J.S.D.C.*, **85** (1969) 355.
8. W Beal, *J.S.D.C.*, **76** (1960) 78.
9. C H Giles, *Rev. Prog. Coloration*, **11** (1981) 89.
10. C H Giles and A H Soutar, *J.S.D.C.*, **87** (1971) 301.
11. B Milicevic, 4th Internat. Congress of Detergency, Section B (4) (1964) 12.
12. R J Ott, U Widmer and H Zollinger, *J.S.D.C.*, **91** (1975) 330.
13. S Lenher and J E Smith, *J. Amer. Chem. Soc.*, **57** (1935) 504.
14. T H Morton, *Trans. Faraday Soc.*, **31** (1935) 262.
15. T Vickerstaff, *The physical chemistry of dyeing*, 2nd Edn (London: Oliver & Boyd, 1954).

16. C Robinson, *Trans. Faraday Soc.*, **31** (1935)245.
17. C Robinson and H E Garrett, *Trans. Faraday Soc.*, **35** (1939) 771.
18. E I Valkó, *J.S.D.C.*, **55** (1939) 173.
19. J B Speakman and H Clegg, *J.S.D.C.*, **50** (1934) 348.
20. F L Goodall, *J.S.D.C.*, **54** (1938) 45.
21. P Alexander and K A Stacey, *Proc. Royal Soc.*, **A212** (1952) 274.
22. J Wegmann, *Amer. Dyestuff Rep.*, **51** (1962) 276.
23. U Baumgarte, *Textilveredlung*, **15** (1980) 413.
24. C C Wilcock, *J.S.D.C.*, **63** (1947) 136.
25. S M Neale and W A Stringfellow, *J.S.D.C.*, **59** (1943) 241.
26. A Schmitz, *J.S.D.C.*, **71** (1955) 910.
27. F H Holmes and H A Standing, *Trans. Faraday Soc.*, **41** (1945) 542.
28. P Gregory and C V Stead, *J.S.D.C.*, **94** (1978) 402.
29. A Müller, *Melliand Textilber.*, **13** (1932) 439, 488.
30. F M Rowe and W G Dangerfield, *J.S.D.C.*, **52** (1936) 48.
31. W J Marshall and R H Peters, *J.S.D.C.*, **68** (1952) 289.
32. L I Belenki and M Kazanskaya, *Tekstil Prom.*, **7** (12) (1947) 18.
33. W Biedermann, *J.S.D.C.*, **87** (1971) 105; **88** (1972) 329.
34. V A Shenai and M C Sadhu, *J. Appl. Polymer Sci.*, **20** (1976) 3141.
35. A S Ghanekar and F Jones, *J.S.D.C.*, **92** (1976) 445.
36. C L Bird, *J.S.D.C.*, **70** (1954) 68.
37. H Schott, *J. Phys. Chem.*, **70** (1966) 2966.
38. J R Aspland and C L Bird, *J.S.D.C.*, **77** (1961) 9.
39. C L Bird and P Harris, *J.S.D.C.*, **73** (1957) 199.
40. E Merian, J Carbonell, U Lerch and V Sanahuja, *J.S.D.C.*, **79** (1963) 505.
41. D Patterson and R P Sheldon, *J.S.D.C.*, **76** (1960) 178.
42. W McDowell and R Weingarten, *J.S.D.C.*, **85** (1969) 589.
43. H F Willis, J O Warwicker, H A Standing and A R Urquhart, *Trans. Faraday Soc.*, **41** (1945) 506.
44. W J Marshall and R H Peters, *J.S.D.C.*, **63** (1947) 446.
45. S Fisichella, G Scarlata and M Torre, *J.S.D.C.*, **94** (1978) 521.
46. J A Fowler, A G H Michie and T Vickerstaff, *Melliand Textilber.*, **32** (1951) 296.
47. K H Meyer, *Melliand Textilber.*, **9** (1928) 573.
48. H H Hodgson, *J.S.D.C.*, **49** (1933) 213.
49. C Robinson and H A T Mills, *Proc. Royal Soc.*, **A131** (1931) 576.
50. K Nishida, *J.S.D.C.*, **82** (1966) 313.
51. H Bach, E Pfeil, W Phillipar and M Reich, *Angew. Chem.*, **75** (1963) 407.
52. C H Giles and A S A Hassan, *J.S.D.C.*, **74** (1958) 846.
53. H R Chipalkatti, C H Giles and D G M Vallance, *J.C.S.*, (1954) 4375.
54. H R Chipalkatti, V B Chipalkatti and C H Giles, *J.S.D.C.*, **71** (1955) 652.
55. D R Lemin and I D Rattee, *J.S.D.C.*, **65** (1949) 217.
56. C H Nicholls, *J.S.D.C.*, **72** (1956) 479.
57. F Townend, *J.S.D.C.*, **61** (1945) 144.
58. J Meybeck and P Galafassi, *Appl. Polymer Symposia*, **18** (1971) 463.
59. C Soiron, *Textilveredlung*, **14** (1979) 169.
60. W Mosimann and H Flensburg, *Taschenbuch für die Textilind.*, (1981) 294.
61. J Hertig and H Scheidegger, *Textilveredlung*, **18** (1983) 325.
62. H R Hadfield and D R Lemin, *J.S.D.C.*, **77** (1961) 97.
63. J A Bone, Proc. 10th Int. Wool Text. Res. Conf. (Aachen) **93** (1984) 170.
64. *The dyeing of synthetic-polymer and acetate fibres*, Ed. D M Nunn (Bradford: Dyers' Company Publications Trust, 1979).
65. A Anton, *Text. Chem. Colorist*, **13** (1981) 31.
66. R J B Marsden and A R Urquhart, *J. Textile Inst.*, **33** (1942) T105.
67. F M Arshid, C H Giles and S K Jain, *J.C.S.*, (1956) 1272.
68. A Cameron, C H Giles and T H MacEwan, *J.C.S.*, (1958) 1224.

69. C H Giles, V G Agnihotri and A S Trivedi, *J.S.D.C.*, **86** (1970) 451.
70. C L Bird and A M Patel, *J.S.D.C.*, **84** (1968) 560.
71. E H Daruwalla, S S Rao and B D Tilak, *J.S.D.C.*, **76** (1960) 418, 680.
72. C L Bird, *J.S.D.C.*, **74** (1958) 688.
73. C L Bird and P Rhyner, *J.S.D.C.*, **77** (1961) 12.
74. K Hoffmann, W McDowell and R Weingarten, *J.S.D.C.*, **84** (1968) 306.
75. J F Dawson, *Rev. Prog. Coloration*, **3** (1972) 18.
76. A Fisher, Chi-Sun Lee and A T Peters, *J. Chem. Tech. Biotechnol.*, **32** (1982) 532.
77. H Gerber, *J.S.D.C.*, **94** (1978) 298.
78. A Urbanik, AATCC Nat. Tech. Conf., (1981) 258.
79. J F Laucius, R A Clarke and J A Brooks, *Amer. Dyestuff Rep.*, **44** (1955) 362.
80. O Glenz and W Beckmann, *Melliand Textilber.*, **38** (1957) 296.
81. W Beckmann, *J.S.D.C.*, **77** (1961) 616.
82. S Rosenbaum, *Text. Research J.*, **33** (1963) 899.
83. K Meldrum and J S Ward, *J.S.D.C.*, **74** (1958) 140.
84. D Balmforth, C A Bowers and T H Guion, *J.S.D.C.*, **80** (1964) 577.
85. S Rosenbaum, *J. Polymer Sci.*, **A3** (1965) 1949.
86. U Mayer, W Ender and A Würz, *Melliand Textilber.*, **47** (1966) 653, 772.
87. U Mayer, *Amer. Dyestuff Rep.*, **56** (1967) 869.
88. J A Koller and M Motter, *Textilveredlung*, **11** (1976) 3.
89. W Beal, *Internat. Dyer*, **157** (1977) 261.
90. S Rosenbaum, *Text. Research J.*, **34** (1964) 159.
91. J Boulton, *J.S.D.C.*, **54** (1938) 268; **60** (1944) 5.
92. C H Giles, A P Montgomery and A H Tolia, *Text. Research J.*, **32** (1962) 99.
93. T L Dawson and J C Todd, *J.S.D.C.*, **95** (1979) 417.
94. C Renard, *Teintex*, **36** (1971) 845.
95. E Waters, *J.S.D.C.*, **66** (1950) 609.
96. A N Derbyshire, *J.S.D.C.*, **90** (1974) 273.
97. T Vickerstaff and E Waters, *J.S.D.C.*, **58** (1942) 116.
98. T Vickerstaff, *J.S.D.C.*, **59** (1943) 92.
99. E Siepmann, *Melliand Textilber.*, **49** (1968) 577.
100. J B Speakman and H Clegg, *J.S.D.C.*, **50** (1934) 348.
101. G A Gilbert, *Proc. Royal Soc.*, **A183** (1944) 167.
102. A B Meggy, *J.S.D.C.*, **66** (1950) 510.
103. B G Ferrini, Y Kimure and H Zollinger, Proc. Internat. Wool Text. Res. Conf., Paris, **3** (1965) 291.
104. S R S Iyer, A S Ghanekar and G S Singh, Symp. on Dye–Polymer Interactions, Bombay, (1971) 13.
105. E Atherton, D A Downey and R H Peters, *Text. Research J.*, **25** (1955) 977.
106. V Kartaschoff and G Farine, *Helv. Chim. Acta*, **11** (1928) 813.
107. T Barr, F M Rowe and J B Speakman, *J.S.D.C.*, **58** (1942) 52.
108. G L Royer, H E Millson and E I Stearns, *Amer. Dyestuff Rep.*, **32** (1943) 288.
109. W H Watkins, G L Royer and H E Millson, *Amer. Dyestuff Rep.*, **33** (1944) 52.
110. J T Voltz, *Text. Chem. Colorist*, **9** (1977) 113.
111. C H Giles, C D Shah, W E Watts and R S Sinclair, *J.S.D.C.*, **88** (1972) 433.
112. N A Evans and I W Stapleton, *J.S.D.C.*, **89** (1973) 208.
113. P Bentley, J F McKellar and G O Phillips, *Rev. Prog. Coloration*, **5** (1974) 33.
114. G E Krichevskii, B Vachobov, Y A Ershov and E B Dovbii, *Text. Research J.*, **45** (1975) 608.
115. C H Giles, D J Walsh and R S Sinclair, *J.S.D.C.*, **93** (1977) 348.
116. G Hallas, *J.S.D.C.*, **95** (1979) 285.
117. C H Giles and E A Strevens, *Text. Research J.*, **49** (1979) 724.
118. R Grecu, M Pieroni and R Carpignano, *Dyes and Pigments*, **2** (1981) 305.
119. N S Allen, *Rev. Prog. Coloration*, **17** (1987) 61.
120. G Baxter, C H Giles, M N McKee and N Macauley, *J.S.D.C.*, **71** (1955) 218.

121. C H Giles, *J.S.D.C.*, **73** (1957) 127.
122. G Baxter, C H Giles and W J Lewington, *J.S.D.C.*, **73** (1957) 386.
123. J Monheim and R Löwenfeld, *Melliand Textilber.*, **53** (1972) 1134.
124. C H Giles and R S Sinclair, *J.S.D.C.*, **88** (1972) 109.
125. D Wegerle, *J.S.D.C.*, **89** (1973) 54.
126. R H Kienle, E I Stearns and P A van der Meulan, *J. Phys. Chem.*, **50** (1946) 363.
127. J Meybeck, P Ruckstuhl and J M Thumann, *Teintex*, **25** (1960) 241.
128. J M Straley and D G Carmichael, *Canadian Textile J.*, **76** (1959) 49.
129. N S Allen, P Bentley and J F McKellar, *J.S.D.C.*, **91** (1975) 366.
130. E Atherton and I Seltzer, *J.S.D.C.*, **65** (1949) 629.
131. E Atherton and R H Peters, *J.S.D.C.*, **68** (1952) 64.
132. M Ahmed and V N Mallet, *J.S.D.C.*, **84** (1968) 313.
133. V N Mallet and B T Newbold, *J.S.D.C.*, **90** (1974) 4.
134. S V Sunthankar and V Thanumoorthy, *Indian J. Chem.*, **8** (1970) 598.
135. M F Sartori, *J.S.D.C.*, **83** (1967) 144.
136. C Muller, *Amer. Dyestuff Rep.*, **59** (3) (1970) 37.
137. C H Giles, D G Duff and R S Sinclair, *Rev. Prog. Coloration*, **12** (1982) 58.
138. O Glenz, W Beckmann and W Wunder, *J.S.D.C.*, **75** (1959) 141.
139. K Grieder, *J.S.D.C.*, **92** (1976) 8.
140. W McDowell, *J.S.D.C.*, **99** (1973) 177.
141. H E Schroeder and S N Boyd, *Text. Research J.*, **27** (1957) 275.
142. F Jones and T S M Leung, *J.S.D.C.*, **90** (1974) 286.
143. J Griffiths and F Jones, *J.S.D.C.*, **93** (1977) 176.
144. J F Dawson, *Rev. Prog. Coloration*, **9** (1978) 25.
145. A T Peters and B A Tenny, *J.S.D.C.*, **93** (1977) 378.

CHAPTER 4

Chemistry of azo colorants

C Vivian Stead

4.1 INTRODUCTION

Shortly after Perkin had produced the first successful synthetic dyestuff, a discovery was made which led to what is now the dominant chemical class of dyestuffs, the azo dyes. This stemmed from the work of Peter Griess, who in 1858 passed 'nitrous fumes' (which correspond to the formula N_2O_3) into a cold alcoholic solution of picramic acid and isolated a strongly cationic product, the properties of which showed it to be a member of a new class of compounds [1]. Griess extended his investigations to other primary aromatic amines and showed his reaction to be generally applicable. He named the products 'diazo' compounds and the reaction came to be known as the diazotisation reaction. The reaction can be represented simply by Scheme 4.1, in which HX stands for a strong monobasic acid and Ar is any aromatic or heteroaromatic nucleus.

$$\mathrm{Ar{-}NH_2 \; + \; HX \; + \; HNO_2 \longrightarrow Ar{-}\overset{+}{N}{\equiv}N \;\; X^- \; + \; 2H_2O}$$

Scheme 4.1

Griess subsequently recognised that the orange compound formed by the reaction of his new 'diazobenzene' with phenol in neutral solution contained two benzene nuclei joined together via an azo (–N=N–) linkage [2] and again the reaction was found to be general and not confined to phenols. This second reaction became known as the diazo coupling reaction and the products as azo compounds. The wide applicability of the reaction opened the way to a vast multitude of compounds produced in generally high yields in simple plant, and led to the azo dyes becoming the largest and most versatile class of synthetic dyestuffs.

The high reactivity of diazonium salts enables them to enter into a great many reactions other than the azo coupling reaction, but these fall outside the scope of the present chapter. Various other methods can be used for the preparation of azo compounds, although these are of minor importance compared with the diazo coupling reaction and they will only be touched on briefly where appropriate.

4.2 MECHANISM OF DIAZOTISATION AND COUPLING

Studies of the mechanism of the diazotisation reaction date back to the end of the nineteenth century and have continued ever since, progressively incorporating the growing knowledge of theoretical chemistry [3,4]. The reaction, which involves nitrosation, has been studied under a wide range of acidities using a variety of amines and nitrosating agents, and numerous effects arising from variations in conditions have been described. In essence, the accepted mechanism is represented by Scheme 4.2, in which X–NO represents one of the many varieties of nitrosating species such as nitrous acid (HO–NO), nitrosyl chloride (Cl–NO), dinitrogen trioxide (O_2N–NO), the nitrosonium ion (NO^+) or the nitrous acidium ion ($H_2NO_2^+$) that may be present in solutions containing nitrous acid. According to the conditions employed, either the formation of the nitrosating agent or its initial attack on the amine is the slowest reaction step and therefore determines the overall rate, the subsequent steps in the reaction sequence being rapid. The concentrations of the various possible nitrosating species can be influenced by other ions, such as chloride ions, or compounds, such as thiourea, present in the reaction mixture and this results in catalytic effects being observed.

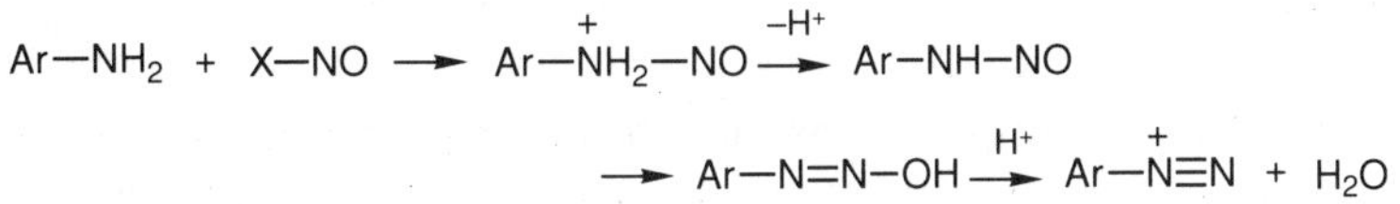

Scheme 4.2

The diazonium ion is a powerful electrophile and, as such, it can readily attack compounds having a nucleophilic centre with the expulsion of a proton, or other entity, from the site of attack [5]. This is the mode of behaviour that leads to the formation of azo compounds by the azo coupling reaction, the compound with the nucleophilic site being termed the coupling component. Suitable components include arylamines, phenols, naphthols and keto–enol compounds. The mechanism with an arylamine coupling component may be represented by Scheme 4.3. In the coupling component illustrated both the *o*- and *p*-positions are activated by the dialkylamino group. However, an approach of the diazonium ion to the *o*-position would cause the electron-donating group to be twisted out of plane and thus to lose its activating influence. For this reason, coupling occurs selectively at the *p*-position.

With a phenol, naphthol or keto–enol coupling component the mechanism is given by Scheme 4.4, in which blocking of the *p*-position forces

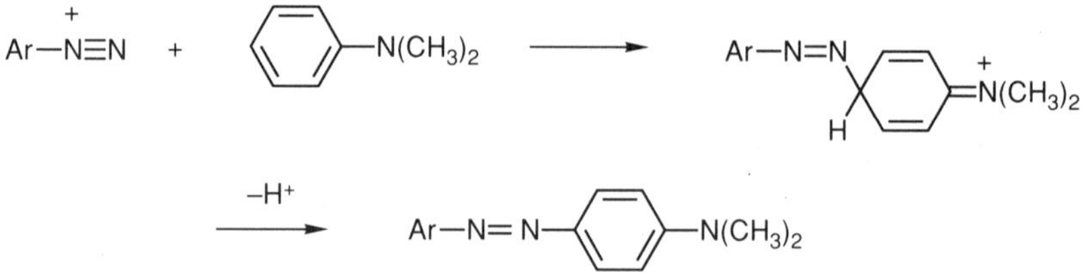

Scheme 4.3

coupling at the *o*-position. In certain cases involving the use of feebly reactive diazonium salts, loss of the proton from the transition state (4.1) in Scheme 4.4 may be slow but may often be facilitated by the addition of a tertiary base, such as pyridine, to the coupling mixture [6,7].

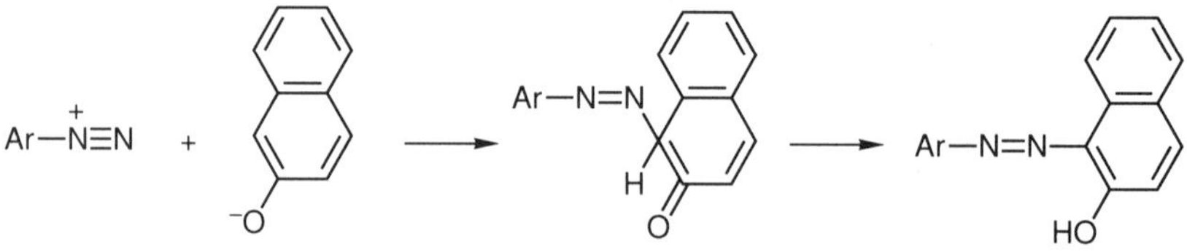

4.1

Scheme 4.4

In these two coupling schemes the arylamine participates in the reaction as the free base, whereas the hydroxylic coupling components react in their ionised forms. Thus whilst raising the pH increases the rate of coupling in both cases, much higher pH values are required to bring about satisfactory coupling with hydroxy compounds than with arylamines. This is taken advantage of in the case of certain aminonaphthol coupling components, particularly H acid (4.2), where one azo group can be selectively directed into the molecule by the amino group at low pH, followed by a second azo group directed by the ionised hydroxy group at higher pH.

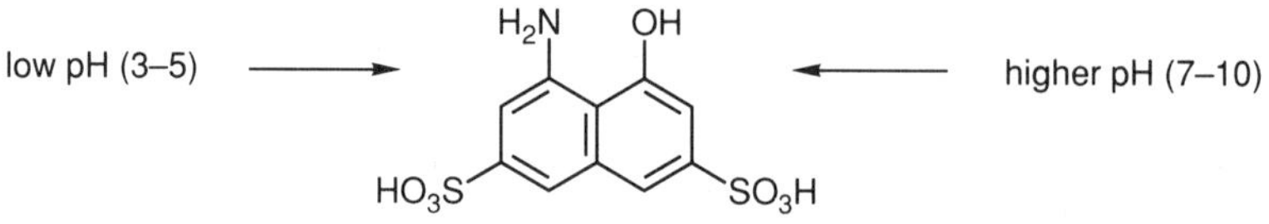

4.2

4.3 DIAZO COMPONENTS AND DIAZOTISATION METHODS

A vast array of arylamines may be converted into the corresponding diazonium salts, and from these 'diazo components' and the host of available coupling components an almost limitless range of azo dyes can be produced. With such a wide range of amines available for diazotisation, a range of diazotisation techniques has been evolved which is designed to cope with the variation in physical properties across the range of amines.

4.3.1 Direct method

The rudiments of the diazotisation technique consist of treating the amine with at least 2.25 equivalents of a mineral acid (usually hydrochloric acid) and one equivalent of sodium nitrite, in accordance with Scheme 4.5. The slight excess of acid specified is used to ensure that the reaction medium remains acid at all times during the diazotisation. In the direct method sodium nitrite is added to a solution of the amine in dilute mineral acid.

$$\mathrm{Ar{-}NH_2} + \mathrm{2HX} + \mathrm{NaNO_2} \longrightarrow \mathrm{Ar{-}\overset{+}{N}{\equiv}N}\ \mathrm{X^-} + \mathrm{NaX} + \mathrm{2H_2O}$$

Scheme 4.5

This method is suitable for aniline and its simple derivatives with groups such as alkyl, alkoxy or chloro as substituents in the aromatic ring, simple naphthylamines and benzidine derivatives. These compounds can be readily diazotised or tetrazotised by dissolving the amine in dilute hydrochloric acid, cooling the solution to 0–5°C with ice and adding the requisite amount of sodium nitrite solution as rapidly as the nitrous acid formed from it is used in the reaction. Spotting on to starch–iodide paper is carried out to test for excess of nitrous acid. On occasions when it is difficult to determine the end point with starch–iodide paper, a dilute acid solution of 4,4′-diaminodiphenylmethane-2,2′-sulphone (4.3), which gives a bright blue colour with nitrous acid, can be used [8]. When a slight excess of nitrous acid has persisted for about 15 minutes the reaction is complete.

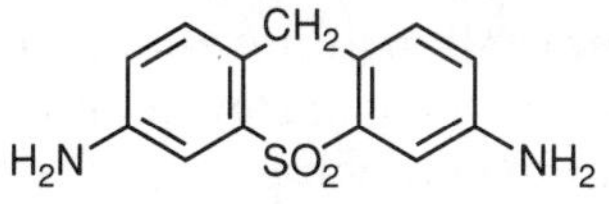

4.3

The excess nitrous acid is then conveniently destroyed by the addition of a little sulphamic acid, which reacts quickly with nitrous acid in accordance with Scheme 4.6, producing innocuous products. The diazonium salt solution is then ready for immediate use in dye preparation.

$$\mathrm{HNO_2} + \mathrm{NH_2SO_3H} \longrightarrow \mathrm{H_2SO_4} + \mathrm{H_2O} + \mathrm{N_2}$$

Scheme 4.6

For the diazotisation reaction to succeed, it is necessary that the amine should be completely converted to the hydrochloride prior to addition of sodium nitrite, since any unprotonated amine can react with the diazonium salt formed yielding a diazoamino compound (Scheme 4.7). This complication needs to be avoided in the case of amines such as the dichloroanilines, which do not dissolve easily in dilute hydrochloric acid [9]. With these compounds it is convenient to dissolve them in a hot acid solution and then cool to 0–5°C. This ensures the absence of free amine, and even if the hydrochloride precipitates on cooling it readily redissolves as the diazotisation proceeds. Similar precautions are required with the nitroanilines, and here an increased amount of mineral acid is advantageously used [10].

$$\mathrm{Ar{-}NH_2} + \mathrm{Ar{-}\overset{+}{N}{\equiv}N} \longrightarrow \mathrm{Ar{-}NH{-}N{=}N{-}Ar} + \mathrm{H^+}$$

Scheme 4.7

Low temperatures are generally necessary for successful diazotisation, except in a few cases where the diazonium salt is exceptionally stable [11]. Temperatures of 50–60°C are used for the diazotisation of 3-aminodibenzofuran (4.4), for example.

4.4

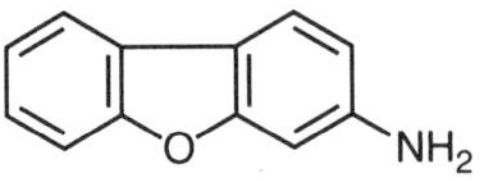

4.3.2 Inverse method

In this method the order of addition of the reactants is changed. The most general application is to the diazotisation of aniline- and naphthylaminesulphonic acids, many of which are sparingly soluble in acidic media and give sparingly soluble diazonium salts. Typically the arylaminesulphonic acid is dissolved as its sodium salt in dilute alkali and sodium nitrite added to the solution. The solution is then run slowly into a mixture of hydrochloric acid and ice, the diazonium salt being formed almost instantaneously.

In a variation on this procedure [12] a small amount of acid is added to a suspension of a sulphonated amine in sodium nitrite solution; the sulphonic acid group on the amine generates nitrous acid, and diazotisation proceeds to completion.

4.3.3 Use of nitrosylsulphuric acid

Very feebly basic amines cannot usually be diazotised in dilute acid media and in these instances the reaction must be carried out in a concentrated acid, normally sulphuric acid. The usual technique is first to dissolve dry sodium nitrite in the concentrated acid, when reaction occurs in two stages

(Scheme 4.8), resulting in the formation of nitrosylsulphuric acid (4.5). Reaction (2) of Scheme 4.8 is slow at room temperature and it is desirable to heat the mixture to 70°C in order to attain equilibrium within a reasonable time. After cooling, the amine is added gradually and after a short time the reaction mixture is poured on to ice, giving an aqueous solution of the diazonium salt [13].

$$(1)\quad H_2SO_4 + NaNO_2 \longrightarrow NaHSO_4 + HNO_2$$

$$(2)\quad H_2SO_4 + HNO_2 \longrightarrow \underset{\textit{4.5}}{HO{-}SO_2{-}O{-}NO} + H_2O$$

Scheme 4.8

This technique is used to diazotise anilines carrying two or more electron-attracting substituents, such as 2,4-dinitro- and 2-cyano-4-nitroaniline, as well as aminoanthraquinones and heteroaromatic amines. Some diazonium salts prepared by this technique, such as those from 6-halogeno-2,4-dinitroanilines, are unstable in water and must be added directly to the coupling component solution.

4.3.4 Other methods

Although not used in dye preparation, three other methods of diazotisation deserve passing mention, one being chiefly of historical interest and the other two sometimes of use in preparative work. The oldest is the method originally used by Griess (section 4.1), who generated the first recognised diazonium salt by passing 'nitrous gases' into a cold ethanolic solution of the amine. This reagent is produced by reducing nitric acid with arsenic(III) oxide or starch; it behaves as if it consisted of dinitrogen trioxide. This method can be useful for the preparation of solid diazonium salts, but a better procedure is that due to Knoevenagel [14]. In this second method an alkyl nitrite is added to a solution of the amine and sulphuric acid in ethanol or other non-aqueous solvent. This allows greater control of the reaction than is possible in the Griess method.

Finally, in a method originated by Witt [15], which is occasionally useful for the diazotisation of stable amines of high relative molecular mass that pose solubility problems, the amine is dissolved in nitric acid and then sodium metabisulphite is added. Nitrous acid is produced in the medium according to Scheme 4.9, enabling diazotisation to proceed.

$$Na_2S_2O_5 + 2HNO_3 \longrightarrow Na_2S_2O_7 + 2HNO_2$$

Scheme 4.9

Diazonium salts vary greatly in reactivity and this affects the ease with which they couple. The presence in the aromatic nucleus of electron-attracting groups, such as nitro and cyano groups, causes difficulties in the diazotisation step but makes the resulting diazonium salt couple extremely readily. Electron-repelling substituents such as methoxy and acetylamino groups, which impart ease of diazotisation, diminish the coupling power. An extreme case is posed by the diazonium salts derived from *o*- and *p*-aminophenols and naphthols, which are very feeble couplers, requiring highly alkaline conditions and often the use of a coupling assistant such as pyridine to bring about satisfactory coupling. This very low level of reactivity is due to the hydroxy substituent interacting structurally with the diazonium group, leading to a mesomeric system in which the limiting forms are represented by Scheme 4.10.

Scheme 4.10

4.4 PREPARATION AND USE OF COUPLING COMPONENTS

As well as being presented with an array of primary amines that can be used for the preparation of azo dyes, the dye chemist has available a multiplicity of coupling components. The coupling components useful in the preparation of azo dyes cover a range of chemical types which can all be seen to contain a keto–enol system or a hydroxy or amino group. Since the use of these compounds is largely restricted to the azo dye field, their preparation is in many cases not covered in the common organic chemistry textbooks. In the discussion of these classes, therefore, a brief mention of their preparation will be given where appropriate.

4.4.1 Acetoacetarylamides

Coupling components of this class have the general structure 4.6 (Scheme 4.11), in which Ar represents an aromatic, usually benzenoid, residue that may carry substituents. Initially they were prepared by reaction of an arylamine with ethyl acetoacetate but now are more conveniently made by treatment of an arylamine with diketene (Scheme 4.11). The electron-attracting effect of the two carbonyl groups flanking the methylene group in structure 4.6 gives the protons on this latter group an acidic character and makes these compounds soluble in alkaline solution, the negative charge produced by the ionisation being spread over the molecule as shown in Scheme 4.12. This provides an electron-rich site on the carbon atom of the methylene group, at which coupling occurs. Coupling of simple diazotised arylamines with acetoacetarylamides produces greenish-yellow pigments and dyes of particular value in the direct and azoic dye fields, under

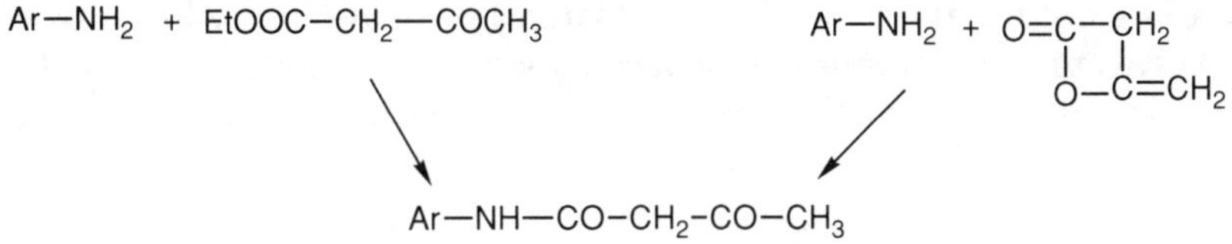

4.6

Scheme 4.11

Scheme 4.12

which headings specific structures will be mentioned. Related to the acetoacetarylamides is barbituric acid (4.7), which has found similar use as a coupling component in synthetic dyes.

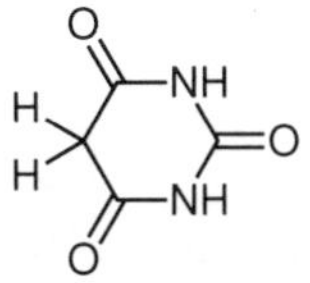

4.7

4.4.2 Pyridones

The pyridone coupling components (4.8), which came into use in the 1960s chiefly for the preparation of greenish-yellow disperse and reactive dyes, are made by the condensation of an alkylamine with malonic ester and ethyl cyanoacetate. Coupling occurs at the position indicated by the arrow in Scheme 4.13.

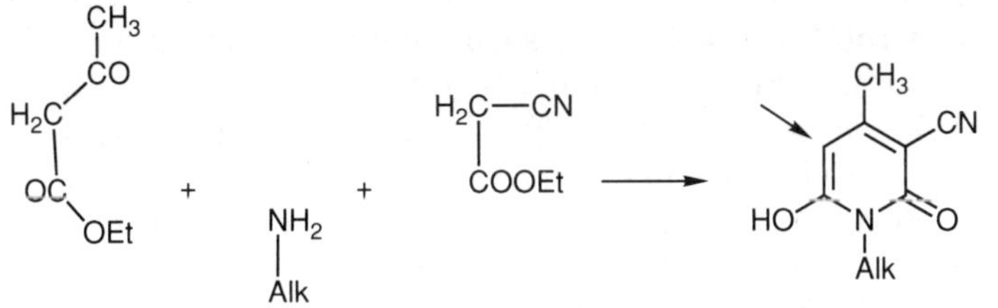

4.8

Scheme 4.13

4.4.3 Pyrazolones

The 1-aryl-5-pyrazolones (4.9), prepared by the two-step condensation of an arylhydrazine with ethyl acetoacetate, are the most commonly used coupling components for the preparation of greenish-yellow azo dyes.

Coupling occurs at the 4-position of the pyrazolone ring which, as in the case of acetoacetarylamides discussed above, is activated towards electrophilic attack by the two flanking unsaturated carbon atoms (Scheme 4.14).

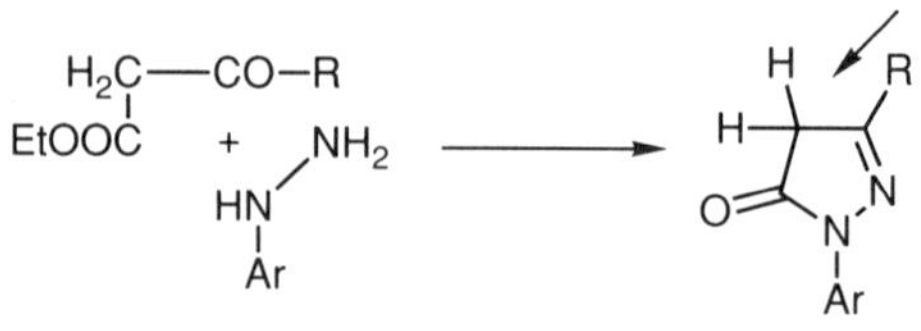

4.9

Scheme 4.14

4.4.4 Aminopyrazoles

Closely related to the pyrazolones, but less commonly used, are the 5-aminopyrazoles (4.10). Again, these give greenish-yellow shades when coupled at the activated 4-position indicated by the arrow in Scheme 4.15. They are prepared by condensation of an arylhydrazine with diacetonitrile, which is itself produced by dimerisation of acetonitrile over a nickel catalyst.

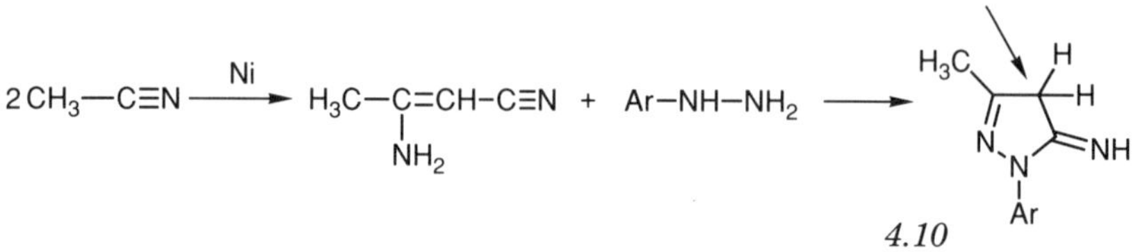

4.10

Scheme 4.15

4.4.5 Phenols

When phenol is treated with one equivalent of a diazonium salt at pH 7–8 coupling occurs mainly in the 4-position of the benzene ring, but small amounts of the 2-azo and 2,4-disazo compounds are formed [16]. With excess diazonium salt under neutral conditions the 2,4-disazo derivative is formed [17], but under strongly alkaline conditions the 2,4,6-trisazo derivative is produced. These factors can lead to impure products and so phenol itself is not often used as a coupling component. 4-Arylazophenols have the disadvantage of being pH-sensitive, ionisation of the hydroxy group under alkaline conditions resulting in a marked change of hue. With 2-arylazophenols hydrogen bonding between the hydroxy group and the azo linkage (4.11) prevents such ionisation. Because of this effect the most commonly employed simple phenols are compounds such as *p*-cresol (4-methylphenol) in which blocking of the 4-position ensures that coupling will occur in the 2-position. With simple diazonium salts these produce yellow azo compounds.

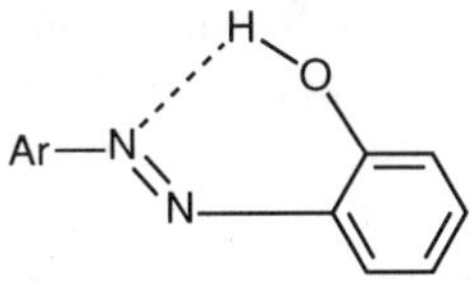

4.11

Resorcinol, in which the two hydroxy groups reinforce each other, is of greater technical importance than phenol not only because it couples much more strongly but also because the presence of two activating hydroxy groups allows more than one azo group to be introduced into the molecule. The preferred first coupling under neutral conditions takes place at the 4-position [16], and when a further equivalent of diazonium salt is employed in the presence of sodium acetate or carbonate a second azo group is introduced into the 2-position; in the presence of sodium hydroxide the second group enters the 6-position. At sufficiently high pH 2,4,6-trisazo compounds are produced [18–20]. The ability to introduce two different azo groups into resorcinol is exploited in the production of brown dyes mainly for use on leather.

The other two dihydric phenols, catechol and hydroquinone, are of little interest in this context since they reduce diazonium salts on attempted coupling, being themselves converted to the corresponding quinones.

4.4.6 Aniline derivatives

Aniline itself couples inefficiently with normal diazonium salts [21], the reaction product of a diazotised amine and aniline often being the diazo-amino compound. However, the presence of electron-donating substituents, such as methyl or methoxy groups, in the benzene nucleus makes coupling much easier and allows the introduction of the azo group into the 4-position of compounds such as *m*-toluidine (3-methylaniline), *p*-xylidine (2,5-dimethylaniline), cresidine (2-methoxy-5-methylaniline) and 2,5-dimethoxyaniline. In the event of the 4-position being blocked, as in 2,5-dimethoxy-4-methylaniline, the azo group may be introduced into the unoccupied *o*-position [22]. Like the 4-arylazophenols, these 4-arylazo-anilines display pH sensitivity, becoming protonated under acid conditions, and they are therefore mainly of use as intermediates in the preparation of disazo dyes.

N-Alkylation of primary aromatic amines increases their nucleophilic character and makes them couple much more readily, the introduction of the azo group occurring in the 4-position. Thus, in contrast to aniline, *N*-methylaniline couples readily and *NN*-dimethylaniline very readily with simple diazonium salts. Diphenylamine also couples, in the 4-position, but less readily than *N*-methylaniline.

The increase in coupling power gained on *N*-alkylation can be exploited to allow the coupling of primary arylamines by warming the base with an aqueous solution of one equivalent of sodium bisulphite and one equivalent of formaldehyde. The complex formed from these two entities reacts with the arylamine, giving a clear solution of the methyl-ω-sulphonate derivative (4.12) which couples readily. The methyl-ω-sulphonate group is labile and can be conveniently removed by acid hydrolysis, affording a facile preparation of aminoazobenzenes not accessible by direct coupling (Scheme 4.16).

$$\mathrm{Ar{-}NH_2 + HOCH_2SO_3Na \longrightarrow \underset{4.12}{ArNHCH_2SO_3Na} + H_2O}$$

Scheme 4.16

The *NN*-dialkylanilines are of paramount importance in the preparation of disperse dyes, the whole range of shades from yellow to greenish-blue being obtainable from them by suitable choice of diazo component. If the 4-position in an *NN*-dialkylaniline is blocked, as for example in *NN*-dimethyl-4-methylaniline, steric hindrance prevents the diazonium ion approaching the *o*-position and no coupling takes place [23].

As with resorcinol, the 1,3-diamines of the benzene series are useful in the preparation of disazo dyes that can form the basis of brown shades. Generally the 2,4-disazo compound is the product of coupling with two equivalents of diazonium salt, but at higher pH values formation of the 4,6-isomer can be significant [24]. Again, the 1,2- and 1,4-diamines reduce diazonium salts on attempted coupling.

4.4.7 Aminophenols

3-Aminophenol is a useful dyestuff intermediate, capable of coupling in either the 2- or the 4-position, and is often used in preference to resorcinol or *m*-phenylenediamine as a final coupling component in the preparation of complex dyes containing more than one azo group, typically trisazo direct dyes. Other aminophenols are of little interest in this context.

4.4.8 Naphthols

In this discussion of coupling components it is convenient to separate naphthalene compounds from their benzenoid analogues since important differences in behaviour exist between the two classes. Naphthalene, although formally containing two fused benzene rings, has much less than twice the resonance stabilisation energy of benzene and therefore has a lower level of aromaticity. The reason for this is that in only one of the three limiting structures (4.13a–c) are both rings 'benzenoid', the other two structures each containing one 'benzenoid' and one 'quinonoid' ring. A consequence is that the contribution of structure 4.13b to the overall structure of naphthalene is greater than that of the other two forms, so that the bonds between adjacent α- and β-positions show a degree of double-bond character. This has an effect on the coupling behaviour of both naphthols and naphthylamines, as will be shown later.

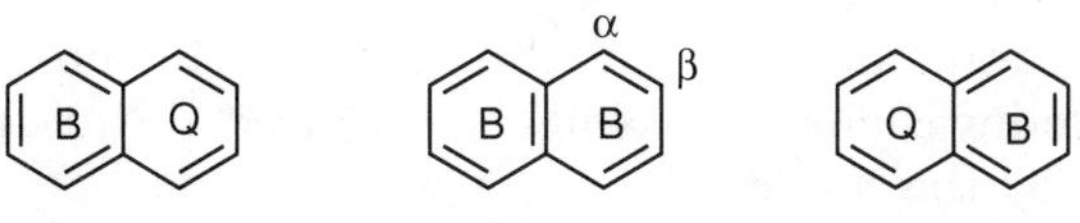

4.13a *4.13b* *4.13c*

The two simple naphthols and their sulphonic acids are very widely used as coupling components in dye manufacture, generally giving orange shades when coupled with simple diazonium salts. 1-Naphthol (4.14) couples in both the 2- and 4-positions and, as with phenol, it is difficult to make the coupling selective. Highly reactive diazonium salts couple chiefly in the 4-position whilst weakly reactive ones, especially those derived from aminophenols and aminonaphthols, couple mainly in the 2-position. The position of coupling in 1-naphthols is, of course, also influenced by substituents; for example, a sulphonic acid group in the 3-, 4- or 5-position inhibits 4-coupling and directs the diazonium ion to attack at the 2-position.

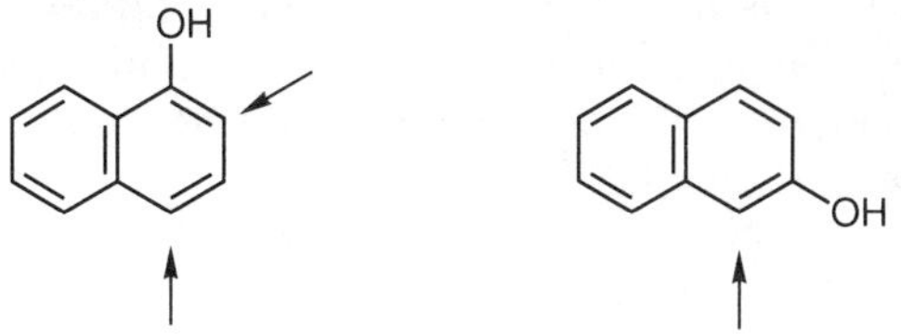

4.14 *4.15*

With 2-naphthol (4.15) the special characteristics of the structure of naphthalene mentioned above come into play and coupling is directed exclusively into the 1-position. When this position is blocked, as with 1-methyl-2-naphthol, no coupling occurs. The reason for this behaviour lies in the nature of the relevant tautomeric forms (4.15a–c) of the naphtholate ion. In structure 4.15a the unsubstituted ring retains its benzenoid character and is therefore vastly preferred over structure 4.15c, in which the unsubstituted ring is forced to assume a quinonoid configuration. This localises the negative charge on the 1-position and leads to selective attack of the diazonium ion at this site.

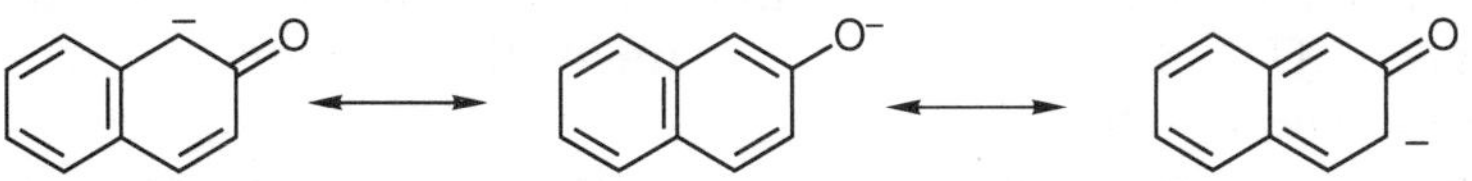

4.15a *4.15b* *4.15c*

4.4.9 Naphthylamines

The aminonaphthalenes couple much more easily than do the amines of the benzene series. This allows the azo compounds to be prepared without the need to reinforce the effect of the amino group with additional electron-repelling substituents. They give hues similar to those obtained from naphthol coupling components and display directive influences similar to those shown by the naphthols in the coupling reaction. Thus with one equivalent of a diazonium salt 1-naphthylamine couples largely in the 4-position [25]. With two equivalents of diazonium salt the 2,4-disazo compound may be obtained [26,27]. 2-Naphthylamine, like 2-naphthol, couples exclusively in the 1-position [24] and derivatives with this position blocked by non-labile substituents fail to couple. When treated with powerful diazonium salts 2-naphthylamines with labile substituents in the 1-position, such as 2-naphthylamine-1-sulphonic acid, couple by introducing the azo group into the 1-position with concomitant expulsion of the labile substituent.

In the past, 2-naphthylamine was an important intermediate for the preparation of dyestuffs, but its recognition as a potent carcinogen has resulted in its total withdrawal by reputable dye manufacturers. Stringent precautions also need to be taken when using 1-naphthylamine to ensure that the material is not contaminated with the 2-isomer. The naphthylaminesulphonic acids, however, remain amongst the most important dyestuff intermediates, finding wide-ranging applications as both diazo and coupling components.

4.4.10 Aminonaphthols

There are 14 isomers carrying both an amino group and a hydroxy group on a naphthalene residue, but only those in which the two groups are located in different rings are of use as coupling components. In these, each group exerts the major directive influence within its own ring, so that at pH 7 or below coupling occurs in the amino-substituted ring, whereas at alkaline pH coupling in the hydroxylated ring becomes important. Often this still leaves a choice of possible coupling sites, leading to mixtures of products. Of the 14 isomers only one or two have found any use in dye manufacture, their usefulness being determined by their ease of manufacture, toxicological considerations and properties of derived dyes.

Of far greater importance are the sulphonated aminonaphthols, in which the versatility conferred by the presence of both an amino and a hydroxy group makes them among the most important group of azo intermediates. Three are outstanding. These are J acid (4.16), γ acid (4.17) and H acid (4.2), which couple under the relevant conditions of pH at the arrowed positions.

Particularly important dyes are produced when the *N*-acylated derivatives are coupled, bright orange shades resulting from the use of *N*-acyl J

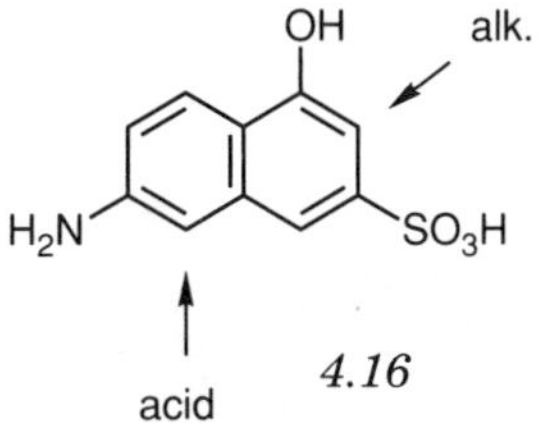

4.16

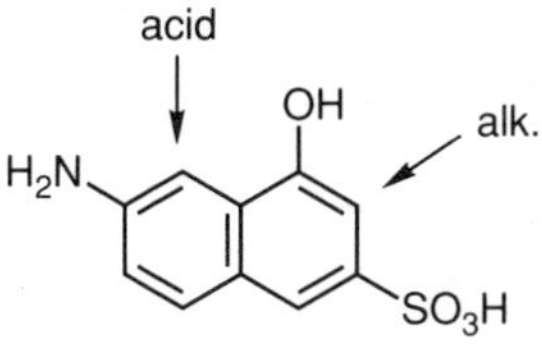

4.17

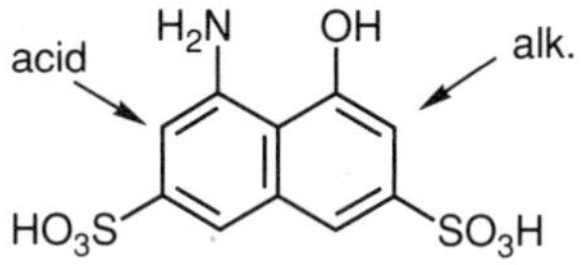

4.2

acids and bright reds from *N*-acyl γ acid and *N*-acyl H acid derivatives. Coupling of γ acid under acid conditions leads to red monoazo dyes that cannot be further coupled and have exceptional fastness to light. These characteristics result from hydrogen bonding of both the amino and hydroxy groups on to the azo linkage (4.18), thus preventing the ionisation of the hydroxy group which is essential for coupling to occur and protecting the azo linkage by involving both its lone pairs of electrons in coordination.

4.18

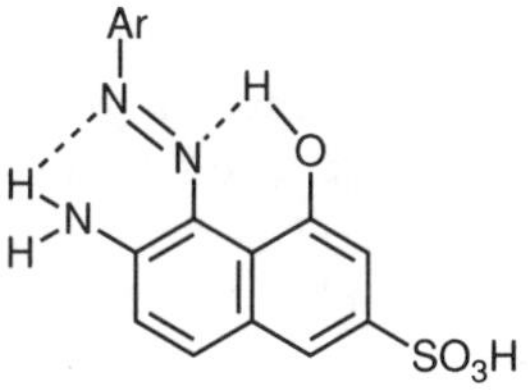

Other important dye classes are formed by coupling aminonaphtholsulphonic acids with two equivalents of diazonium salt, giving a disazo dyestuff. In the case of H acid it is essential that the acid coupling stage is carried out first, the products being navy in shade. With J acid, which again gives dark blue disazo dyes, the initial coupling should preferably be carried out under alkaline conditions.

4.5 STRUCTURE OF AZO DYES

4.5.1 Stereochemical isomerism

The simplest azo compound, azobenzene, exists as a mixture of a stable *trans* (4.19) and an unstable *cis* (4.20) form [28,29]. Formation of the *cis* isomer is induced by exposure to light, the quantum yield of the process depending upon the wavelength of the light employed [30]. The proportion of the *cis* isomer can be appreciable in an equilibrium mixture. Thus a concentration of 24% of this unstable form builds up within a few hours when an acetic acid solution of azobenzene is exposed to sunlight in shallow white trays. Reversion to the *trans* form occurs readily on heating and is

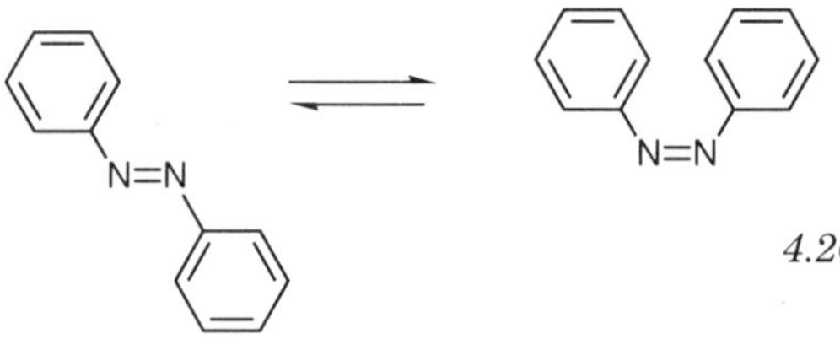

4.19 4.20

catalysed by a variety of substances that can function as electron donors or acceptors [31].

The two isomers display differing properties and can be separated by exploiting differences in their distribution between immiscible solvents. The molecule of the *trans* form is very nearly planar, whereas in the *cis* isomer the benzene rings are inclined at an angle of about 56° to each other [32]. The *trans* isomer forms monoclinic crystals, has no dipole moment and melts at 68°C whereas the *cis* isomer forms orthorhombic bipyramidal crystals, has a dipole moment of 3.0 debye and melts at 71.4°C. The two isomers also differ in hue. Some substituted azobenzenes display similar stereoisomerism but these are mainly simple structures, increasing molecular complexity causing steric factors to control the configuration exclusively in the *trans* form. Some simple substituted azobenzenes serve as disperse dyes, but in these the difference in hue between the *cis* and *trans* forms gives rise to the phenomenon of phototropism, manifested as a change in hue on exposure to light; this is an undesirable feature of these dyes.

4.5.2 Tautomerism

Whilst azo compounds prepared from diazonium salts and phenolic or keto–enol coupling components are usually depicted in the azophenol form (4.11), an alternative tautomeric structure can be drawn for these compounds. This is the quinonehydrazone form (4.21) and, in cases where the azo and hydroxy groups are located on adjacent carbon atoms, hydrogen bonding can occur between the two groups as shown. Similar pairs of structures, but without hydrogen bonding, can be drawn for *p*-hydroxyazo compounds.

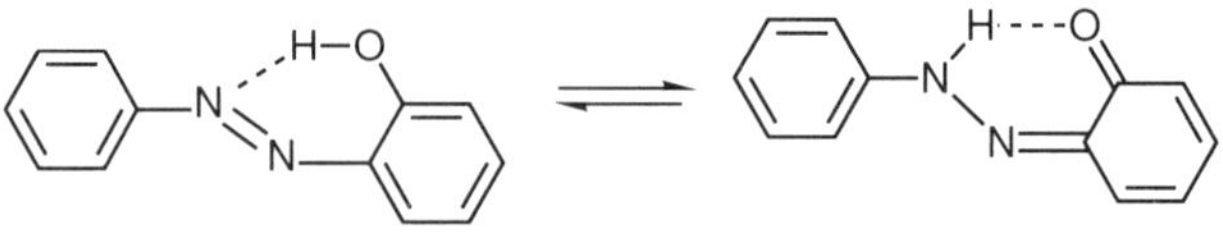

4.11 4.21

The actual structure of any particular azo compound will depend on the relative energy levels of the two tautomers that can be drawn for that compound. Where one form has a much lower energy level than the other, the lower-energy form will predominate; a tautomeric mixture will result where the energy levels of the two forms are similar. In the case of azo compounds derived from keto–enol coupling components (acetoacetarylamides, pyridones, pyrazolones and aminopyrazoles) the quinonehydrazone form has a much lower energy level than the azophenol and represents the true structure of these compounds. Physical measurements support this conclusion. For example, spectroscopic evidence shows azo compounds derived from acetoacetarylamides to have structure 4.22 [33]. This is confirmed by n.m.r. evidence, which also indicates pyrazolone-based azo compounds to have the structure 4.23 [34,35].

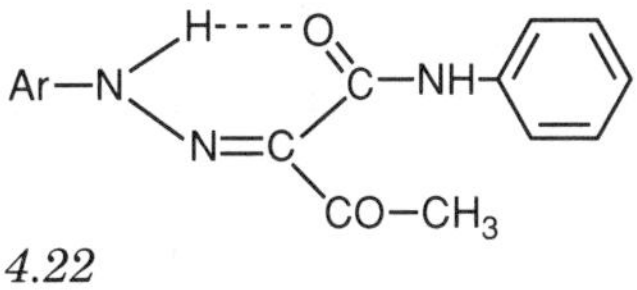

4.22

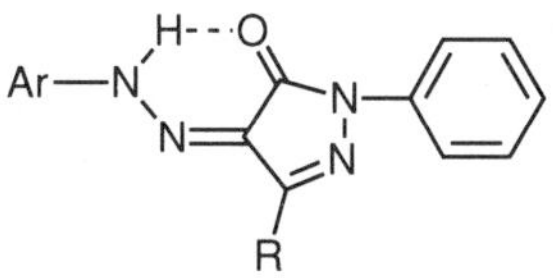

4.23

The converse of this situation is presented by diazotised arylamines coupled with phenols, where assumption of a quinonehydrazone structure would necessitate loss of the aromatic character of the phenolic ring. Consequently these compounds exist solely in the azophenol form.

The phenylazonaphthols present a particularly interesting series. Despite the partial loss of aromaticity in the hydrazone structures that can be drawn for 1-phenylazo-2-naphthol, 2-phenylazo-1-naphthol and 4-phenylazo-1-naphthol, these structures are at most only 8 kJ/mol higher in energy than the azo forms [36], and thus there should be no significant preference for one form over the other. When the visible absorption spectra curves of 4-phenylazo-1-naphthol are measured in a variety of solvents they all pass through a common point (isosbestic point) (Figure 4.1, overleaf) [37]. 1-Phenylazo-2-naphthol shows similar behaviour. This is clear evidence that in solutions of these compounds both tautomers are present. Tautomerism both in the solid state and in solution is confirmed by infra-red spectroscopic measurements on all three compounds [38,39].

A totally different picture is presented by 3-phenylazo-2-naphthol. This unusual isomer cannot be prepared by a normal coupling procedure but has been obtained by reaction of 3-amino-2-naphthol with thionyl chloride to give the *N*-sulphinylamine (4.24, overleaf), condensation of which with *N*-phenylhydroxylamine yields the desired product [40]. Here assumption of a quinonehydrazone form would entail loss of aromatic character in both rings of the naphthalene residue and the energetic unfavourability of this ensures that the compound exists solely in the azophenol form.

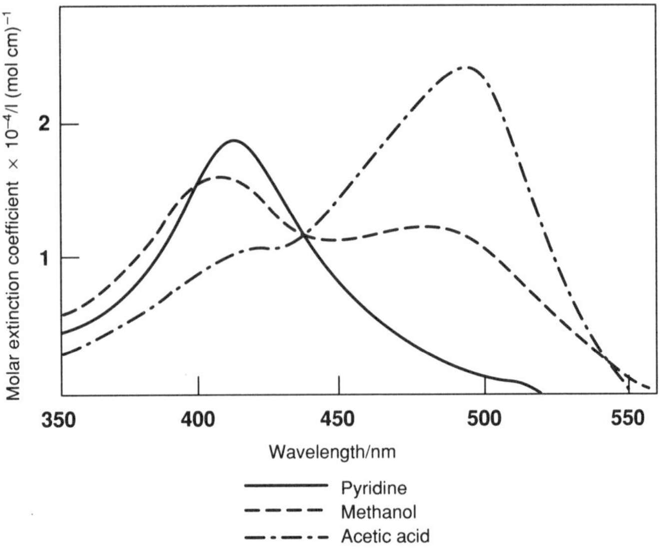

Figure 4.1 – Effect of solvent on 4-phenylazo-1-naphthol spectrum

4.24
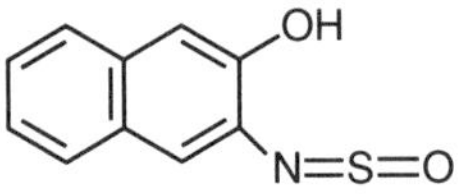

4.6 PREPARATION AND IMPORTANCE OF NAPHTHALENE INTERMEDIATES

4.6.1 Importance of naphthalene

When purely benzenoid intermediates are employed in the preparation of azo dyes, the products exist solely in the azophenol form since the energy level of this form is significantly lower than that of the quinonehydrazone form. The position changes when a naphthalene intermediate is used, for here the two tautomeric forms are of comparable energy and so coexist. The quinonehydrazone form generally possesses a higher extinction coefficient than does the azophenol form; the practical consequence is that the former has a greater tinctorial strength – in crude terms, more colour value per unit mass. The participation of the quinonehydrazone form in the naphthalene-based dyes allows them to gain this strength advantage which, together with the ready availability of naphthalene and its convenient processing to ideally substituted derivatives, has made naphthalene of great importance as a primary starting material in the manufacture of intermediates for the dyestuffs industry.

Naphthalene intermediates [41] are always built up by substitution reactions with the cheap and plentiful hydrocarbon using, in the main, only seven basic reactions. Most of these are reactions that will be generally familiar from benzene chemistry but which are slightly modified, since naphthalene has two different possible positions of substitution. These are often designated the α- and β-positions, the four α-positions being *ortho* and the four β-positions *meta* to the nearest carbon atom of the central bond. A further modifying influence is the lower level of aromaticity of naphthalene compared with benzene.

The main reactions used in the manufacture of dyestuff intermediates from naphthalene are described in sections 4.6.2 to 4.6.8.

4.6.2 Sulphonation

Treatment of naphthalene with concentrated sulphuric acid [42] results in substitution at one of the α-positions (4.25) if the temperature is below 60°C, but the reaction is reversible and at higher temperatures rearrangement occurs yielding the stable β-isomer (4.26).

Under more forcing conditions, such as the use of higher operating temperatures or of oleum (sulphur trioxide dissolved in anhydrous sulphuric acid), more than one sulphonic acid group can be introduced. This can lead to mixtures of isomers and, in practice, the reaction conditions must be carefully defined if the desired isomer is to be produced in maximum yield.

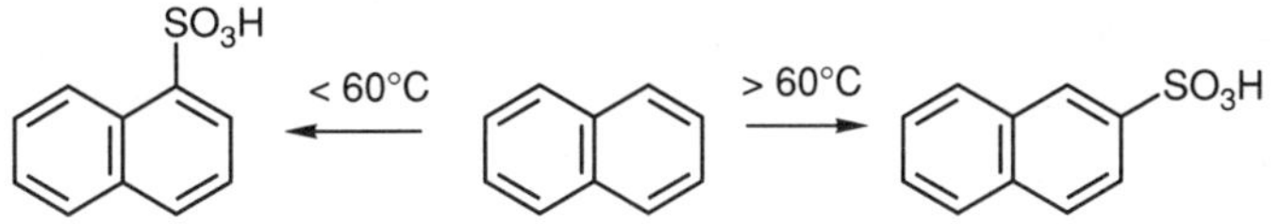

4.25 *4.26*

4.6.3 Nitration

This reaction [43], unlike sulphonation, is not reversible. It results in mainly α-substitution, yielding 1-nitronaphthalene (4.27).

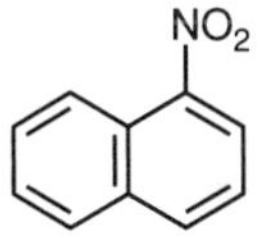

4.27

4.6.4 Reduction

Reduction of a nitro compound to the corresponding naphthylamine proceeds exactly as in the benzene series. The process was formerly carried out using iron and hydrochloric acid, but catalytic hydrogenation is now more generally used.

4.6.5 Replacement of a sulphonic acid group by a hydroxy group

This is accomplished by heating the sulphonated derivative at a high temperature with sodium or potassium hydroxide [44]. Typical is the preparation of 2-naphthol (4.15) from naphthalene-2-sulphonic acid (Scheme 4.17). Displacement of the sulphonic acid group occurs more readily when it is located at the α- rather than the β-position, the former requiring a fusion temperature of about 200°C and the latter one of about 250°C. This difference can be exploited to prepare naphtholsulphonic acids by fusion of suitable naphthalenedisulphonic acids.

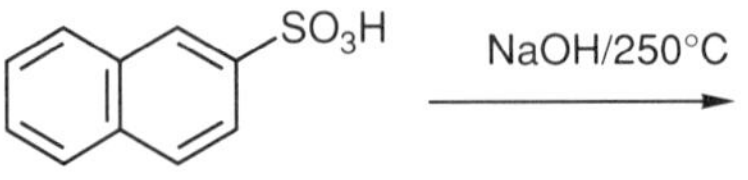

4.26 *4.15*

Scheme 4.17

4.6.6 Bucherer reaction

This reaction [45–47] is only rarely encountered in the benzene series but is extremely useful in the naphthalene series, where the mechanism of the reaction has been extensively investigated. It allows a hydroxy group to be exchanged for an amino group or vice versa. When a hydroxy group is to be converted into an amino group, the naphthol is heated under pressure with ammonium bisulphite (often produced *in situ* by introduction of ammonia liquor and sulphur dioxide into a sealed autoclave) at a temperature of 100–150°C; the naphthol is thus converted into the corresponding naphthylamine. The mechanism of the reaction is shown in Scheme 4.18.

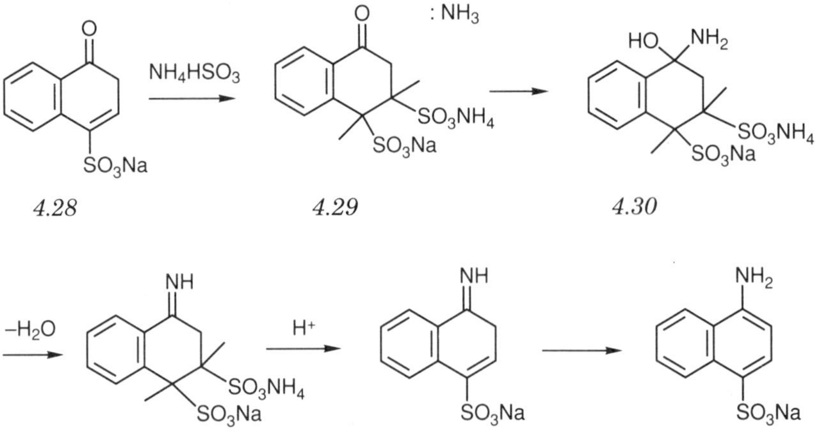

Scheme 4.18

In the initial step the naphthol reacts in its tautomeric form (4.28), adding ammonium bisulphite across the isolated double bond in the sulphonated ring. Nucleophilic attack by ammonia on the tetralonesulphonic acid (4.29) is followed by rearrangement of a proton from the nitrogen atom to the oxygen atom producing an intermediate (4.30) which then loses the elements of water, giving the imine (4.31). After cooling and discharging the autoclave, the reaction mixture is acidified and boiled. The added bisulphite is eliminated from the molecule and rearrangement of the imine tautomer (4.32) so produced completes the conversion to the naphthylamine (4.33).

The reverse of this reaction, which allows a naphthylamine to be converted into the corresponding naphthol, is similarly carried out by heating the amine with sodium bisulphite and finally acidifying the mixture; the mechanism is shown in Scheme 4.19.

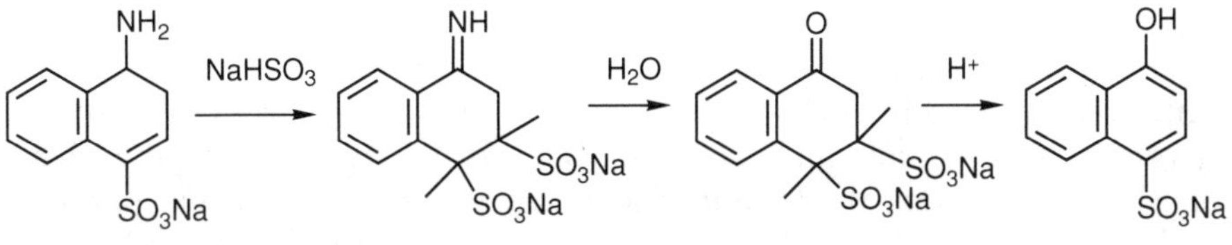

Scheme 4.19

The Bucherer reaction can also be used in the preparation of substituted naphthylamines. Carrying out the reaction on a naphthol but employing an alkylamine in place of ammonia results in the production of the corresponding *N*-alkylnaphthylamine. Alternatively a naphthylamine may be used as starting material, when the excess alkylamine employed results in the amino group being displaced by the alkylamino group. Of greater importance is the preparation of *N*-arylnaphthylamines by heating either a naphthol or a naphthylamine with sodium bisulphite and an arylamine, whereupon the amino or hydroxy group in the starting material is exchanged for the arylamino group.

In the above reaction mechanisms it is noteworthy that the sulphonic acid group introduced has been shown to enter at the 3-position and not the 4-position as previously postulated. A consequence of this is that attempted Bucherer reactions on naphthols and naphthylamines carrying a sulphonic acid group located *meta* to the hydroxy or amino group would require a second sulphonic acid group to be introduced at this position. Since it is impossible to locate two sulphonic acid groups on the same carbon atom, these compounds cannot undergo the transformation.

4.6.7 Arylation

In some cases where it is difficult to carry out the Bucherer reaction successfully it is easier to prepare *N*-arylnaphthylamines by heating together a naphthylamine and an arylamine. In particular, this reaction is useful in the preparation (Scheme 4.20) of 1-phenylaminonaphthalene-8-sulphonic acid (4.34; *N*-Phenyl Peri acid) and its *N*-4-methylphenyl analogue (Tolyl Peri acid), both of which intermediates are valuable components for the production of navy blue dyes.

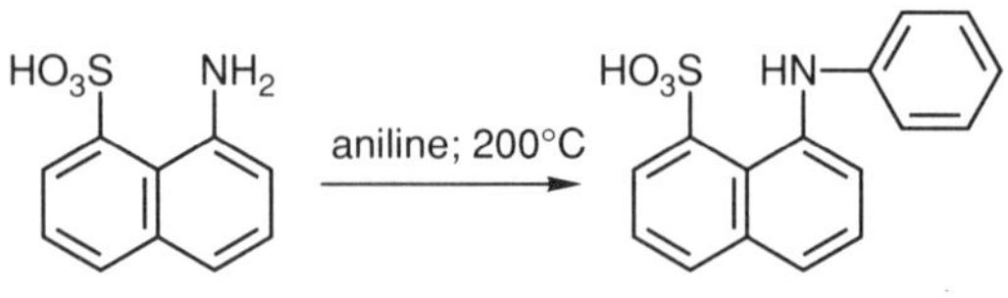

4.34

Scheme 4.20

4.6.8 Kolbe–Schmitt reaction

This reaction is of importance only for the manufacture of BON acid (the parent acid of compound 4.35) from 2-naphthol. It is accomplished by heating the sodium salt of 2-naphthol under pressure in an atmosphere of carbon dioxide (Scheme 4.21). Under these conditions an equilibrium mixture of the naphthol and its carboxylated derivative is established, the mixture containing about 30% of the carboxylic acid. The desired carboxylic acid is extracted from the reaction mixture and the unchanged 2-naphthol recycled.

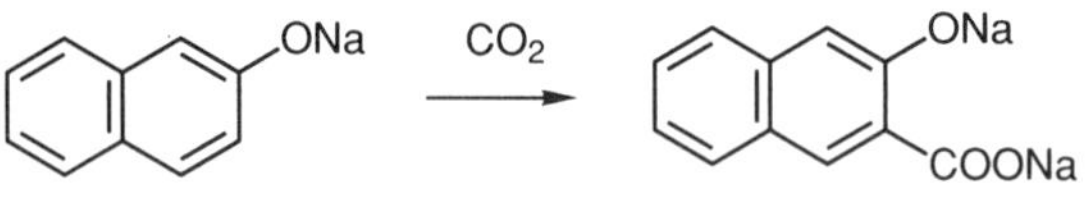

4.35

Scheme 4.21

4.6.9 Preparation of specific intermediates

These reactions can be used to prepare a variety of multi-purpose intermediates that are of great value in dyestuffs manufacture. Many of these intermediates were prepared and found to be of value in dyestuff preparation before the orientation of their substituents was fully worked out. This early lack of basic knowledge resulted in these intermediates being given trivial names that have persisted in the dyestuffs industry.

In the preparation of naphthalene intermediates the reactions must be employed in the correct order to achieve a desired orientation in the final product. A further crucial consideration can be the need to avoid steps that would result in the formation of carcinogenic materials. These points are illustrated in the following examples.

Naphtholsulphonic acids and γ acid
The preparation of 2-naphthol by high-temperature sulphonation of naphthalene followed by fusion of the resulting naphthalene-2-sulphonic acid has been mentioned previously. Further sulphonation of 2-naphthol produces several useful naphtholsulphonic acids, and conditions can be chosen to make one or other of these the main product. The initial product is the unstable 2-naphthol-1-sulphonic acid, which readily rearranges to 2-naphthol-6-sulphonic acid (4.36; Schaeffer's acid), further sulphonation of which leads to 2-naphthol-6,8-disulphonic acid (4.37; G acid) at low temperature and 2-naphthol-3,6-disulphonic acid (4.38; R acid) at higher temperature.

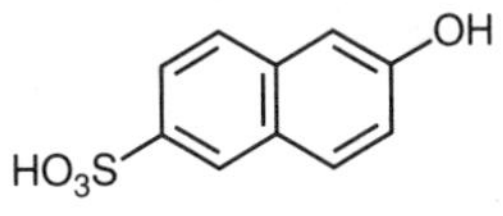

4.36

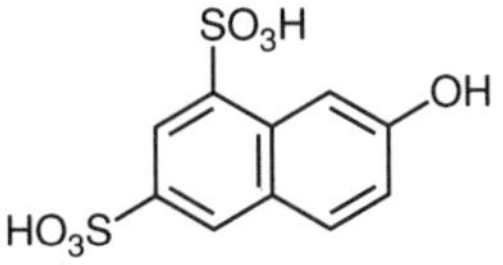

4.37

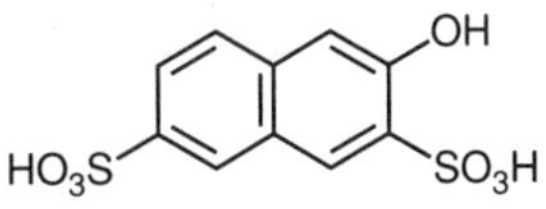

4.38

Fusion of 2-naphthol-6,8-disulphonic acid with sodium hydroxide yields naphthalene-2,8-diol-6-sulphonic acid (4.39), the more readily displaced α-sulphonic acid group being replaced. When this compound is subjected to a Bucherer reaction only the 2-hydroxy group is exchanged for an amino group, yielding γ acid (4.40) (Scheme 4.22). This selectivity arises because the 8-hydroxy group is unable to form the intermediate addition product necessary for the Bucherer reaction to proceed, the sulphonic acid group being located in the *meta* position.

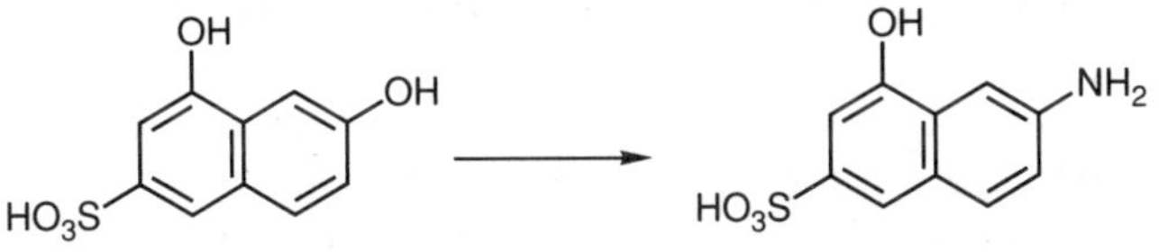

4.39 *4.40*

Scheme 4.22

H acid
H acid (4.2) is possibly the most important single naphthalene-based intermediate. The preparation of this intermediate starts with a high-temperature sulphonation of naphthalene using 65% oleum (anhydrous

sulphuric acid in which 65% by weight of sulphur trioxide has been dissolved) to give mainly naphthalene-1,3,6-trisulphonic acid, the nitration product from which is purified by selective isolation. Reduction of the nitro group followed by hydrolysis of the 1-sulphonic acid group by heating with sodium hydroxide solution at 180°C completes the process (Scheme 4.23).

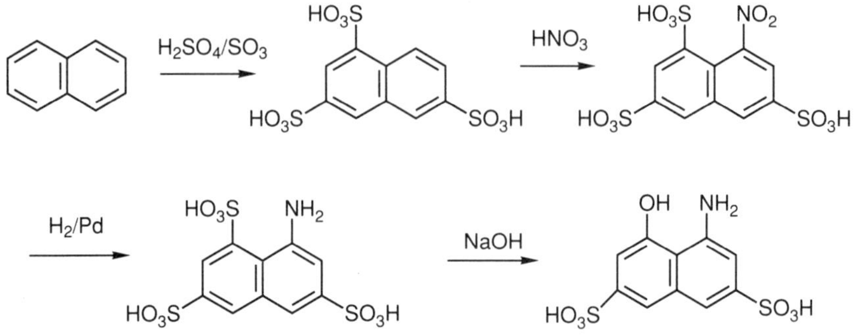

4.2

Scheme 4.23

Use of a higher temperature (280°C) at the final stage results in more extensive hydrolysis, with the amino group also being replaced by a hydroxy group, leading to the formation of naphthalene-1,8-diol-3,6-disulphonic acid (chromotropic acid).

J acid

Of particular interest is the chain of reactions leading to J acid (4.43) since several of the compounds that feature as intermediate stages in the chain are themselves useful as dyestuff intermediates. The starting point for this chain is 2-naphthol which, in early syntheses, was converted by means of a Bucherer reaction into 2-naphthylamine, this then being sulphonated to yield 2-naphthylamine-5,7-disulphonic acid (4.42; Amido J acid). The recognition that 2-naphthylamine is a potent carcinogen caused this route to be abandoned. In the method of preparation now in use (Scheme 4.24) a sulphonic acid group is introduced into the 1-position of the naphthalene nucleus and is carried through the early stages, so that 2-naphthylamine-1-sulphonic acid (4.41; Tobias acid) rather than the unsulphonated amine figures in the preparation.

After sulphonation to 2-naphthylamine-1,5,7-trisulphonic acid the labile 1-sulphonic acid substituent, which has now served its purpose, is eliminated by diluting the sulphonation mixture and heating. Fusion of the

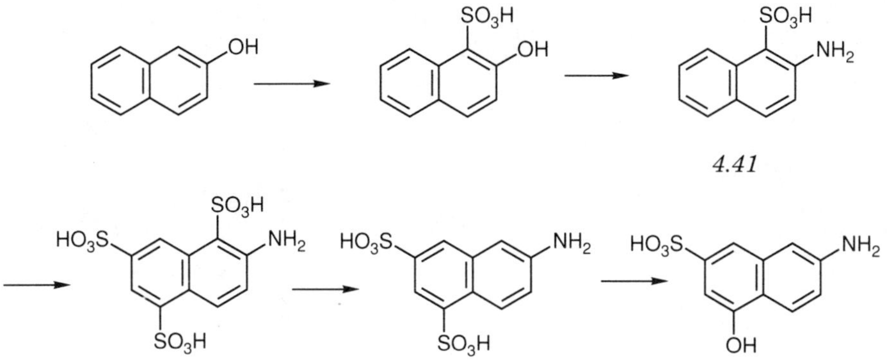

4.41 4.42 4.43

Scheme 4.24

resulting 5,7-disulphonic acid with sodium hydroxide replaces the more labile 5-sulphonic acid group by a hydroxy group, forming J acid.

2-Naphthylamine-5,7-disulphonic acid and 2-naphthylamine-1-sulphonic acid, which are intermediate products in the above scheme, as well as 2-naphthylamine-1,5-disulphonic acid (obtained on careful low-temperature sulphonation of the latter compound), are all used in the preparation of azo dyestuffs.

1-Naphthylaminesulphonic acids

Nitration of naphthalene-1-sulphonic acid produces two isomeric nitronaphthalenes that have very similar solubilities. It is convenient to reduce the mixture without separating, giving a mixture of 1-naphthylamine-8-sulphonic acid (4.44; Peri acid) and 1-naphthylamine-5-sulphonic acid (4.45; Laurent's acid) (Scheme 4.25). These two compounds can be readily separated since the former has a much lower solubility. The isomers also differ markedly in their value as intermediates, Peri acid being a very useful dye component whilst Laurent's acid finds little use.

In all the above sequences single isomers are produced by careful control of reaction conditions combined with purification by selective isolation at various points in the synthesis. Occasionally two isomers are produced which give dyestuffs that have very similar properties; in these cases it is often quite acceptable and economically beneficial not to separate the individual components but to use the total mixture in dye preparation. An example is the mixture of 1-naphthylamine-6- and 7-sulphonic acids (4.46; mixed Cleve's acids), which arises by nitration and reduction of naphthalene-2-sulphonic acid (Scheme 4.26).

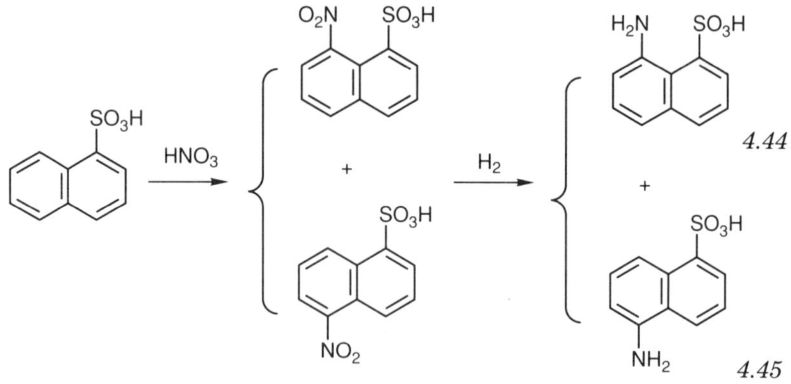

Scheme 4.25

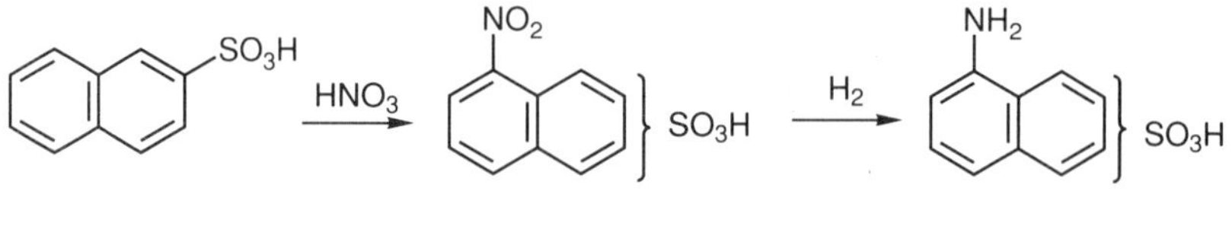

Scheme 4.26

4.7 SCHEMATIC REPRESENTATION OF COUPLING

The simple process of diazotising an amine and coupling the resulting diazonium salt with a coupling component produces a monoazo dye, that is, a dye containing a single azo group. Such dyes are often represented as A → E dyes, A standing for the amine, E the end coupling component and the arrow symbol meaning 'diazotised and coupled with'. The certainty with which diazotisation and coupling proceed, together with the wide variety of components available, allows more complex dye structures to be built up by combining two or more diazotisation and coupling sequences. A convenient shorthand way of classifying these more complex products has gained common acceptance. This shorthand, known as Winther symbols, uses the arrow symbol combined with the following symbols to indicate the nature of the component:

- A to represent an amine that is diazotised
- E to represent an end coupling component
- D to represent a diamine that is tetrazotised (both amino groups diazotised)
- M to represent an amine that is first coupled with a diazotised amine and then diazotised
- Z to represent a coupling component that can couple twice.

Using this nomenclature disazo dyes (dyes containing two azo groups) fall into one or other of the three classes represented thus:

$$E_1 \leftarrow D \rightarrow E_2 \qquad A_1 \rightarrow Z \leftarrow A_2 \qquad A \rightarrow M \rightarrow E$$

4.8 SULPHONATED AZO DYES

In the majority of dyeing processes the dye is transferred from aqueous solution on to the fibre, so adequate aqueous solubility is a desirable property in a dye. The sulphonic acid group provides the cheapest way of solubilising a dyestuff and does not generally cause any diminution in the light fastness of the chromogen to which it is attached. Consequently dyes containing sulphonic acid groups find use in the dyeing of both amide and cellulosic fibres (see sections 3.2.1 and 3.2.2). In the ranges of dyes available for the dyeing of these fibres, sulphonated monoazo structures provide most of the dyes in the yellow to red sector as well as a few of the more bathochromic hues. As indicated previously, the coupling component is of prime importance in determining the hue, with the diazo component merely modifying it, although when the diazo component contains powerful electron-attracting or -repelling substituents this modification can be profound. Pyridone and pyrazolone coupling components are generally used in yellow dyes and naphthol, naphthylamine, aminonaphthol or acylaminonaphthol coupling components to cover the orange to red region. Disazo dyes cover the spectrum from yellow to greenish-blue and find use on cellulosic and amide fibres. Trisazo and polyazo dyes are restricted to use on the cellulosic fibres.

The actual structures chosen vary according to the particular fibre as well as the end use to which the dyed material will be put, and it is appropriate to divide the discussion of the structures of these dyes according to the fibres on which they are used.

4.8.1 Acid dyes for wool

Simple monoazo dyes containing one or two sulphonic acid groups are used to dye wool, the dye being retained on the fibre by electrostatic bonding. Pyrazolones, naphthols, naphthylamines and acylaminonaphthols are commonly used coupling components. These 'levelling acid' dyes are typified by tartrazine (4.47; C.I. Acid Yellow 23) and C.I. Acid Orange 20 (4.48). These simple dyes are markedly hydrophilic in character and this necessitates them being applied to the substrate from a strongly acid dyebath. Under these conditions they readily penetrate the wool fibre and distribute themselves evenly, but their highly hydrophilic character makes them easily desorbed and thus limits their fastness to washing. A higher level of wash fastness can be achieved by using disazo dyes of larger molecular size, usually carrying two sulphonic acid groups. These 'acid milling' dyes are typified by C.I. Acid Yellow 42 (4.49), C.I. Acid Red 151 (4.50) and C.I. Acid Blue 116 (4.51).

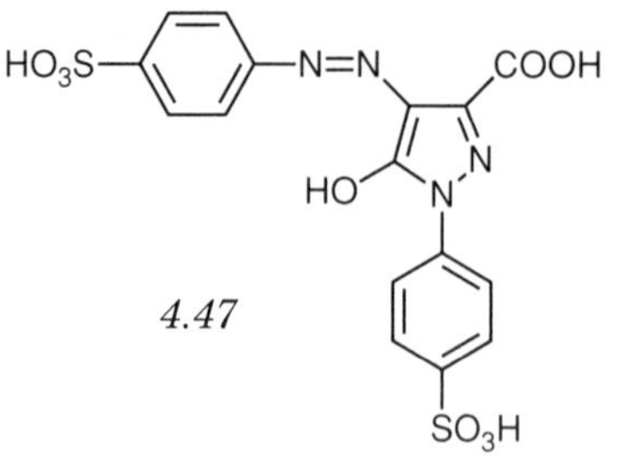

4.47

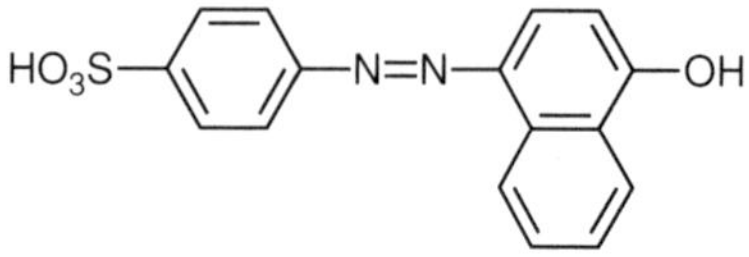

4.48

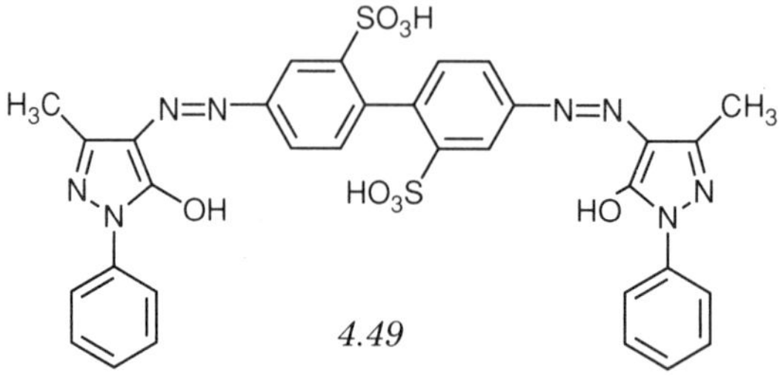

4.49

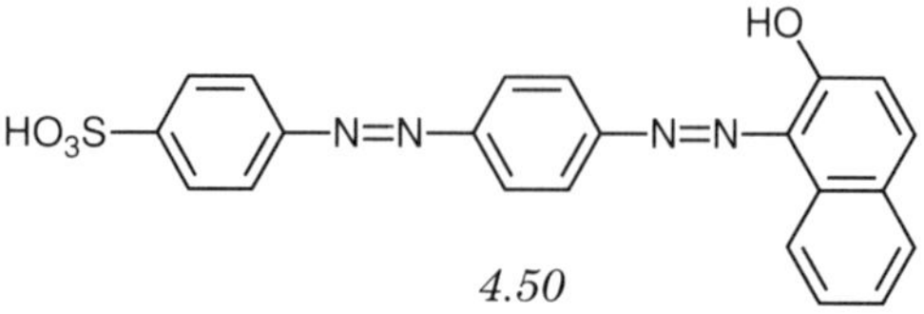

4.50

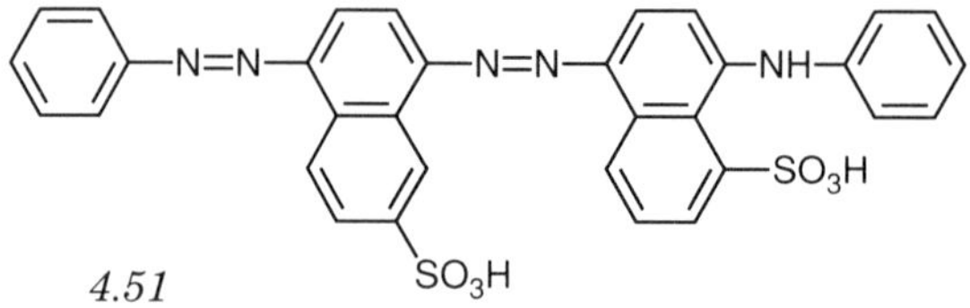

4.51

Much superior wet fastness can be achieved by incorporating a highly hydrophobic weighting group into the dye molecule [48]. Non-polar bonding between this group and hydrophobic side chains in wool imparts neutral-dyeing affinity to these 'super milling' dyes (see section 3.2.2). A typical structure [48] is C.I. Acid Red 138 (4.52).

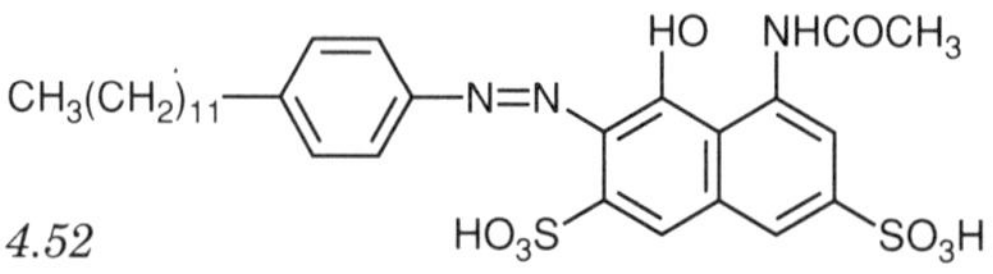

4.52

Of particular interest are the light-fast red dyes obtained by coupling on to γ acid under acid conditions. In this latter type, an example of which is C.I. Acid Red 32 (4.53), both the amino group and the hydroxy group are hydrogen-bonded on to the azo linkage [49]. This double hydrogen bonding prevents the tautomerism that is normally a feature of naphthalene-based azo dyes by locking the dye in the azo form, leading both to the red hue and to the high level of light fastness.

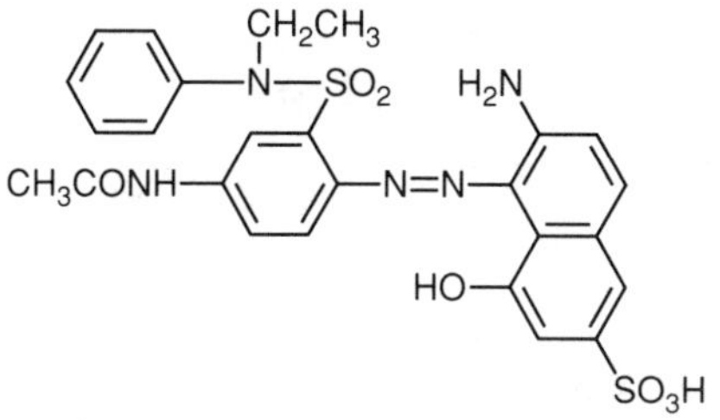

4.53

Strongly electron-attracting or -repelling substituents in the diazo component can have a profound effect on the hue, producing a large bathochromic shift (see section 1.4). The use of diazo components of the naphthalene series produces a similar but less marked effect. These effects are exploited in C.I. Acid Blue 92 (4.54), prepared by coupling diazotised *O*-tosyl H-acid with *N*-Phenyl Peri acid and hydrolysing. Here the proximity of the *peri*-sulphonic acid group to the secondary amino group prevents any marked pH sensitivity, removing the need to stabilise the hue by acylation. Again the hydroxy group on the diazo component is hydrogen-bonded on to the azo linkage, resulting in excellent light fastness.

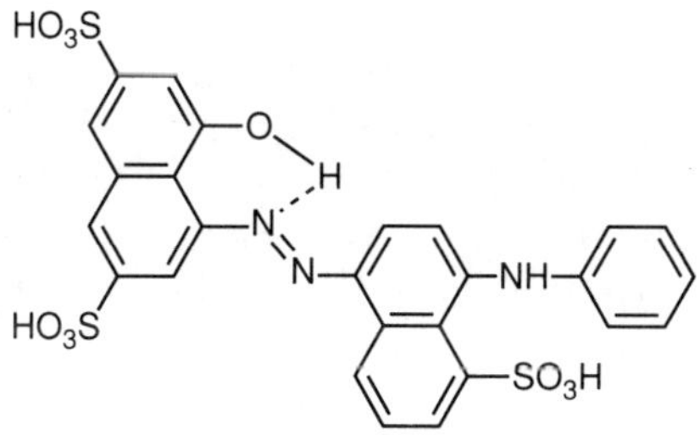

4.54

4.8.2 Acid dyes for nylon

The synthetic polyamide fibre nylon can be dyed with either acid or disperse dyes, the former giving better fastness and the latter better levelling properties. The types of acid dye used are broadly similar to those for wool, that is, monosulphonated levelling acid dyes with relative molecular mass values in the 300–500 region and disulphonated milling acid dyes with relative molecular mass of 600–900. Dyes of small molecular size are often deficient in wet fastness and those of r.m.m. higher than about 800 tend to give unlevel dyeings. Within these crude guidelines certain structural features exert specific effects; for instance, hydrogen-bonding substituents such as hydroxy and acylamino groups boost wash fastness but may impair level dyeing behaviour.

Monoazo dyes containing a single sulphonic acid group conveniently fall into the levelling acid group. A typical structure is C.I. Acid Red 266 (4.55).

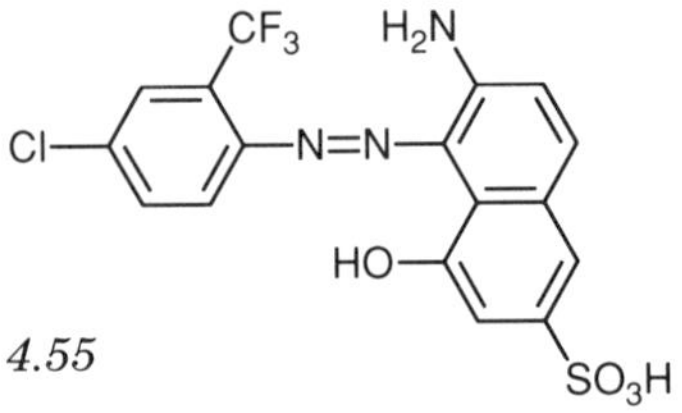

4.55

Monosulphonated levelling acid dyes are mainly yellow to red in colour, whilst the larger molecular size that can be tolerated with a disulphonated acid milling dye allows the use of disazo compounds, such as C.I. Acid Blue 118 (4.56), to cover more bathochromic hues such as navy blue.

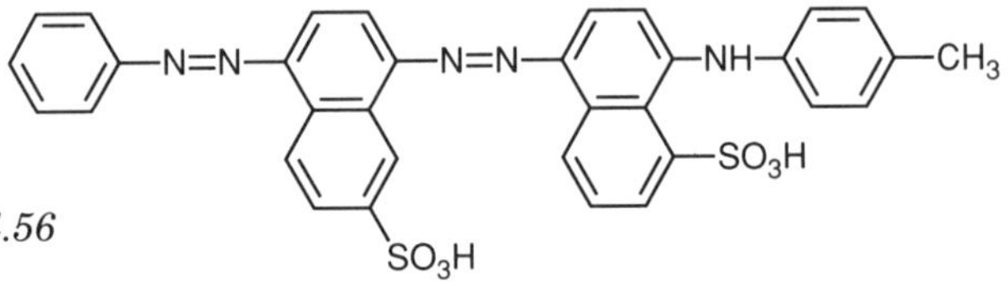

4.56

4.8.3 Direct dyes

Certain water-soluble dyes are directly adsorbed on to cotton that has not been pretreated in any way (see Chapter 1). The first dye in which this phenomenon was observed was Congo red (4.57; C.I. Direct Red 28), discovered in 1884 by Böttiger. The extreme pH sensitivity of this dye now restricts its use to that of an indicator, but it deserves mention as the forerunner of the direct dye class.

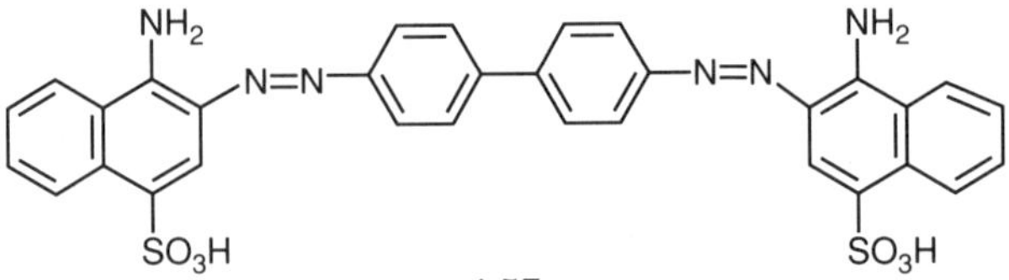

4.57

If the solubility of dyes of this type is kept to the minimum, an acceptable level of wash fastness can also be achieved. This low solubility requirement is met by the use of the minimum number of solubilising groups consistent with the need to prepare an aqueous dyebath. Hydrogen-bond formation with the fibre depends upon having numerous groups containing electron-rich atoms, such as oxygen and nitrogen, positioned at intervals along an extended planar molecule. Suitable features conferring substantivity include amino and acylamino groupings, hetero atoms in aromatic rings and the azo group itself.

The simplest monoazo dyes fail to meet these requirements, but by choosing intermediates known to confer substantivity and by building up the molecule to provide the necessary length and coplanarity (see Chapter 3), direct dyes can be produced from this class. Thus the highly substantive character of the benzothiazole nucleus is exploited in C.I. Direct Yellow 8 (4.58), as is the alignment of the azo, ureido and acylamino groups in the substituted J acid coupling component of C.I. Direct Red 65 (4.59).

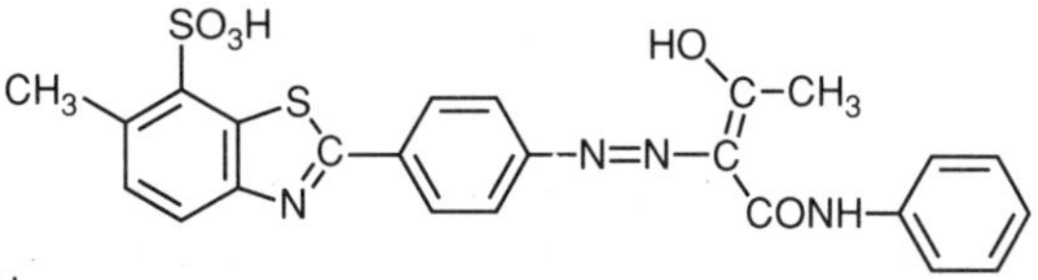

4.58

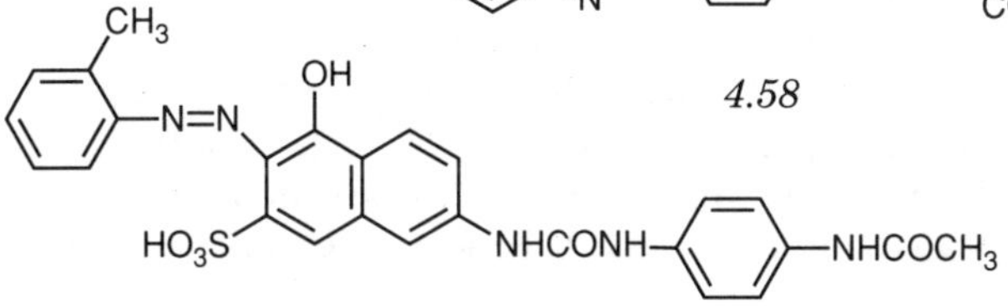

4.59

The required long planar shape is more readily supplied by simple disazo structures such as C.I. Direct Blue 1 (4.60). Copper complexes of such disazo compounds are important and will be dealt with elsewhere (see Chapter 5). A widely used method of producing A → Z ← A disazo dyes relies on treating J acid with phosgene ($COCl_2$) to give the bis-coupling component, carbonyl J acid (4.61).

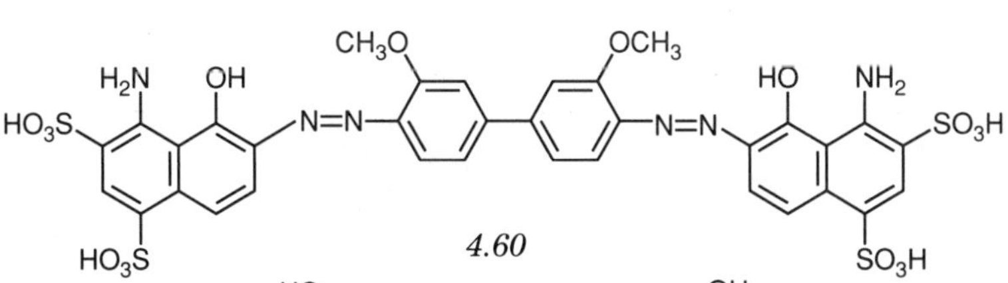

4.60

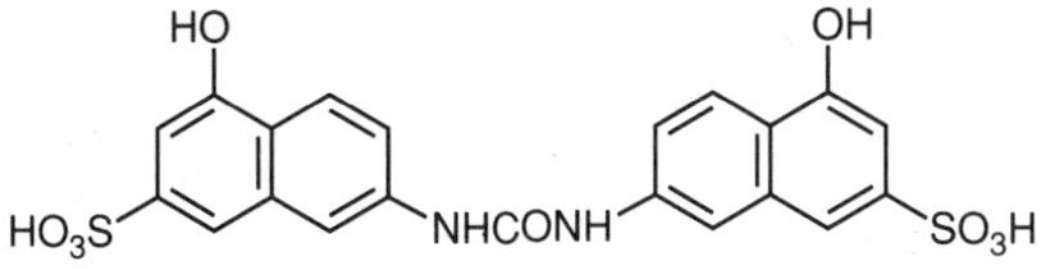

4.61

Disazo dyes can also be prepared by phosgenating suitable amino monoazo dyes. The phosgenation technique for building up direct dye molecules by joining together sub-units generally produces symmetrical structures. However, the use of cyanuric chloride (4.62) allows different sub-units to be joined together and this intermediate was employed to link non-substantive anthraquinone structures to substantive azo moieties in green direct dyes, such as C.I. Direct Green 28 (4.63).

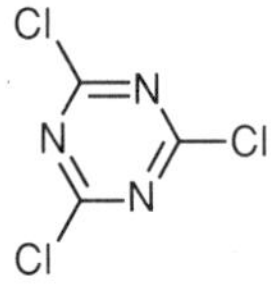

4.62

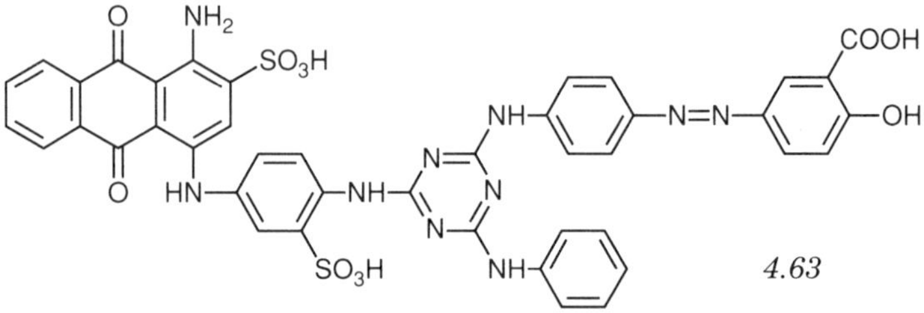

4.63

It is in the direct dye class that the more complex polyazo dyes come into their own, and trisazo structures such as C.I. Direct Blue 78 (4.64), C.I. Direct Brown 222 (4.65) and C.I. Direct Black 38 (4.66) are classic examples of this type of dye. Generally, the $A \rightarrow M_1 \rightarrow M_2 \rightarrow E$ type (such as 4.64) afford reasonably bright blue shades whilst dyes prepared according to other schemes, such as the $E \leftarrow D \rightarrow Z \leftarrow A$ type (for example, 4.65 and 4.66) yield drab shades.

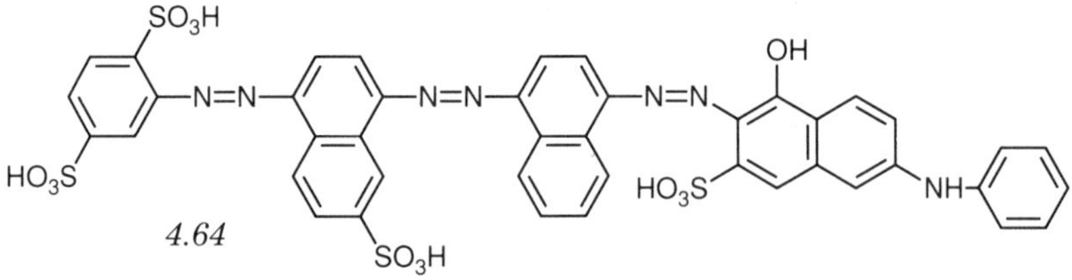

4.64

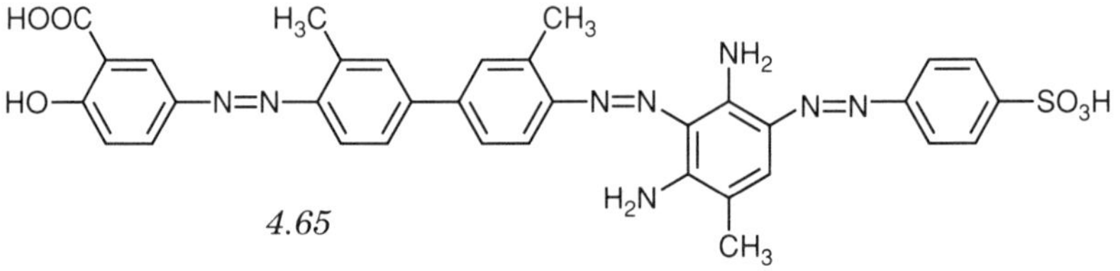

4.65

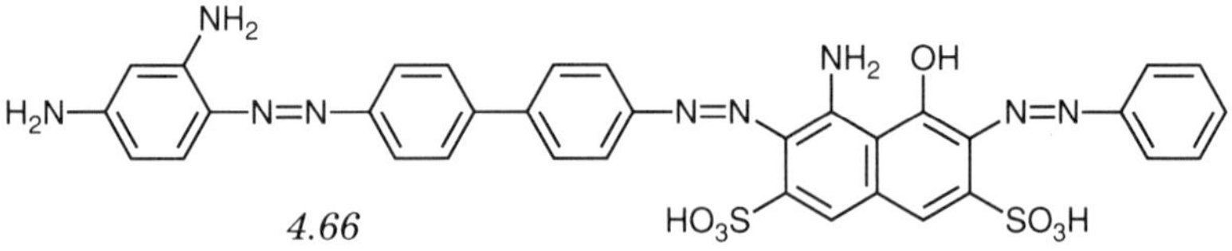

4.66

The last of this trio highlights an area of concern in direct dye chemistry. Many dyes of this type relied on the extremely useful intermediate benzidine. In the tetrazonium salt prepared from this diamine, the two diazonium groups are conjugated and strongly activate each other. This makes the first coupling proceed very readily, but after this has been completed the second diazonium group is no longer subject to the powerful activating influence of a conjugated diazonium group and couples far more sluggishly. The two diazonium groups can thus be coupled selectively in a stepwise manner to produce a host of useful products. In the preparation of C.I. Direct Black 38, for example, one diazonium grouping in tetrazotised benzidine was coupled at low pH with H acid and diazotised aniline then added. The pH was raised to cause the benzenediazonium chloride to couple next to the hydroxy group on the H acid unit. Finally *m*-phenylenediamine was added and the second benzidinediazonium grouping allowed to couple slowly to produce the third azo linkage. The withdrawal of the notorious carcinogen benzidine by major dyestuff manufacturers has led to an extensive search for replacements for these dyes. One product that has emerged from this effort is C.I. Direct Black 166 (4.67), in which 4,4′-diaminobenzanilide is used as a benzidine replacement.

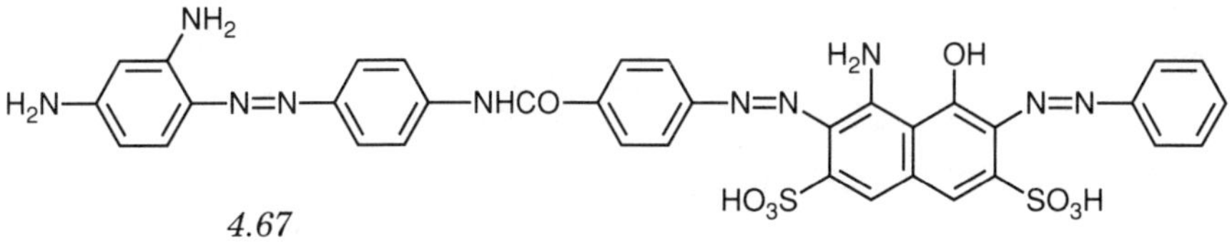

4.67

Tetrakisazo dyes (dyes containing four azo groups) are less common but some do find use, these products often arising from the phosgenation of an amino disazo dye.

Direct dyes have only modest fastness to washing, which may be improved by aftertreatments such as metal-complex formation (see Chapter 5) or diazotisation of the dye on the fibre and further coupling of the diazonium salt with an insoluble coupling component (see section 1.6.13). In addition to their use on cotton and viscose, direct dyes are important in the dyeing of leather. The cheapest members of this class also find use in the coloration of paper, since for this purpose fastness properties are largely irrelevant and price is all-important.

4.9 UNSULPHONATED MONOAZO DYES

4.9.1 Solvent-soluble dyes

These dyes, already mentioned in section 2.9, require adequate solubility for the coloration of various organic solvents and must be cheap [50]. The simplest and least polar dyes of this class are used for the coloration of petrol and ball-pen inks, with more polar types being used in lacquers, stains and varnishes. Some products of lower solubility are used in mass coloration.

These unexacting requirements make the simplest unsulphonated azo structures, often monoazo types, quite acceptable. Typical of the least polar members of this class are C.I. Solvent Yellow 1 (4.68), C.I. Solvent Orange 1 (4.69) and C.I. Solvent Red 17 (4.70). Simple azo structures carrying sulphonamide, sulphone or carboxylate ester groups are used where a somewhat more polar, less soluble dye is needed. Simple disazo compounds (4-aminoazobenzene→2-naphthol, for example) are used as red solvent dyes. Probably the only structural feature worthy of note in this class is the occasional adoption of structures carrying long alkyl chains to enhance solubility, as in the case of the disazo dye C.I. Solvent Yellow 107 (4.71).

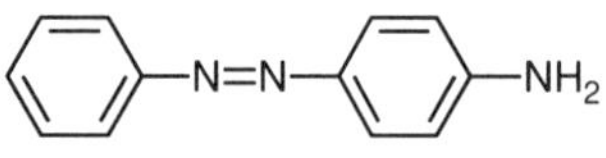

4.68

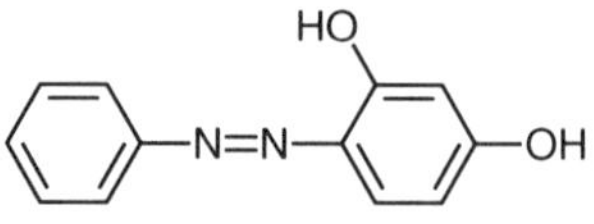

4.69

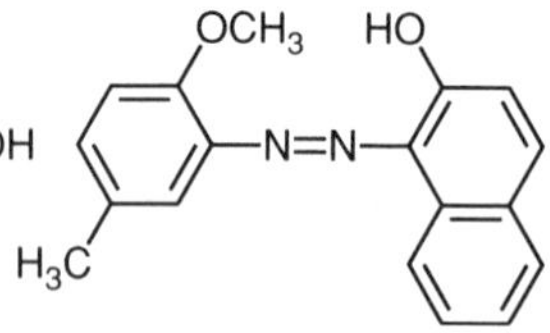

4.70

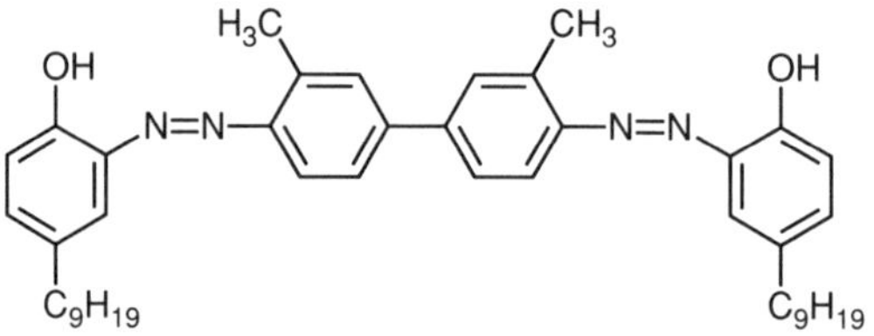

4.71

4.9.2 Disperse dyes

The introduction of the man-made fibre cellulose acetate in the early 1920s posed a new problem for the dyestuff chemist, an initial solution to which was supplied by Green and Saunders in 1923 with the launch of the Ionamine range [51,52]. These dyes carried, as sole solubilising group, an ω-methanesulphonate grouping, which was incorporated into the molecule by coupling on to an arylamine ω-methanesulphonate. During the dyeing process this grouping hydrolysed, leaving a fine dispersion of a water-insoluble dye that was taken up by the fibre. It was later realised that there was no real need for this temporary solubilisation of the dye and that satisfactory dye uptake could be achieved by merely using an aqueous dispersion of a water-insoluble dye as the dyebath. This type of dye, known as a disperse dye [53,54], is now used for the dyeing of cellulose acetate, triacetate and polyester fibres as well as offering an alternative to the acid dyes for the dyeing of nylon.

In order to facilitate satisfactory dye uptake the molecular size of a disperse dye must be kept small; monoazo structures are therefore exceptionally important, particularly in the coloration of polyester and cellulose triacetate. In the yellow shade area this generally poses no problem and the various available coupling components can all be used without making the molecule too large. A very simple example of the type of structure employed using a phenolic coupling component is C.I. Disperse Yellow 3 (4.72).

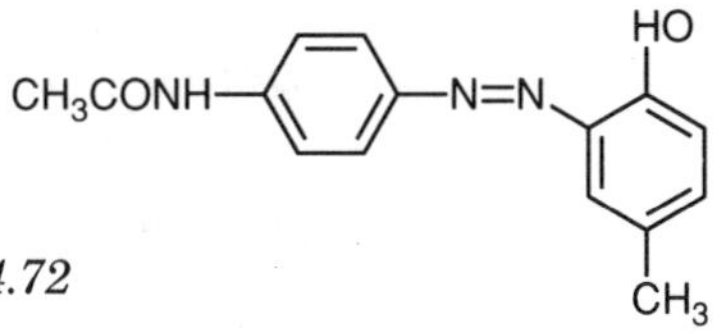

4.72

Heterocyclic coupling components are widely used in the disperse dye field for the production of yellow dyes. Numerous conventional dyes are based on simple pyrazolones, often combined with an *o*-nitroaniline diazo component, the *o*-nitro group being particularly helpful in ensuring good light fastness. C.I. Disperse Yellow 8 (4.73), which uses a very simple pyrazolone coupling component, is an example.

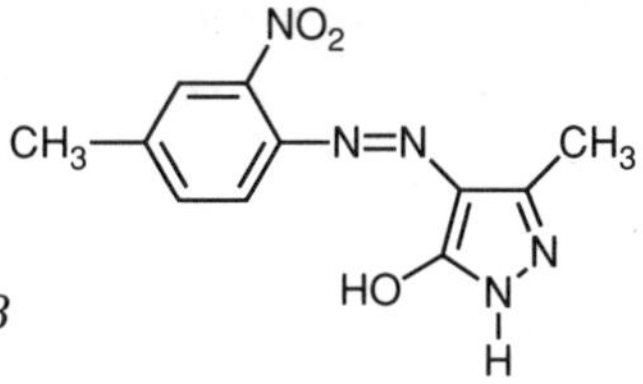

4.73

A very noticeable feature has been the use of a wider range of heterocyclic coupling components than in water-soluble azo dyes. An early example of this trend was C.I. Disperse Yellow 5 (4.74), which uses 4-hydroxy-1-methylcarbostyril as coupling component. It was in this field that the pyridones made their first impact (for example, 4.75). These compounds are now the most important yellow-producing coupling components, giving exceptionally bright greenish-yellow hues. Their success has stimulated the investigation of other compounds, with the particular aim of extending the range of hues available from such simple coupling

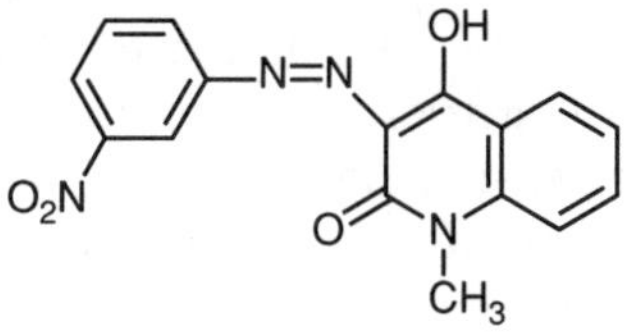

4.74

components. Thus derivatives of 2,6-diaminopyridine with simple diazo components yield orange dyes, such as 4.76.

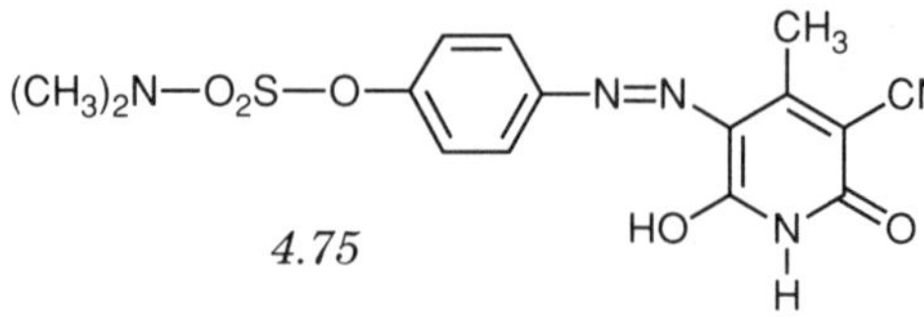

4.75

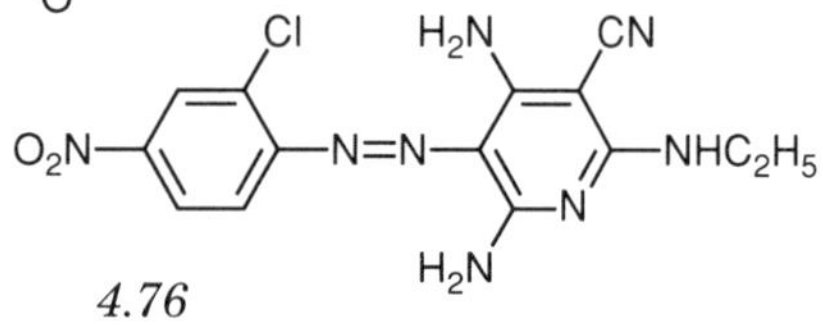

4.76

With the more bathochromic hues the restriction posed by the requirement for small molecular size can only be overcome by using ω-substituted *NN*-dialkylaniline coupling components and allying the powerful dialkylamino auxochrome with electron-attracting groups in the diazo component. This arrangement, with electrons being repelled from one end of the molecule and attracted to the other, produces highly polar structures, with hues varying from yellow to red to blue as the polarity increases. Examples are C.I. Disperse Red 90 (4.77) and C.I. Disperse Blue 183 (4.78). In this latter dye the polarity is further increased by incorporating an amide group in the coupling component to reinforce the electron-repelling effect of the dialkylamino group.

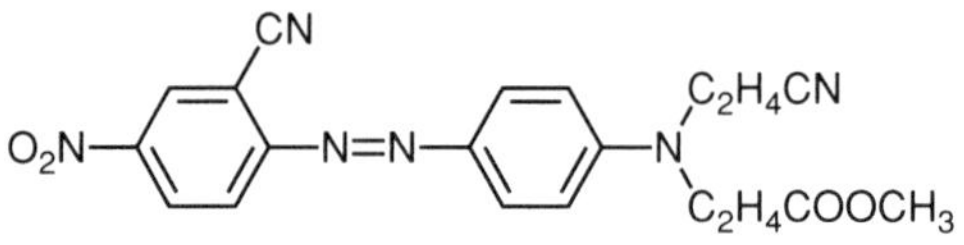

4.77

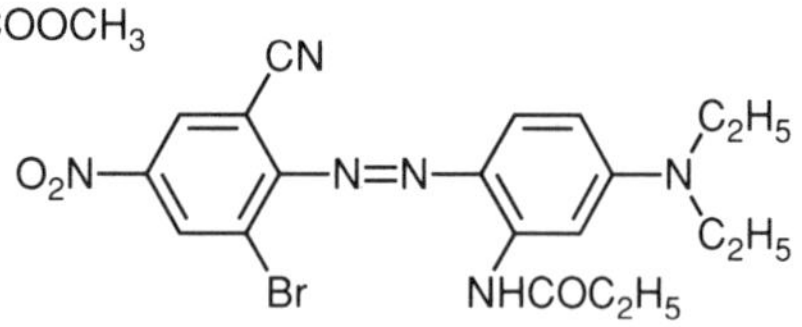

4.78

The coupling component in C.I. Disperse Red 90 is a complex tertiary amine, carrying cyano and carboethoxy groups on the alkyl chains. Such a coupling component is prepared by reaction of the primary arylamine with acrylonitrile ($H_2C{=}CHCN$) and methyl acrylate ($H_2C{=}CHCOOCH_3$), the amino group adding across the activated double bond. The reason for incorporating these substituents is to enhance the fastness properties of the dye. Thus the use of a cyanoethyl substituent in preference to a simple ethyl group in the tertiary amine coupling component generally improves fastness to light and sublimation.

In addition to benzenoid diazo components, diazotised heterocyclic amines in which the amino group is attached to a nitrogen- or sulphur-containing ring figure prominently in the preparation of disperse dyes, since these can produce marked bathochromic shifts. The most commonly used of these are the 6-substituted 2-aminobenzothiazoles (4.79), prepared by the reaction of a suitable arylamine with bromine and potassium thiocyanate (Scheme 4.27); using these compounds red dyes such as C.I. Disperse Red 156 (4.80) can be produced easily from the simplest coupling components.

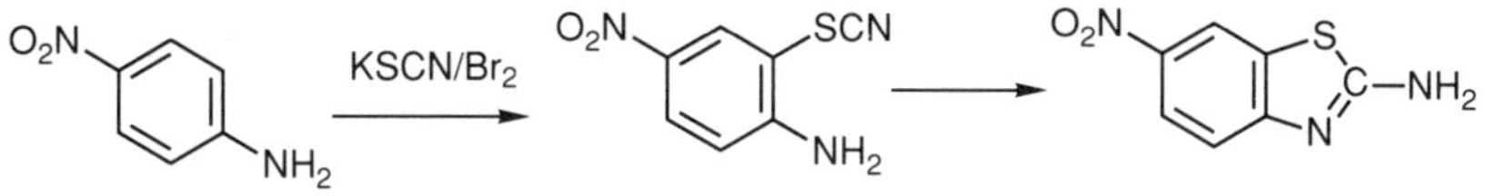

4.79

Scheme 4.27

4.80

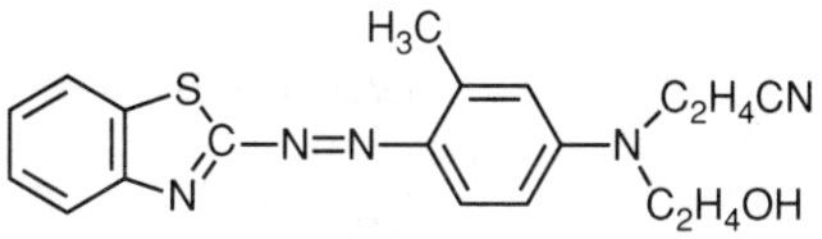

The use of heterocyclic diazo components is particularly important in the preparation of blue disperse dyes. It is often difficult to prepare, diazotise and/or couple the highly electronegatively substituted diazo components of the benzene series, needed to achieve blue hues, and feebly basic aminoheterocycles offer a convenient alternative. The small molecular size of these heterocyclic amines, compared with that of heavily substituted anilines, is another factor favouring their use.

One of the earliest diazo components of this type to be introduced was 2-amino-5-nitrothiazole (4.81), prepared by condensation of thiourea with chloroacetaldehyde and nitration of the resultant 2-aminothiazole (Scheme 4.28). This component yields bright dischargeable blues, such as dye 4.82, which have outstanding build-up, very high extinction coefficients and good fastness to burnt gas fumes. Use of diazo component 4.81 with coupling component 4.83 yields a greenish-blue dye.

Other examples include 3-aminobenzisothiazoles and especially their 5-nitro derivatives (4.84) [55] (synthesised as shown in Scheme 4.29) and electronegatively substituted 2-aminothiophenes (Scheme 4.30). Both of these are capable of giving blues with arylamine coupling components. For example, 5-acetyl-2-amino-3-nitrothiophene (4.85) used in dye 4.86 gives very attractive greenish-blue dyeings [56].

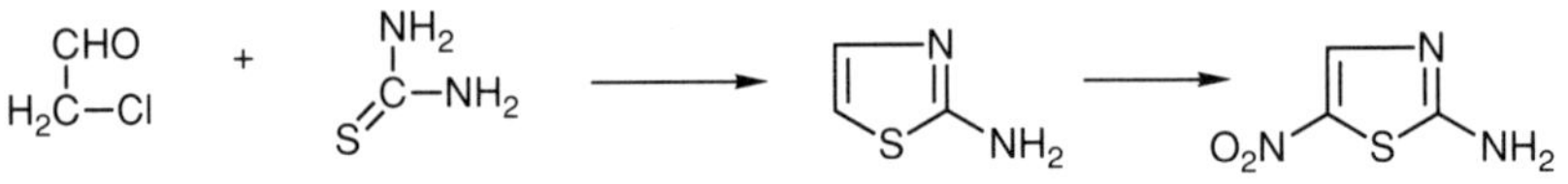

4.81

Scheme 4.28

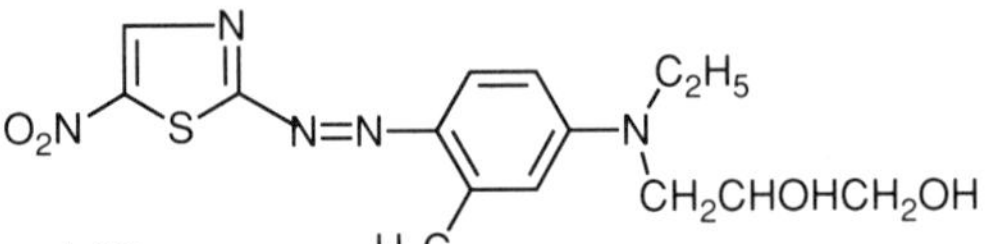

4.82

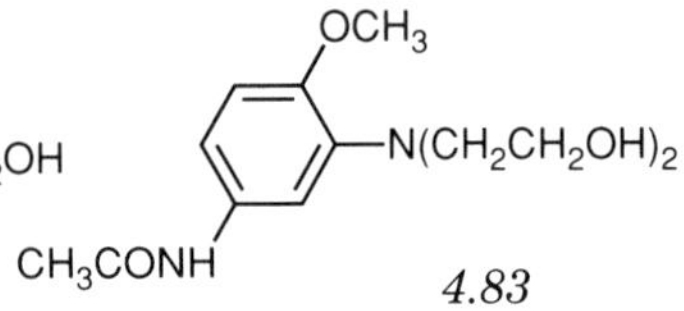

4.83

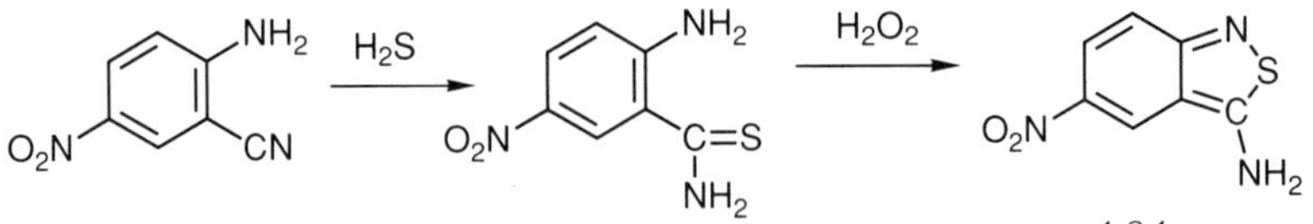

4.84

Scheme 4.29

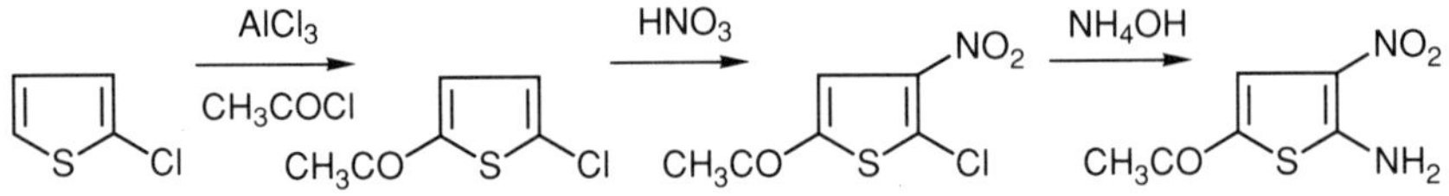

4.85

Scheme 4.30

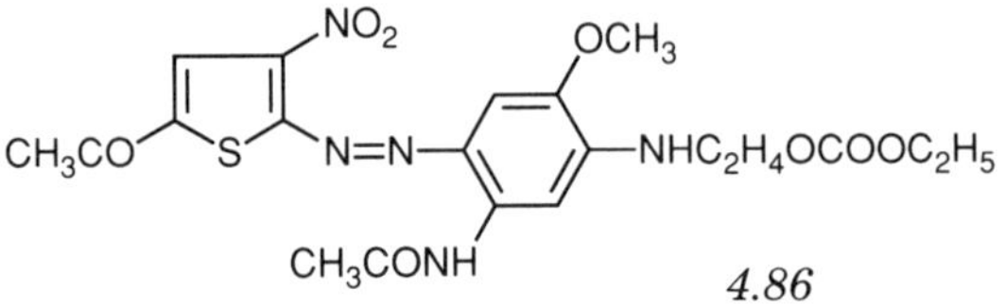

4.86

Whilst monoazo dyes of small molecular size dominate the disperse dye field, the use of high-temperature dyeing and printing methods for the coloration of polyester has permitted some relaxation in the necessity for minimal molecular size. This has allowed the use of monoazo dyes such as dye 4.87, employing a typical azoic coupling component, particularly as disperse dyes in processes for the dyeing of polyester/cotton blends. A novel development in the dyeing of such blends has been the introduction of disperse dyes such as dye 4.88, in which the tertiary amine coupling component contains ester groupings. These can be hydrolysed in an alkaline clearing bath to facilitate removal of loose dye at the end of the dyeing process, thus minimising staining of the cellulosic portion of the blend.

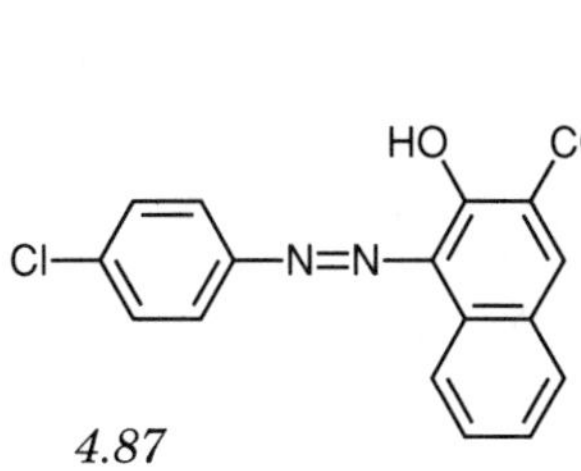

4.87

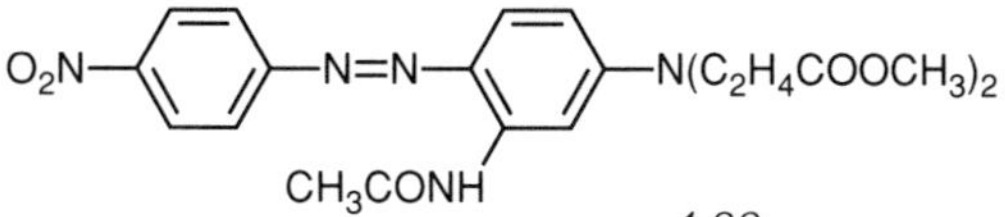

4.88

The introduction of high-temperature dyeing methods has also allowed the use of simple disazo structures, such as C.I. Disperse Orange 13 (4.89) and C.I. Disperse Orange 29 (4.90), as economic dyes giving chiefly yellow and orange hues.

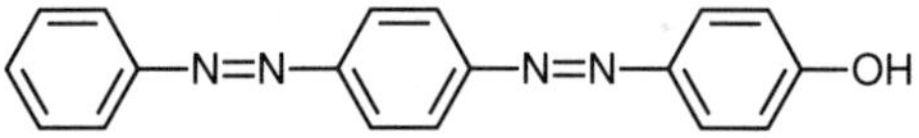

4.89

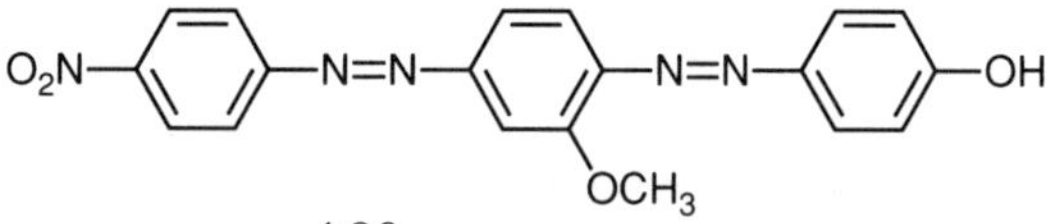

4.90

4.10 BASIC AZO DYES

This class of dyes [57] includes some old-established products such as chrysoidine (4.91; C.I. Basic Orange 2) that have been of importance in the past due to their bright, strong, economical colours. Today the main area of application of basic azo dyes is on acrylic fibres, and the more advanced dyes developed for this purpose differ from the older ones in that the positive charge is supplied by a quaternary ammonium group rather than merely by a protonated amine. In the dyeing of acrylic fibres problems arise because the levelling properties of the adsorbed dye are poor, since the electrostatic attraction of a basic dye to the acidic centres in the fibre provides extremely strong binding, resulting in excellent fastness properties.

The structures used fall into two main classes, which vary in the way in which the basic centre is built into the molecule. Either it is incorporated as a cyclammonium group that is part of the chromogen or a tertiary amino group is attached pendently, usually via a short alkyl chain. Both types often bear a marked resemblance to the structures found in disperse dyes, with similar intermediates being employed, except that in the case of the basic dye a nitrogen atom in one of the intermediates has been quaternised.

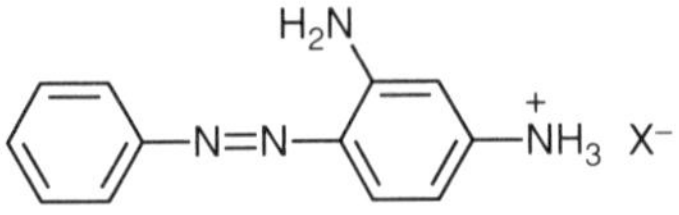

4.91

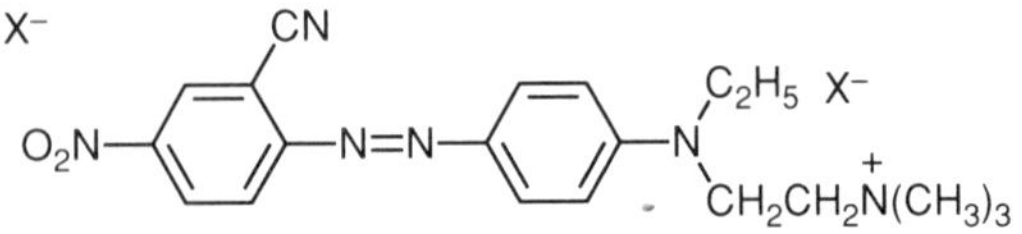

4.92

Thus the typical C.I. Basic Red 24 (4.92) has a structure reminiscent of a disperse dye, except that a quaternary ammonium group is carried on the pendent alkyl chain in the coupling component. This coupling component is prepared by reaction of *N*-ethylaniline with ethylene oxide followed by conversion of the resulting β-hydroxyethyl derivative to the β-chloroethyl compound, which is then treated with trimethylamine. By contrast, in C.I. Basic Orange 24 (4.93) the basic site is located in the diazo component.

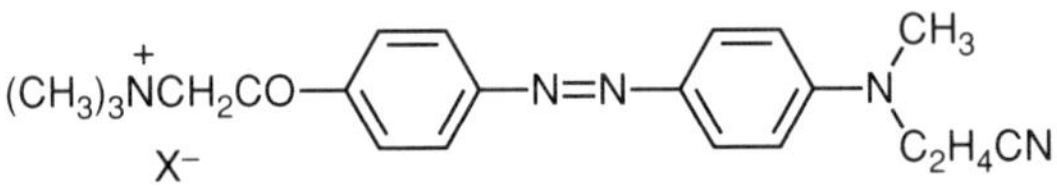

4.93

Basic dyes in which the positive charge is supplied by a quaternised heterocyclic ring that is part of the chromophoric system often have particularly bright, strong shades. The intermediates used here are again typical disperse dye components, quaternisation being carried out after coupling by treating the dye with an alkylating agent such as dimethyl sulphate. Using typical tertiary amine coupling components, bright blue shades, such as C.I. Basic Blue 41 (4.94), are readily obtained.

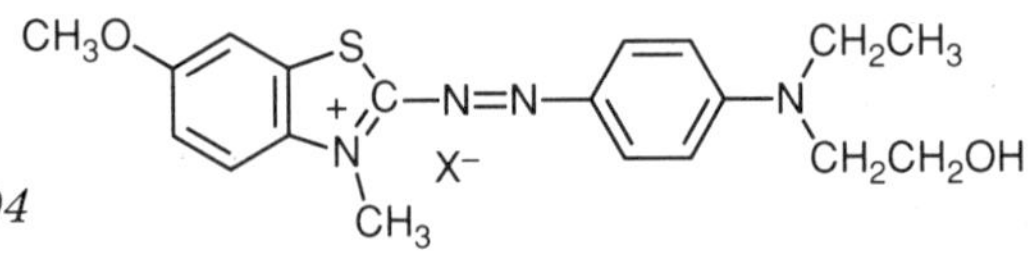

4.94

Considerable difficulties are often experienced in the preparation of basic dyes from quaternised heterocyclic diazo components. An alternative technique is to use an oxidative coupling procedure, in which a mixture of a hydrazone with a coupling component is treated with a mild oxidising agent, such as a hexacyanoferrate(III) [58]; an azo compound is then produced as shown in Scheme 4.31 [59]. Examples of dyes prepared by this technique are C.I. Basic Red 22 (4.95) and C.I. Basic Red 30 (4.96).

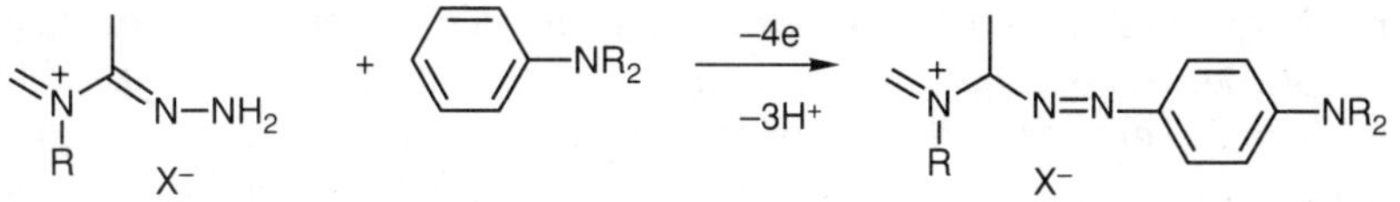

Scheme 4.31

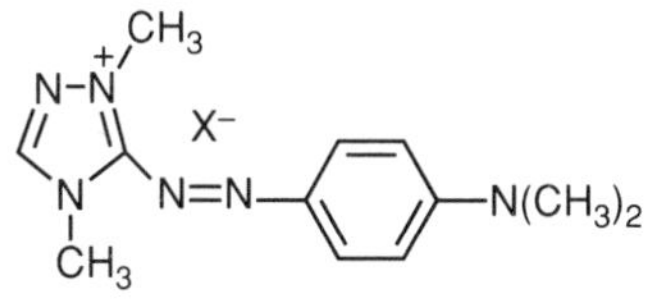

4.95

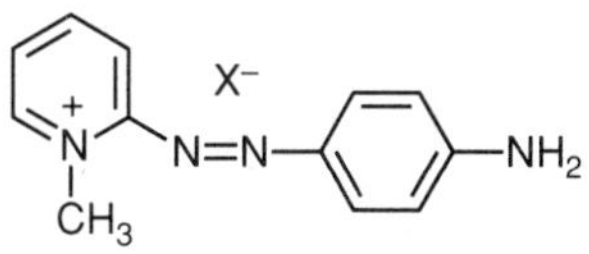

4.96

It is convenient to include in this class certain tautomeric structures that can exist either in the azo form or in the alternative hydrazone form. These are the diazatrimethinecyanines, which can be viewed as being derived from the trimethinecyanines (R–CH=CH–CH=R) by replacement of two of the CH units by nitrogen atoms (R–N=N–CH=R). These dyes are important in achieving yellow and red shades and they are often most conveniently prepared by an oxidative coupling procedure using coupling components peculiar to basic dyes. C.I. Basic Red 29 (4.97) and C.I. Basic Yellow 24 (4.98) are typical of this group.

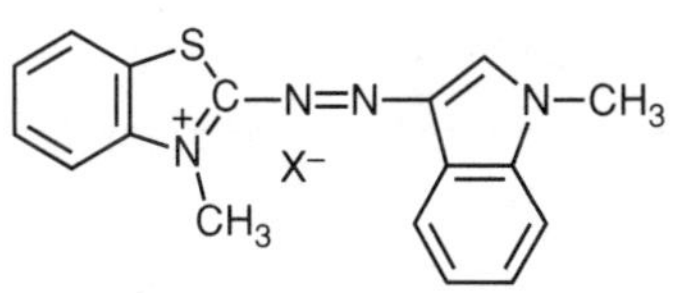

4.97

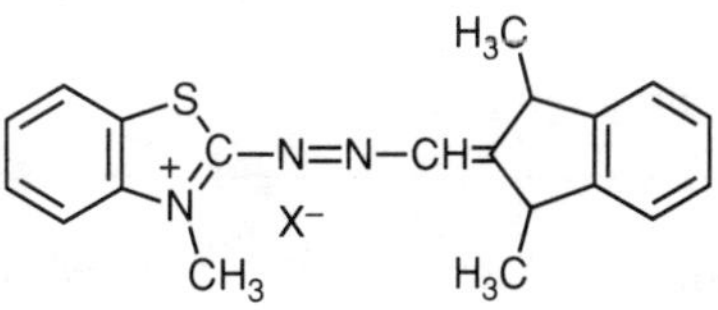

4.98

4.11 AZOIC DIAZO AND COUPLING COMPONENTS

In 1880 a process for the coloration of cotton by initial impregnation with an alkaline solution of 2-naphthol and subsequent treatment with a solution of a diazonium salt was patented by Thomas and Robert Holliday [60]. In this process a water-insoluble monoazo dye of good fastness to washing was formed within the fibre pores. The dyer's need to diazotise the arylamine component at low temperature led to these dyes becoming known as 'ice colours', and the product obtained using diazotised 4-nitroaniline became known as para red. Over the next thirty years minor improvements in the process were introduced, but a major step forward was made in 1912 when Griesheim–Elektron introduced a new type of coupling component, 3-hydroxy-2-naphthoic anilide (4.99), under the trade name Naphtol AS (HOE, C.I. Azoic Coupling Component 2). The guiding principle behind this advance was that the new intermediate was substantive towards cellulose and therefore was held in place on the fibre prior to and during the treatment with diazonium salt. Furthermore, the increased substantive character and diminished solubility of the resulting azo combination gave a marked improvement in the level of wash fastness obtained.

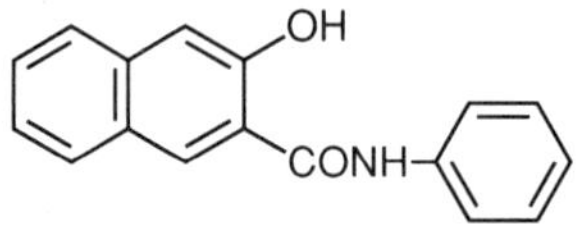

4.99

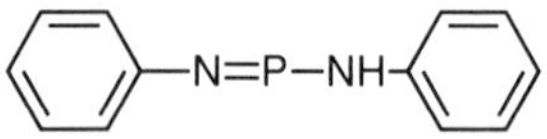

4.100

Naphtol AS is best prepared by heating together BON acid, aniline and phosphorus trichloride in an inert solvent, the initially formed intermediate (4.100) reacting further with the naphthoic acid to give the desired anilide. A range of similar products followed, with a variety of arylamines being used in place of aniline. A selection of these is given in Table 4.1.

TABLE 4.1

Azoic coupling components

Commercial name[a]	Chemical name
Naphtol AS-TR (HOE) (8)	3-hydroxy-2-naphthoic 4′-chloro-2′-methylanilide
Naphtol AS-OL (HOE) (20)	3-hydroxy-2-naphthoic 2′-methoxyanilide
Naphtol AS-D (HOE) (18)	3-hydroxy-2-naphthoic 2′-methylanilide
Naphtol AS-PH (HOE) (14)	3-hydroxy-2-naphthoic 2′-ethoxyanilide
Naphtol AS-ITR (HOE) (12)	3-hydroxy-2-naphthoic 5′-chloro-2′,4′-dimethoxyanilide

[a] C.I. Azoic Coupling Component number in parentheses

Ranges of diazotisable amines were supplied, designed to go with these coupling components and known as Fast Bases, despite being marketed as the hydrochlorides. Whilst hues obtainable at first were limited to the orange to red region of the spectrum, the later introduction of amines heavily substituted with electron-repelling groups allowed extension into the violet to blue region. Examples of Fast Bases are given in Table 4.2.

TABLE 4.2

Fast Bases

Commercial name[(a)]	Chemical name
Fast Orange GC Base (HOE) (2)	3-chloroaniline hydrochloride
Fast Orange RD Base (HOE) (49)	2-chloro-5-trifluoromethylaniline hydrochloride
Fast Scarlet GG Base (HOE) (3)	2,5-dichloroaniline hydrochloride
Fast Red TR Base (HOE) (11)	4-chloro-2-methylaniline hydrochloride
Fast Red KB Base (HOE) (32)	5-chloro-2-methylaniline hydrochloride
Fast Red RC Base (HOE) (10)	5-chloro-2-methoxyaniline hydrochloride
Fast Red ITR Base (HOE) (42)	2-methoxyaniline-5-sulphondiethylamide hydrochloride
Fast Red RL Base (HOE) (34)	2-methyl-4-nitroaniline hydrochloride
Fast Violet B Base (HOE) (41)	4-benzoylamino-2-methoxy-5-methylaniline hydrochloride
Fast Blue BB Base (HOE) (20)	4-benzoylamino-2,5-diethoxyaniline hydrochloride

(a) C.I. Azoic Diazo Component number in parentheses

Extension into other hues was not possible using anilides of BON acid, and to meet this requirement coupling components of differing structure but similar levels of substantivity were introduced (4.101 to 4.104; the arrows indicate coupling positions). Yellow dyeings were obtained using the bis-acetoacetarylamide derived from 3,3′-dimethylbenzidine, marketed as Naphtol AS-G (4.101; HOE, C.I. Azoic Coupling Component 5). Browns were supplied using the 2′-methylanilide, Naphtol AS-BR (4.102, X = CH_3, Y = H; HOE, C.I. Azoic Coupling Component 3), and the 4′-chloroanilide, Naphtol AS-LB (4.102, X = H, Y = Cl; HOE, C.I. Azoic Coupling Component 15), of 2-hydroxycarbazole-1-carboxylic acid. Blacks were obtained from the 4″-methoxy-2″-methylanilide of 3′-hydroxy-1,2-benzocarbazole-3′-carboxylic acid, Naphtol AS-SR (4.103; HOE, C.I. Azoic Coupling Component 25). Duller greens were obtained from the 2′-methylanilide of 3-hydroxyanthracene-2-carboxylic acid, Naphtol AS-GR (4.104;

HOE, C.I. Azoic Coupling Component 36), whilst bright greens were eventually provided by Naphtol AS-FGGR (HOE, C.I. Azoic Coupling Component 108), the reaction product of a metal phthalocyanine with an aminoarylpyrazolone whose exact constitution has not been disclosed [61,62].

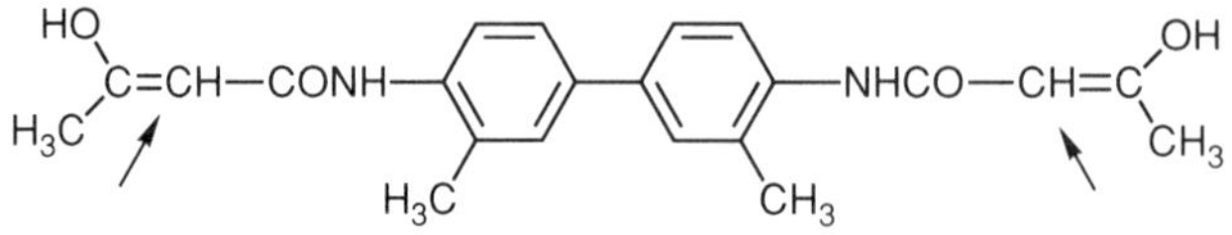

4.101

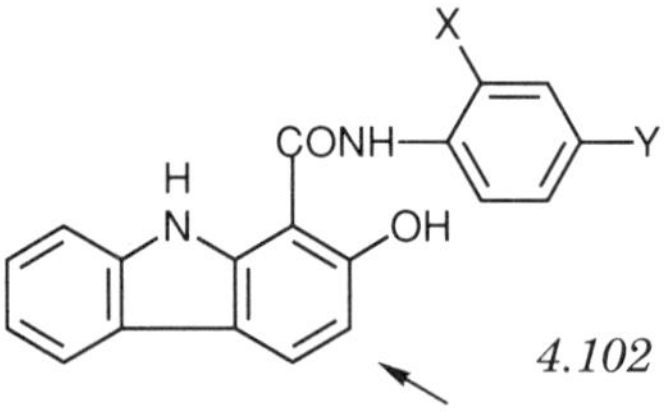

4.102

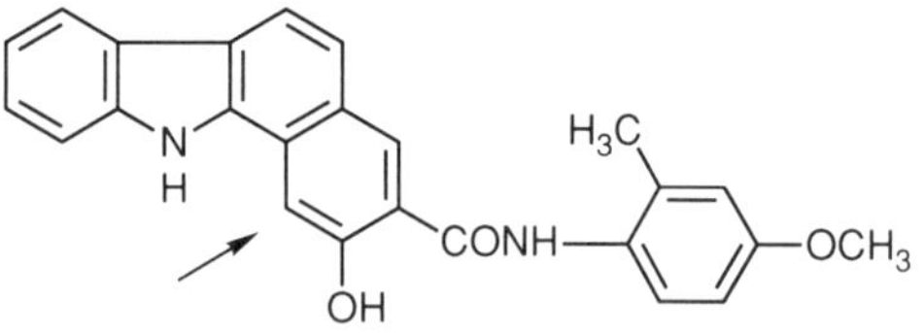

4.103

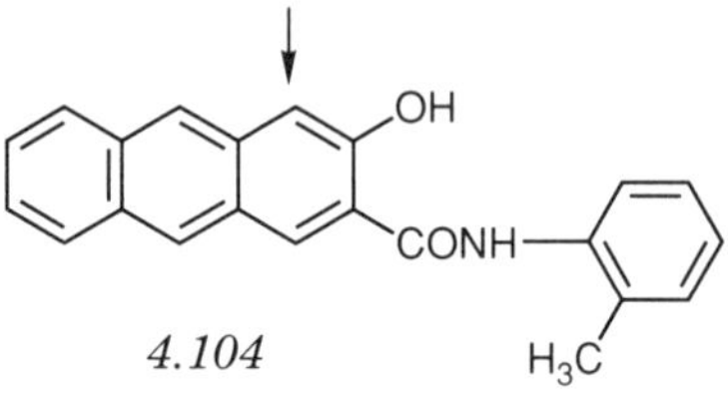

4.104

4.12 STABILISED DIAZONIUM SALTS AND AZOIC COMPOSITIONS

4.12.1 Fast Salts

As the range of components available for use in the azoic dyeing process expanded, research was simultaneously targeted on improvements designed to make the process more attractive to the commercial dyer. The necessity for the dyer to diazotise the Fast Base was removed with the introduction of stabilised diazonium salts [63], known as Fast Salts. Stabilisation was achieved by a judicious selection of the counterion to the diazonium ion; various anions have found use in commercial Fast Salts and a selection is listed in Table 4.3. Particularly effective is the diazonium tetrachlorozincate, which can be readily prepared by adding excess zinc chloride solution to a solution of the diazonium salt. The precipitated complex diazonium salt is usually admixed with an inert diluent, which enhances its stability, and in use the dyer only needs to dissolve the powder in water to prepare the necessary diazonium salt solution.

TABLE 4.3

Stabilised diazonium salts

Commercial name[a]	Parent base
Tetrachlorozincates	
Fast Black K Salt (38)	4-amino-2,5-dimethoxy-4′-nitroazobenzene
Fast Blue B Salt (48)	3,3′-dimethoxybenzidine (dianisidine)
Fast Blue BB Salt (20)	4-benzoylamino-2,5-diethoxyaniline
Fast Bordeaux GP Salt (1)	4-methoxy-2-nitroaniline
Fast Orange GR Salt (6)	2-nitroaniline
Fast Red AL Salt (36)	1-aminoanthraquinone
Fast Red 3GL Salt (9)	4-chloro-2-nitroaniline
Fast Red ITR Salt (42)	2-methoxyaniline-5-sulphondiethylamide
Fast Scarlet GG Salt (3)	2,5-dichloroaniline
Fast Scarlet R Salt (13)	2-methoxy-5-nitroaniline
Chloride	
Variamine Blue B Salt (35)	4-amino-4′-methoxydiphenylamine
Naphthalene-1,5-disulphonates	
Fast Red B Salt (5)	2-methoxy-4-nitroaniline
Fast Red TR Salt (11)	4-chloro-2-methylaniline
Tetrafluoroborates	
Fast Orange GC Salt (2)	3-chloroaniline
Fast Red GG Salt (37)	4-nitroaniline

[a] C.I. Azoic Diazo Component number in parentheses

4.12.2 Rapid Fast colours

In addition to these 'active' stabilised diazonium salts that give a diazonium salt solution immediately on dissolving in water, there are certain derivatives of diazonium compounds from which the diazonium salt can be readily regenerated. One group of such derivatives is the class of compounds known as *anti*-diazotates [64]. Diazotates are formed when the pH of a diazonium salt solution is raised to between 9 and 12 with an alkali metal hydroxide. The *syn*-diazotates initially formed are somewhat unstable and isomerise into the stable *anti*-diazotate form. The evidence overwhelmingly points to the two forms being stereoisomers (4.105 and 4.106 respectively).

Stable mixtures of these *anti*-diazotates and Naphtols were marketed by IG Farbenindustrie as Rapid Fast colours for printing on to fabric with

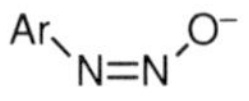

4.105

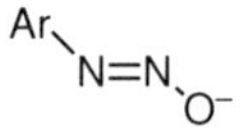

4.106

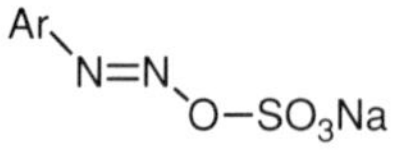

4.107

development of the azoic dye by steaming. The *anti*-diazosulphonates (4.107) [65], which are prepared by treatment of a diazonium salt with sodium sulphite and which regenerate the diazonium ion on treatment with an oxidising agent, found similar use. Both groups are now of only historical interest.

4.12.3 Rapidogens and Neutrogenes

Superior 'passive' stabilised diazo compounds are afforded by the diazoamino compounds (triazenes) that arise by reaction of diazonium salts with a variety of tertiary amines [66]. Typically sarcosine (CH_3NHCH_2COOH) (which gives products such as compound 4.108), *N*-methyltaurine ($CH_3NHCH_2CH_2SO_3H$) and *N*-methylaniline-4-sulphonic acid have been used for this purpose. Again these were marketed admixed with a coupler for application to cotton and colour development by acid steaming, under the trade name Rapidogen by IG Farbenindustrie in the 1930s [67]. A disadvantage was the corrosive effects of the acid (usually acetic or formic) on equipment, and this was eventually overcome by CFMC with the Neutrogene range [68], which used an *N*-substituted anthranilic acid (such as compound 4.109) as the tertiary amine required for triazene formation.

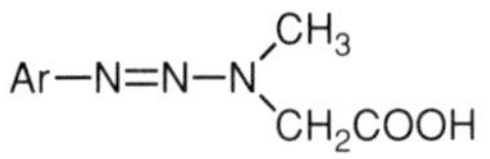

4.108

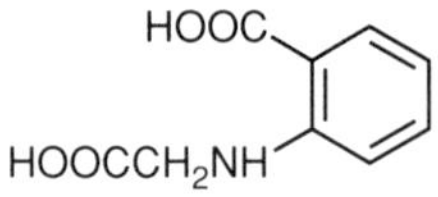

4.109

These products, which have been copied by other manufacturers under their own trade names, are designed specifically for printing applications. Azoic dyeing applications were catered for in 1967 when Hoechst introduced the Azanil salts [69–71], a limited range of mixtures of substantive triazenes and Naphtols that can be adsorbed on the fabric and the colour developed by acidification of the dyebath.

4.13 AZO PIGMENTS PRODUCED BY FINAL COUPLING

Insoluble azo compounds produced by a final coupling stage are in the main made from conventional azoic dye intermediates and find wide use as general-purpose pigments where the highest fastness is not essential. For yellow pigments acetoacetarylamide coupling components are the most commonly employed, but the monoazo combinations derived from these, such as C.I. Pigment Yellow 1 (4.110) and C.I. Pigment Yellow 98 (4.111), are generally of poor fastness to heat and solvents. Superior properties are shown by disazo pigments of the $E_1 \leftarrow D \rightarrow E_2$ type, which are often based on the use of 3,3′-dichlorobenzidine as tetrazo component. A typical disazo pigment is C.I. Pigment Yellow 171 (4.112).

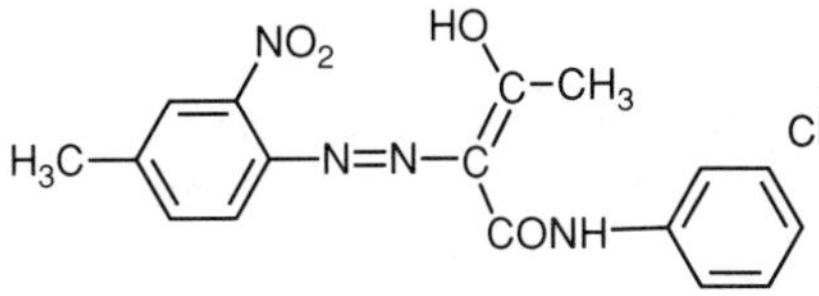

4.110

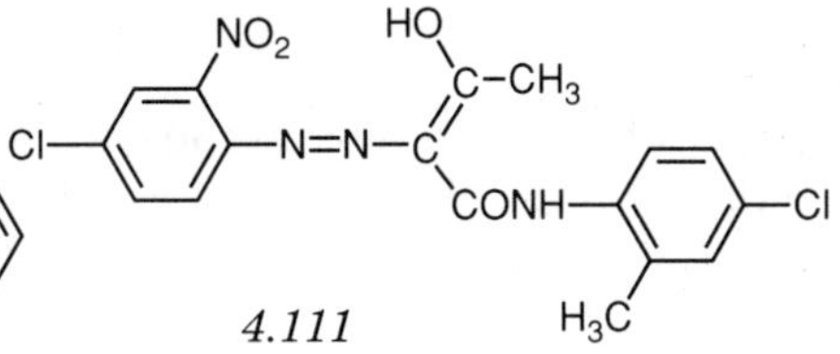

4.111

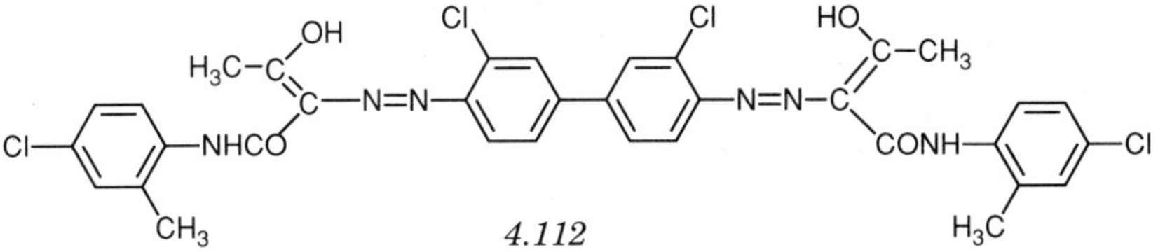

4.112

For orange colours simple pyrazolones are the most commonly used coupling components as, for instance, in C.I. Pigment Orange 6 (2-nitro-4-methylaniline→1-phenyl-3-methyl-5-pyrazolone). 2-Naphthol is also useful, as in C.I. Pigment Orange 5 (2,4-dinitroaniline→2-naphthol). As in the yellow series, superior disazo pigments can be prepared using 3,3′-dichlorobenzidine as tetrazo component with 1-phenyl-3-methyl-5-pyrazolone as coupler.

In the red to violet area Naphtol AS and its derivatives are widely used. Simple monoazo combinations such as 2,5-dichloroaniline and Naphtol AS serve where high heat and solvent fastness are not essential, but for better properties polar features, such as amide linkages, are incorporated. C.I. Pigment Red 245 (4.113), C.I. Pigment Red 188 (4.114) and C.I. Pigment Violet 50 (4.115) are typical of this group.

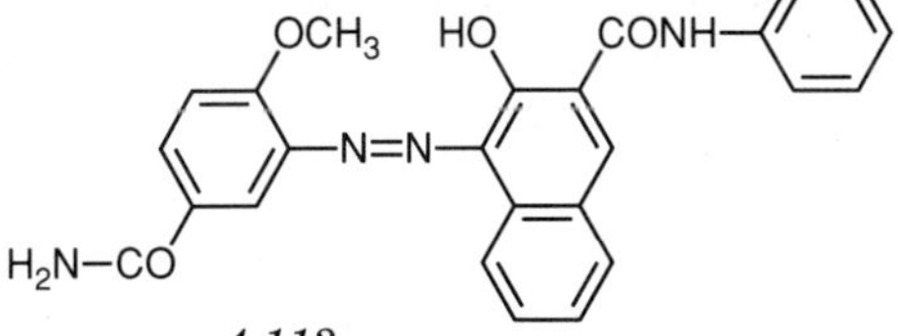

4.113

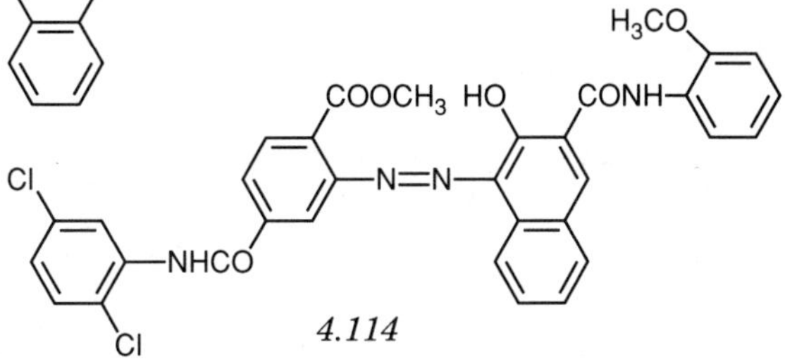

4.114

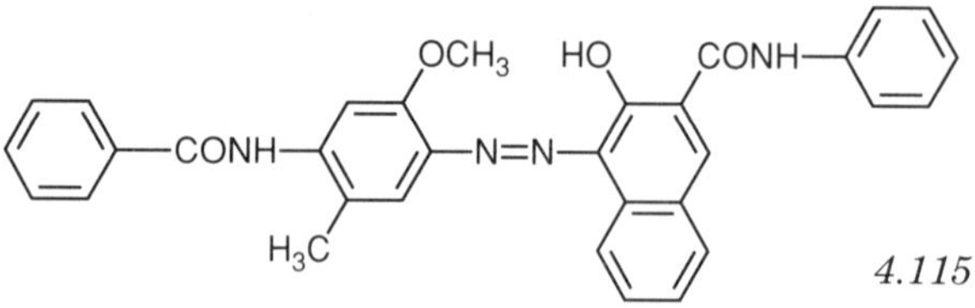

4.115

In the blue region conventional azoic combinations are used as pigments and the adoption of dianisidine as tetrazo component allows the disazo type to be produced, such as C.I. Pigment Blue 26 (4.116).

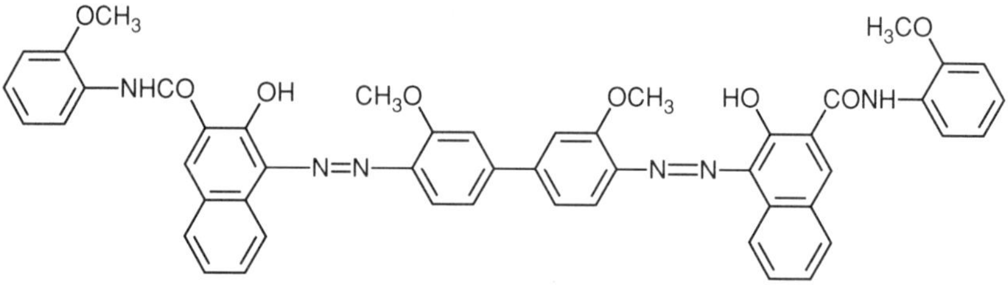

4.116

The objective in preparing a pigment is to produce a highly insoluble organic compound and in order to approach this ideal it is necessary to resort to intermediates that themselves have only very limited solubility. These inevitable difficulties in preparation limit the fastness achieved from a final coupling process and have led to the development of alternative procedures for the production of high-grade azo pigments; these alternatives are discussed elsewhere (see section 2.3).

The wholly organic azo compounds discussed above have the soft texture and high tinctorial strength that are desired pigmentary properties. Somewhat inferior in these properties are the 'lakes' of sulphonated azo compounds, which are also widely used as pigments. These 'lakes' are insoluble salts, usually calcium or barium salts of sulphonated arylamines linked to cheap coupling components. They rely on the polar character of the ionic bond to provide their low solubility. Examples are C.I. Pigment Red 57 (the calcium lake of compound 4.117) and C.I. Pigment Red 53 (the barium lake of compound 4.118).

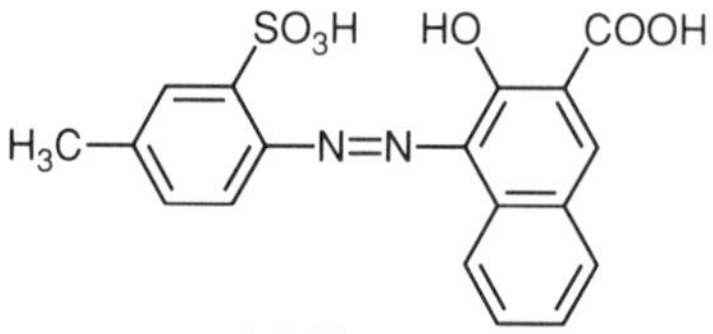

4.117

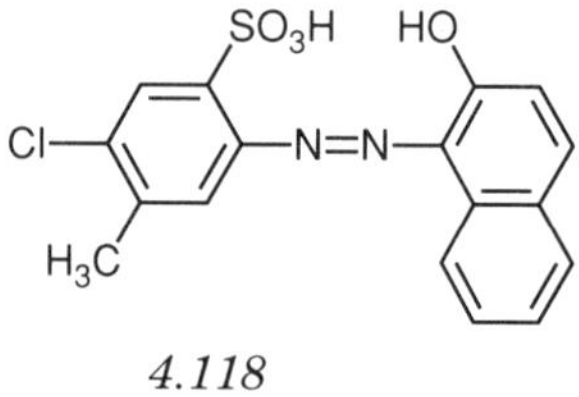

4.118

4.14 IMPLICATIONS OF NEW TECHNOLOGY IN DIAZOTISATION AND COUPLING

The manufacture of azo dyestuffs has been for many years a batch operation, individual batches of dye being made in multipurpose plant, isolated by filtration techniques, stove-dried and standardised. The traditional processes [72] are now steadily being replaced [73,74] by more efficient production methods that offer greater cost-effectiveness and improved working conditions. Although most azo dye production is still carried out by batch operation in vastly improved plants, more dedicated continuous units for the manufacture of high-volume products are being introduced. In these plants the diazotisation and coupling steps are carried out continuously by computer-controlled metering of the intermediates and reactants into a tube, flow rates being chosen to allow the reaction time to be less than the dwell time of the reaction mixture in the tube. The product is still usually isolated by filtration but, even here, automatic discharging has removed much of the heavy physical effort and dirty conditions previously required. Spray drying of pre-standardised pastes with careful attention to particle form provides much improved products for the user. The supply of dyes as concentrated solutions allows the automatic metering-in of dyestuff by the user, making automation easier and allowing greater control over dyeing conditions. These improvements in manufacturing technique, together with the availability of a much wider variety of analytical methods, serve to ensure that the user is supplied with a high-quality product at an acceptable price.

REFERENCES

1. P Griess, *Ann.*, **106** (1858) 123; **113** (1860) 201; *Compt. Rend.*, **49** (1859) 77.
2. P Griess, *Ann.*, **137** (1866) 85; *Ber.*, **9** (1876) 627.
3. J H Ridd, *J.S.D.C.*, **75** (1959) 285.
4. J H Tedder, *The chemistry of synthetic dyes*, Vol. 3, Ed. K Venkataraman (New York: Academic Press, 1970).
5. W Bradley, *J.S.D.C.*, **75** (1959) 289.
6. IG, BP 248 230.
7. V Chmátal and Z J Allan, *Coll. Czech. Chem. Comm.*, **30** (1965) 1205.
8. O Stein, *Ber.*, **27** (1894) 2806.
9. E Noelting and E Kopp, *Ber.*, **38** (1905) 3506.
10. C Schwalbe, *Z. Farb. Text. Ind.*, **4** (1905) 433.
11. IG, BP 426 403.
12. N V Philips, BP 659 897.
13. ICI, BP 429 936.
14. E Knoevenagel, *Ber.*, **23** (1890) 2994; **28** (1895) 2048.
15. O N Witt, *Ber.*, **42** (1909) 2953.
16. T S Gore and K Venkataraman, *Proc. Indian Acad. Sci.*, **34A** (1951) 368.
17. H H Hodgson and D E Nicholson, *J.C.S.*, (1943) 379.
18. W R Orndorff and B J Ray, *Ber.*, **40** (1907) 3211.
19. F Muhlert, *Z. Angew. Chem.*, **21** (1908) 2611.
20. IG, BP 455 229.
21. H V Kidd, *J. Org. Chem.*, **2** (1937) 206.
22. IG, BP 465 484.

23. W Schwarin and O Kaljanov, *Ber.*, **41** (1908) 2056.
24. F Muzik and Z J Allan, *Chem. Listy*, **49** (1955) 212.
25. E R Ward, D B Pearson and P R Wells, *J.S.D.C.*, **75** (1959) 484.
26. C Krohn, *Ber.*, **21** (1888) 3241.
27. V Chmátal and Z J Allan, *Coll. Czech. Chem. Comm.*, **27** (1962) 1835.
28. G S Hartley, *Nature*, **140** (1937) 281.
29. G S Hartley, *J.C.S.*, (1938) 633.
30. G Zimmerman, L-Y Chow and U-J Paik, *J. Amer. Chem. Soc.*, **80** (1958) 3528.
31. D Schulte-Frohlinde, *Ann.*, **612** (1957) 131.
32. D J W Bullock, C W N Cumper and A I Vogel, *J.C.S.*, (1965) 5316; D L Beveridge and H H Jaffé, *J. Amer. Chem. Soc.*, **88** (1966) 1948.
33. Y Yagi, *Bull. Soc. Chim. Japan*, **36** (1963) 487.
34. R Jones, A J Ryan, S Sternhill and S E Wright, *Tetrahedron*, **19** (1963) 1497.
35. S Toda, *Nippon Kagaku Zasshi*, **80** (1959) 402.
36. E Fischer and Y F Frei, *J.C.S.*, (1959) 3159.
37. P F Gordon and P Gregory, *Organic chemistry in colour* (Berlin: Springer Verlag, 1983) 100; A Burawoy, A G Salem and A R Thompson, *J.C.S.*, (1952) 4793.
38. K J Morgan, *J.C.S.*, (1961) 2151.
39. D Hadzi, *J.C.S.*, (1956) 2143.
40. H E Fierz-David, L Blangey and E Merian, *Helv. Chim. Acta*, **34** (1951) 34, 846.
41. N Donaldson, *The chemistry and technology of naphthalene compounds* (London: E Arnold, 1958).
42. N N Vorozhtsov, *The chemistry of synthetic dyes*, Vol. 3, Ed. K Venkataraman (New York: Academic Press, 1970) 93.
43. N N Vorozhtsov, *The chemistry of synthetic dyes*, Vol. 3, Ed. K Venkataraman (New York: Academic Press, 1970) 110.
44. N N Vorozhtsov, *The chemistry of synthetic dyes*, Vol. 3, Ed. K Venkataraman (New York: Academic Press, 1970) 146.
45. H T Bucherer, *J. Prakt. Chim.* (2), **69** (1904) 49.
46. N L Drake, *Organic Reactions*, **1** (1942) 105.
47. H Seeboth, *Angew. Chem.* Internat. Edn, **6** (1967) 307.
48. ICI, BP 443 835.
49. D L Ross and E Reissner, *J. Org. Chem.*, **31** (1966) 2571.
50. Chi-Kang Dien, *The chemistry of synthetic dyes*, Vol. 8, Ed. K Venkataraman (New York: Academic Press, 1978) 81.
51. A G Green and K H Saunders, *J.S.D.C.*, **39** (1923) 10.
52. A G Green and K H Saunders, *J.S.D.C.*, **40** (1924) 138.
53. J F Dawson, *Rev. Prog. Coloration*, **9** (1978) 26; **14** (1984) 90.
54. J M Straley, *The chemistry of synthetic dyes*, Vol. 3, Ed. K Venkataraman (New York: Academic Press, 1970) 423.
55. *Mod. Textile Mag.*, **50** (1969) 52.
56. ICI, BP 1 351 382.
57. U Mayer and E Siepmenn, *Rev. Prog. Coloration*, **4** (1973) 65.
58. S Hünig, *Angew. Chem.* Internat. Edn, **7** (1968) 335.
59. H Zollinger, *Azo and diazo chemistry* (New York: John Wiley, 1961) 243.
60. T and R Holliday, BP 2757.
61. F Gund, *J.S.D.C.*, **76** (1970) 151.
62. BAY, BP 811 221.
63. R L M Allen and K H Saunders, *Aromatic diazo compounds* (London: E Arnold, 1985) 138.
64. R L M Allen and K H Saunders, *Aromatic diazo compounds* (London: E Arnold, 1985) 214.
65. R L M Allen and K H Saunders, *Aromatic diazo compounds* (London: E Arnold, 1985) 248.
66. R L M Allen and K H Saunders, *Aromatic diazo compounds* (London: E Arnold, 1985) 368.

67. R L M Allen and K H Saunders, *Aromatic diazo compounds* (London: E Arnold, 1985) 409.
68. B Jomain, *J.S.D.C.*, **69** (1953) 661.
69. HOE, BP 1 077 777.
70. HOE, BP 1 088 694.
71. HOE, BP 1 104 364.
72. H E Fierz-David and L Blangey, *Fundamental processes of dye chemistry* (New York: Interscience, 1949).
73. G Booth, *Rev. Prog. Coloration*, **8** (1977) 1; **14** (1984) 114.
74. G Booth, *The manufacture of organic colorants and intermediates* (Bradford: SDC, 1988).

CHAPTER 5

The chemistry and properties of metal-complex dyes

Francis Jones

5.1 INTRODUCTION

Transition-metal complexes of suitably substituted organic compounds are of increasing importance both academically and industrially. Although their use is well known in catalysis, biological systems, metal recovery and sequestration, this chapter is primarily concerned with complexes formed by the interaction of organic chromophoric systems with transition metals to give dyes for wool, silk, nylon and cellulosic materials. They also find use as high-quality solvent dyes, as lakes in pigment and printing ink technology, as charge-control agents in photocopying, as dye developers in instant colour photography and as reagents in colorimetric analytical chemistry.

It is not proposed to deal with these non-textile uses. Admirable reviews of metal-complex dyes, referring to many patents, are available [1–3]. The fundamentals of coordination chemistry from a student viewpoint have not been emphasised in these reviews; however, a basic understanding is necessary to complete the picture of present-day thinking in the technology of metal complexes, with particular reference to dyes.

The concept of a metal complex originated from the work of Alfred Werner at the University of Zürich in 1891. At that time the first azo dye capable of forming a metal complex, C.I. Mordant Orange 1 (5.1), had already been synthesised (although, of course, naturally occurring mordant dyes applied to premetallised fibres had been known long before this). The emphasis during the next twenty years was on the synthesis of other azo dyes suitable for mordant application to wool.

By 1908 Werner's coordination theories were fully developed and were being applied to these mordant dyes [4]. He had concluded that a metal ion was characterised by two types of valency, which he called primary and

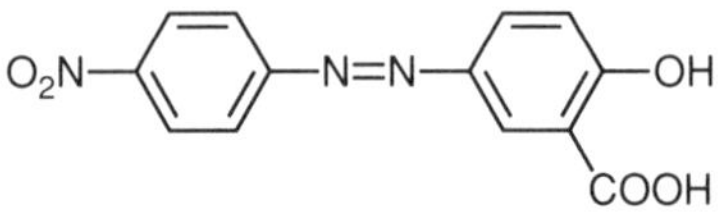

5.1

secondary. The primary valency satisfied the chemical equivalence (that is, the number of positive charges on the metal cation) and the secondary valency partook in coordination. We now refer to the primary valency as the oxidation state (number) of the central metal atom and the secondary valency as the number of associated ligand atoms required to complete the coordination number. It must be remembered that the concept of directional valency, such as the tetrahedral bond angles about the carbon atom in methane, had not then been developed, and knowledge of the mode of attachment of the ligand atoms was rather vague.

The first metal-complex dyes to be prepared in substance, rather than within the fibre, were discovered by René Bohn (BASF) in 1912. These were sulphonated 1:1 chromium–dye complexes [5] and the first patent for applying them directly to wool appeared three years later [6]. They can be regarded as the forerunners to the Palatine Fast (BASF) and Neolan (CGY) dyes that came into general use during the following twenty years. In 1949 the first water-soluble 1:2 metal–dye complex, which contained no ionisable sulphonic acid groups, was described by Guido Schetty [7] for application to wool under neutral conditions. This dye was C.I. Acid Black 58; more recent developments have included unsymmetrical 1:2 metal-complex dyes containing from one to three sulphonate groups and complexes based on formazans (see section 5.4.2).

5.2 BASIC CONCEPTS

Metal-complex formation involves the interaction of ligand systems with metal cations; this brings about changes in the characteristics of both.

5.2.1 Ligand systems

A ligand system is any charged or uncharged single entity or atomic component of a substituent group containing an excess electron or electron pair. The nature of the bond between the ligand atom and the metal is dependent upon the ligand atom acting as an electron-pair donor and the metal as an electron-pair acceptor. The atomic size and the electron distribution in the outermost shells of the transition elements of the first series are most favourable or conducive to the formation of metal complexes. The donation of an electron pair from ligand atom to metal gives rise, in molecular orbital terminology, to a sigma (σ) or covalent bond. It is sometimes convenient to distinguish between conventional covalent bonds arising from the sharing of electrons, denoted by a continuous line, and coordinated covalent bonds arising by electron donation from a ligand system, denoted by an arrow, although there is no real difference between them once the bonds are formed.

In organic metal-complex dye synthesis the ligand atoms may form part of a basic or acidic functional group. A basic group contains an atom carrying a lone pair of electrons, which may interact with the metal ion (or indeed with a proton). Examples include $-NH_2$ (amino), =NH (imino), –N= (nitrogen in azo or azomethine), =O (oxygen in carbonyl) and –S– (sulphur in thioether). Acidic functional groups lose a proton during metallisation to give a formal negative charge such as $-O^-$ (from carboxylic, phenolic or enolic hydroxy groups), $-S^-$ (from thiophenolic or thioenolic mercapto groups) or $-N^-H$ (from amino groups).

In the case of a ligand group such as ammonia or water, only one pair of electrons is utilised in forming a σ-bond with the metal. This type of ligand group is said to be mono- or uni-dentate (literally 'one-toothed'). Monodentate ligand groups are to be found in many metal-complex dyes, satisfying sites not sterically occupied by coordinated dye molecules. Dye molecules used as intermediates in making metal complexes have at least two ligand donor groups arranged so that it is geometrically possible to coordinate at two or more positions in the coordination shell of a single metal ion. Examples are to be found throughout this chapter in which from two to six ligand groups may be bound to the one metal ion. Where two groups originating in the same molecule combine with the single metal ion, a heteroatomic ring is formed.

Scheme 5.1 shows such ring formation where X and Y are donor groups and M is a metal. This process was originally postulated by Morgan and Drew [8] who termed it 'chelation' from the Greek *chela* meaning 'claw' (for example, of a lobster or a crab). Chelation profoundly affects the chemical and physical properties of both metal ion and ligand groups. Although the formation of the single heterocyclic ring results in an increase in the stability of the complex, it is not sufficiently stable to be directly useful for dyeing. Scheme 5.1 is a reversible reaction, which implies that demetallisation can take place, to an extent that depends on the severity of the conditions. The metal chelate may be thermodynamically stable but kinetically labile. The chromophoric ligand groups may become replaced by others, such as water. This applies to certain iron complexes based on nitrosonaphthols, such as C.I. Pigment Green 12 (5.2), a green colorant formerly used on wool, which may be regarded as one of the forerunners of the Palatine Fast and Neolan dye ranges. On the other hand, chromium and cobalt form complexes containing two condensed heteroatomic ring systems (see section 5.4.2). These are much more stable, both thermodynamically and kinetically, and undergo ligand exchange reactions extremely slowly.

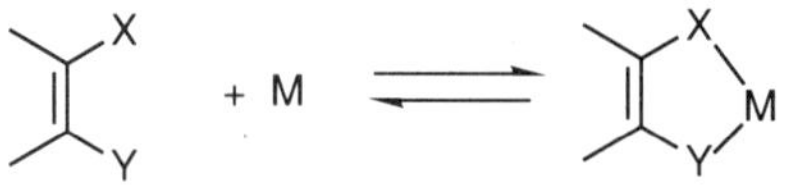

Scheme 5.1

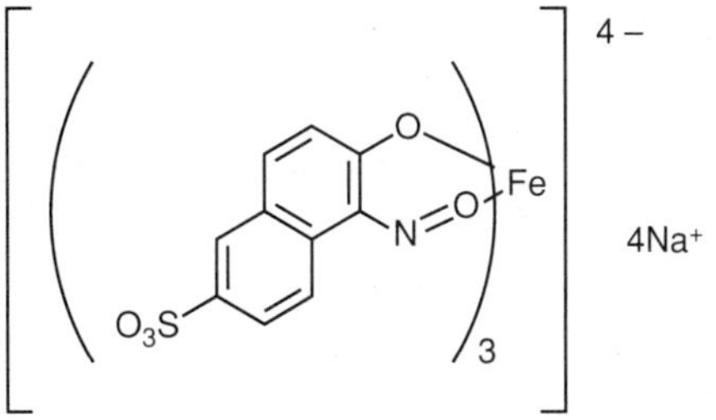

5.2

5.2.2 Coordination number

The number of ligand systems associated with a metal atom in a complex is known as the coordination numberof the metal, often abbreviated to CN. For many metals in their lower oxidation states (1+, 2+, 3+) the CN is usually 4 or 6. The metals used in dye complexes are of this type. In other less common complexes the number can range from 2 to 10. The CN depends on the size and geometry of the ligand system and on the size of the metal ion: the larger the central cation, the greater the number of ligand groups that can be accommodated around it. However, the charge on the cation also influences the CN. Metals in a high valency state, such as 6+ or 7+, generally have low CN values since the removal of six or seven electrons reduces the ionic size and accommodation of so many ligand groups becomes sterically impossible.

5.3 ELECTRONIC STRUCTURE OF TRANSITION-METAL IONS

The formation of metal-complex dyes, which must have a high degree of stability during their preparation or application, is limited to some elements of the first transition series, notably chromium, cobalt and nickel: except for iron, used in nitrosonaphthol pigments, and divalent copper in metallised direct dyes, the other members of the transition series form relatively unstable complexes. The following descriptions of the influence of electronic structure are applicable, however, to all members of the series.

The principal characteristic of the transition elements is that of an incomplete electronic subshell which confers specific properties on the metal concerned. Ligand entities may partake in coordination not only by electron donation to the 3d levels in the first transition series but also by donation to incomplete outer 4s and 4p shells.

Figure 5.1 shows that the difference in orbital energy levels between the 4s, 4p and 3d orbitals is much smaller than, for example, the difference between inner 2s and 2p levels. Consequently transitions between the 4s, 4p and 3d levels are easily effected and coordination is readily achieved. The manner in which ligand groups surround the metallic central ion and their orientation are both determined by the number and energy levels of the electrons in the incomplete subshells.

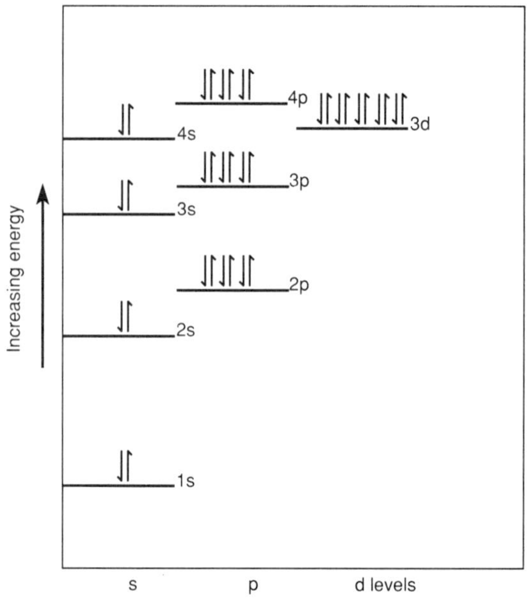

Figure 5.1 – Relative energy levels up to atomic number 36

The first transition series of elements (scandium, titanium, vanadium, chromium, manganese, iron, cobalt, nickel and also copper(II)) is built up by the addition of electrons to the outer 3d shell, which can accommodate a total of ten electrons in five subshells of equal energy, each capable of containing two electrons. The influence of the electrical fields originating from the ligand atoms as they approach the electrons of these outermost 3d, 4s and 4p shells is described by ligand field theory. On approach, hybridisation between available ligand electrons and metal electrons takes place with bond formation. It is this and the strength of the ligand field that determines the final disposition of the ligand entities around the central ion.

For elements of CN 6, such as trivalent chromium or cobalt, the most stable spatial arrangement of six identical ligand groups is that of an octahedron with four of the groups situated in equatorial positions and the other two respectively above and below the central metal atom (Figure 5.2(a)).

For elements of CN 4 there are two possibilities: a square planar distribution on the equatorial plane, as in Figure 5.2(b), or the tetrahedral configuration shown in Figure 5.2(c). It is possible to obtain a distorted octahedral (i.e. tetragonal) configuration by an increase in the distance of the ligand groups in the pole positions of the octahedron. In the extreme, when the two pole ligand groups are completely removed, this leads to a

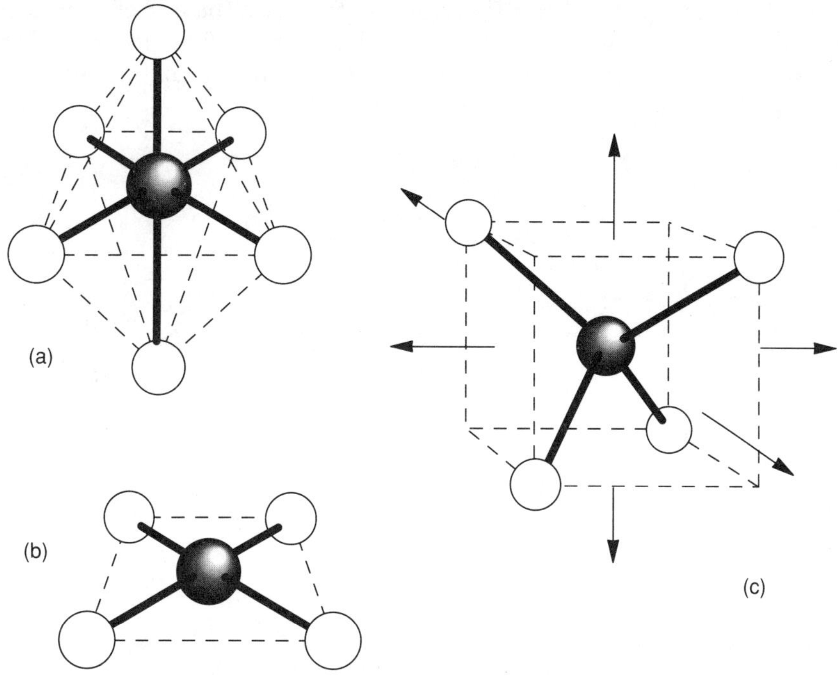

Figure 5.2 – Ligand arrangements about a central nucleus

square planar arrangement. In early objections to the electrostatic theory of bonding in metal complexes it was argued that the symmetrical arrangement of four entities about a spherical central ion was best explained in terms of a tetrahedral structure. It was not appreciated that the d shell electrons of the central atom are, in fact, non-symmetrical and stable square planar arrangements are possible.

As described above, in each of the transition metals there are five d orbitals, each of which can accommodate two electrons. These orbitals are arranged spatially in two groups, one acting between the main x-, y- and z-axes of the metal ion and the other acting along these axes.

5.3.1 Octahedral arrangement

In an octahedral weak ligand field environment, the influence of the six polar or ionic ligand atoms is to cause the ligand electrons to avoid the regions where the charge density associated with the d electrons is greatest. This charge density distribution is shown in Figure 5.3. The line of closest approach is therefore along a line of minimum charge repulsion, that is, along the main x-, y- and z-axes for the d_{xy}, d_{xz} and d_{yz} orbitals. The $d_{x^2-y^2}$ and d_{z^2} orbitals have charge densities that are at a maximum along these main axes. They therefore conflict with the line of approach of the ligand atoms and are consequently destabilised. The five orbitals are thus split into two groups of different energy: an upper doublet referred to as e_g and a lower triplet level known as t_{2g}.

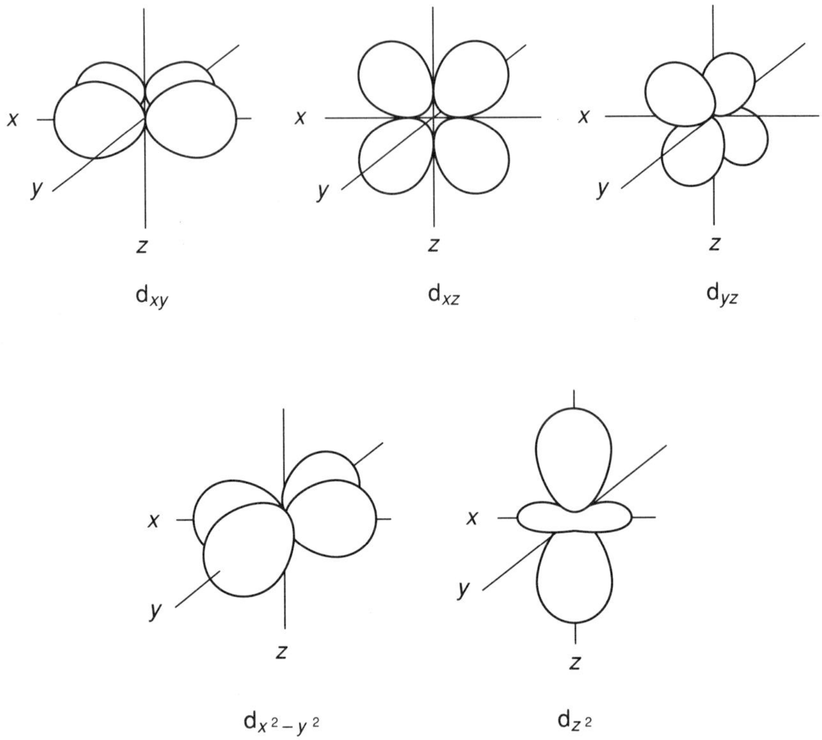

Figure 5.3 – Atomic d orbitals in transition elements

The manner in which these d electrons are distributed between these orbitals is determined by the difference in energy levels (Δ) between them. This energy difference is directly proportional to the electric field strength of the ligand entities partaking in complex formation. Since transition-metal (or any other) ions do not normally exist in isolation, they have to be considered in association with counter-ions or solvents such as water and other ligand systems. In isolation the electron distribution in Cr^{3+}, described in Table 5.1(a), consists of three unpaired electrons in the d shell.

TABLE 5.1

Electron distribution in some transition-metal ions (1s, 2s, 2p, 3s and 3p shells are complete in all cases)

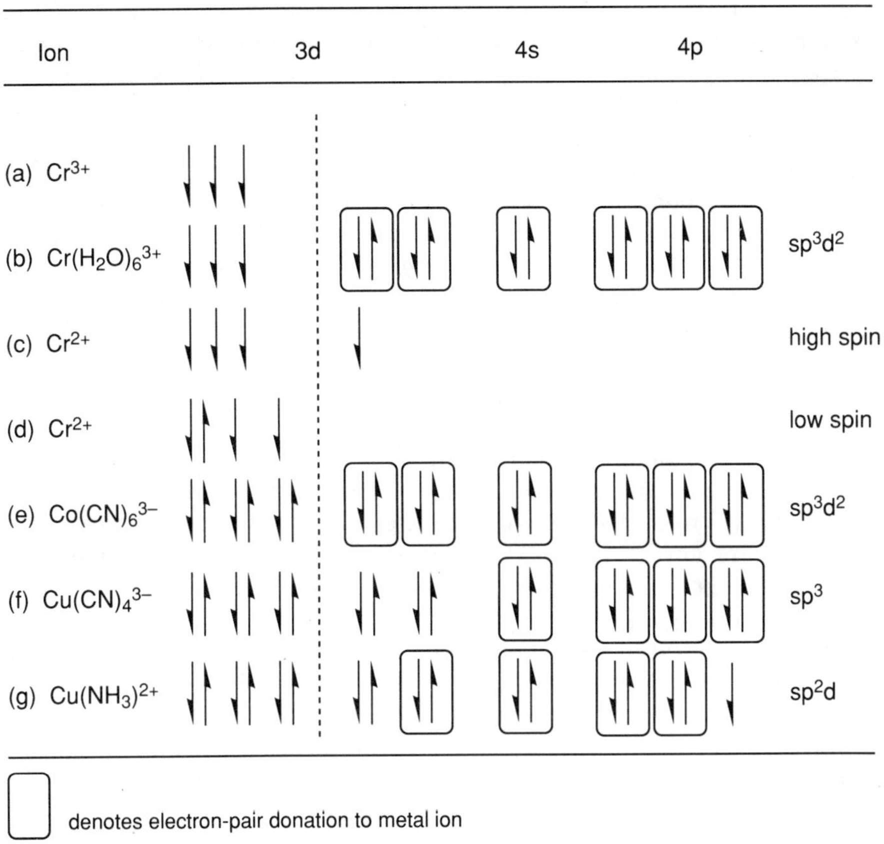

In the Table 5.1(b) arrangement the electron pairs donated from the oxygen atoms give rise to sp^3d^2 hybridisation. This is characteristic of an octahedral structure. A similar situation arises with trivalent cobalt complexes in Table 5.1(e), where each of the three t_{2g} levels is doubly occupied by an electron pair. With divalent chromium there are two distinct possibilities, shown in Table 5.1(c) and (d). With an ion having four or more electrons the fourth electron may occupy an e_g level (c) or a t_{2g} level (d), depending on whether the energy required to cause spin pairing is less or greater than Δ; these result in a 'high-spin' and a 'low-spin' complex respectively; the field strength of the ligand entities determines which is actually formed. The unpaired electrons in Cr^{2+} are unstable and divalent chromium cannot be utilised in preparing dye complexes. Invariably a d^4

electron is lost and oxidation to Cr^{3+} occurs, with concurrent reduction of the ligand chromophoric system [9].

The presence of unpaired electrons in a complex can be detected by the study of its magnetic properties. A single spinning electron has an associated magnetic moment. When a complex is weighed in a magnetic field any unpaired electrons can be detected by an apparent increase in mass compared with the mass in the absence of the field. The magnitude of attraction of the complex to a magnet is a measure of the number of unpaired electrons. The complex is said to be paramagnetic. Conversely, where there is no apparent difference in mass the substance contains no unpaired electrons and is referred to as diamagnetic. Thus this data is of value in determining the electronic distribution around the central nucleus in the complex. The oxidation state and electron distribution in the 3d shell at least partially accounts for the fact that most metal-complex dyes for wool and nylon are based on chromium or cobalt.

5.3.2 Tetrahedral arrangement

Depending on the ligand field strength and the number of ligand systems that can be accommodated, hybridisations other than octahedral are possible. Perhaps the most familiar is the tetrahedral arrangement exemplified by sp^3 hybridisation in saturated carbon atoms. Similarly the cyanocuprate ion in Table 5.1(f) has a tetrahedral configuration. The copper(I) ion possesses a completely filled d shell and the introduction of the four cyanide ions occurs in the 4s and 4p levels.

In a tetrahedral arrangement, the $d_{x^2-y^2}$ and d_{z^2} orbitals (that is, the e_g orbitals) are more stable than the t_{2g} orbitals, since the former are situated at the maximum distance from the influence of the four ligand groups. Conversely the d_{xy}, d_{xz} and d_{yz} orbitals become destabilised. The ligand groups are best visualised as occupying alternate corners of a cube whose main axes represent those of the metal ion (Figure 5.2(c)).

5.3.3 Square planar arrangement

The square planar arrangement in Figure 5.2(b) can be considered as an octahedral structure in which any two electrons – on the *z*-axis, for example – are completely removed from the sphere of influence of the central metal atom. If two ligand groups on the *z*-axis are moved so that they become slightly more distant from the central metal atom than the groups in the *xy* plane, the result is a distorted octahedral (that is, a tetragonal) structure. This outward movement allows the *xy* equatorial groups to move in closer. Consequently, the d orbitals in this plane experience a greater repulsion than in a regular octahedral structure and the $d_{x^2-y^2}$ and d_{xy} orbitals (Figure 5.3) increase in energy. At the same time the outward removal of the ligand groups along the *z*-axis reduces the repulsion effects on the d orbitals along the *z*-axis and in the *xz* and *yz* planes. Thus the energy of the d_{z^2} orbital is reduced, the energies associated with the d_{yz} and d_{xz} also being reduced compared with their energy in a regular octahedral environment. When the two groups on the z-axis are completely removed to give the square planar complex these effects are further enhanced.

The splitting of d orbital energy levels into t_{2g} and e_g levels under the influence of ligand fields for different geometries is summarised in Figure 5.4. The energy difference between e_g and t_{2g} orbitals in a tetrahedral structure Δ_t is approximately one-half of the energy difference Δ_o for that of the octahedron. These effects favour formation of an octahedral rather than a tetrahedral structure. Further reading useful to this basic theory of metal-complex formation can be found elsewhere [10–13].

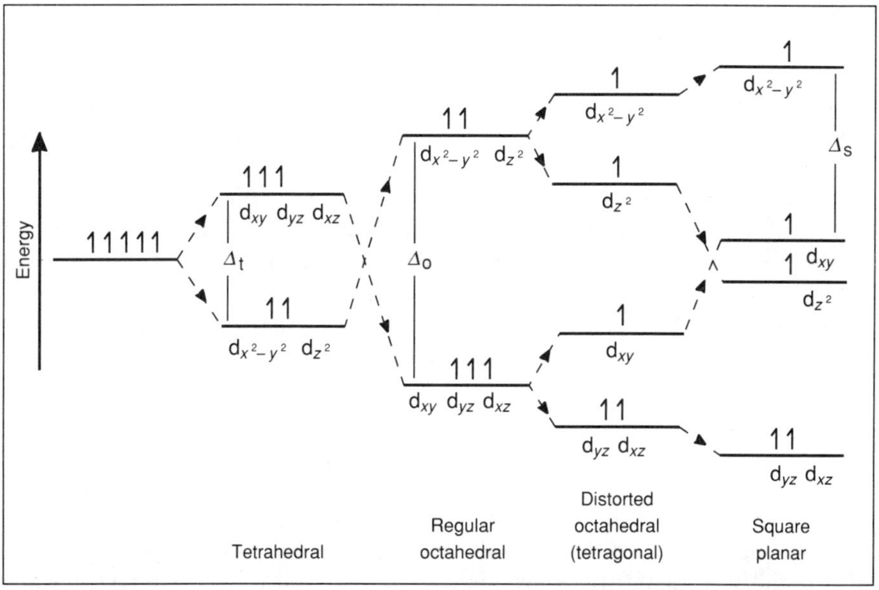

Figure 5.4 – Field splitting of d orbitals with different geometries

5.4 STRUCTURAL CHARACTERISTICS NECESSARY FOR COMPLEX FORMATION

5.4.1 Bidentate ligand entities

Bidentate ligand systems in which three molecules of a dye containing either a terminal salicylic acid residue (5.1) or an *o*-nitrosonaphthol (5.2) can chelate with a transition-metal cation such as Cr^{3+} or Fe^{3+} of CN 6 to form a 1:3 metal–dye complex. Perhaps one of the best-known bidentate ligand systems is that based on 1,2-dihydroxyanthraquinone or alizarin

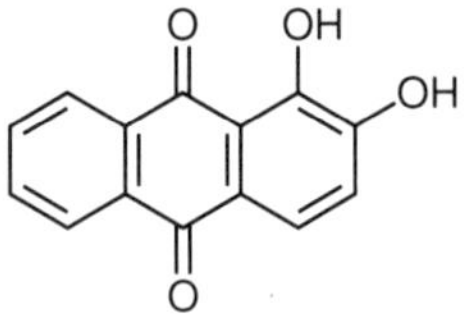

5.3

(5.3). Natural dyes such as madder, fustic, Persian berries, kermes and cochineal, which contain similar derivatives, can be fixed on textile materials by complexing the ionised *o*-dihydroxy groups with transition-metal oxides or salts used as mordants. Although mordanted wool dyed with synthetic alizarin had excellent fastness, reproducibility especially in bulk dyeings was difficult due to the shade changes that occurred on metallisation. The famous Turkey red, in which alizarin was applied to aluminium-mordanted wool in the presence of calcium salts, consisted of a metallised structure the nature of which is still in considerable doubt. It has been suggested that chelation involves interaction between the metal atom and the *peri* 1-hydroxy and keto groups of the anthraquinone residue, with possible further involvement between the metal atom and the 2-hydroxy group.

Most metal-complex dyes of commercial significance are based on the azo chromophore; in practice, tridentate ligand systems incorporating the azo group are necessary to achieve adequate stability in application and satisfactory fastness. Each nitrogen atom in an azo (–N=N–) or azomethine (–CH=N–) grouping possesses a lone electron pair that can participate in coordination. The linkage, when formed, is relatively weak and there is no evidence that the azo group in azobenzene, for example, even in the *cis* configuration can act as a donor group with a first-series transition element. On the other hand, azo compounds that also contain an *o*-hydroxy or *o*-amino group can act as bidentate ligand groups in which the azo group does participate in coordination. Several complexes formed from these derivatives with salts of copper, nickel and cobalt have been obtained, although their structures were not conclusively established until the work of Drew and Landquist in 1938 [14]. These workers were able to show that both *o*-hydroxyazo (5.4) and *o*-aminoazo compounds formed 1:2 metal complexes with divalent copper, nickel or cobalt, with a proton loss from each molecule, thereby giving a neutral species. In addition, by comparing the *o*-substituted azobenzene complexes with the corresponding azomethine derivatives (such as compound 5.5), they were able to show that the β-nitrogen of the azo group acted as the donor atom. Interestingly compound 5.5 is currently used as a catalyst for crosslinking reactions in polyesters.

In general, metal complexes formed from bidentate azo chromophores are little used as dyes but do find applications as pigments. There are exceptions, however, such as the nickel complex of *p*-nitroaniline → BON acid (5.6). This bordeaux complex gives light-fast prints on cotton. It should be noted that the *o*-hydroxyazo grouping is not the only chelating system present.

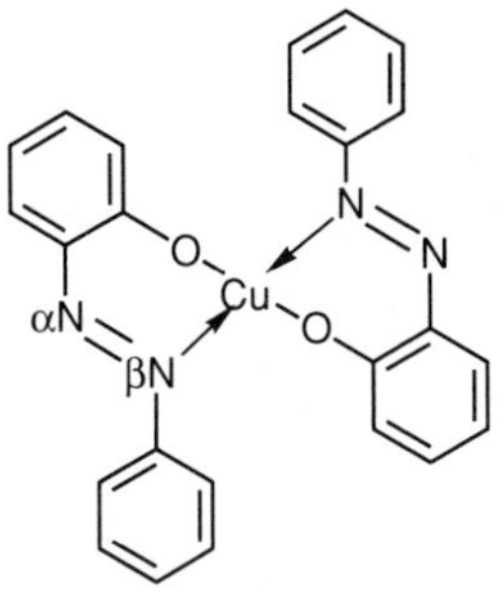

5.4

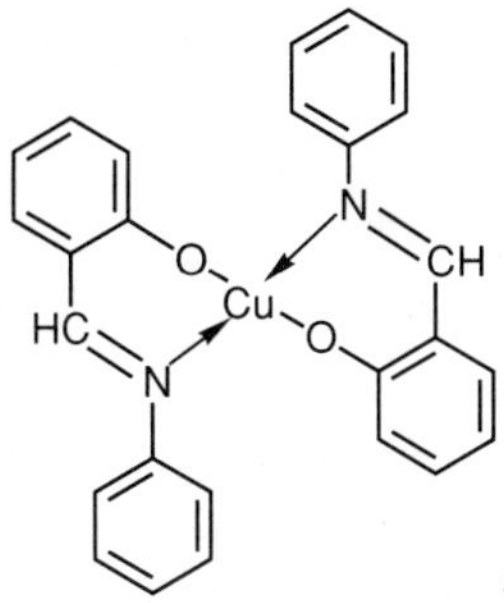

5.5

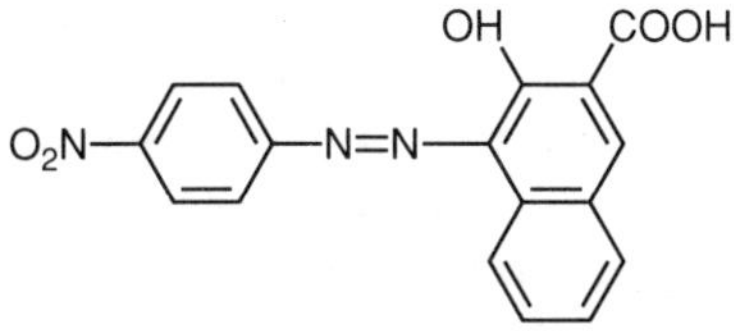

5.6

Terminal metallisation

In compound 5.6 the hydroxy and carboxyl groups may be ionised and in this state coordinate with a metal ion. This structural feature is also present in azo dyes that contain salicylic acid residues in either a median or a terminal position. Salicylic acid itself undergoes metallisation, and ammonium chromisalicylate (5.7) is used as an intermediate in metal–dye complex production. C.I. Mordant Orange 1 (5.1) and C.I. Mordant Yellow 5 (5.8) are examples of dyes containing terminal salicylic acid residues.

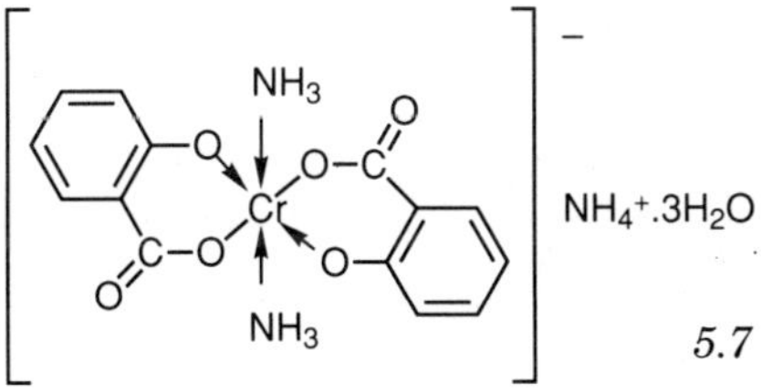

5.7

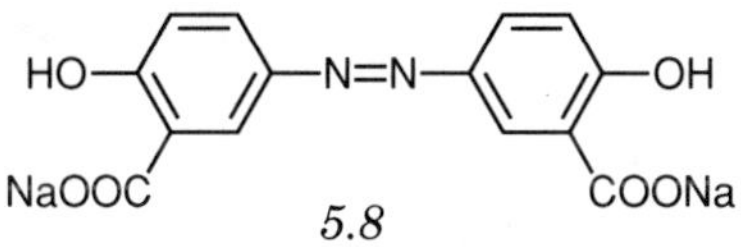

5.8

Since both ligand groups in the salicylic acid residue are electronically associated with the azo chromophore, changes in shade become apparent on metallisation. Difficulties in shade matching are experienced on afterchroming these mordant dyes. The change may be minimised by introducing an insulating group, such as $-SO_2-$, $-SO_2NH-$ or $-CH_2-$, between the salicylic acid residue and the chromophoric system. C.I. Mordant Red 26 (5.9) is typical of this kind of dye.

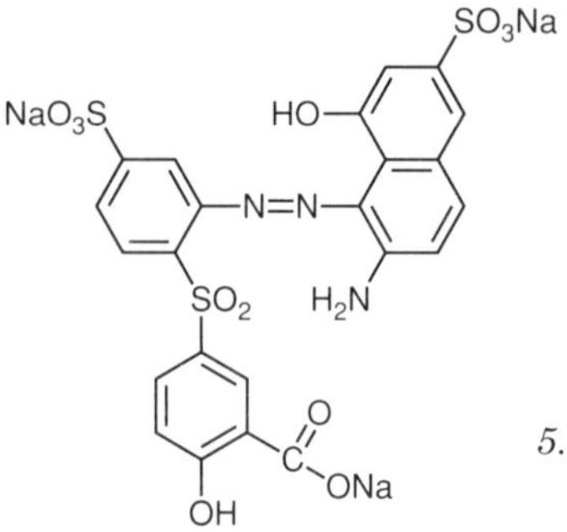

5.9

As might be anticipated, dyes in which the azo group does not participate in metallisation are generally brighter in hue. Other terminal bidentate groups include 8-hydroxyquinoline. The dyes are disazo dyes that owe their aqueous solubility to the presence of two terminal 8-sulphatoquinoline groups (5.10); they can be applied as direct dyes to cellulosic fibres to give bright hues. On aftercoppering, which improves fastness properties, metal complexes can be formed via hydrolysis catalysed by the copper(II) ion. It has been suggested [15] that an intermediate complex (5.11) is formed as a precursor, the electron withdrawal from sulphur in the O–S bond facilitating hydrolysis (Scheme 5.2). The presence of two sulphated or phosphonated 8-hydroxyquinoline groupings in terminal positions leads to the formation of a polymeric metal complex (5.12) with lower aqueous solubility and improved resistance to wet treatment on metallisation.

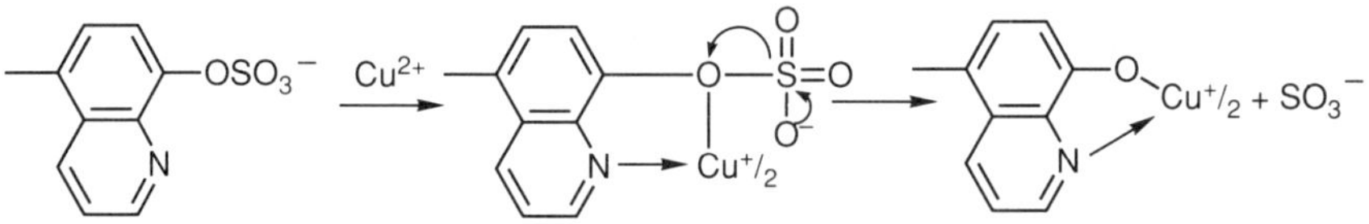

5.10 5.11

Scheme 5.2

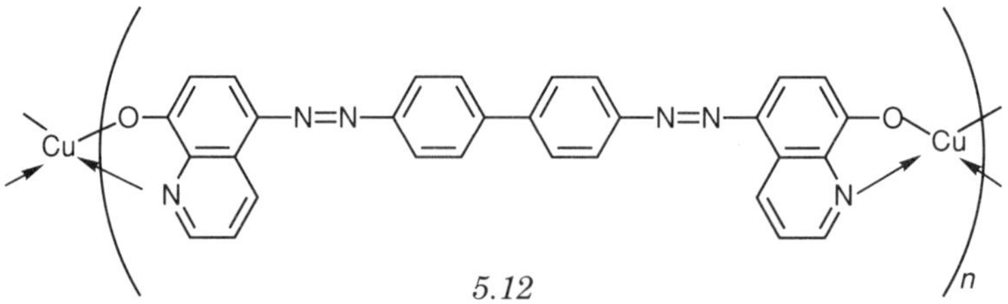

5.12

o-Nitrosonaphthols can also be considered as terminally substituted bidentate ligand systems (as in structure 5.2). On replacing the ionised sulphonic acid group in each of the ligand entities by a non-ionised modified solubilising group, such as an alkylsulphonyl (R–SO_2–), sulphonamido (H_2N–SO_2–) or dialkylsulphonamido (R_2N–SO_2–) group, where R is usually a methyl group, the overall negative charge on the 1:3 metal–dye complex is reduced from –4 to –1. Consequently dyeing can take place from neutral or weakly acidic dyebaths. Similarly green to olive shades can be obtained on wool or nylon using divalent iron complexes of 5-nitroso-6-hydroxy-indazoles (5.13): the amphoteric nature of the three heterocyclic rings in the metal complex accounts for its solubility. Protonation of the –N= in each ring in the complex depends on the pH, and the charge on the complex can vary from –1 for the unprotonated form to +2 for the fully protonated form (5.14) [16]. Thus cationic or basic metal complexes are a possibility. In their preparation the use of iron(III) salts causes initial partial oxidation of the nitrosonaphthol with consequent reduction to iron(II), which then interacts with the nitrosonaphthol to form the 1:3 metal–dye complex.

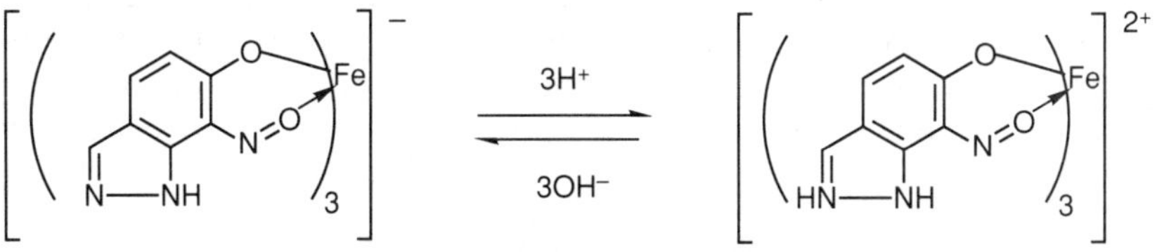

5.13 *5.14*

5.4.2 Tridentate ligand entities

Following earlier developments undertaken in BASF, in 1915 the Society for Chemical Industry (now incorporated in Ciba–Geigy) introduced a range of *o,o'*-dihydroxyazo compounds containing chromium which were applicable to wool by a single-bath method. Typical amongst these were the chromium complexes Neolan Yellow GR (C.I. Acid Yellow 99) (5.15) and Acidol Bordeaux M-B (C.I. Acid Violet 90) (5.16). Similar ranges were introduced for the direct dyeing of cotton by converting suitably substituted dyes to their copper complexes, which possessed enhanced resistance, particularly to light.

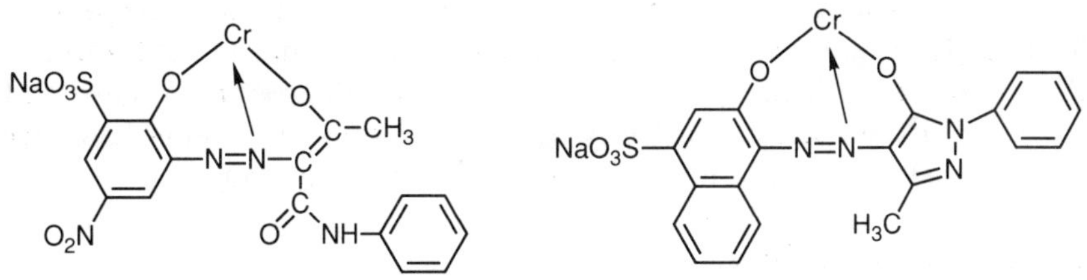

5.15 *5.16*

For some time after their introduction, possibly because of the independent studies of Morgan and Drew [8], these dyes were formulated as bidentate systems, utilising only one of the hydroxy groups in conjunction with the azo group. Drew and Fairbairn [17] were able to confirm that C.I. Acid Violet 90 (5.16) was in fact a 1:1 chromium–dye complex in which both hydroxy groups and the azo group were coordinated with the metal, the remaining three sites being taken up by monodentate ligand groups.

Of all the multidentate chromophoric ligand systems available for metal-complex dyes, the tridentate azo systems have proved to be the most important. Those capable of coordination with metals of CN 6 (trivalent chromium or cobalt) or of CN 4 (such as divalent copper) are mainly limited to *o*-hydroxy-*o*′-substituted diarylazo or azomethine derivatives containing the structures 5.17a and 5.17b respectively. The more important dyes for wool and nylon are the derivatives of trivalent chromium or cobalt in which X is usually a hydroxy group but may be an amino, carboxyl or other ligand group from which a proton is lost prior to or during complex formation. Coordination of these and the azo groups with the metal leads to a fused-ring system with one shared bond, constituting a bicyclic or annelated ring system. If the *o*-substituent is amino than one ring is five-membered (5.18a), but if it is carboxyl both rings are six-membered (5.18b).

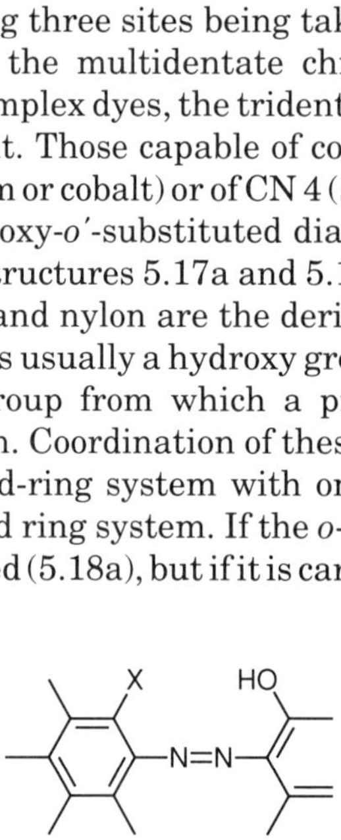

5.17a

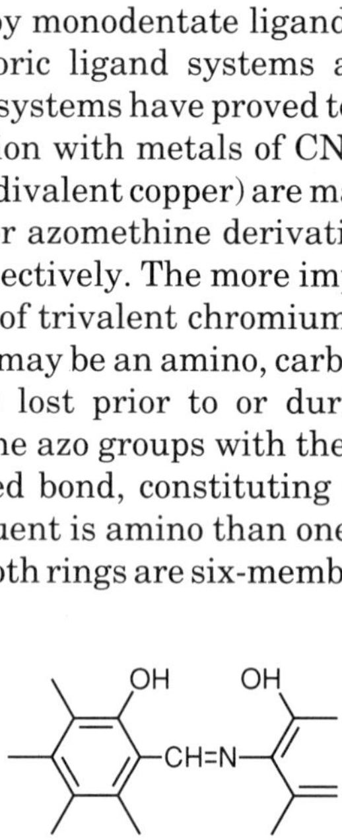

5.17b

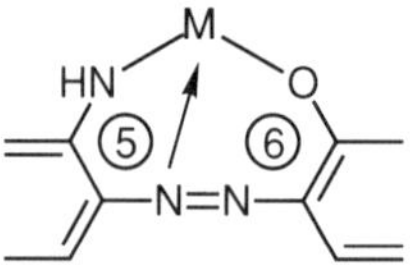

5.18a

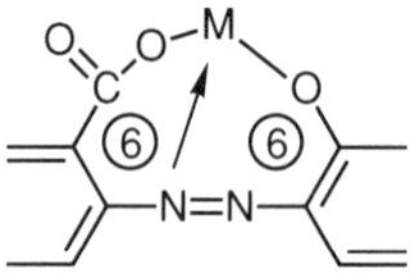

5.18b

This difference has an important bearing on the physical properties of the complexes (see section 5.6.1). Fused-ring formation enhances the stability of the resultant metal–dye complex to a degree that its use under dyeing conditions is commercially possible; bidentate systems, in which a fused-ring system is not formed, are much less important.

1:1 Metal–dye complexes

The earlier Palatine Fast (BASF) and Neolan (CGY) dyes were 1:1 complexes of trivalent chromium in which the three remaining sites were occupied by colourless monodentate ligand groups. With metals of CN 6 there is the possibility of forming both 1:1 and 1:2 metal–dye complexes. The factors determining the type of complex obtained include pH, the structure of the tridentate chromophoric ligand groups and the stoichiometric ratio between the dye and the metal. 1:1 Chromium–dye complexes in which the three monodentate ligand groups are water molecules are stable at low pH values, although similar cobalt complexes are not. However, 1:1 cobalt–dye complexes have been obtained [18] by heating a tridentate dye with a cobalt(II) salt in the presence of excess ammonia, the monodentate ligand groups being three ammonia molecules. The reaction involves oxidation of the Co^{2+} ion to Co^{3+} prior to complex formation. Even so, the resultant complexes were insufficiently stable to find use as dyes. Stable 1:1 cobalt–dye complexes of use in coloration can, however, be prepared by replacing the colourless monodentate ligand groups with colourless tridentate groups such as diethylenetriamine, as in compound 5.19. By varying the number of ionised sulphonic acid groups from zero to three the net charge on the complex can be varied from +1 (in a cationic complex suitable for dyeing acrylic fibres) to zero (for pigment usage) through to −1 and −2 (for nylon and wool dyeing).

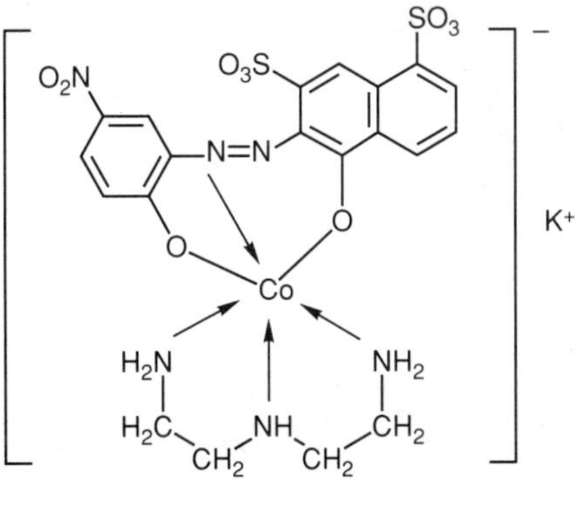

5.19

1:1 Chromium complexes were the predominant class of metal-complex dyes prior to the 1940s. The presence of sulphonic acid groups in their structures conferred solubility and their anionic nature allowed salt linkages with protonated sites in the wool during dyeing. Strongly acidic conditions were necessary to achieve high substantivity, but could cause degradation of the wool. The presence of these ionised groups in the complexes gave rise to anomalous results in structure elucidation by elemental analysis, since simple metal salts could also be formed between these negatively charged sulpho groups and the metal cation.

1:2 Metal–dye complexes

The coordination sites in 1:1 metal–dye complexes of trivalent chromium can be occupied by colourless mono-, bi- and tri-dentate ligand groups. Such groups can generally be replaced by a second tridentate chromophore, as was shown by Drew and Fairbairn [17] as early as 1939. This replacement of labile ligand groups results in a more stable configuration consisting of two bicyclic ring systems as in structure 5.20, in which either or both of groups X and X′ may be –N= or –CH=. Colour, solubility, and dyeing and fastness properties can be controlled by selection from a wide range of sulphonated or unsulphonated azo or azomethine tridentate ligand systems. By further extension to include dis- and tris-azo systems browns, blacks and navy blues become possible. With the use of sulphonated intermediates, however, unlevel dyeing remained a problem because of the low pH necessary to achieve satisfactory exhaustion. This difficulty was partially overcome by the use of non-ionic additives but the most important factor was the introduction in 1954 of modified water-solubilising groups which did not develop a formal negative charge [19,20]. These included the C_1–C_3 alkylsulphonyl (R–SO_2–), *NN*-mono- and di-alkylsulphonamides (R_1R_2N–SO_2–) and cyclic sulphones. In these modified solubilising groups the hydrogen-bonding capacity of the sulphonyl oxygen atoms with surrounding water molecules is not as great as that between the oxygen atoms and water in the ionised sulphonic acid groups. 1:2 Metal–dye complexes of this type are Perlon Fast Orange RRS (IG) (5.21) and C.I. Acid Violet 78 (5.22).

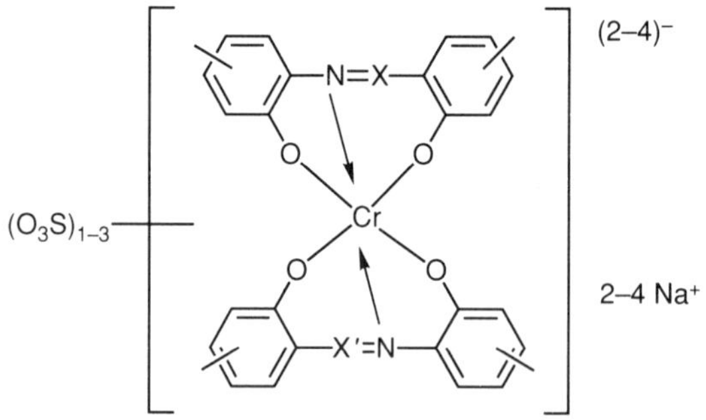

5.20

Similarly, symmetrical 1:2 cobalt complexes can be produced. These readily give level dyeings on wool or nylon under neutral conditions, and heavy shades are attainable. The cobalt complexes are generally faster to light and absorb at shorter wavelengths than the corresponding chromium complexes.

The reaction of two moles of a tridentate chromophore (A) with one equivalent of M leads only to a symmetrical AMA structure. Reaction with two different chromophoric structures (A and B) leads to the possibility of

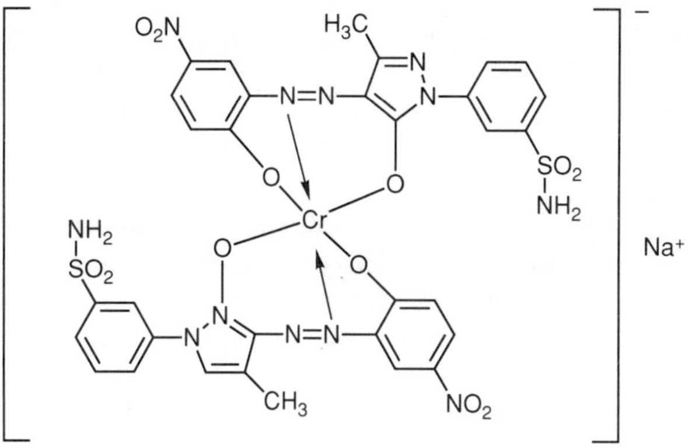

5.21

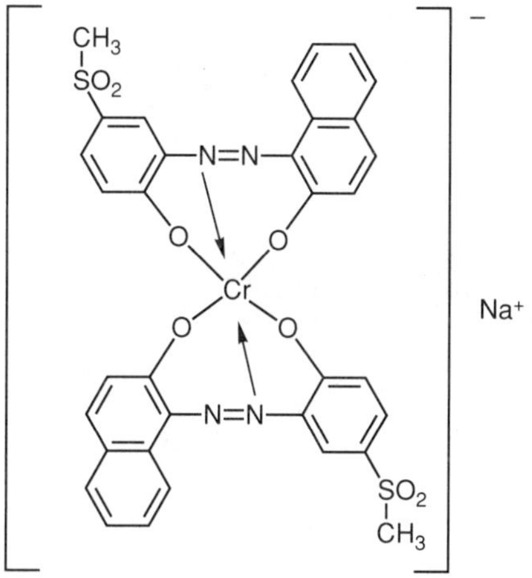

5.22

a mixture of structures (AMA, AMB and BMB). Mixtures of this type are industrially important, since they are less expensive to produce than the pure unsymmetrical complexes (AMB), which can only be obtained in a pure form by the reaction of a pure 1:1 complex (AM, for example) with the other ligand entity (B). Monosulphonated unsymmetrical chromium complexes with two negative charges are of particular importance. The variable position of the sulphonate group leads to differences in colour properties and only weakly acid conditions are required for dyeing.

Tridentate formazans

Formazans may be obtained by the coupling of a diazonium salt in alkaline medium with a component containing a reactive methylene group at which two diazonium ions may react. Compounds such as nitromethane or acetone may be used for this purpose, as illustrated in Scheme 5.3. Thus by coordination of both azo groups 1:2 complexes with metals such as nickel,

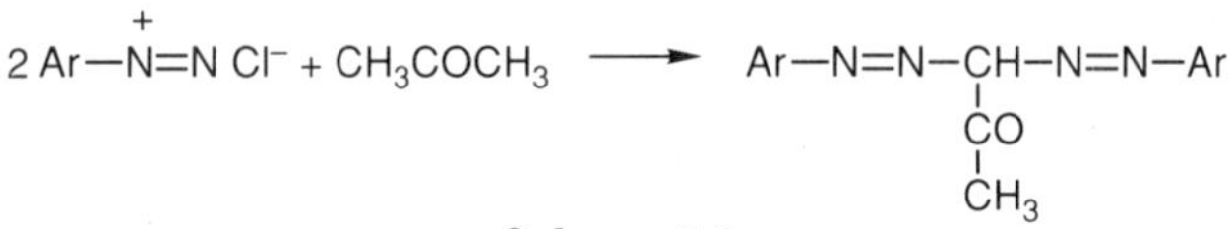

Scheme 5.3

cobalt and copper may be obtained [21]. Such monocyclic chelates have low stability but formazans containing an *o*-hydroxy or *o*-carboxy ligand group in the aryl ring [22] can be used to produce intensely coloured tridentate, bicyclic, neutral 1:1 metal–dye complexes. Metallisation with divalent copper, for example, gave a change in colour from red to violet (5.23). On completing the CN 4 by the addition of a monodentate ligand group such as pyridine, a further colour change can be observed. With metals of CN 6 tridentate formazans, like tridentate azo compounds, give 1:2 complexes with trivalent chromium and copper, such as structure 5.24.

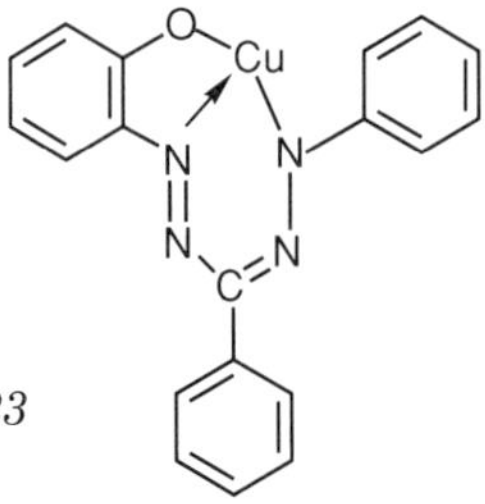

5.23

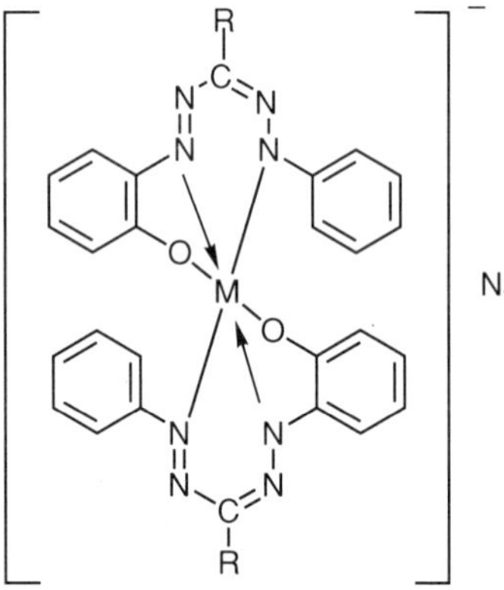

5.24

5.4.3 Quadridentate ligand entities

Quadridentate formazans

Among quadridentate ligand systems is to be found a relatively small group of metal-complex formazan dyes in which four ligand substituent groups originating in the one formazan molecule can coordinate with a metal to give a complex possessing a condensed tricyclic ring system. As already indicated, these metal complexes are more stable than their equivalent bicyclic systems based on tridentate formazans, and several technical dye and pigment applications have been suggested. Quadridentate formazans can be metallised with salts of nickel, copper, chromium or cobalt. Only a 1:1 metal–dye ratio is possible, and for metals of CN 6 the coordination sphere is completed by the use of colourless ligand groups such as pyridine, ammonia or water. Solubility and dyeing properties may be varied by the presence of a sulpho group or an uncharged solubilising group as in structure 5.25. This copper complex dyes nylon in blue shades from a neutral bath [23].

Similar formazan–metal complexes that are sufficiently solubilised by the introduction of sulphonic acid groups, together with a labile or reactive residue attached via the aryl rings in the 1-, 3- (meso) or 5-positions of the tetrazine ring, can be used as reactive dyes for cellulose; an example is structure 5.26, which in the form of its copper complex gives fast dyeings on cellulose.

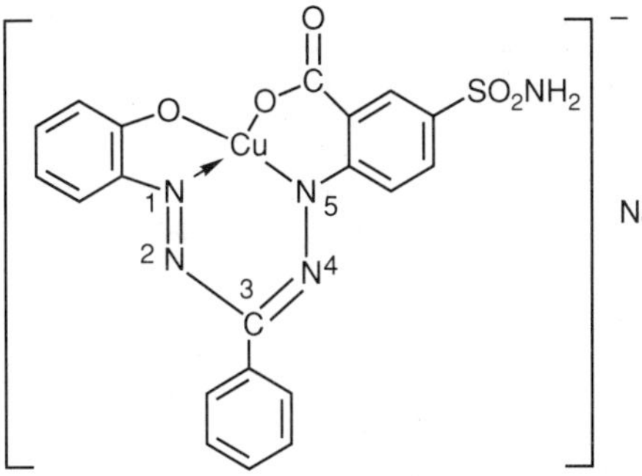

5.25

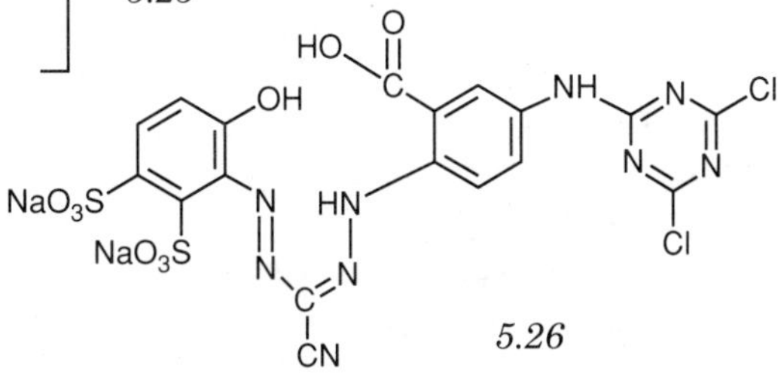

5.26

Phthalocyanine metal complexes

Although not generally classified under the heading of metal-complex dyes, copper phthalocyanine is also a representative of a quadridentate ligand system. Its chemistry, properties, and its use as a pigment are described in section 2.4 [24].

Copper phthalocyanine possesses a quadridentate tetracyclic condensed ring system that has high kinetic and thermal stability (5.27). It can be sublimed without decomposition *in vacuo* at temperatures as high as 550°C and the solid may undergo polymorphic change at 250°C under atmospheric conditions. In addition to copper many other transition metals and rare-earth elements can form 1:1 complexes with phthalocyanine, although the characteristic colour and stability of the copper complex make this particular pigment commercially valuable.

It may be solubilised for use as a direct dye on cellulosic materials by the introduction of sulphonic acid groups, modified solubilising groups or a combination of both. Similarly other ionised or non-ionised groups may be incorporated to give dyes for many different substrates. The presence of halogen atoms causes a slight colour change to greenish-blue. The 18 π-electrons of the central chromophore system are, however, insulated from

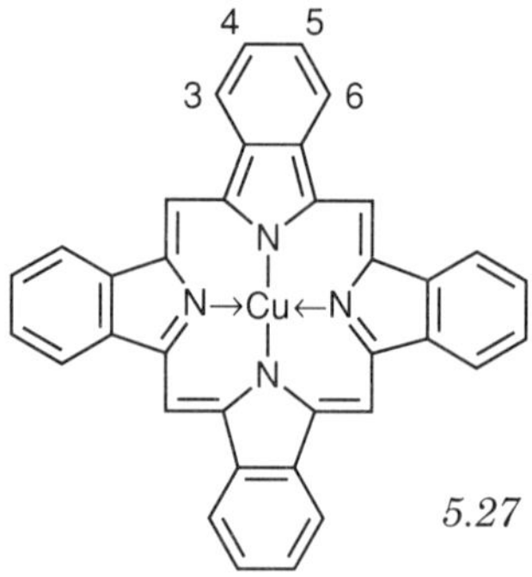

5.27

the outermost 3-, 4-, 5- and 6-positions in each aromatic or quinonoid ring, so that the colour range anticipated from the presence of auxochromes is extremely limited.

5.5 PREPARATION OF METAL COMPLEXES

There are numerous ways of introducing a metal atom into a dye molecule to form a metal complex. Most attention has been paid to chromium and cobalt for this purpose; there are essential differences in the preparation conditions of their complexes.

5.5.1 Chromium complexes

Table 5.1 (in section 5.3.1) shows that the trivalent hexa-aquo chromium ion possesses three low-energy t_{2g} orbitals, each singly occupied. The ligand field stabilisation energy is considerable as a consequence. Since this energy has to be overcome, the replacement of the water ligand groups by other entities is a slow process, and the most successful practical methods are designed to take this into account.

The replacement reactions may be represented generally by a two-stage process (Schemes 5.4 and 5.5, in which D^{2-} represents a tridentate azo dye system, assuming prior ionisation). The reaction in Scheme 5.4 is slower than that in Scheme 5.5 and is the rate-controlling step. Where a tridentate *o,o′*-dihydroxyazo dye is used in excess with a hydrated chromium salt under conditions favouring 1:2 metal–dye complex production, no 1:1 complex is formed, indicating that the reaction in Scheme 5.5 is more rapid than that in Scheme 5.4. In order to obtain a 1:1 complex, therefore, it is necessary first to obtain the 1:2 complex (favoured at pH > 9) and then to treat this complex under strongly acid conditions; the reaction in Scheme 5.5 then proceeds to the left.

$$D^{2-} + [Cr(H_2O)_6]^{3+} \xrightleftharpoons{\text{slow}} [DCr(H_2O)_3]^{+} + 3H_2O$$

Scheme 5.4

$$[DCr(H_2O)_3]^+ + D^{2-} \underset{HCl}{\overset{rapid}{\rightleftharpoons}} [D_2Cr]^- + 3H_2O$$

Scheme 5.5

The chromium salt used can be the acetate, chloride or sulphate, the reaction with metallisable dyes being carried out in aqueous media at the boil at atmospheric pressure or greater. Where the metallisable dye is insoluble in water a solvent such as ethanol or ethylene glycol may be added.

Other solvents such as formamide or molten urea have been recommended and it is probable that such solvents will displace water molecules and coordinate with the chromium(III) ion as the first step in the reaction. By using organic chelates of chromium, such as those derived from oxalic or tartaric acid, in place of or in addition to hydrated chromium(III) salts, the difficulty of replacing the strongly coordinated water molecules in the first stage of reaction is removed and the reaction can be carried out at high pH without the precipitation of contaminating chromium hydroxide. The use of the complex ammonium chromisalicylate (5.7) in this connection should be noted.

5.5.2 Cobalt complexes

Like trivalent chromium, trivalent cobalt complexes contain three low-energy t_{2g} orbitals, this time each doubly occupied. In general trivalent cobalt complexes are easier to prepare than the chromium analogues, although their properties are similar. Direct use of cobalt(III) salts is avoided since the Co^{3+} ion is strongly oxidising and slow to react. Most 1:2 metal–dye complexes are prepared by the reaction between the metallisable dye and a cobalt(II) salt at a relatively high pH. The product is invariably the diamagnetic cobalt(III) complex, the oxidation of the cobalt(II) ion taking place through reduction of some azo dye molecules. This can be avoided by adding a less stable oxidising agent. Unlike the corresponding 1:1 chromium complexes, 1:1 cobalt–dye complexes are unstable and cannot be obtained in acceptable yield at low pH. However, advantage can be taken [25] of the strong affinity of nitrogen electron donors for cobalt. The reaction of the cobalt(II) ion with a tridentate dye molecule in an excess of ammonia yields a 1:1 complex in which the coordination sphere is completed by three ammonia molecules. This again is diamagnetic, indicating the absence of unpaired electrons necessary to achieve an sp^3d^2 octahedral structure.

Although 1:1 metal–dye complexes have little importance in direct application to textile substrates, they are widely used as intermediates in the production of unsymmetrical 1:2 metal–dye complexes, particularly when the asymmetry arises from the presence of one or more sulphonic acid groups in one of the metallisable dye ligands partaking in complex formation.

5.5.3 Copper complexes

The only copper complexes of tridentate azo compounds are 1:1 complexes, since copper has a CN of 4; they are little used on wool or nylon. They can be prepared by the reaction of the azo compound with a copper salt in an aqueous medium at 60°C.

The major application for copper complexes is in the prior metallisation or aftertreatment of direct dyes containing at least one *o,o'*-dihydroxyazo or *o*-methoxy-*o'*-hydroxyazo chromophore system. In the manufacture of these dyes it is sometimes easier to synthesise the latter system simply because the *o*-methoxyarylamines when diazotised are more efficient couplers than are the corresponding *o*-hydroxyarylamines. Once the mono- or dis-azo chromophore is obtained, metallisation can be achieved with cuprammonium sulphate in the presence of an alkanolamine such as diethanolamine. The best-known example is the disazo dye C.I. Direct Blue 1, the greener copper complex of which contains the system shown in structure 5.28.

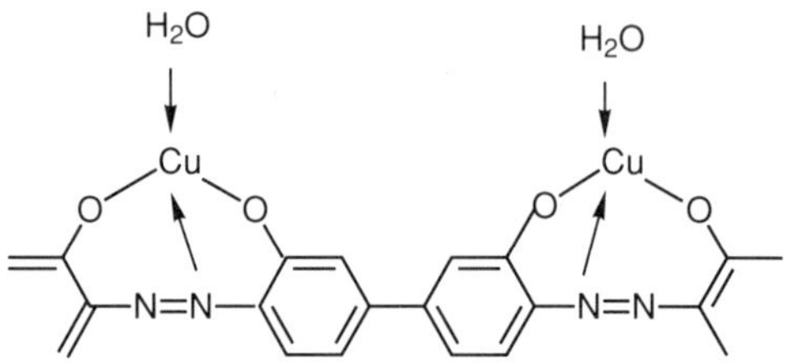

5.28

During metallisation the two methoxy groups originating in the central tetrazotised dianisidine component are demethylated. It is probable that coordination occurs between the copper(I) ion and the electron-rich oxygen of the methoxy group with simultaneous or subsequent loss of the methyl group as a carbonium ion, according to Scheme 5.6. The removal of the methyl group is probably assisted by the coordination of the methoxy oxygen, particularly in the presence of electronegative groups in *p*-positions.

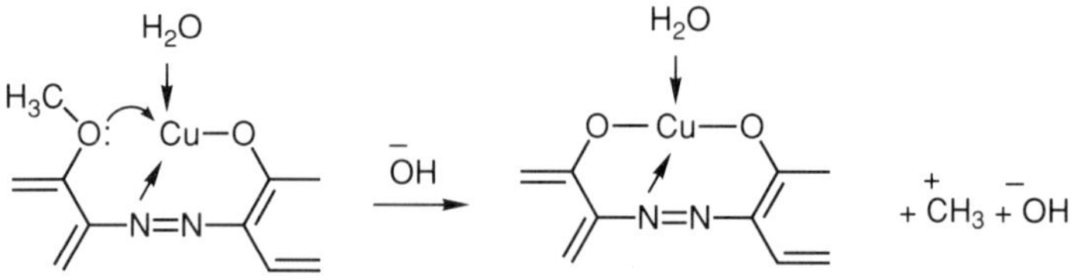

Scheme 5.6

Oxidative coppering

Other methods of introducing copper into dye structures include the process of oxidative coppering. In this method an azo compound containing a single *o*-hydroxy group is treated with a copper salt and an oxidant in an aqueous medium at 40–70°C and pH 4.5–7.0. Pfitzner and Baumann [26] have reviewed applications of this method. Sodium peroxide, sodium perborate, hydrogen peroxide or other salts of peroxy acids may be used as oxidants, the function of which is to introduce a second hydroxy group in the *o*′-position. The situation is similar to that investigated earlier by Morgan and Main-Smith, who found that C.I. Acid Red 14, containing a single *o*-hydroxyazo grouping, did not form a chromium complex with chromium salts but could do so on oxidation with chromic acid. The resultant complex was later found to be identical with that obtained by treating C.I. Mordant Black 3 with chromium salts. The latter dye differs from C.I. Acid Red 14 only in the presence of a second *o*′-hydroxy group.

Replacement of labile halogen

The halogen atom in an *o*-halogeno-*o*′-hydroxyazo compound may be replaced by a hydroxy group under mild conditions provided that the halogen atom is activated by the presence of electron-attracting substituents in the *o*- or *p*-positions [27]. More recent observations [28] suggest the formation of an intermediate copper complex of general structure 5.29, in which chlorine coordinates with copper. Coordination of chlorine with copper facilitates nucleophilic attack by the hydroxide ion at the carbon atom to which the chlorine is attached, leading to the formation of a *o,o*′-dihydroxyazo derivative.

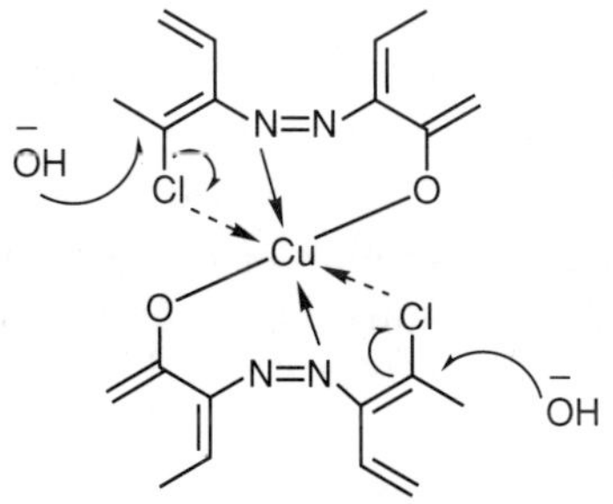

5.29

5.6 PHYSICAL PROPERTIES OF METAL-COMPLEX DYES

5.6.1 Isomerism in metal-complex dyes

The fact that ligand systems are distributed in the space surrounding a central metal atom allows the possibility that metal complexes can form spatial isomers (stereoisomers). Metals of CN 4 may partake in either a tetrahedral or planar arrangement. With two symmetrical bidentate ligand groups (two A–A or two B–B), a tetrahedral configuration gives rise to a single structure (Figure 5.5(a)). However, unsymmetrical bidentate ligand groups (two A–B) can exist as enantiomers, possessing non-superimposable mirror images, as in Figure 5.5(b), and exhibiting optical isomerism.

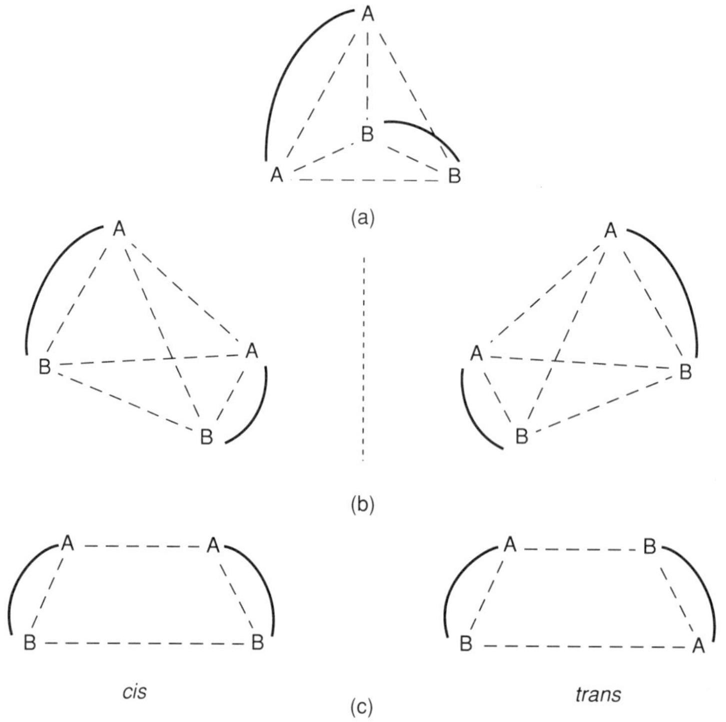

Figure 5.5 – Possible bidentate isomeric structures

Equatorial planar arrangements of unsymmetrical bidentate entities exist in *cis* and *trans* forms (Figure 5.5(c)). 1:2 Copper complexes of salicylic acid have the *trans* planar configuration; the existence of the *cis* form has yet to be shown.

When we consider complexes containing unsymmetrical tridentate structures in conjunction with metals of CN 6 such as trivalent chromium or cobalt, the possible number of stereoisomers becomes much greater. The requirement that the tridentate dye molecules retain a high degree of coplanarity in forming the complex does, however, restrict this number.

As early as 1939 Drew and Fairbairn [17], and later Pfitzner [19], concluded that in 1:2 complexes of *o,o′*-dihydroxyazo or *o*-carboxy-*o′*-hydroxyazo dyes with chromium the major planes of the two dye molecules are arranged about the central atom in a mutually perpendicular fashion.

Figure 5.6(a) shows such a structure, which became known as the Drew–Pfitzner model or as a meridial ('mer') complex. These complexes were not found to exhibit isomerism, but Pfeiffer and Saure [29] isolated two isomers of the 1:2 chromium complex of an *o*-carboxy-*o*′-hydroxyazo derivative and concluded that the three donor ligand atoms of each dye molecule were situated at the apices of an equilateral triangle forming the face of a regular octahedron. In an extensive series of papers Schetty and co-workers [30] confirmed the existence of these isomers. In complexes based on *o,o*′-dihydroxyazo dyes ('mer' isomers) the angle subtended by the three coordinating groups ABC and A′B′C′ (Figure 5.6(a)) is 90°, and is best fitted by the Drew–Pfitzner model. On metallisation these dyes each formed a 5,6′-membered bicyclic ring system. On the other hand, *o*-carboxy-*o*′-hydroxyazo dyes formed a 6,6′-membered ring system, with an angle of 60° between groups ABC and also between A′B′C′. Such a configuration could be more readily accommodated at the corners of the equilateral faces of the octahedron. This Pfeiffer–Schetty 'sandwich' structure or 'fac' (facial) configuration, shown in Figure 5.6(b), can exhibit a minimum of three isomers, each isomer existing as an enantiomeric mirror-image pair. The possibility that compounds forming two 6-annelated rings may coordinate to give a 'mer' isomer is not excluded, however, although the necessary angular distortion required will reduce resonance and cause a change in absorbance maxima, as shown by spectral differences.

Other studies extended to *o*-amino-*o*′-hydroxy-azo and -azomethine derivatives, as well as to *o*-carboxy- and *o*-hydroxy-formazans [31], confirmed Schetty's general conclusions that all diarylazo dyes that form annelated 5,6′-membered rings give the Drew–Pfitzner ('mer') structure and those dyes forming a 6,6′-membered ring system give isomers of the Pfeiffer–Schetty ('fac') type. An elegant confirmation of the 'fac' structure by X-ray crystallography has been given by Jaggi [32].

Stereoisomers of this type, having the same elemental analysis, differ in solubility and absorption spectra and may be separated by chromatography. The activation energy of conversion from one form to another is low and a solution of one isomer rapidly changes to an equilibrium mixture of all four. Attempts to demonstrate differences in dyeing properties are therefore difficult to assess and, while no definite relationship can be established, complexes based on *o*-carboxyarylazopyrazolone in the 'fac' structures are claimed [33] to have significantly inferior wet fastness and marginally better light fastness on wool than the 'mer' configuration.

5.6.2 N_α/N_β isomerism

A second type of isomerism in metal-complex dyes, originally suggested by Zollinger [34], has become known as N_α/N_β-isomerism. It occurs in 1:1 copper(II) and nickel(II) complexes and also in 1:2 chromium(III) and

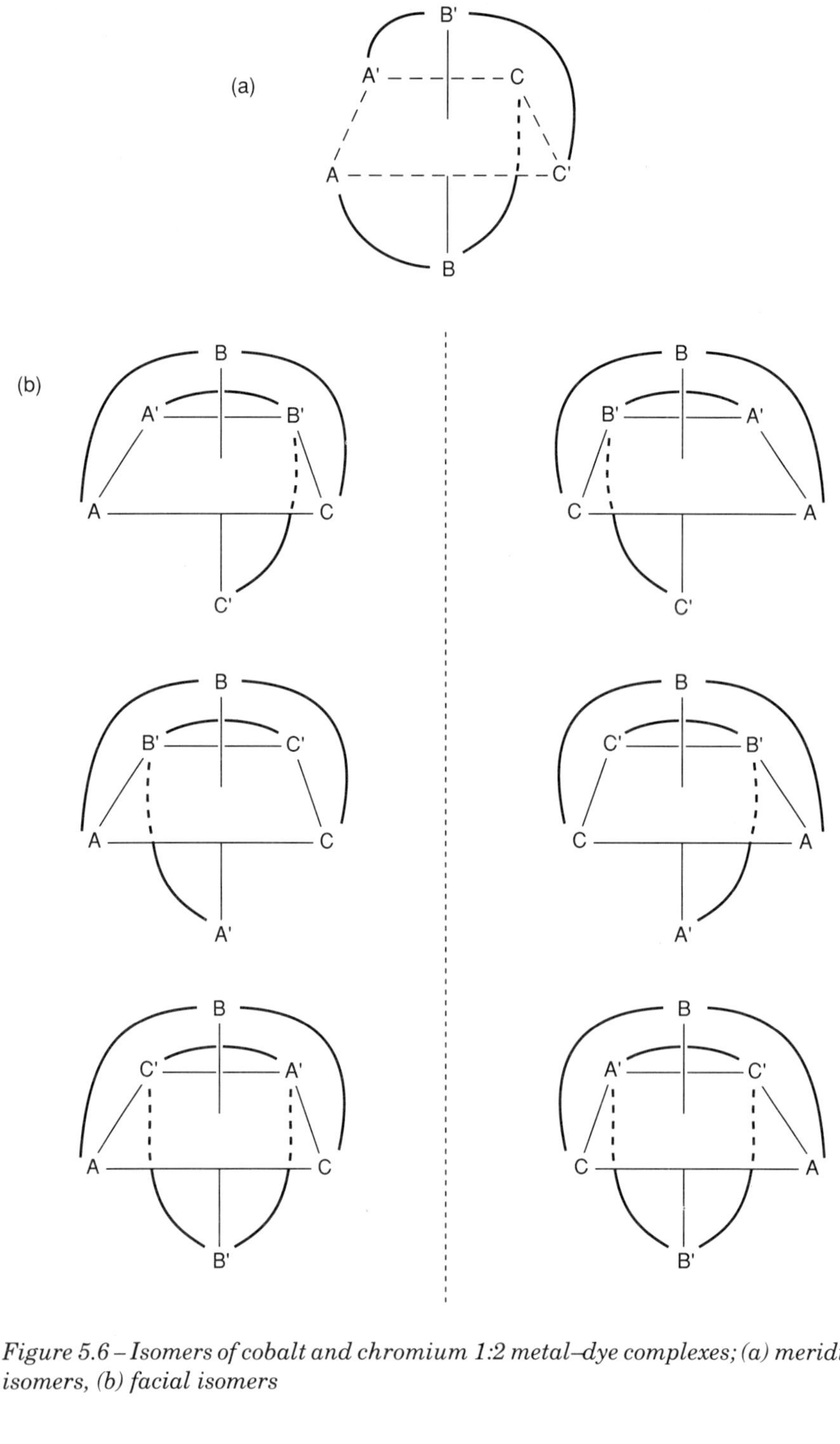

Figure 5.6 – Isomers of cobalt and chromium 1:2 metal–dye complexes; (a) meridial isomers, (b) facial isomers

cobalt(III) complexes, provided that the *o*-hydroxyazo chelating dye molecules are unsymmetrical and in the azo form on metallisation. These isomers (5.30a–c) differ in the positions of the nitrogen atoms taking part in coordination. Their existence was conclusively proved by Schetty and Steiner [35]. Where group R is alkyl, differences in the proton magnetic resonance signals, associated with the alkyl group protons, depend on the relative positions of the alkyl group and the nitrogen atom that coordinates with the metal. From the data, and by comparison with similar azomethine derivatives in which the coordinating nitrogen atom can be identified unambiguously, the proportions of the three isomers in the mixture can be calculated.

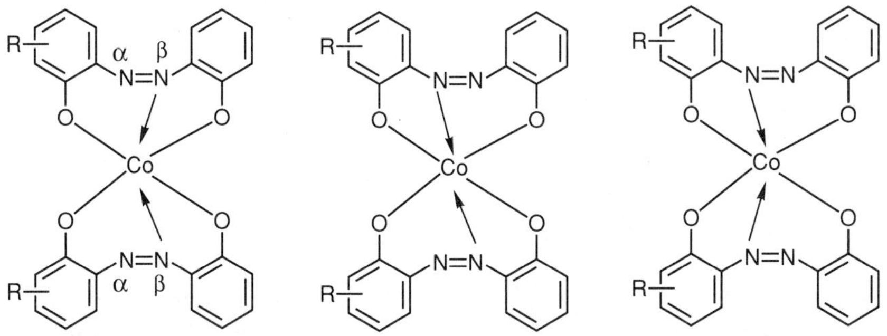

a β,β   b α,β   c α,α

5.30

5.7 STABILITY OF METAL-COMPLEX DYES

The formation of any metal complex is a reversible reaction so that, at equilibrium, the complex is always partially dissociated into its various components. The thermodynamic stability constant *K* is a measure of the extent of this dissociation. It is defined in general terms as the ratio of the factor of the concentrations of the end products to the factor of the concentrations of the starting materials (Scheme 5.7, in which *K* is defined according to Eqn. 5.1, and M represents the metal and L the ligand). The square brackets indicate the concentration terms, such as moles per litre. In practice concentration terms are used, since in dilute solutions the solutes approximate to ideal behaviour; otherwise activity coefficients should be included. A high value of *K* represents a relatively high concentration of the complex ML – that is, the complex is thermodynamically stable.

$$M^+ + L^- \overset{K}{\rightleftharpoons} ML$$

Scheme 5.7

$$K = \frac{[\mathrm{ML}]}{[\mathrm{M}^+][\mathrm{L}^-]} \tag{5.1}$$

Stability constants are closely related to the ionisation constants of acids and bases. For example, in the simple case of a monobasic acid (Scheme 5.8) the ionisation constant K_i can be expressed by Eqn 5.2. The term K_i then represents a measure of the ease with which the acid yields up a proton. In other words, K_i is an instability constant if we regard K, the stability constant of a metal complex, as a measure of the ease of binding between the metal ion and the ligand system.

$$\mathrm{HCl} \overset{K_i}{\rightleftharpoons} \mathrm{H^+ + Cl^-}$$

Scheme 5.8

$$K_i = \frac{[\mathrm{H}^+][\mathrm{Cl}^-]}{[\mathrm{HCl}]} \tag{5.2}$$

In the formation of a copper complex of an *o*-hydroxyazo ligand entity, for example, the reversible reactions shown in Scheme 5.9 can take place. Thus acceptance of a copper(II) ion by the ionised entity is analogous to its acceptance of a proton, or the reverse of ionisation. Expressing the above in terms of ionisation and stability constants gives Schemes 5.10 and 5.11 (and hence Eqns 5.3 and 5.4), where HL represents the non-ionised entity. Combining Schemes 5.10 and 5.11 gives Scheme 5.12 (and hence Eqn 5.5, where K_s represents the overall stability constant for the reaction). Eqn 5.5 includes the term [H^+] and the solution becomes more acidic during complex formation through removal of the ionised ligand entity L^-. In this example further interaction between Cu^+L and a second ionised entity leads to a 1:2 copper complex.

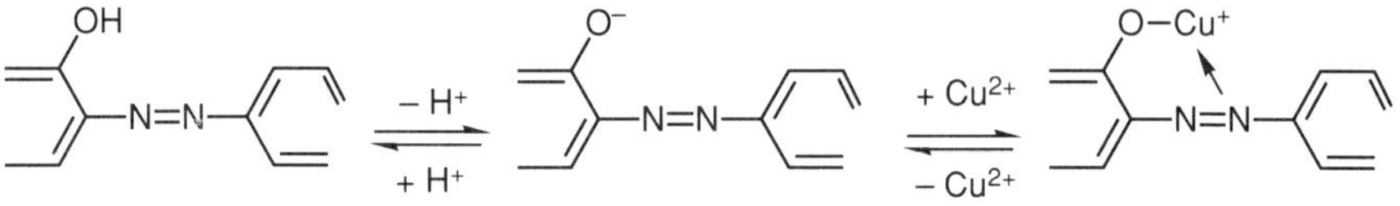

Scheme 5.9

$$\mathrm{HL} \overset{K_i}{\rightleftharpoons} \mathrm{H^+ + L^-}$$

Scheme 5.10

$$Cu^{2+} + L^- \overset{K'}{\rightleftharpoons} Cu^+L$$

Scheme 5.11

$$K_i = \frac{[H^+][L^-]}{[HL]} \tag{5.3}$$

$$K' = \frac{[Cu^+ L]}{[Cu^{2+}][L^-]} \tag{5.4}$$

$$Cu^{2+} + HL \overset{K_s}{\rightleftharpoons} Cu^+L + H^+$$

Scheme 5.12

$$K_s = K_i\, K'$$

$$\text{i.e.} \quad K_s = \frac{[Cu^+ L][H^+]}{[Cu^{2+}][HL]} \tag{5.5}$$

For a particular complex to have a stability sufficient to warrant its use as a dye, the stability constant must be as large as possible – in practice, 10^{15} or greater. The corresponding value of the ligand proton acceptance (reverse dissociation) is usually of the order of 10^7–10^9, giving a formation or stability constant at least some 10^6 times greater.

As exemplified throughout this chapter, factors influencing the stability of complexes include the size and number of annelated ring systems formed on metallisation. Chelates with five- and six-membered rings are more stable than complexes formed from terminally substituted salicylic acid, which give only a single chelate ring. Those with 6,7-annelated rings, such as those formed with *o,o'*-dicarboxyazo ligand groups, show stabilities closer to those of bidentate systems.

Metal-complex stability is also related to the basic strength of the ligand entity. It has been shown [36] that for a series of ligand structures (5.31) the acid dissociation constants pK_a are linearly related to the stability constants log K of their 1:2 copper complexes, the more acidic groups forming the less stable complexes. Where group X in structure 5.31 is a nitro group, pK_a is 8.1 and log K is 17.2; where X is methoxy, pK is 8.5 and log K is 19.2. The relationship, however, is valid only for a series of strictly comparable ligand groups and is dependent on the nature of the solvent.

5.31 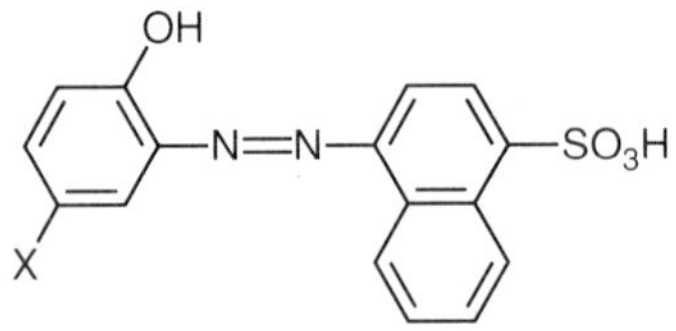

Similarly *o,o′*-dihydroxyazo dyes have been investigated by Snavely et al. [37]. The first and second acid dissociation constants related to the two hydroxy groups in the azo derivative 5.32, where X = H, were found to be 11.0 and 13.75 respectively. When X = Cl or NO_2 the first dissociation constant pK'_a decreased from 11.0 to 10.55 and 7.67 respectively. The stability constant log *K* depends on the metal concerned, being particularly high for nickel at 19.6 and copper at 23.3.

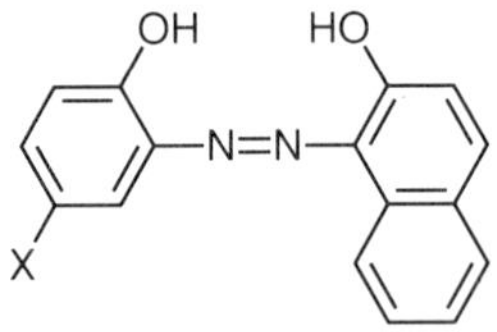

5.32

It has been suggested [38] that the partially ionised form of the dihydroxyazo ligand group obtained at lower pH values may be subject to association in solution but that the fully ionised form is not, because of greater repulsion effects. This possibility may be related to the type of solvent used in preparation. In comparing the data obtained from a series of ligand groups of general structure 5.31, where group X is directly conjugated with the azo group, Johnson et al. [36] showed that the stability constants did not change with group X. It was proposed that such entities could exist as azo–hydrazone tautomers, the proportion of each depending on the solvents used. Thus metal-complex formation could proceed according to Scheme 5.13 by ionisation of the –NH– group of the hydrazone

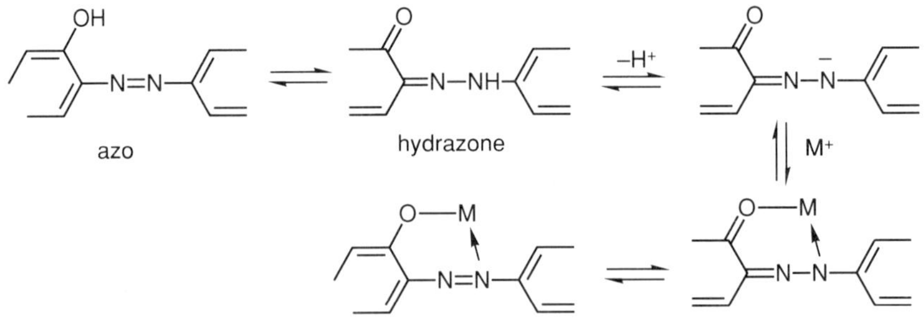

Scheme 5.13

tautomer, although such metallisation is relatively slow. Also there is no reason why tautomeric forms of the metal complex should not exist.

Finally, the nature of the donor atom in a coordinating group plays a part in the resultant stability of the complex. We have already seen that stable 1:1 cobalt(III) complexes can be obtained when the coordination number is completed by three nitrogen atoms. Iron(III) complexes tend to form more stable complexes with oxygen donor atoms, and chromium(III) complexes show equal stability with either nitrogen or oxygen donors. In a series of *o*-substituted-*o*′-hydroxyazo compounds based on the same 5-pyrazolone coupling component, where the *o*-substituent was a methoxy, methylthio or dimethylamino group, Snavely et al. [39] showed that even though the

TABLE 5.2

Commercial metal-complex dyes for wool and nylon

Name and producer	Solubilising group(s)	Comments
(1) *No solubilising groups*		
Palatine Fast (BASF)		First true 1:1 metal–dye
Neolan (CGY)		complexes for wool
Neopalatine (BASF)		
Vialon Fast (BASF)		1:2 complexes, good
Amichrome (ICI)		fastness on nylon, barry
Avilon (CGY)		on filament fabrics
(2) *Modified solubilising groups*		
Irgalan (CGY)	Alkylsulphonyl,	Most common for
Isolan (BAY)	alkylsulphonamide,	use on wool
Lanasyn (S)	*NN*-dialkylsulphonamide,	
Ortolan (BASF)	etc.	
(3) *One solubilising group*		
Isolan S (BAY)	Sulphonic acid	Fastness properties on
Lanacron (CGY)		wool and silk better
Lanasyn S (S)		than group (2)
Neutrichrome S (ICI)		
(4) *Two solubilising groups*		
Elbelan (LBH)	Carboxylic acid	Highly soluble
Acidol M (BASF)	Sulphonic acid	Used for wool/nylon blends and acid-dyeable polypropylene

acid dissociation constants of the oxygen- and sulphur-containing groups were practically identical, those metal complexes containing sulphur donors possessed greater stability constants than those containing oxygen donors.

5.8 COMMERCIAL METAL-COMPLEX DYES

Although the original Palatine Fast, Neolan and later Neopalatine (BASF) 1:1 metal complexes yield bright colours, their current importance is as intermediates in making 1:2 unsymmetrical complexes in which the position and number of solubilising groups can differ in the two chelating systems. This asymmetry results in a wider range of solubility and dyeing properties. Since they are invariably 1:2 metal–dye complexes, their classification can be based on the number and types of solubilising groups present. The most recent survey by Schneider [40] is given in summarised form in Table 5.2. All these dyes are classified in the *Colour Index* as acid dyes and are 1:2 metal–dye complexes, and almost all are based on chromium. There are one or two exceptions: Avilon Golden Yellow 2R (CGY, C.I. Acid Yellow 119) and Lanasyn Yellow LNW (S) are 1:2 cobalt complexes, and Isolan Orange K-RLS (BAY, C.I. Acid Orange 107) is a bidentate 1:2 copper complex.

REFERENCES

1. F Beffa and G Back, *Rev. Prog. Coloration*, **14** (1984) 33.
2. R Price in *Chemistry of synthetic dyes*, Ed. K Venkataraman, Vol. 3 (New York: Academic Press, 1970) 303.
3. G. Schetty, *J.S.D.C.*, **71** (1955) 705.
4. A Werner, *Ber.*, **41** (1908) 1062.
5. R Bohn (BASF), German P 282 987 (1912).
6. BASF, German P 284 856 (1915).
7. G Schetty, *Textilrundschau*, **5** (1950) 399.
8. G T Morgan and H D K Drew, *J.C.S.*, **117** (1920) 1456.
9. R Price, *Chimia*, **28** (1974) 221.
10. *The chemistry of coordination compounds*, Ed. J C Bailar Jr (New York: Reinhold, 1956).
11. F Basolo and R C Johnson, *Coordination chemistry*, Revised Edn (New York: Benjamin, 1964).
12. C F Bell, *Principles and applications of metal chelation* (Oxford: Clarendon Press, 1977).
13. E Coates, *J.S.D.C.*, **99** (1983) 382.
14. H D K Drew and J K Landquist, *J.C.S.*, (1938) 292.
15. R W Hay and J A G Edmonds, *Chem. Commun.*, (1967) 969.
16. R F M Sureau, *Chimia*, **19** (1965) 254.
17. H D K Drew and R E Fairbairn, *J.C.S.*, (1939) 823.
18. P A Mack and R Price, *Ind. Chim. Belge Suppl.*, (1967) 31.
19. H Pfitzner, *Melliand Textilber.*, **35** (1954) 649.
20. G Schetty, *J.S.D.C.*, **71** (1955) 705.
21. L Hunter and L B Roberts, *J.C.S.*, (1941) 820.
22. R Wizinger, *Chimia* (Suppl.), (1968) 82.
23. DUP, USP 2 662 075 (1952).

24. F H Moser and A L Thomas, *Phthalocyanine compounds* (New York: Reinhold, 1963).
25. ICI, BP 1 094 746 (1965).
26. H Pfitzner and H Baumann, *Angew. Chem.*, **70** (1958) 232.
27. B I Stepanov in *Recent progress in the chemistry of natural and synthetic colouring matters and related fields*, Ed. T S Gore et al. (New York: Academic Press, 1962) 451.
28. R Price in *Chemistry of synthetic dyes*, Ed. K Venkataraman, Vol. 3 (New York: Academic Press, 1970) 348.
29. P Pfeiffer and S Saure, *Ber.*, **74** (1941) 935.
30. G Schetty and W Kuster, *Helv. Chim. Acta*, **44** (1961) 2193; G Schetty, *Helv. Chim. Acta*, **45** (1962) 1095, 1473; **46** (1963) 1132; **47** (1964) 921.
31. L Beffa, P Leinhard, E Steiner and G Schetty, *Helv. Chim. Acta*, **46** (1963) 1369.
32. H Jaggi, *Helv. Chim. Acta*, **51** (1967) 580.
33. G Schetty, *Textilveredlung*, **1** (1961) 3.
34. H Zollinger, *Chemie der Azofarbstoffe* (Basel: Verlag Birkhäuser, 1958) 238.
35. G Schetty and E Steiner, *Helv. Chim. Acta*, **57** (1974) 2149.
36. A Johnson, M S Mort, R H Peters and M J Wada, *de Tex*, **21** (1962) 453.
37. F A Snavely, W C Fernelius and B E Douglas, *J.S.D.C.*, **73** (1957) 491.
38. E Coates and B Rigg, *Trans. Faraday Soc.*, **64** (1968) 3255.
39. F A Snavely, B D Krecker and C G Clark, *J. Amer. Chem. Soc.* **81** (1959) 2337.
40. K R Schneider, *Dyer*, **172** (9) (1987) 13.

CHAPTER 6

Chemistry of anthraquinonoid, polycyclic and miscellaneous colorants

Geoffrey Hallas

6.1 ANTHRAQUINONE ACID, DISPERSE, BASIC AND REACTIVE DYES

It was pointed out in Chapter 1 that, after the azo class, anthraquinone derivatives form the next most important group of organic colorants listed in the *Colour Index*. The major application groups are vat dyes, disperse dyes and acid dyes (Table 1.1).

Of the various anthracenedione isomers, only the 9,10-compound is used for the synthesis of dyes; it is usually referred to simply as anthraquinone (6.1). The parent compound is pale yellow in colour, having a weak absorption band in the visible region (n→π* transition). The presence of one or more electron-donating substituents leads to significant bathochromic effects so that relatively simple derivatives are of commercial importance as dyes. The colour of such compounds, which usually contain amino or hydroxy groups, can be attributed to the existence of a charge-transfer absorption band [1].

From a historical viewpoint, the advent of synthetic anthraquinone dyes can be traced to the elucidation of the structure of the important naturally occurring compound alizarin (1,2-dihydroxyanthraquinone) (6.2) by Graebe and Liebermann in 1868 [2]. Natural alizarin is the colouring matter of madder and was the source of the then valuable mordant dye Turkey red (C.I. Mordant Red 11), an aluminium–calcium complex. In the following year Caro and Perkin independently discovered a commercially viable route to alizarin from anthraquinone via the 2-sulphonic acid (6.3). Treatment of the sodium salt of anthraquinone-2-sulphonic acid (called 'silver salt', from its appearance) with aqueous sodium hydroxide under pressure at an elevated temperature in the presence of an oxidising agent gives the disodium salt of alizarin as the main product (Scheme 6.1). Perkin soon devised a more efficient and cheaper route that involved chlorination of

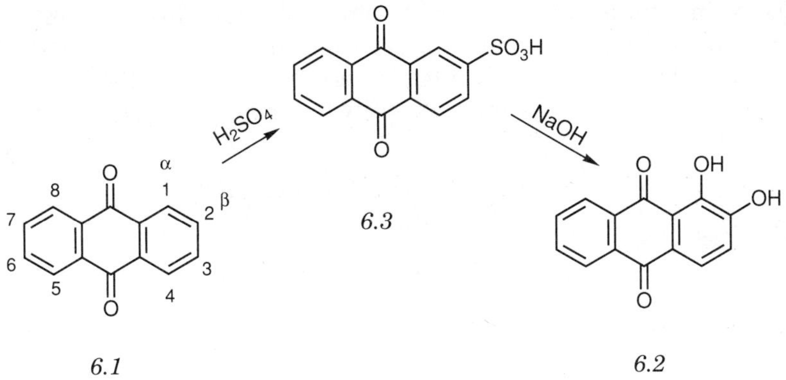

Scheme 6.1

anthracene to give 9,10-dichloroanthracene, which was then sulphonated, oxidised and fused with sodium hydroxide to yield alizarin [3].

6.1.1 Methods of synthesis

Anthraquinone itself is traditionally available from the anthracene of coal tar by oxidation, often with chromic acid or nitric acid; a more modern alternative method is that of air oxidation using vanadium(V) oxide as catalyst. Anthraquinone is also produced in the reaction of benzene with phthalic anhydride (6.4) using a Lewis acid catalyst, typically aluminium chloride. This Friedel–Crafts acylation gives *o*-benzoylbenzoic acid (6.5) which undergoes cyclodehydration when heated in concentrated sulphuric acid (Scheme 6.2). Phthalic anhydride is readily available from naphthalene or from 1,2-dimethylbenzene (*o*-xylene) by catalytic air oxidation.

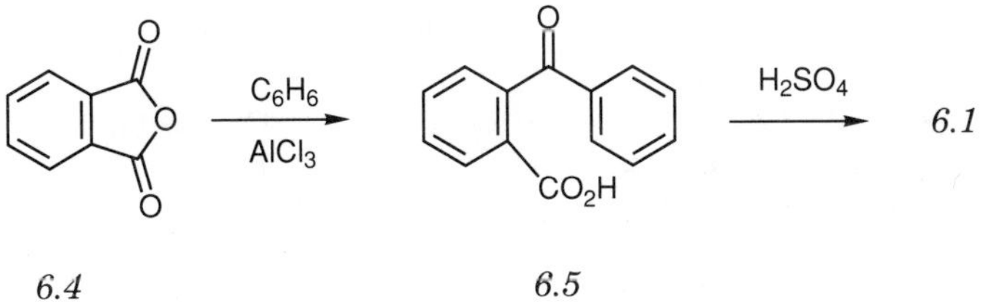

Scheme 6.2

Although anthraquinone is the starting point for the preparation of many derivatives, involving substitution and replacement reactions, certain compounds are obtained directly by varying the components in the above synthesis. Thus, for example, replacement of benzene with methylbenzene (toluene) leads to the formation of 2-methylanthraquinone. A particularly important variation on the phthalic anhydride route is the synthesis of quinizarin (1,4-dihydroxyanthraquinone) (6.6) using 4-chlorophenol with sulphuric acid and boric acid as catalyst (Scheme 6.3). The absence of aluminium chloride permits hydrolysis of the chloro substituent to take place.

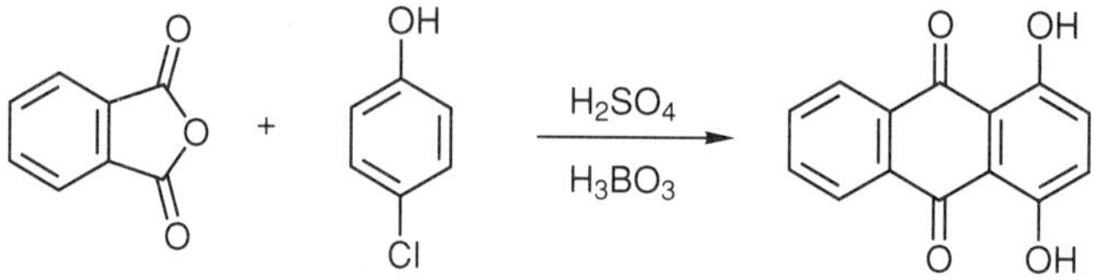

6.6

Scheme 6.3

Electrophilic substitution at the anthraquinone ring system is difficult due to deactivation (electron withdrawal) by the carbonyl groups. Although the 1-position in anthraquinone is rather more susceptible to electrophilic attack than is the 2-position, as indicated by π-electron localisation energies [4], direct sulphonation with oleum produces the 2-sulphonic acid (6.3). The severity of the reaction conditions ensures that the thermodynamically favoured 2-isomer, which is not subject to steric hindrance from an adjacent carbonyl group, is formed. However, the more synthetically useful 1-isomer (6.7) can be obtained by sulphonation of anthraquinone in the presence of a mercury(II) salt (Scheme 6.4). It appears that mercuration first takes place at the 1-position followed by displacement. Some disulphonation occurs, leading to the formation of the 2,6- and 2,7- or the 1,5- and 1,8-disulphonic acids respectively. Separation of the various compounds can be achieved without too much difficulty. Sulphonation of anthraquinone derivatives is also of some importance.

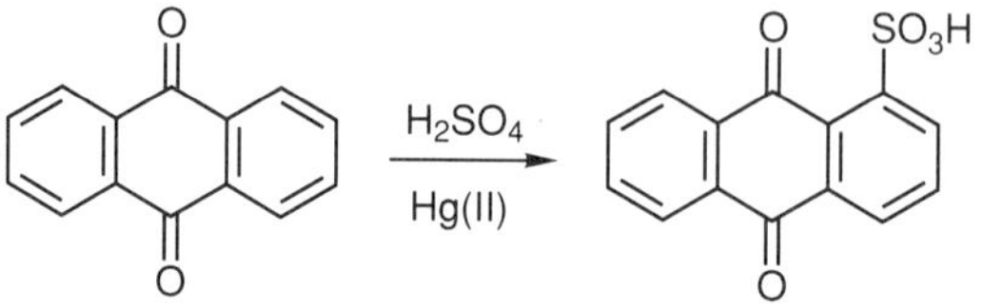

6.1 *6.7*

Scheme 6.4

Anthraquinone-1-sulphonic acid is the traditional precursor of 1-aminoanthraquinone (6.8), the most important anthraquinone intermediate. Since it is expensive to eliminate mercury(II) ions from waste water, an alternative route via 1-nitroanthraquionone has been examined in recent years. Nitration of anthraquinone gives, as well as the desired 1-nitro derivative, significant amounts of the 2-isomer together with 1,5- and 1,8-dinitroanthraquinones. Nevertheless, chemists at Sumitomo in Japan and at Bayer in West Germany have optimised the nitration procedure with respect to both yield and purity of the 1-nitro compound [5]. Nitration of anthraquinone derivatives is also of some significance.

Direct halogenation is normally carried out only on derivatives of anthraquinone. The importance of chloro or bromo derivatives lies mainly in the displacement of the halogen by nucleophiles, especially amines. Simple derivatives, such as 1-chloroanthraquinone (6.9), are available from the appropriate sulphonic acid by replacement (Scheme 6.5). 1-Amino-4-bromoanthraquinone-2-sulphonic acid (bromamine acid) (6.10) is a very useful intermediate in the synthesis of dyes derived from 1,4-diaminoanthraquinone. Bromamine acid is obtained from 1-aminoanthraquinone (6.8) by sulphonation to give the 2-sulphonic acid (6.11), which is then brominated (Scheme 6.6). Direct bromination of 1-aminoanthraquinone affords the 2,4-dibromo derivative. Hydroxyanthraquinones are also susceptible to halogenation. In the case of alizarin (6.2), for example, the 1-hydroxy group is involved in intramolecular hydrogen bonding with the adjacent carbonyl group so that bromination takes place at the 3-position (6.12), which is both *ortho* to the 2-hydroxy group and *meta* to the other carbonyl group (Scheme 6.7).

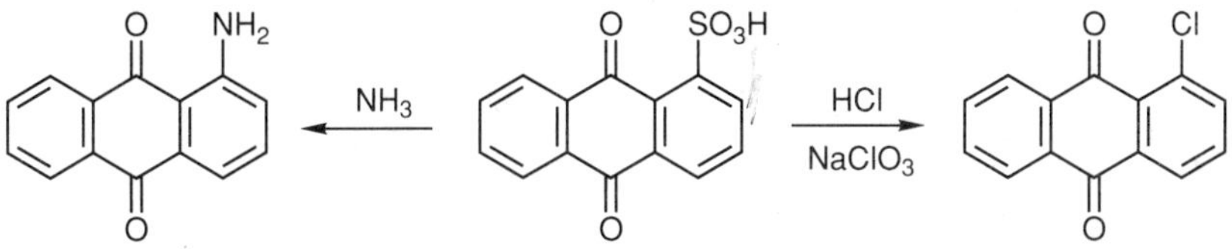

6.8 *6.7* *6.9*

Scheme 6.5

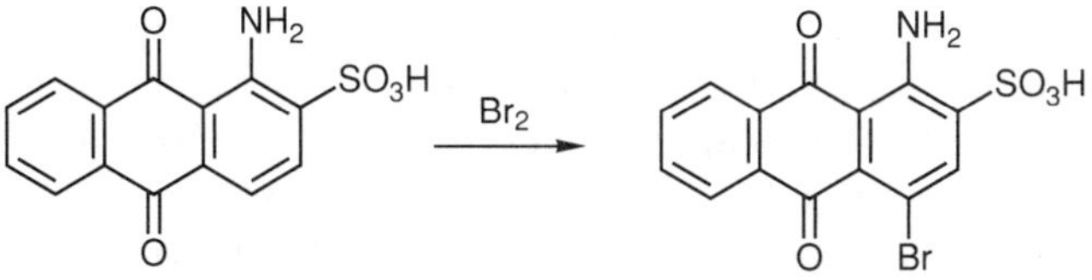

6.11 *6.10*

Scheme 6.6

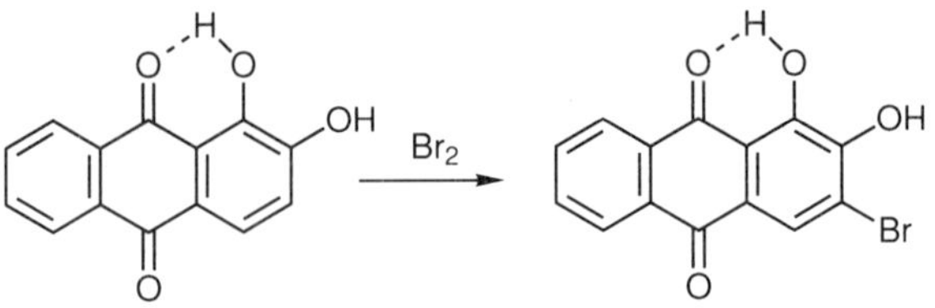

6.2 *6.12*

Scheme 6.7

Although some hydroxyanthraquinones, such as quinizarin (6.6), are available by direct synthesis, most hydroxy derivatives are produced by nucleophilic displacement of sulphonic acid groups or halogen atoms. Thus 1,5-dihydroxyanthraquinone (6.13; anthrarufin) and the 1,8-isomer (6.14; chrysazin) are obtained by alkaline hydrolysis of the corresponding disulphonic acids using calcium hydroxide or dilute sodium hydroxide. Under such relatively mild conditions additional hydroxy groups are not introduced. In the case of anthraquinone-2-sulphonic acid (6.3), as pointed out earlier, hydroxylation as well as displacement can take place to produce alizarin (6.2). From a mechanistic point of view it seems likely that the reaction first involves hydroxylation to produce a hydroquinone (6.15), which is then oxidised to 1-hydroxyanthraquinone-2-sulphonic acid (6.16) before nucleophilic displacement of the sulphonic acid group takes place (Scheme 6.8) [5].

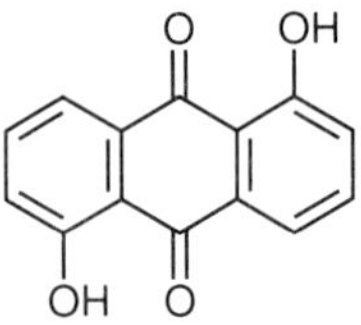

6.13

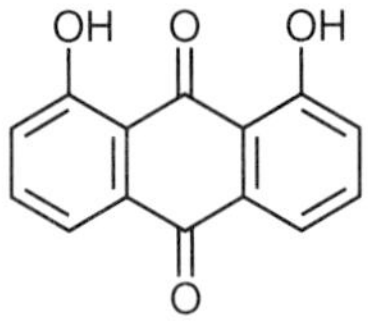

6.14

Polyhydroxyanthraquinone derivatives can be obtained by means of the Bohn–Schmidt reaction, in which hydroxyanthraquinones containing at least one α-hydroxy group react with fuming sulphuric acid; thus, for example, alizarin can be converted into 1,2,5-trihydroxyanthraquinone (6.17). The presence of boric acid leads to the formation of mixed anhydrides with sulphuric acid and to cyclic intermediates. The substitution mechanisms are complex but it has been established [6] that sulphur trioxide can interact with the 1-hydroxy group and the adjacent carbonyl group to form esters containing six-membered rings, exemplified by structure 6.18. The sulphuric acid esters are readily hydrolysed to hydroxy compounds during work-up. Sulphonation and subsequent desulphonation can also take place.

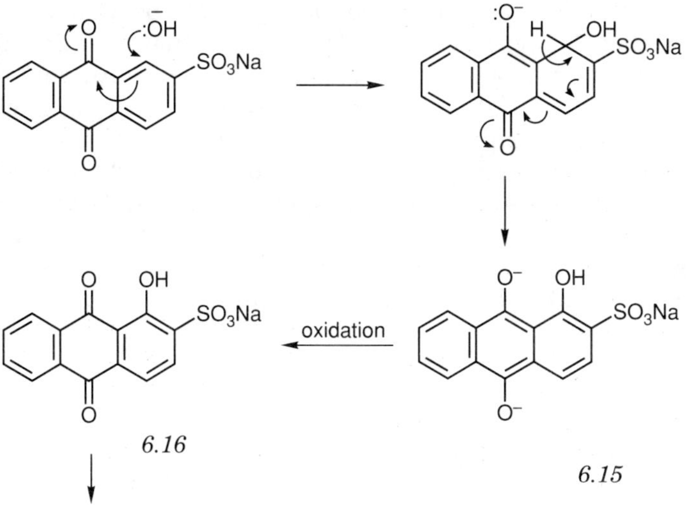

6.16

6.15

6.2

Scheme 6.8

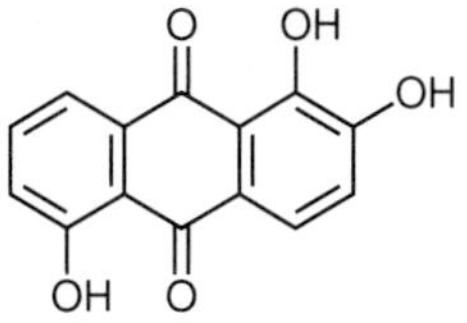

6.17

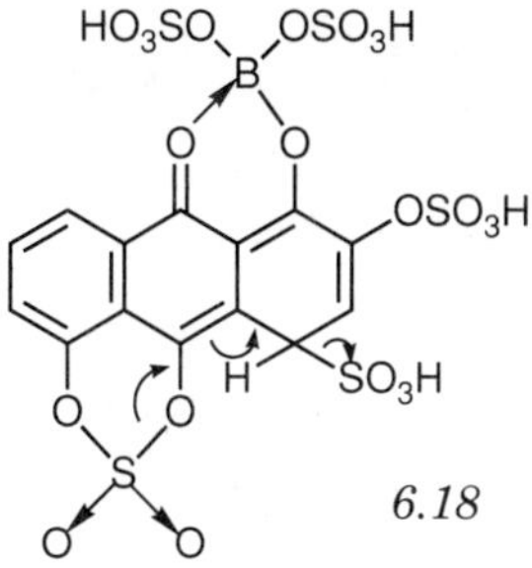

6.18

The Bohn–Schmidt procedure is also applicable to aminoanthraquinones. Thus 1,4-diaminoanthraquinone (6.19) can be converted into 1,4-diamino-5,8-dihydroxyanthraquinone (6.20).

Nucleophilic displacement of halogen by amines is an important method of introducing amino groups into the anthraquinone ring system. In the Ullmann reaction the displacement is catalysed by metallic copper or by

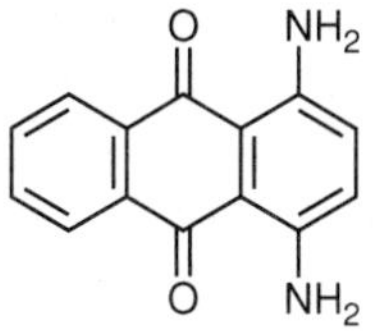

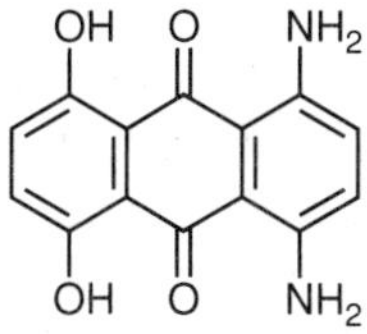

6.19

6.20

copper ions so that relatively mild conditions can be used. Recent mechanistic studies [5] suggest that it is copper(I) ions that exert a catalytic effect via complex formation. Derivatives of 1,4-diaminoanthraquinone are of considerable industrial significance. Many compounds are prepared from the reduced form of quinizarin (6.6). Leucoquinizarin can be represented by various tautomers such as structures 6.21 and 6.22; however, ^{1}H- and ^{13}C-n.m.r. have shown the dominant species to be the 2,3-dihydro tautomer (6.22) [7]. Alkylamines react with leucoquinizarin in a stepwise manner to give the corresponding 1,4-dialkylamino derivative (6.23) after air oxidation (Scheme 6.9). The less nucleophilic arylamines do not react unless

6.21

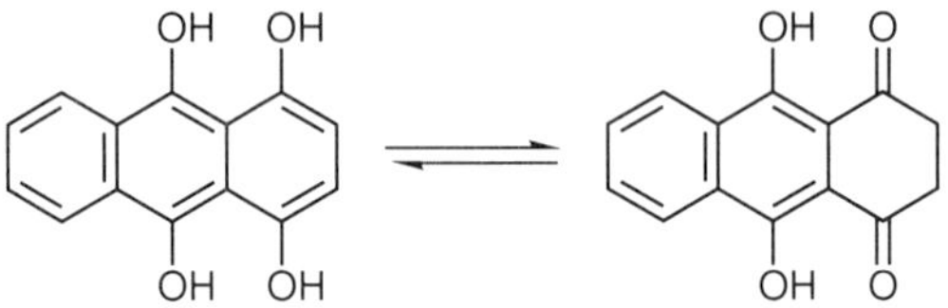

6.22

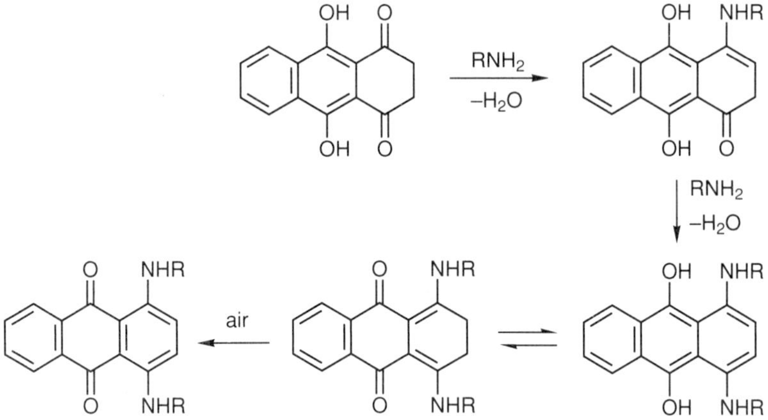

6.23

Scheme 6.9

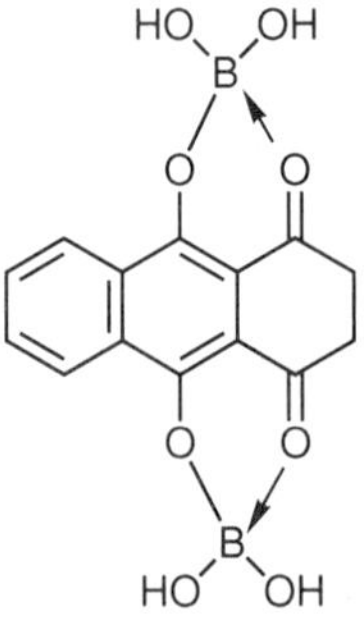

6.24

boric acid is present as a catalyst. The complex thus formed (6.24) is more susceptible to nucleophilic attack at the 1- and 4-positions than is leucoquinizarin. Unsymmetrical 1,4-diaminoanthraquinones can be synthesised by using two different amines; the corresponding symmetrical dyes are also formed to a certain extent.

6.1.2 Acid, disperse, basic and reactive dyes

Anthraquinone dyes tend to predominate in the violet, blue and green hue sectors. Although they have the advantages of brightness and chemical stability, they are more expensive to manufacture than are azo dyes, which are also tinctorially stronger.

In general, 1-substituted derivatives of anthraquinone are more bathochromic than the corresponding 2-substituted isomers, in accordance with PPP–MO calculations [1]. Intramolecular hydrogen bonding is not possible between the carbonyl group and a 2-substituent and, in the 2-substituted series, the bathochromic shift increases with electron-donor power in the sequence OH < OCH_3 < $NHCOCH_3$ < NH_2 < $NHCH_3$ < $N(CH_3)_2$. An enhanced bathochromic effect is observed when a 1-substituent, such as hydroxy or methylamino (6.25), is able to hydrogen-bond with the adjacent carbonyl group and thereby assist the conjugation of the donor lone pair of electrons with the anthraquinone ring system; for example, a methylamino group is a more effective donor than is a dimethylamino group in the 1-substituted series. Moreover, the latter group is unable to conjugate fully due to steric hindrance between the bulky substituent and the adjacent carbonyl group, which decreases the bathochromic effect and reduces the tinctorial strength (intensity). Steric interaction is minimised by rotation of the dimethylamino group out of the plane of the anthraquinone ring system (6.26). Some illustrative spectral data are given in Table 6.1 [8]. Significant crowding effects can also arise from the presence of *ortho* groups, as in 1,2-disubstituted anthraquinones.

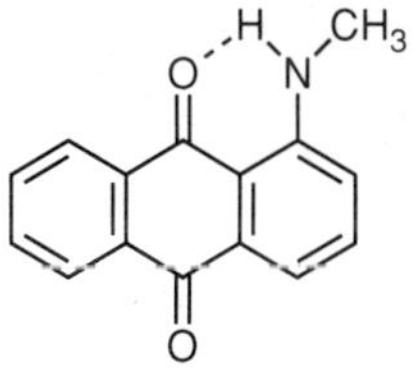

6.25

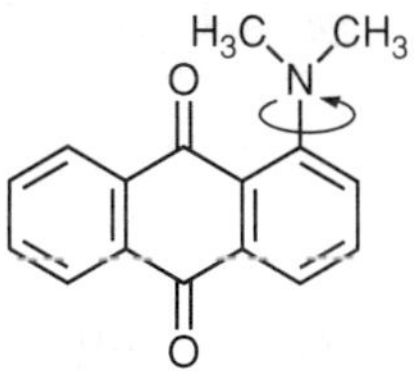

6.26

Anthraquinone itself can be regarded approximately as consisting of two *ortho*-interlinked benzoyl groups [9]. Thus the introduction of a second group into the unsubstituted ring of a monosubstituted anthraquinone effectively doubles the extinction coefficient (compare ε_{max} values for α-hydroxy and α-amino substitution in Tables 6.1 and 6.2). However, the

TABLE 6.1

Spectral data for some monosubstituted anthraquinones in methanol

	1-position		2-position	
Substituent	λ_{max}/nm	ε_{max}	λ_{max}/nm	ε_{max}
OH	402	5500	368	3900
OCH_3	378	5200	363	3950
$NHCOCH_3$	400	5600	367	4200
NH_2	475	6300	440	4500
$NHCH_3$	503	7100	462	5700
NHC_6H_5	500	7250	467	7100
$N(CH_3)_2$	503	4900	472	5900

TABLE 6.2

Spectral data for some disubstituted anthraquinones in methanol

Substituents	λ_{max}/nm	ε_{max}
1,4-diOH	470	17 000
1,5-diOH	425	10 000
1,8-diOH	430	10 960
1,4-diNH_2	550, 590	15 850, 15 850
1,5-diNH_2	487	12 600
1,8-diNH_2	507	10 000

presence of two donor groups in the 1,4-positions gives rise to a pronounced increase in intensity together with a significant bathochromic shift (Table 6.2) [8]. These enhanced effects have been widely exploited in commercial dyes. The order of bathochromicity for amino-substituted anthraquinones is 2 < 1 < 1,5 < 1,8 < 1,2 < 1,4 < 1,4,5,8. This sequence can be explained by VB theory [9] as well as by PPP–MO calculations [1], but only the latter approach can account for variations in tinctorial strength. The twin absorption peaks of 1,4-diaminoanthraquinones are thought to be caused by vibrational fine structure associated with one electronic transition in the visible region. 1,4-Bis(alkylamino)anthraquinones are generally blue. The introduction of arylamino substituents into the 1,4-positions gives rise to an additional absorption band in the 400 nm region that imparts a yellow component to the dominant blue colour. A good example is Alizarine Cyanine Green G (6.27; C.I. Acid Green 25), which in aqueous solution absorbs at 410, 608 and 646 nm [10].

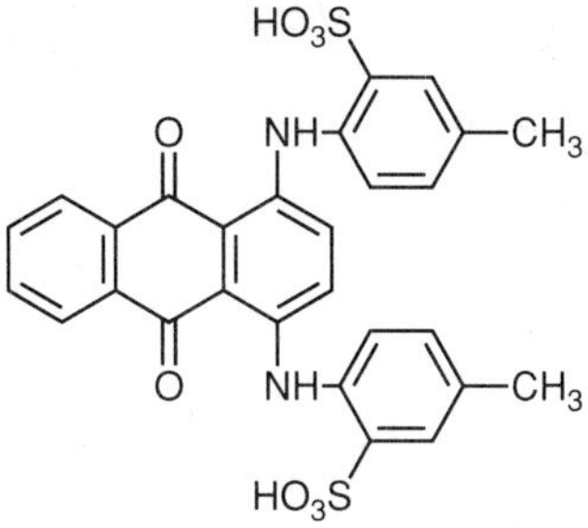

6.27

Acid dyes
In 1871 Graebe and Liebermann discovered that alizarin (6.2) could be applied to wool by mordant dyeing after sulphonation to produce the 3-sulphonic acid (6.28). This dye is commonly referred to as Alizarine Red S (C.I. Mordant Red 3) and is still listed in the latest revision of the *Colour Index* as a commercial product [11]. Although many sulphonated polyhydroxyanthraquinones have been examined, few remain in current use. Another, and more important, classic dye that continues in commercial use as an acid dye is Alizarine Saphirol B (6.29; C.I. Acid Blue 45). This dye was discovered in 1897 by Schmidt and can be made from anthrarufin (6.13) by disulphonation, subsequent dinitration and reduction. The dye gives an attractive blue on wool with good all-round fastness properties.

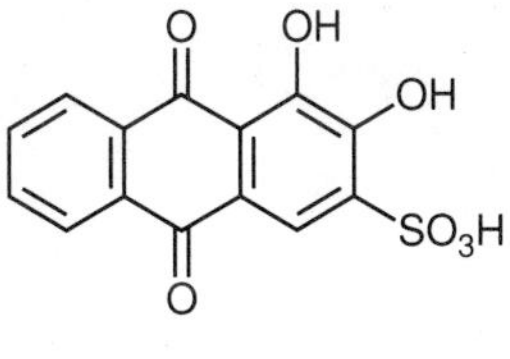

6.28

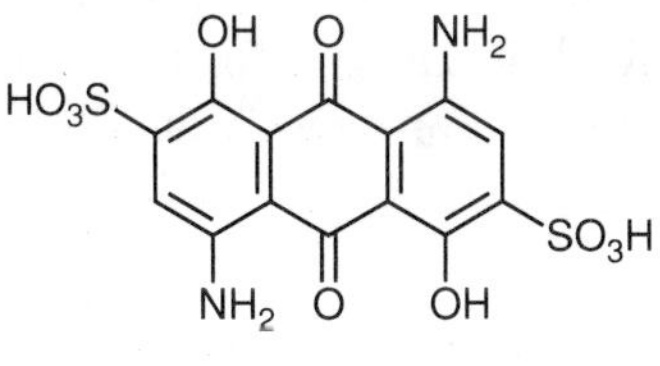

6.29

Many blue acid dyes have been derived from bromamine acid (6.10) by nucleophilic displacement of the halogen atom using arylamines of varying complexity in the presence of a copper catalyst. Simple examples include C.I. Acid Blue 25 (6.30; R = H) and the analogous cyclohexyl derivative C.I. Acid Blue 62 (6.31). Reaction with *p*-butylaniline gives C.I. Acid Blue 230 (6.30; R = butyl), which possesses good washing fastness owing to hydrophobic interaction with the substrate. Somewhat more complex structures are illustrated by C.I. Acid Blue 40 (6.32) and C.I. Acid Blue 129 (6.33). C.I. Acid Blue 40 (6.32) is greener than the parent dye (6.30; R = H) due to the electronic effect of the acetylamino substituent.

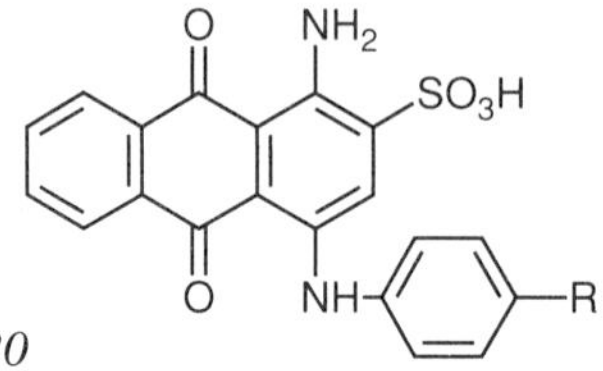

6.30

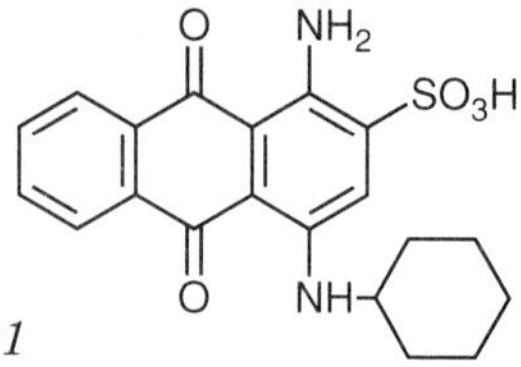

6.31

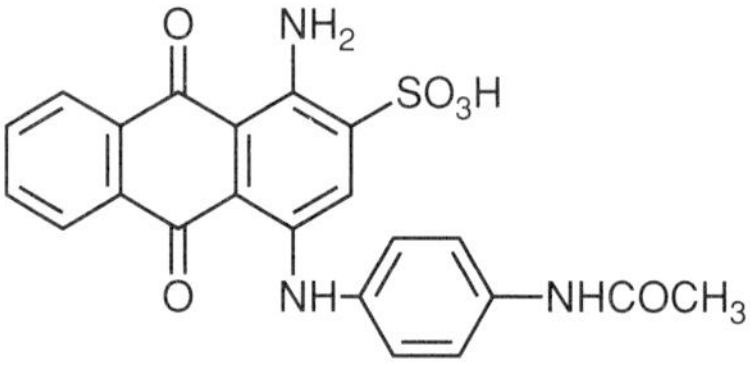

6.32

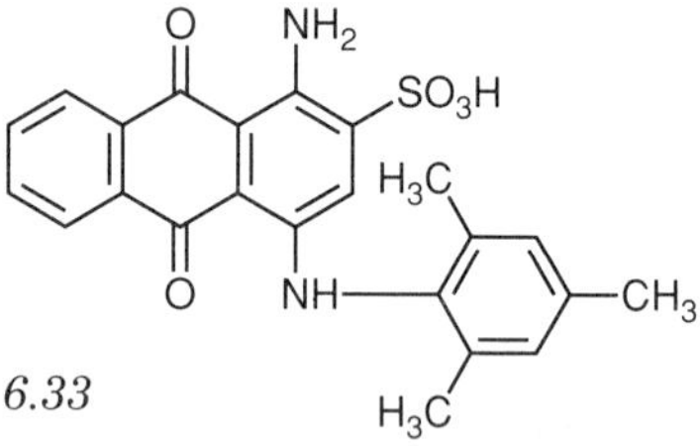

6.33

Steric crowding in the case of C.I. Acid Blue 129 (6.33) reduces the conjugation of the 4-substituent with the remainder of the system and results in a reddish-blue hue. The non-aromatic cyclohexyl group in C.I. Acid Blue 62 (6.31) has a similar effect on the hue.

Use of the intermediates 1-amino-2,4-dibromoanthraquinone and 1-amino-4-bromo-2-methylanthraquinone leads to similar dyes, exemplified by C.I. Acid Blue 78 (6.34; X = Br) and C.I. Acid Blue 47 (6.34; X = CH_3) respectively; in the production of these two dyes sulphonation is the final step.

Milling dyes with very good wet fastness are obtained by the reaction of one mole of a diamine with two moles of bromamine acid, as in the case of C.I. Acid Blue 127 (6.35). The dye is suitable for dyeing wool bright blue from a neutral or weakly acid bath.

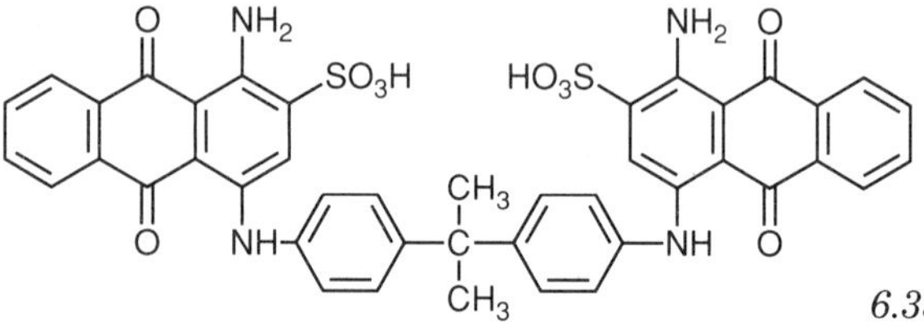

6.35

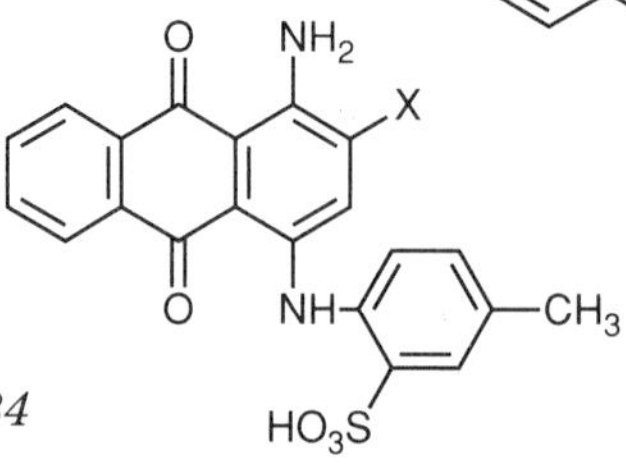

6.34

Condensation of one mole of leucoquinizarin (6.22) with two moles of an arylamine in the presence of boric acid, followed by oxidation and sulphonation, is the route to some symmetrically substituted acid dyes; these are blue to green according to the structure of the amine. The classical, but still important, dye C.I. Acid Green 25 (6.27) has already been mentioned. A related dye of commercial significance is C.I. Acid Blue 80 (6.36), which can be obtained either from leucoquinizarin or from 1,4-dichloroanthraquinone. Unsymmetrically substituted 1,4-diaminoanthraquinone derivatives bearing only one sulphonic acid group are better for the level dyeing of nylon than are their disulphonated analogues.

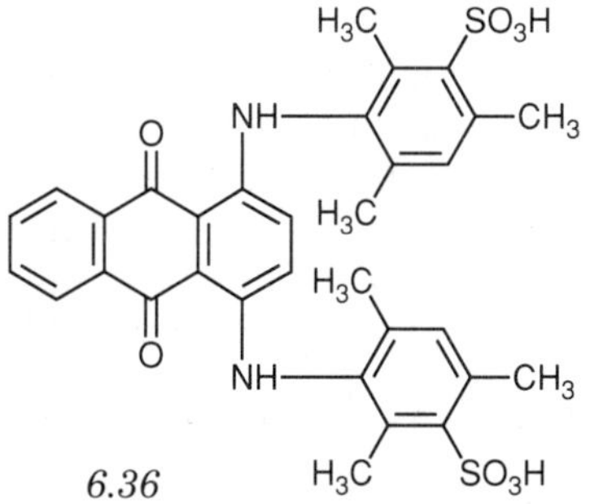

6.36

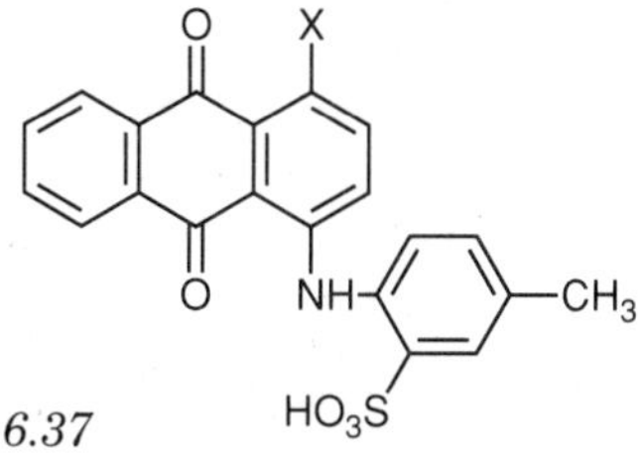

6.37

A suitable example is C.I. Acid Blue 27 (6.37; X = $NHCH_3$), which is made from 4-bromo-1-methylaminoanthraquinone by nucleophilic displacement followed by sulphonation. Some commercial acid dyes based on 1,4-diaminoanthraquinone contain substituents in both the 2- and the 3-positions. Thus, for example, C.I. Acid Violet 41 (6.38) is produced by the condensation of 1,4-diamino-2,3-dichloroanthraquinone with phenol in the presence of sodium sulphite and manganese dioxide.

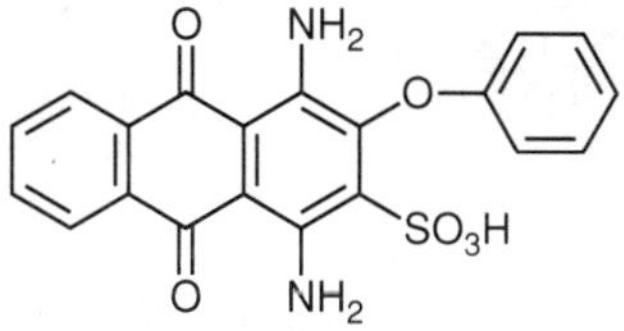

6.38

Aminohydroxyanthraquinone derivatives are available from leucoquinizarin by heating with an appropriate arylamine and boric acid in aqueous ethanol, followed by oxidation. The use of *p*-toluidine and subsequent sulphonation gives C.I. Acid Violet 43 (6.37; X = OH).

Disperse dyes

Anthraquinone dyes are second only to azo dyes in importance as disperse dyes and are predominant in the red, violet, blue and blue-green ranges [12]. Because anthraquinone disperse dyes are relatively expensive to manufacture, attempts have been made in recent years to replace some of them with technically equivalent and more economical products [13]. The replacement process has been most successful in the red region using, for example, heterocyclic azo dyes and novel chromophores. The brilliance of the anthraquinones with their narrow spectral absorption bands is difficult to attain with other structures, however, as is their high light fastness and chemical stability. The development of anthraquinone disperse dyes is included in a recent review by Dawson [14].

Dyes for cellulose acetate are relatively simple molecules, typified by C.I. Disperse Red 15 (6.39; X = OH), C.I. Disperse Violet 4 (6.39; X = $NHCH_3$) and C.I. Disperse Blue 3 (6.40), the last-named being manufactured from leucoquinizarin and the appropriate amines. The unsymmetrically substituted product inevitably contains significant amounts of the related symmetrical compounds. Bright red 2-alkoxy-1-amino-4-hydroxyanthraquinones, such as C.I. Disperse Red 4 (6.41), can be obtained from 1-amino-2,4-dibromoanthraquinone by hydrolysis to give 1-amino-2-bromo-4-hydroxyanthraquinone (C.I. Disperse Violet 17), which is then condensed with the appropriate alcohol.

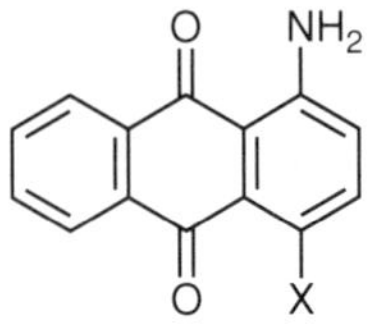

6.39

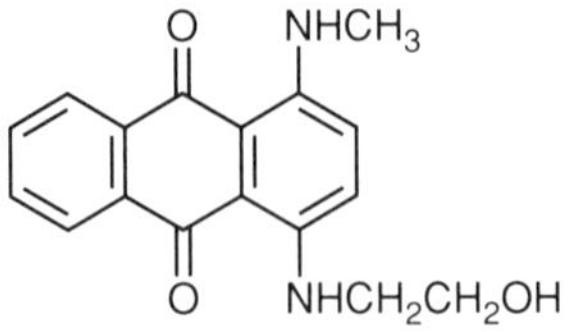

6.40

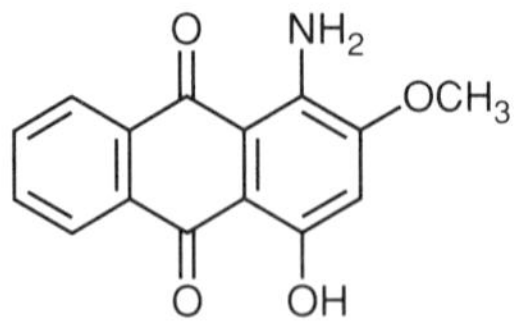

6.41

Suitable disperse dyes for polyester require good sublimation fastness and generally contain additional or more hydrophobic substituents compared with acetate dyes. Thus, for example, C.I. Disperse Red 60 (6.42) is important for the dyeing of polyester fabrics but has only moderate sublimation fastness. It is, however, the most important red dye for transfer printing. Improved sublimation fastness can be achieved if substituents are introduced into the phenoxy ring. Exceptional fastness is achieved by chlorosulphonation of C.I. Disperse Red 60 followed by condensation with 3-methoxypropylamine to give compound 6.43 [14].

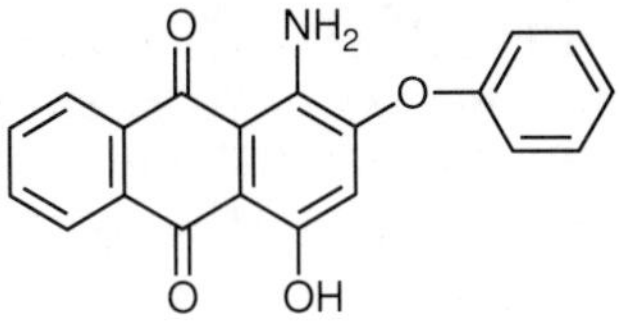

6.42

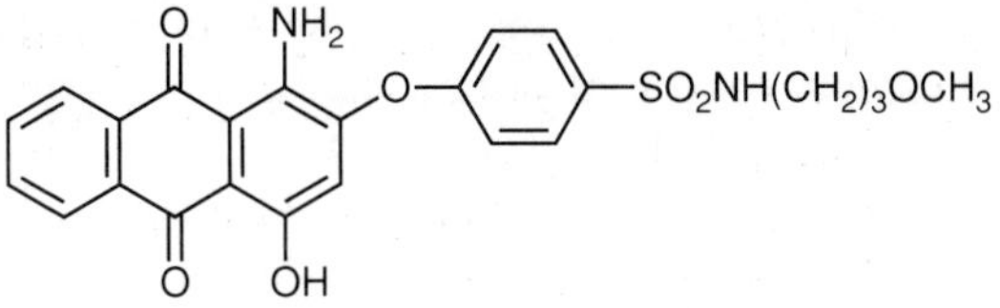

6.43

Monoarylation of 1-amino-4-hydroxyanthraquinone (6.39; X = OH) results in violet dyes such as C.I. Disperse Violet 27 (6.44; R = H) and the bluish-violet C.I. Disperse Blue 72 (6.44; R = CH_3); the latter dye is also important as C.I. Solvent Violet 13. Chlorination of 1,4-diaminoanthraquinone with sulphuryl chloride gives the 2,3-dichloro derivative (C.I. Disperse Violet 28), which on condensation with phenol yields C.I. Disperse Violet 26 (6.45). Monoaryl or dialkyl derivatives of 1,4-diaminoanthraquinone (6.19; C.I. Disperse Violet 1) are blue. Typical examples include C.I. Disperse Blue 19 (6.46) and C.I. Disperse Blue 23 (6.47).

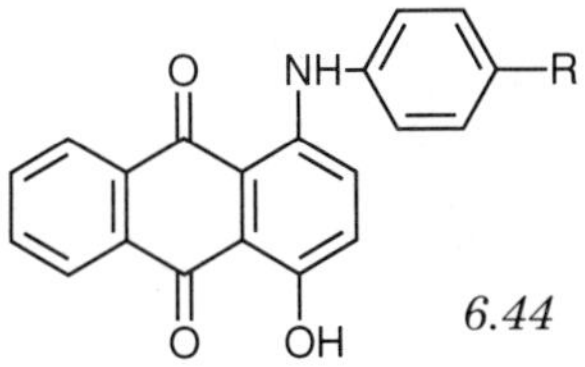

6.44

6.45

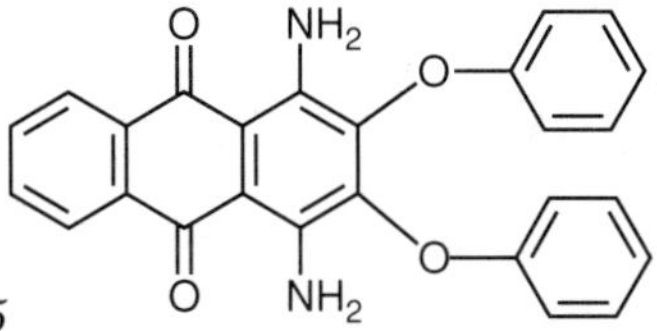

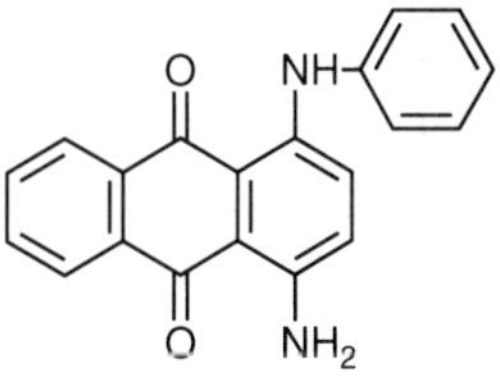

6.46

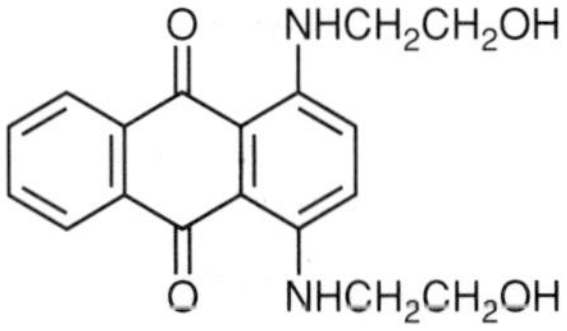

6.47

Many commercially important blue dyes are derived from tetra-α-substituted anthraquinones, especially the diamino derivatives of anthrarufin (6.13) and chrysazin (6.14). Examples include C.I. Disperse Blue 26 (6.48) and C.I. Disperse Blue 56 (6.49). The presence of amino groups is often associated with inferior fastness to burnt gas fumes. This drawback can be overcome by the use of dyes such as C.I. Disperse Blue 27 (6.50), which, although relatively expensive, has very good all-round fastness properties on both cellulose acetate and polyester. This dye can be obtained from chrysazin (6.14) by nitration and subsequent condensation with *p*-(2-hydroxyethyl)aniline. Bright blue dyes for polyester can be manufactured from the well-known acid dye C.I. Acid Blue 45 (6.29) by the *C*-arylation of its boric ester with, for example, a mixture of phenol and anisole, followed by removal of the sulphonic acid groups. This procedure gives C.I. Disperse Blue 73 (6.51; R = H and CH_3). 1,4,5,8-Tetra-aminoanthraquinone (C.I. Disperse Blue 1) is a cellulose acetate dye of long standing, as is the product of partial methylation, C.I. Disperse Blue 31.

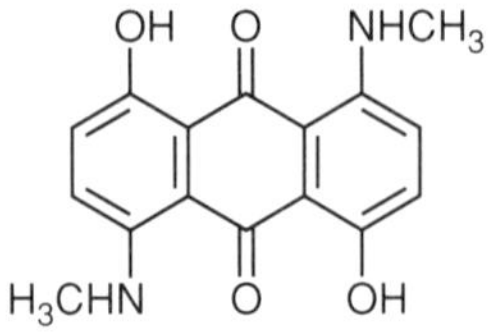

6.48

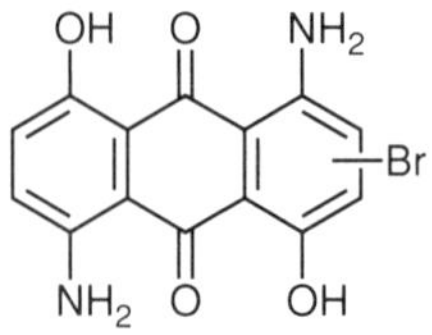

6.49

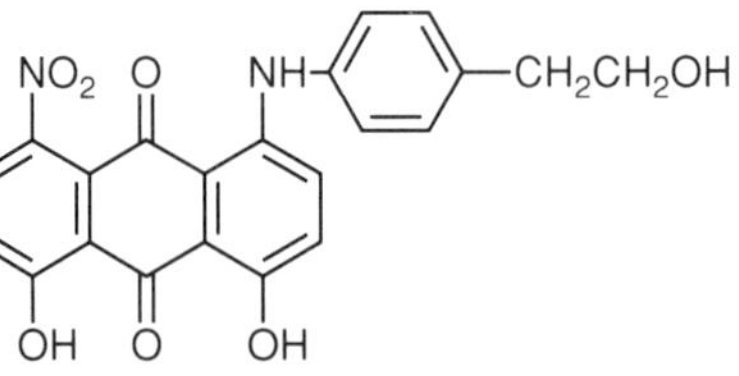

6.50

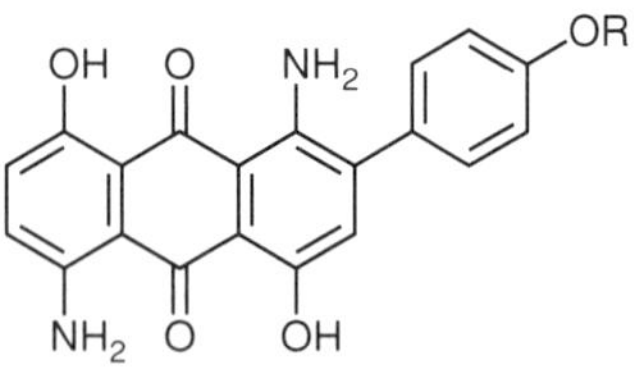

6.51

Few heterocyclic anthraquinone analogues are of established importance as disperse dyes. An interesting exception is the system 6.52, which provides bright turquoise blue dyes of very good light fastness on polyester [15]. A route to C.I. Disperse Blue 60 (6.52; R = $CH_3OCH_2CH_2CH_2$) involves hydrolysis of 1,4-diamino-2,3-dicyanoanthraquinone followed by alkylation of the resulting imide; several closely related products can be obtained by varying the alkyl group.

Not many green anthraquinone disperse dyes are still available commercially. One example is C.I. Disperse Green 6:1 (6.53), which has wider use as C.I. Solvent Green 3 and is the precursor of C.I. Acid Green 25 (6.27).

6.52

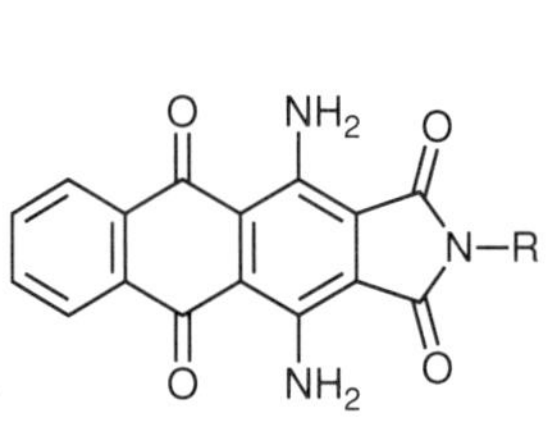

6.53

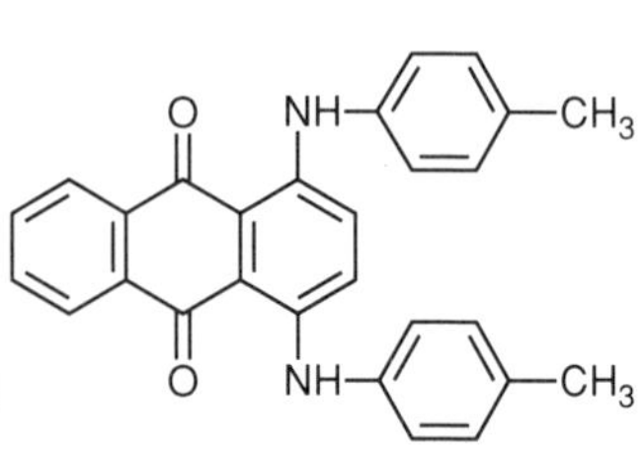

Basic dyes

Some anthraquinone dyes with pendent cationic groups are used commercially on acrylic and modacrylic fibres [16]. Only two disclosed structures are included in the latest revision of the *Colour Index*, the dyes being the reddish-blue C.I. Basic Blue 47 (6.54) and the greenish-blue C.I. Basic Blue 22 (6.55).

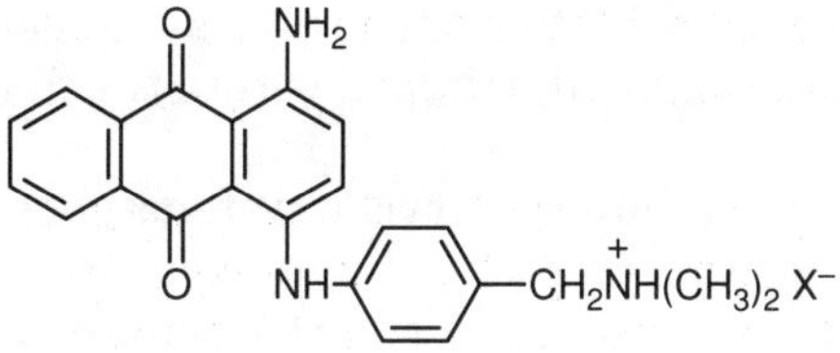

6.54

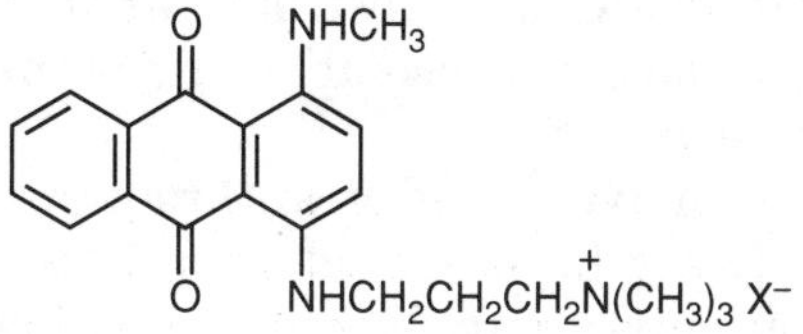

6.55

Reactive dyes

The chemistry of reactive dyes is discussed in detail in Chapter 7. However, it is appropriate to point out in this section that the anthraquinone system provides several commercially valuable bright blue reactive dyes. Bromamine acid (6.10) is the key intermediate for the synthesis of the vinylsulphone dye Remazol Brilliant Blue R (6.56; HOE, C.I. Reactive Blue 19), which finds application in dyeing and printing. The dichlorotriazinyl dye Procion Blue MX-R (6.57; ICI, C.I. Reactive Blue 4) is derived from the same intermediate. Procinyl Blue R (6.58; ICI, C.I. Reactive Blue 6), obtained from the reaction of 1,4-diaminoanthraquinone with epichlorohydrin, was a reactive disperse dye suitable for nylon. In recent years anthraquinone blues have been replaced to some extent by derivatives of the intrinsically bright triphenodioxazine chromophore [17].

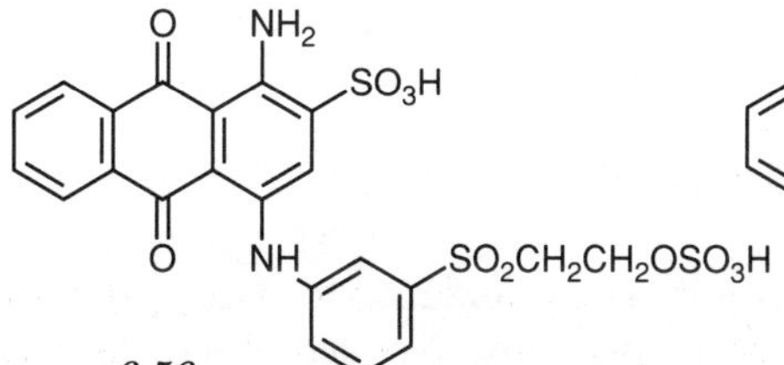

6.56

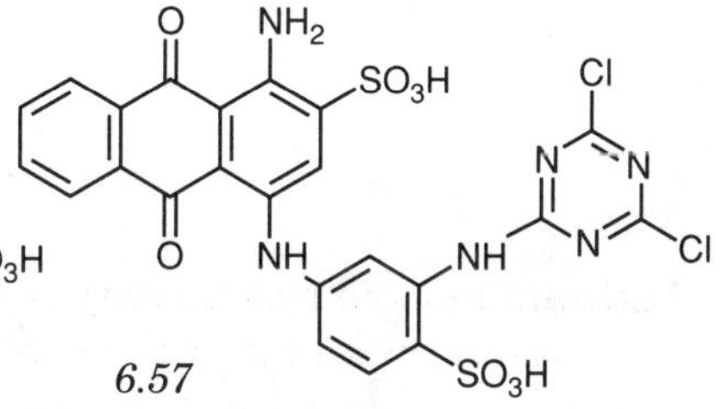

6.57

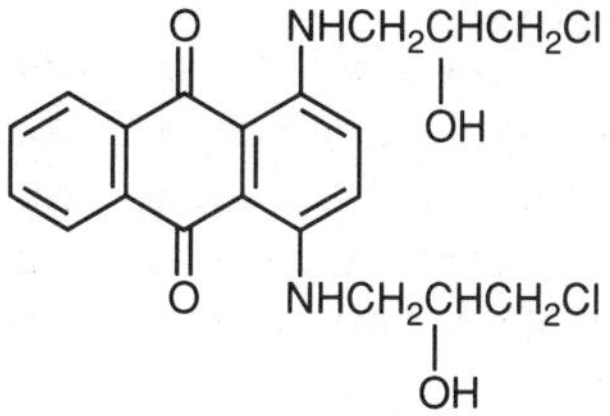

6.58

6.2 POLYCYCLIC VAT DYES

Vat dyes are insoluble in water and are applied to cellulosic fibres, usually with sodium dithionite under alkaline conditions, by a vatting process involving reduction to produce a water-soluble leuco form which possesses affinity for the substrate. Air or chemical oxidation then results in regeneration of the insoluble form of the dye (pigment) trapped within the fibre structure.

Vat dyes based on anthraquinone and related polycyclic systems generally exhibit outstanding fastness properties. Most vat dyes contain two carbonyl groups linked by a conjugated (quinonoid) system (Scheme 6.10). They can be applied as the water-soluble disulphuric esters of the leuco form, which have low affinity for cellulose; these can be obtained by esterifying the leuco compound with chlorosulphonic acid. Reduction and esterification may be carried out without isolating the leuco compound by using pyridine and iron together with chlorosulphonic acid [18].

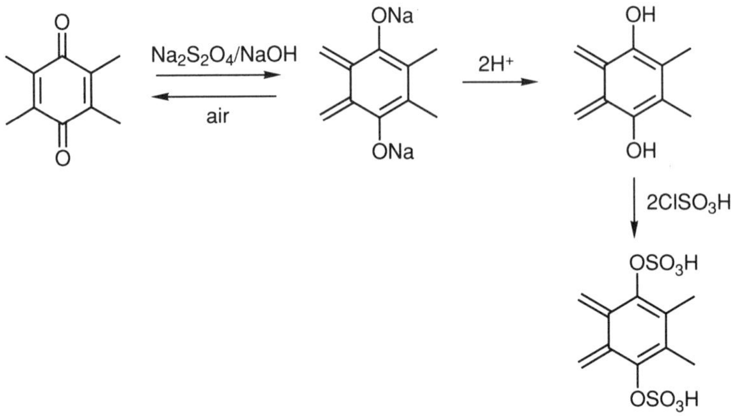

Scheme 6.10

Anthraquinonoid systems are the most important vat dyes, providing hues covering the whole of the visible spectrum. Nevertheless, the main commercial products are found within the blue, green, brown and black ranges. Some of the relatively simple yellow and orange dyes cause phototendering of the dyed substrate, bringing about adverse structural changes on exposure to light.

Most of the currently available vat dyes have been known for many years, and very few new dyes have been marketed during the last twenty years [19]. Recent research efforts have concentrated on improved methods of manufacture and more economical finishing of traditional vat dyes.

The general stability of the polycyclic vat dyes permits many of them to be used as pigments.

6.2.1 Anthraquinones

In 1909 Deinet observed that the product of *N*-benzoylation of 1-aminoanthraquinone could be applied as a vat dye; whereas the leuco form of 1-aminoanthraquinone has no affinity for cotton, the corresponding benzoyl derivative does have a moderate affinity.

Acylaminoanthraquinones are structurally the simplest of the vat dyes, and are used for yellow, orange, red and violet hues. Numerous acylaminoanthraquinones have been examined for use as vat dyes but only a few have survived competition from more cost-effective types. One example is C.I. Vat Yellow 26 (6.59), which is made from 1-aminoanthraquinone and isophthaloyl chloride. Another yellow dye of commercial importance is C.I. Vat Yellow 33 (6.60), which is made by reducing 4′-nitrobiphenyl-4-carboxylic acid with glucose and sodium carbonate. The resulting azo compound is acylated with thionyl chloride before condensing with 1-aminoanthraquinone. It is of significance that the azo bridge remains intact during the vatting process; in this dye the azo group is unable to tautomerise into a hydrazone and consequently resists reduction [20].

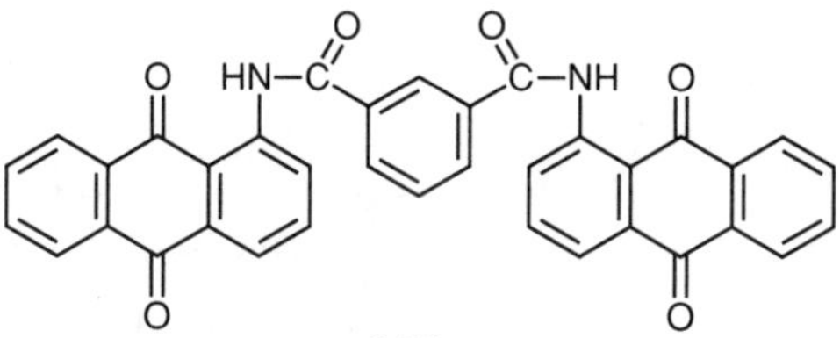

6.59

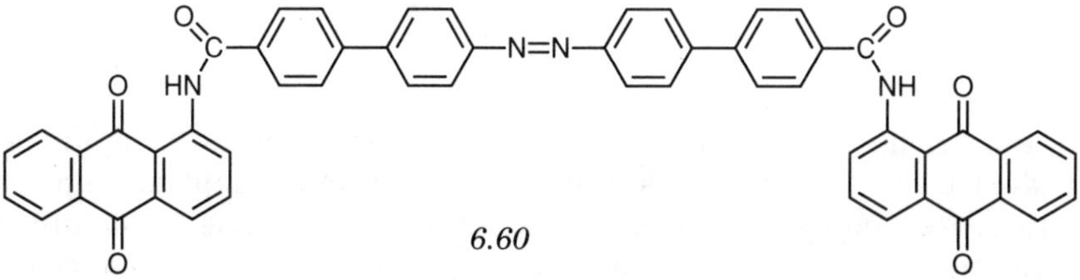

6.60

Derivatives of 1,4- and 1,5-diaminoanthraquinone provide some important examples in this class. Relatively simple dyes include C.I. Vat Yellow 3 (6.61; X = H) and C.I. Vat Violet 15 (6.61; X = OH).

It is generally found that dyes based on the 1,4-diaminoanthraquinone system are more bathochromic than isomers derived from 1,5-diaminoanthraquinone. This difference can be related to the relative stabilities of the dipolar excited states of the dyes. Qualitatively, the 1,4-disubstituted derivatives are aromatic (6.62) and therefore at a lower energy, whereas the 1,5-disubstituted isomers (6.63) are not. Consequently, the energy difference between the ground state of the molecule and its first excited state is smaller for 1,4-diaminoanthraquinones than for the 1,5-isomers. Thus, for example, the 1,4-isomer of dye 6.61 (X = H), C.I. Vat Red 42, is a yellowish-red (it is no longer in commercial use). More complex examples include C.I. Vat Red 21 (6.64), which contains amino groups in the acyl ring systems.

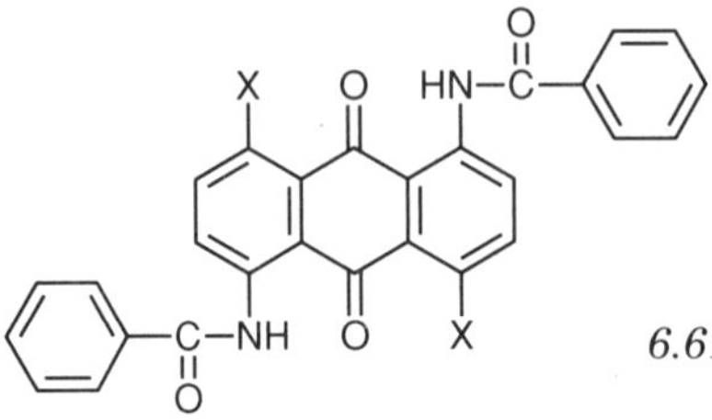

6.61

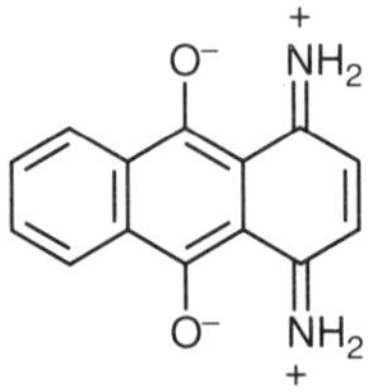

6.62

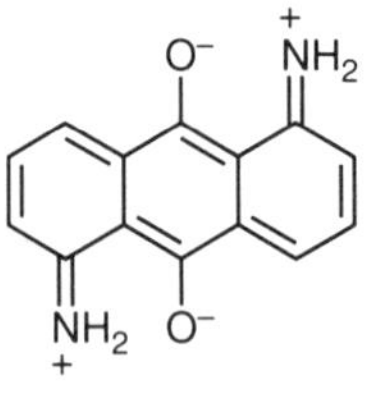

6.63

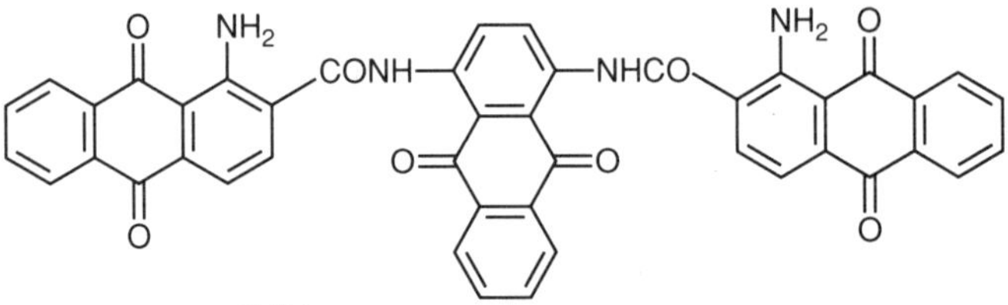

6.64

6.2.2 Indanthrones

In 1901 Bohn discovered the first vat dye based on anthraquinone, a landmark in the history of synthetic dyes. In an attempt to prepare an anthracene analogue of indigo (6.112), Bohn applied the Heumann synthesis to 2-aminoanthraquinone by fusing the condensation product of this amine and chloroacetic acid with potassium hydroxide. The brilliant blue product was named indanthrene, a name coined from those of indigo and anthracene. Bohn showed that the same product resulted from the alkali fusion of 2-aminoanthraquinone itself so that the dye could not be an analogue of indigo. The new vat dye was marketed by BASF as Indanthren Blue R (C.I. Vat Blue 4) and was subsequently given the chemical name indanthrone (6.65). The structure was proposed on the basis of analysis ($C_{28}H_{14}O_4N_2$, suggesting the loss of four atoms of hydrogen from two molecules of 2-aminoanthraquinone), the absence of azo and of primary amino groups and the exceptional stability of the compound, and also by synthesis: for example, Bohn obtained indanthrone by heating 1-amino-2-bromoanthraquinone in nitrobenzene solution under Ullmann conditions.

Indanthrone has low fastness to bleaching because the corresponding yellowish-green azine (6.66) is formed. Its importance at the time of its discovery lay in the dearth of fast dyes for cotton. Indanthrone is also used as a pigment (C.I. Pigment Blue 60).

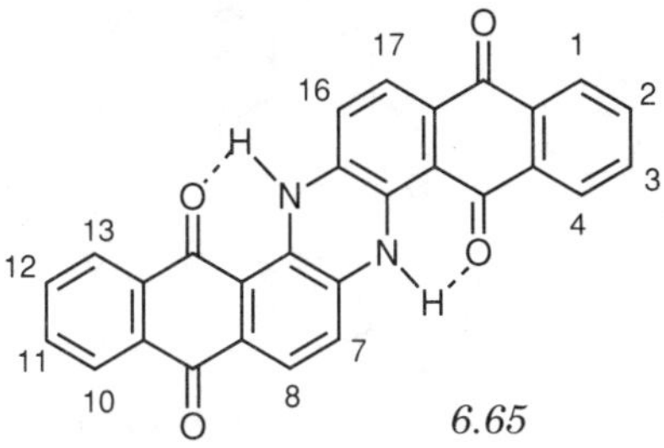

6.65

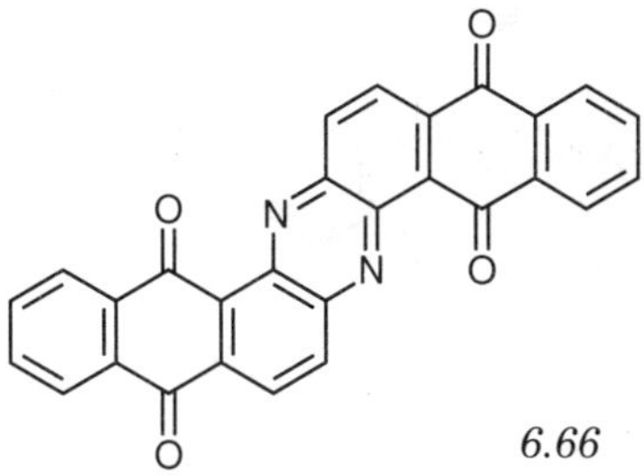

6.66

Very careful control of the fusion conditions is necessary in order to minimise the formation of by-products. Alkali fusion at approximately 220°C gives mainly the blue alkali salt of leuco indanthrone; air oxidation is then required to produce the dye. Fusion at lower temperatures results in the formation of alizarin (6.2) whereas reaction at a higher temperature (above 300°C) affords significant amounts of a yellow dye as well as the blue indanthrone. The yellow product was first marketed under the name Flavanthren and later as Indanthren Yellow G (BASF, C.I. Vat Yellow 1), and was subsequently given the chemical name flavanthrone (6.67). Today flavanthrone is used mainly as a pigment (C.I. Pigment Yellow 24). Flavanthrone is readily reduced to a blue leuco form; hydrogen abstraction to generate the leuco form can also take place on cellulose, leading to fibre damage. The leuco compound is believed, on the basis of spectroscopic evidence, to have the structure 6.68 [21]. The yellow to blue colour change is made use of in testing for the presence of reducing agents under alkaline conditions. The structure of flavanthrone was confirmed by an Ullmann synthesis using the Schiff's base (6.69) obtained from 2-amino-1-chloro-anthraquinone, which gave a bianthraquinonyl precursor (6.70) (Scheme

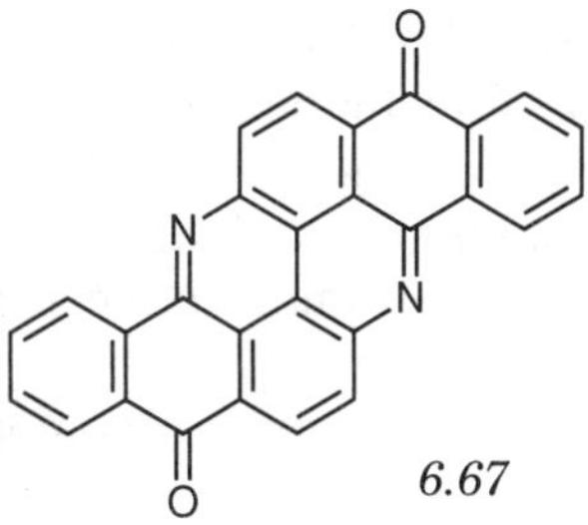

6.67

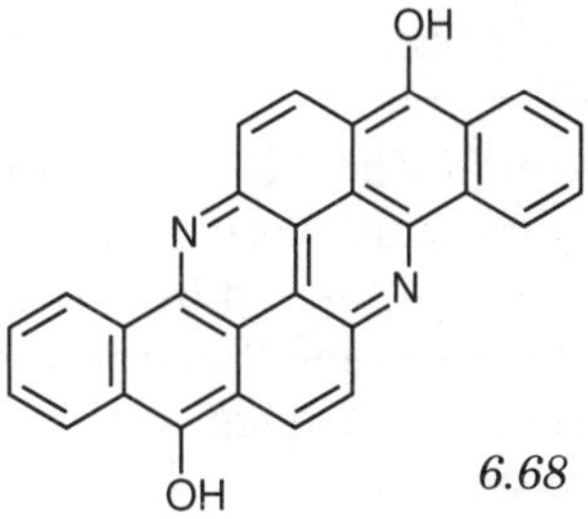

6.68

6.11). Hydrolysis of compound 6.70 gives the corresponding diamine, which subsequently undergoes cyclodehydration to flavanthrone. In a related industrial synthesis, the diphthalimide of 2-amino-1-chloroanthraquinone is converted into the corresponding bianthraquinonyl, which undergoes hydrolysis and ring closure in dilute sodium hydroxide.

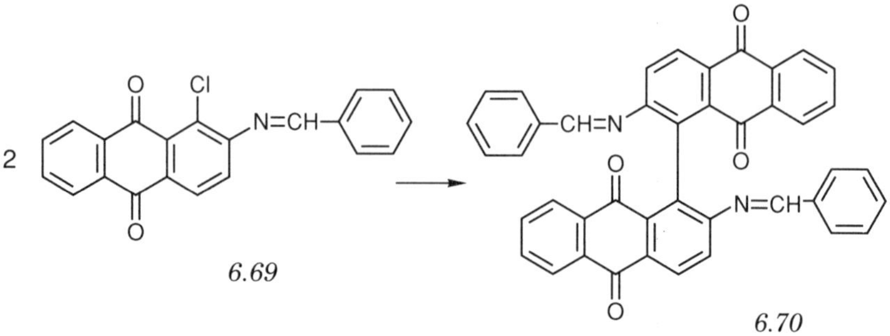

6.69

6.70

Scheme 6.11

The formation of indanthrone and flavanthrone, as well as alizarin, during the alkali fusion of 2-aminoanthraquinone can be explained mechanistically on the basis of the initial loss of a proton. The resulting anionic species can be represented as a resonance hybrid and is also tautomeric (Scheme 6.12). Primary 1-hydroxylation of 2-aminoanthraquinone is probably the first step in the formation of the alizarin by-product (cf. Scheme 6.8). Such an attack may initiate the formation of flavanthrone [22]. It is also possible to envisage the formation of all three species by a radical mechanism [23].

The normal blue vat dye obtained from indanthrone corresponds to a dihydroanthraquinone (compare flavanthrone) and yields a sparingly soluble disodium salt. It seems likely that strong hydrogen bonding between the inner hydroxy groups and the azine nitrogen atoms accounts for the ready formation of the planar molecule 6.71. The infra-red spectrum of the disodium salt indicates the absence of carbonyl groups. Treatment of the salt with dimethyl sulphate produces the corresponding dimethyl ether.

Halogenated indanthrones provide improved resistance to hypochlorite treatments and many derivatives have been examined. Direct halogenation of indanthrone leads to substitution at the 7-, 8-, 16- and 17-positions. The commercially important 7,16-dichloroindanthrone (6.72; C.I. Vat Blue 6) can be synthesised by a rather lengthy route from phthalic anhydride and chlorobenzene (Scheme 6.13).

Polycyclic vat dyes of this type also find use as electrophotographic photoreceptors.

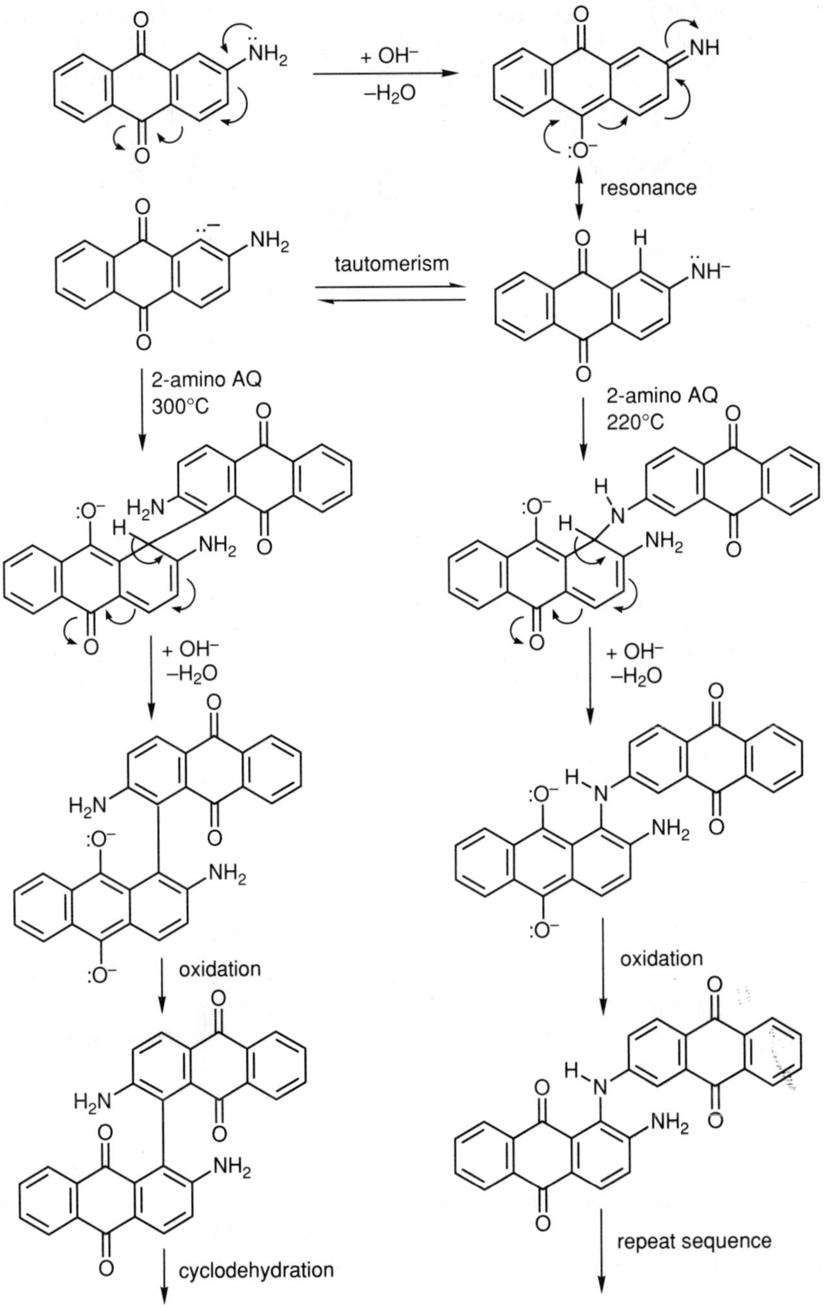

6.67

6.65

Scheme 6.12

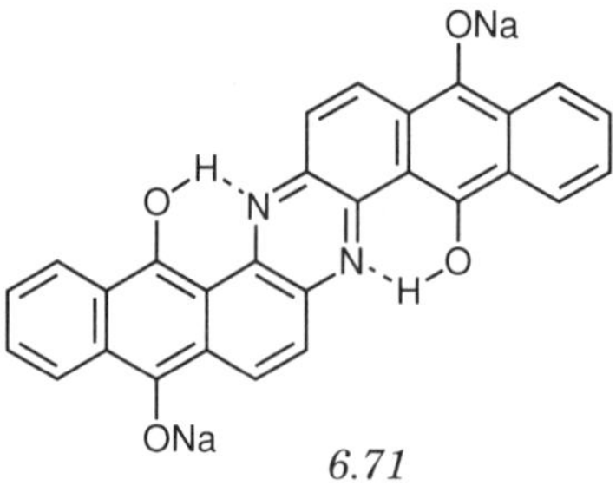

6.71

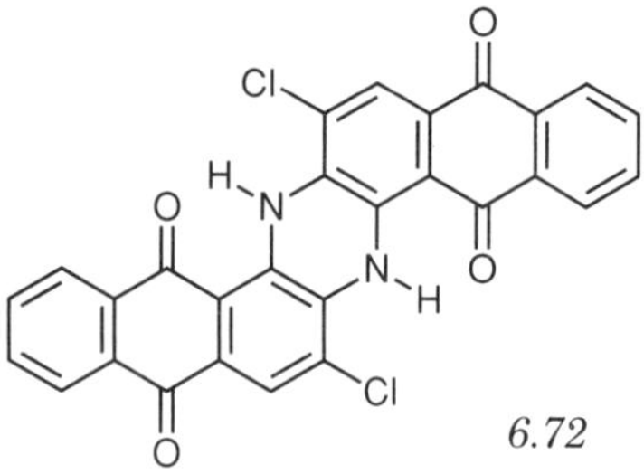

6.72

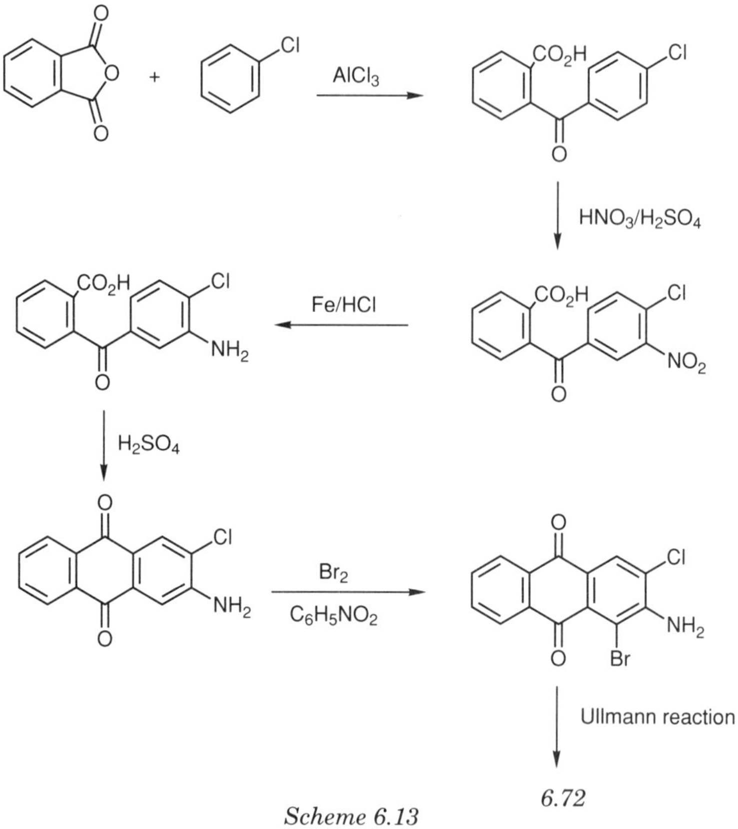

Scheme 6.13

6.2.3 Benzanthrone derivatives

Benzanthrone (6.73) is the source of various commercially important violet, blue and green vat dyes. This tetracyclic system can be prepared from a mixture of anthraquinone and glycerol (propane-1,2,3-triol) by heating with iron powder in concentrated sulphuric acid. The reaction involves reduction of anthraquinone to anthrone (6.74) followed by condensation with acrolein (Scheme 6.14), the latter compound being generated from glycerol by dehydration. The structure of benzanthrone was confirmed by Scholl who prepared it by cyclising 1-benzoylnaphthalene with aluminium chloride.

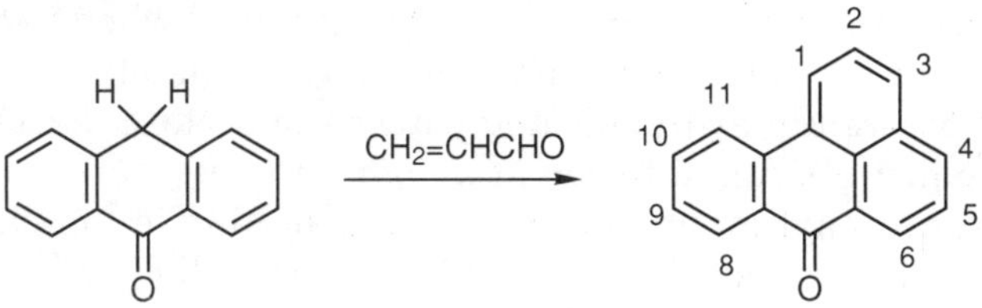

6.74 *6.73*

Scheme 6.14

The benzanthrone system is susceptible to both electrophilic and nucleophilic attack. The most reactive sites towards electrophiles are the 3- and 9-positions, which can be compared with the 4- and 4′-positions in biphenyl. The 9-position is somewhat deactivated by the carbonyl group, however. Thus, for example, monobromination takes place at the 3-position and further substitution gives 3,9-dibromobenzanthrone. Nitration and benzoylation similarly give rise to the 3-substituted product. The 3-position is in fact *peri*-hindered (compare with naphthalene) so that sulphonation yields the 9-sulphonic acid. Electron withdrawal by the carbonyl group activates the 4- and 6-positions towards nucleophilic attack: for example, hydroxylation occurs at these sites.

In 1904 Bally obtained a blue-violet solid by alkali fusion of benzanthrone at approximately 220°C. Two isomeric compounds were isolated by vatting the reaction mixture and filtering off a sparingly soluble sodium salt. Oxidation of the filtrate gave a blue vat dye, violanthrone (6.75; C.I. Vat Blue 20), as the main component, and the residue similarly afforded a violct product, isoviolanthrone (6.76; C.I. Vat Violet 10). The formation of isoviolanthrone can be suppressed by carrying out the fusion in a solvent such as naphthalene or a polyethylene glycol in the presence of sodium acetate and sodium nitrite. Dyes of this type are often referred to as dibenzanthrones.

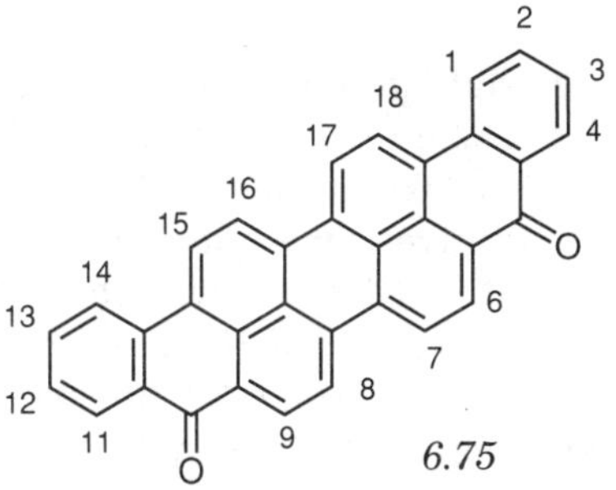

6.75

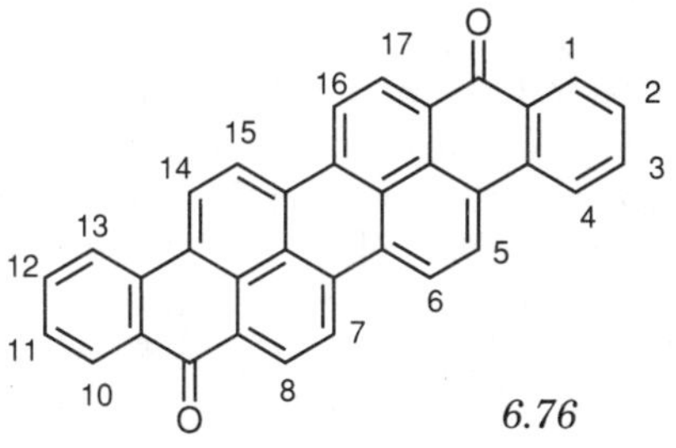

6.76

It was established in 1929 by Lüttringhaus and Neresheimer that 4,4′-bibenzanthronyl (6.77) is an intermediate in the formation of violanthrone. Thus compound 6.77 results when an Ullmann reaction is carried out on 4-chlorobenzanthrone; the same product is obtained when benzanthrone reacts under relatively mild conditions (approximately 110°C) with a mixture of potassium hydroxide and potassium acetate in isobutanol (2-methylpropan-1-ol). Alkali fusion at a higher temperature then converts 4,4′-bibenzanthronyl into violanthrone. Use of aluminium chloride also leads to ring closure (Scholl reaction). Mechanistically it is likely that the resonance-stabilised anion resulting from deprotonation at the 4-position of benzanthrone couples with non-ionised benzanthrone (Scheme 6.15). Linkage across the 3- and 3′-positions under more strongly alkaline conditions may also involve loss of a proton, coupling and oxidation. Chlorination of benzanthrone followed by an Ullmann reaction produces 3,3′-bibenzanthronyl (6.78), which readily undergoes 4,4′-coupling under mild conditions. 3,3′-Bibenzanthronyl is available from benzanthrone directly by treatment with manganese dioxide in sulphuric acid.

Isoviolanthrone (6.76) was first made by heating 3-chlorobenzanthrone with ethanolic potassium hydroxide at 150°C. The reaction probably involves deprotonation at the 4-position to form a carbanion which then displaces the chloro substituent in another molecule of 3-chlorobenzanthrone. Repeating the sequence in the intermediate 3,4′-bibenzanthronyl derivative then gives isoviolanthrone. In an improved, but mechanistically obscure, method of synthesis 3-chlorobenzanthrone is treated with sodium sulphide to produce 3,3′-bibenzanthronyl sulphide; on refluxing with potassium hydroxide in isobutanol, this is converted into isoviolanthrone. Isoviolanthrone itself is no longer used as a vat dye but some halogenated derivatives are commercially available. For example, dichlorination of isoviolanthrone in nitrobenzene gives C.I. Vat Violet 1, which is also useful as a pigment (C.I. Pigment Violet 31). The greenish-blue 6,15-dimethoxy derivative (C.I. Vat Blue 26) is also of some importance.

Nitration of violanthrone gives C.I. Vat Green 9, the main component of which is the 16-nitro derivative. Vatting gives the corresponding amino compound, which gives a bluish-black hue when oxidised on the fibre. Several halogenated derivatives, such as the trichloro derivative C.I. Vat Blue 18, are of industrial significance.

Of special importance is the outstanding bright green vat dye originally marketed as Caledon Jade Green X (6.79; C.I. Vat Green 1), discovered in 1920 by Davies, Fraser-Thomson and Thomas. Oxidation of violanthrone with manganese dioxide and sulphuric acid gives 16,17-dihydroxyviolanthrone. Methylation of this diol yields the dimethoxy derivative (6.79). The dye has excellent fastness to light, alkali and chlorine. Its constitution was confirmed by synthesis from 2-methoxybenzanthrone. An alternative method of manufacture involves the oxidation of 4,4′-bibenzanthronyl

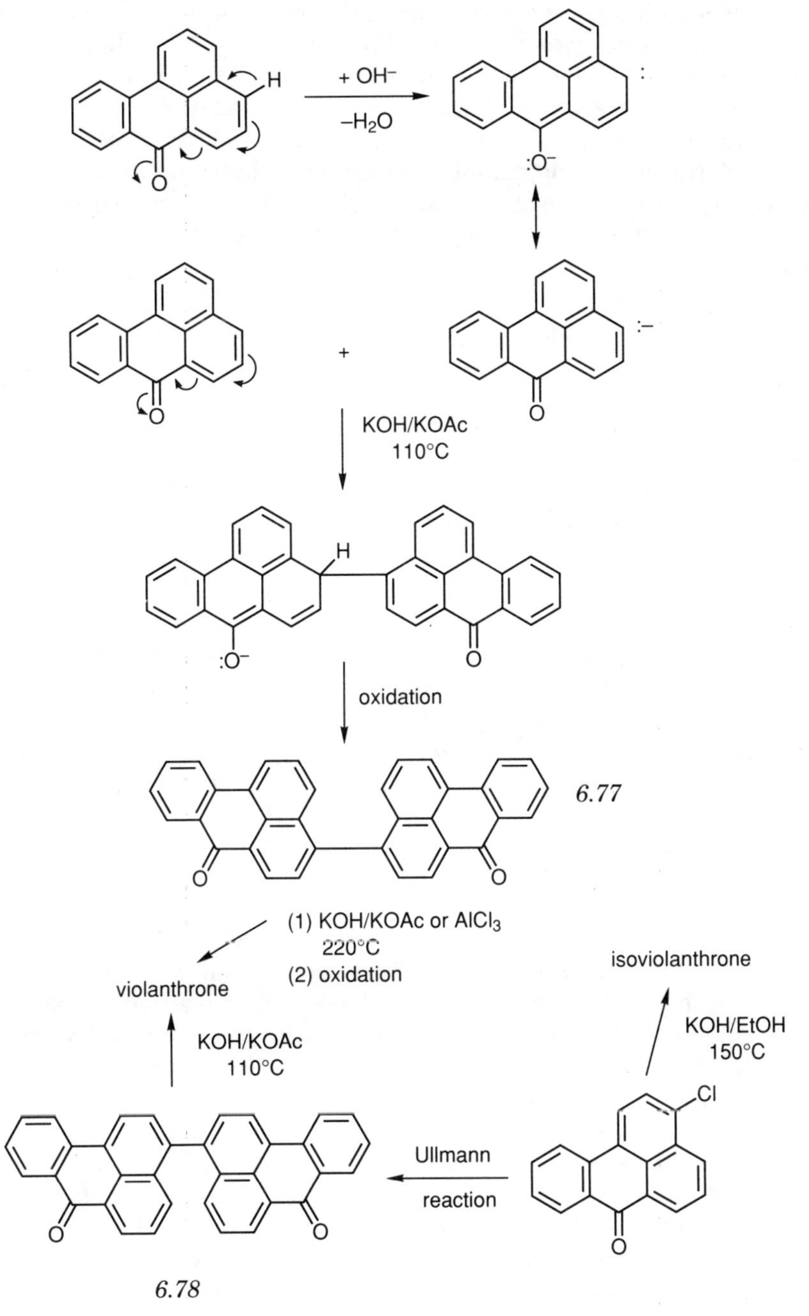
+ OH⁻
−H₂O
:O⁻
+
KOH/KOAc
110°C
H
:O⁻
oxidation
6.77
(1) KOH/KOAc or AlCl₃
220°C
(2) oxidation
violanthrone
isoviolanthrone
KOH/KOAc
110°C
KOH/EtOH
150°C
Cl
Ullmann
reaction
6.78

Scheme 6.15

(6.77) with manganese dioxide and sulphuric acid to give violanthrone-16,17-dione (6.80), which on reduction with sodium bisulphite gives an excellent yield of the 16,17-diol. Treatment of C.I. Vat Green 1 with bromine in oleum gives the 3,12-dibromo derivative (C.I. Vat Green 2). The substitution pattern was confirmed by the n.m.r. spectrum of the dimethyl ether of the leuco compound. Many other alkyl derivatives of 16,17-dihydroxyviolanthrone have been examined but few remain of commercial significance. Of some interest is the navy blue vat dye C.I. Vat Blue 16, in which the 16- and 17-positions are connected by an ethylenedioxy bridge (6.81).

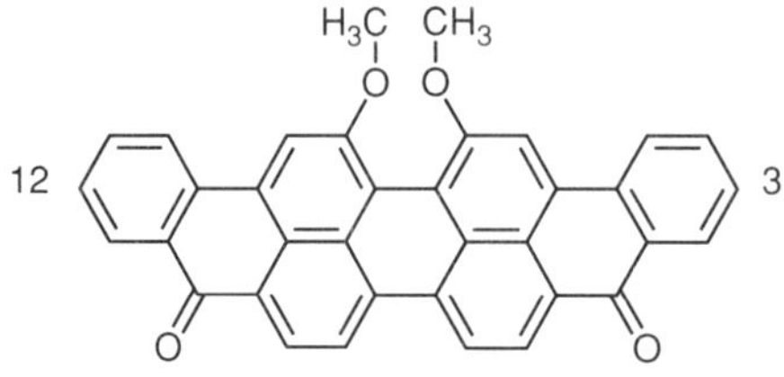

6.79

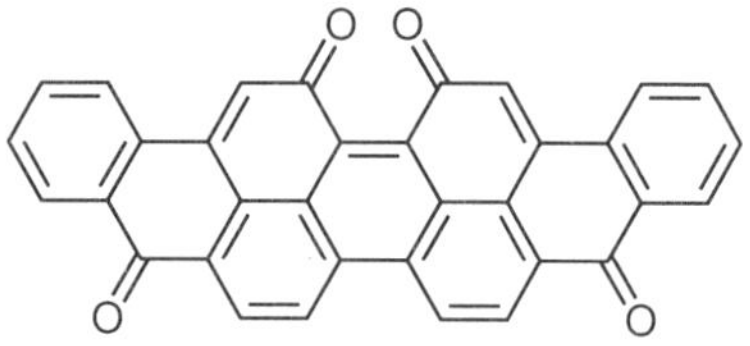

6.80

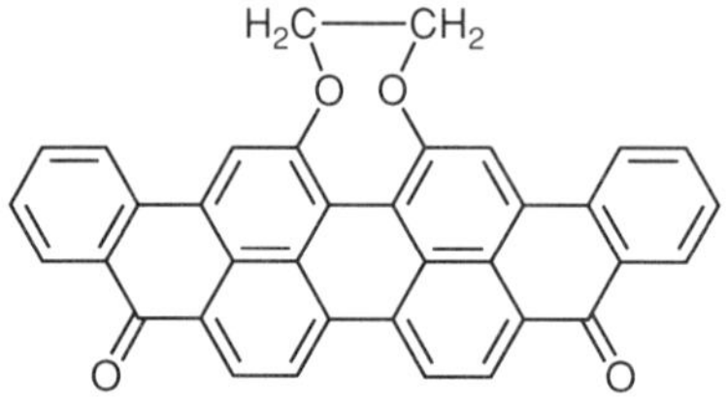

6.81

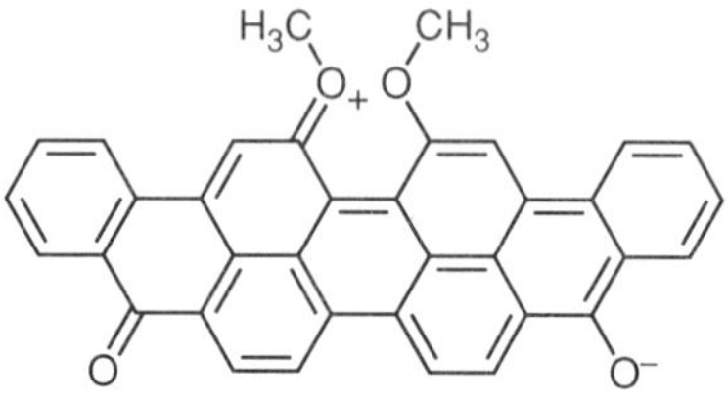

6.82

Violanthrone and isoviolanthrone are effectively *cis* and *trans* isomers. The important *cis* compound absorbs more intensely and at a longer wavelength than the *trans* isomer (Table 6.3). These differences are found to be general for this type of vat dye and can be accounted for theoretically by the PPP–MO approach [24]. It is of interest to compare the absorption maxima of violanthrone and some of its derivatives (Table 6.3). Steric crowding at an essential single bond [25] normally results in a hypsochromic shift of the absorption band as shown by the 16,17-dimethyl derivative and, to a lesser extent, by the 16-methoxy compound. In contrast, however, the blue 16,17-ethylenedioxyviolanthrone exhibits a bathochromic effect that is somewhat greater in the green dimethoxy analogue. Since the steric requirements of the two methoxy groups are not significantly less than those of the equivalent methyl groups, it must be concluded that this bathochromic shift is electronic in origin. A possible explanation is the formation of a stabilised excited state in the case of the

TABLE 6.3

Spectral data for some dibenzanthrones in dimethylformamide

Compound	λ_{max}/nm	ε_{max}
Isoviolanthrone (6.76)	588	41 700
Violanthrone (6.75)	600	60 260
6.75; 16-OCH_3	598	38 900
6.79	636	41 700
6.81	618	
6.75; 16,17-diCH_3	575	

dimethoxy dye (6.82) in which some attraction exists between the adjacent oxygen atoms; any steric clash can be minimised by rotation of the methoxy groups away from the side of the ring system. A pronounced bathochromic effect is also exhibited by the 7,8-dimethoxy isomer, in which the substituents are similarly situated, but not by the 3,12-dimethoxy compound [23]. It is likely that the constraint imposed by the puckered eight-membered ring in the ethylenedioxy dye reduces the interaction between the oxygen lone-pair orbitals and the π orbitals of the polycyclic system.

6.2.4 Carbazole derivatives

Diphthaloylcarbazole derivatives constitute an important group of vat dyes, possessing good all-round fastness properties and providing hues ranging from yellow to black.

The precursors of the carbazoles are the anthrimides, in particular the 1,1′-dianthrimides (1,1′-dianthraquinonylamines). These intermediates were originally used as vat dyes themselves, but very few give bright colours and the vatting process also brings about reduction to 1-aminoanthraquinone. Only one dianthrimide is currently active in the *Colour Index* [11]; this grey vat dye is C.I. Vat Black 28 (6.83). 1,1′-Dianthrimides are easily prepared by heating a 1-aminoanthraquinone with a 1-chloroanthraquinone in nitrobenzene or another solvent in the presence of sodium carbonate and a copper catalyst.

In 1910 Mieg found that 1,1′-dianthrimides undergo carbazolisation (ring closure) when heated with potassium hydroxide at approximately 220°C or with aluminium chloride at a somewhat lower temperature (Scholl reaction). The simplest dye of this type was originally marketed as Indanthren Yellow FFRK (6.84; MLB, C.I. Vat Yellow 28). Cyclisation

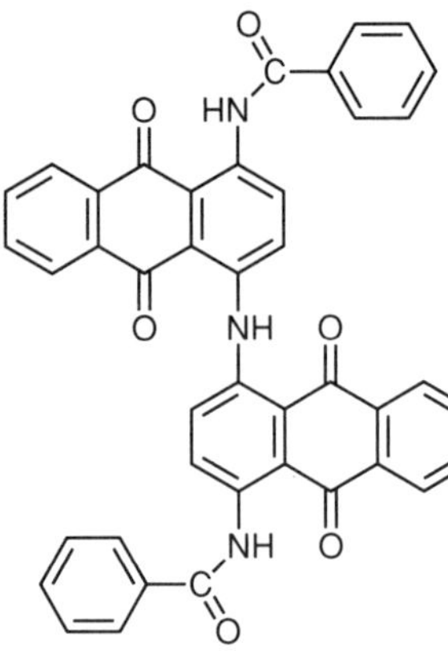

6.83

under strongly basic conditions is probably initiated by deprotonation (Scheme 6.16); an oxidation step is necessary after ring closure. The characteristic structure of this type of vat dye, which is stabilised by intramolecular hydrogen bonding, was established by an alternative synthesis based on the Graebe route to carbazole and involving an Ullmann reaction with 2-chloro-1-nitroanthraquinone, followed by reduction and cyclisation.

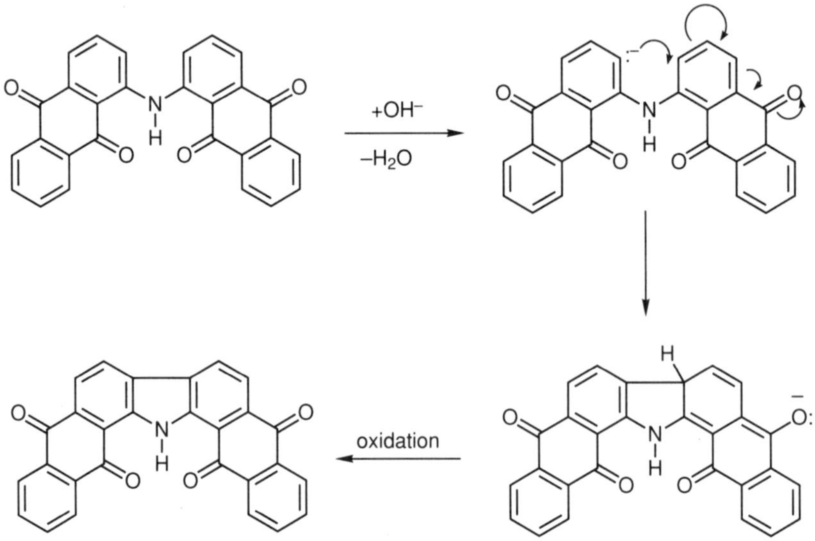

6.84

Scheme 6.16

The parent dye, which has a phototendering action on cellulose, is now obsolete but several relatively simple benzoylamino derivatives are of commercial importance. Thus C.I. Vat Orange 15 (6.85; $R_1 = R_4 = NHCOC_6H_5$,

$R_2 = R_3 = H$) is a 1,5-diaminoanthraquinone type whereas the more bathochromic olive vat dye C.I. Vat Black 27 (6.85; $R_1 = R_4 = H$, $R_2 = R_3 = NHCOC_6H_5$) is a 1,4-diaminoanthraquinone derivative. The unsymmetrically substituted isomer (6.85; $R_1 = R_3 = H$, $R_2 = R_4 = NHCOC_6H_5$) is a useful brown vat dye (C.I. Vat Brown 3).

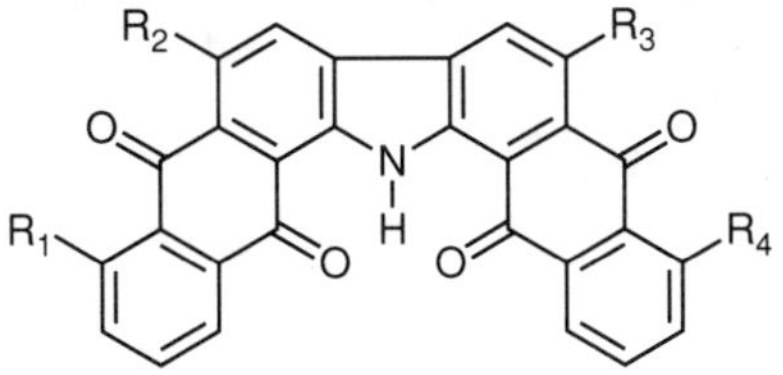

6.85

The presence of two benzoylamino groups in the dianthrimide system permits carbazole ring closure to take place relatively easily in concentrated sulphuric acid at approximately 30°C. It is likely that the benzoylamino groups assist protonation at the adjacent carbonyl group and thereby aid the subsequent cyclisation process. The final oxidation step can take place with sulphuric acid alone. It is possible that the cyclisation occurs by a radical mechanism [23].

Various more complex dyes of industrial significance contain more than one carbazole system. For example, C.I. Vat Orange 11 (6.86) contains two carbazole components and a central unit derived from 1,5-diaminoanthraquinone. This dye is prepared by carbazolisation of the trianthrimide produced when two moles of 1-chloroanthraquinone react with one mole of 1,5-diaminoanthraquinone. The equivalent isomeric dye obtained from 1,4-diaminoanthraquinone is reddish-brown (6.87; C.I. Vat Brown 1). The interesting, symmetrically substituted tetracarbazole dye C.I. Vat Green 8 (6.88) was first synthesised in 1911 by Hepp from 1,4,5,8-tetrachloroanthraquinone and marketed as Indanthren Khaki GG (HOE). Not surprisingly, the product ($C_{70}H_{28}N_4O_{10}$; relative molecular mass 1084) is of very low solubility. The structure was confirmed in 1957 by Jayaraman [23], who found no evidence of uncyclised anthrimides in the u.v. spectrum of the dye solution in concentrated sulphuric acid.

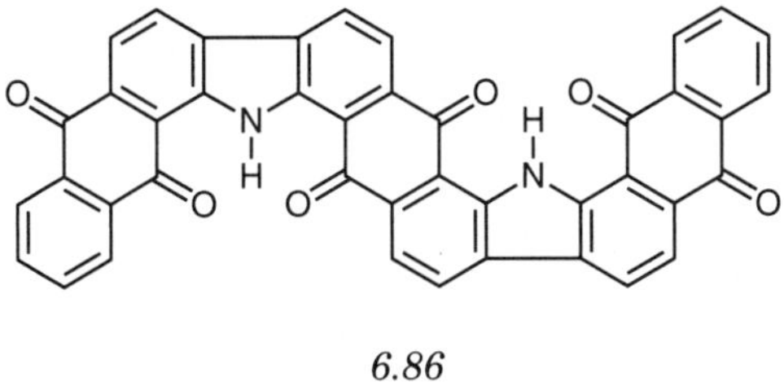

6.86

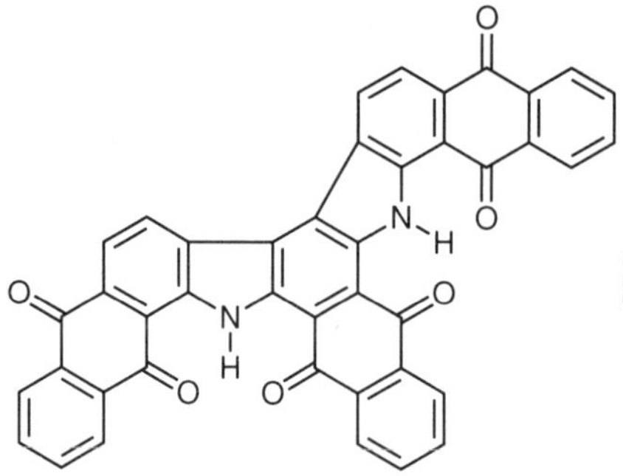

6.87

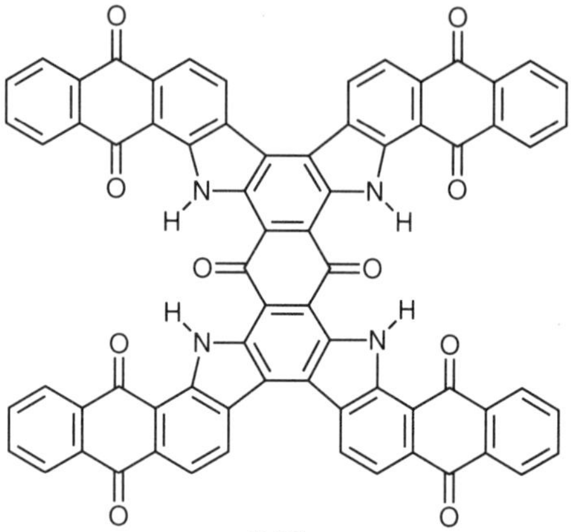

6.88

6.2.5 Acridone derivatives

The first anthraquinone vat dye containing an acridone ring system was synthesised in 1909 by Ullmann. The parent compound (6.89) can be made by condensation of 1-chloroanthraquinone with anthranilic acid in the presence of copper (Ullmann reaction), followed by cyclodehydration in concentrated sulphuric acid (Scheme 6.17).

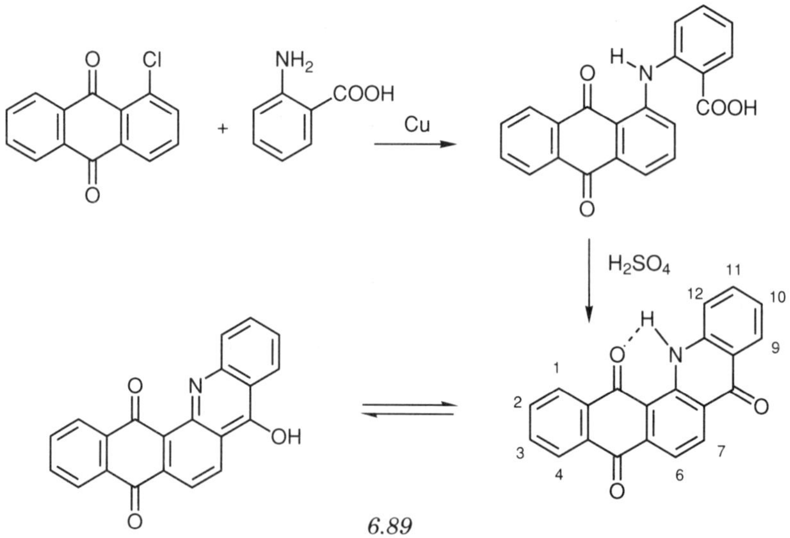

Scheme 6.17

Alternatively, the synthesis may begin by condensing aniline with the 1-chloro-2-carboxy intermediate. Acridone vat dyes of this type have excellent light fastness but only moderate resistance to alkali due to the keto–enol equilibrium. It is interesting that this pentacyclic dye is approximately 30 nm more bathochromic than the closely related tetracyclic 1-amino-2-benzoylanthraquinone.

The parent acridone is not useful as a vat dye, although some halogenated derivatives are commercially important. Chlorination of dye 6.89 with sulphuryl chloride gives the reddish-violet C.I. Vat Violet 14, which consists mainly of the 6,10,12-trichloro compound. The colour of acridone derivatives is very sensitive to the number and position of substituents; for example, the 6-chloro derivative is violet whereas the 9,12-dichloro compound is red and its 9,11-isomer is orange. The presence of a substituted amino group at the 6-position leads to blue dyes such as C.I. Vat Blue 21 (6.90), obtained from bromamine acid (6.10) and 4-trifluoromethylanthranilic acid followed by ring closure, during which desulphonation take place, and benzoylation. The relatively simple bisacridone C.I. Vat Violet 13 (6.91) is the original dye made in 1909 by Ullmann from 1,5-dichloroanthraquinone; this product is still of industrial importance.

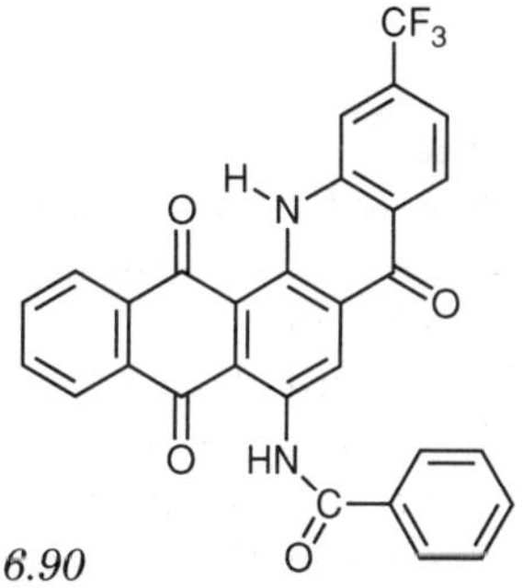

6.90

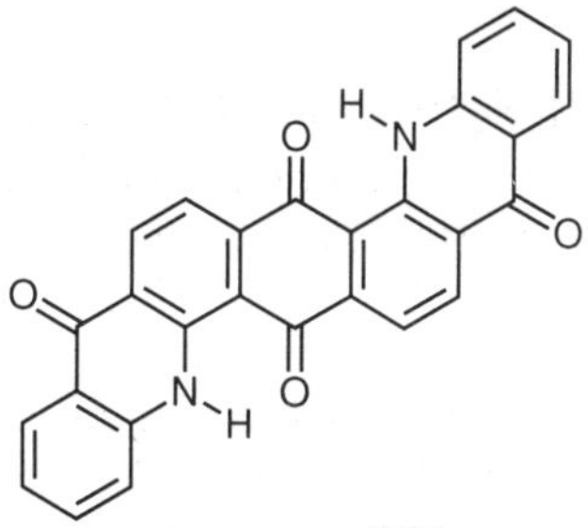

6.91

Various complex carbazole–acridones have been used to obtain brown, khaki, grey and olive hues. The fused carbazole–acridone C.I. Vat Brown 55 (6.92), which possesses excellent all-round fastness properties, is currently available. This compound can be obtained from C.I. Vat Violet 14 by condensation with two equivalents of 1-amino-5-benzoylaminoanthraquinone followed by carbazolisation.

The important olive-green dye C.I. Vat Green 3 (6.93; R = H), which has outstanding light fastness, can be regarded as an acridone analogue. This dye is readily available by condensing 3-bromobenzanthrone with 1-aminoanthraquinone under Ullmann conditions followed by ring closure with potassium hydroxide. Its 1-anthraquinonylamino derivative is a useful brownish-grey dye (C.I. Vat Black 25), which is prepared in a similar manner from 3,9-dibromobenzanthrone.

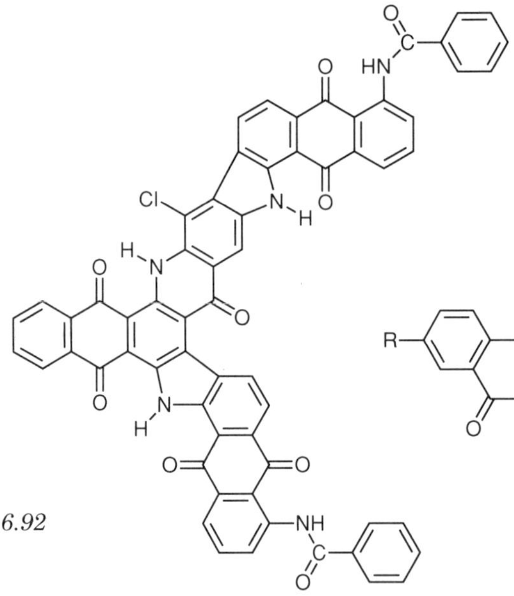

6.92

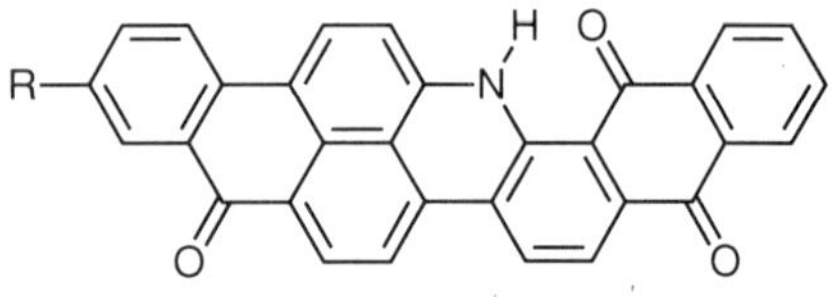

6.93

6.2.6 Miscellaneous polycyclic quinones

The carbocyclic analogue of flavanthrone was first prepared in 1905 by Scholl. Chlorination of 2-methylanthraquinone with sulphuryl chloride yields the 1-chloro derivative, which under Ullmann conditions gives 2,2′-dimethyl-1,1′-bianthraquinonyl (6.94). Treatment of the latter compound with ethanolic potassium hydroxide produces pyranthrone (6.95; C.I. Vat Orange 9) (Scheme 6.18). The cyclisation reaction was the first example of its kind and probably proceeds via carbanion formation. As well as being an anthrone derivative, pyranthrone is a fused benzanthrone and also contains a central pyrene ring system. Pyranthrone, a powerful phototenderer, gives rise to brilliant orange hues of good fastness properties. Bromination of pyranthrone in chlorosulphonic acid gives mainly the reddish-orange 4,12-dibromo derivative, C.I. Vat Orange 2.

An alternative route to pyranthrone, involving baking 1,6-dibenzoylpyrene (6.96) with aluminium chloride, was also devised by Scholl (Scheme 6.18). The Scholl reaction is a key step in the synthesis of several polycyclic quinones; the cyclisation of 1-benzoylnaphthalene to give benzanthrone has already been mentioned. The mechanism of this cyclodehydrogenation reaction may involve an initial protonation step, if traces of water are present, or complexation with aluminium chloride. Electrophilic substitution is thereby promoted, followed by an oxidation step (Scheme 6.19). It has been suggested that radical cations are formed as intermediates [26].

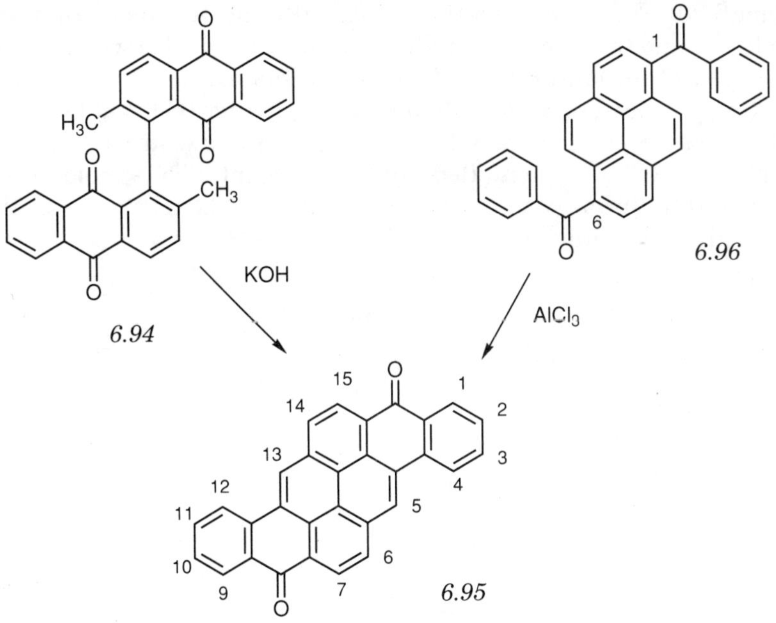

Scheme 6.18

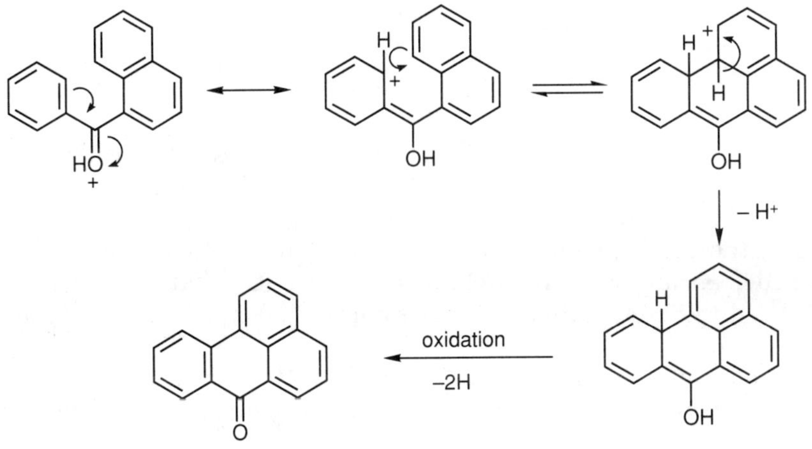

6.73

Scheme 6.19

The anthanthrone system (6.97) was discovered in 1912 by Kalb. In the latter stages the synthesis of anthanthrone involves hydrolysis of naphthostyril (6.98), diazotisation and biaryl formation under Gattermann conditions to give 1,1′-binaphthyl-8,8′-dicarboxylic acid before cyclodehydration

(Scheme 6.20). The leuco derivative of anthanthrone itself has little affinity for cellulose and the oxidation product has low tinctorial power. However, bromination of anthanthrone or of the intermediate dicarboxylic acid leads to the bright reddish-orange 4,10-dibromo derivative (C.I. Vat Orange 3), which possesses good general fastness properties and which also finds use as a pigment (C.I. Pigment Red 168). Treatment of one mole of 4,10-dibromoanthanthrone with two moles of 1-amino-4-benzoylaminoanthraquinone gives the bluish-grey dye C.I. Vat Black 29.

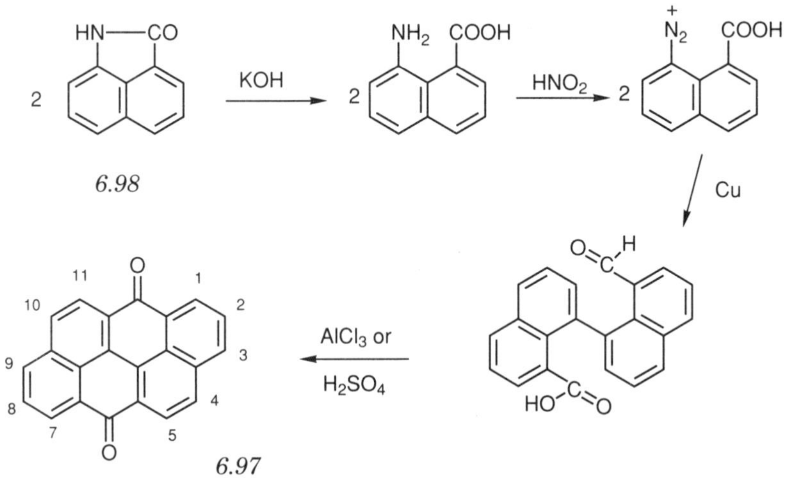

6.98

6.97

Scheme 6.20

There are several isomeric dibenzopyrenequinones, of which only C.I. Vat Yellow 4 (6.99) is of commercial importance. This compound is synthesised from 1,5-dibenzoylnaphthalene using the Scholl cyclisation procedure; the same product can be obtained in a similar manner from 3-benzoylbenzanthrone. Its dibromo derivative (C.I. Vat Orange 1), prepared by direct bromination, possesses improved fastness to light, washing and bleaching.

The reddish-brown dye C.I. Vat Brown 45 (6.100) is a dichloro derivative of the unusual acedianthrone system. The dye is prepared by condensing

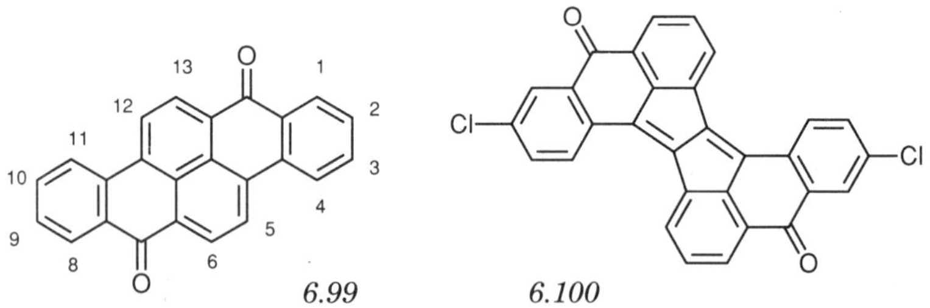

6.99 6.100

two moles of 2-chloroanthrone with one mole of glyoxal followed by oxidative cyclisation, which probably takes place in a stepwise manner [22].

The 1,9-heterocyclic derivative flavanthrone (6.67) has already been considered. Several related heterocyclic compounds are of importance as vat dyes. The 1,9-pyrimidanthrone derivative 6.101 is of commercial significance both as a vat dye (C.I. Vat Yellow 20) and also as a pigment (C.I. Pigment Yellow 108). The pyrimidine ring system, which prevents phototendering on cellulose, can be obtained from the appropriately substituted 1-aminoanthraquinone with dimethylformamide and either thionyl chloride ($SOCl_2$) or phosgene ($COCl_2$). 1,9-Pyrazolanthrones, like the anthrapyrimidines, can be regarded as heterocyclic analogues of benzanthrone. The parent tautomeric pyrazolanthrone system (6.102) is available from 1-aminoanthraquinone by diazotisation, reduction to the hydrazine with sodium bisulphite and ring closure in sulphuric acid. As with benzanthrone, dimerisation takes place on fusion with alkali. Improved resistance to alkali is achieved by alkylation of the product. The useful bluish-red dye C.I. Vat Red 13 (6.103) is obtained by ethylation of the parent dimer. Several other valuable dyes also contain a pyrazolanthrone system, such as the navy blue dye C.I. Vat Blue 25 (6.104), which is prepared by condensing 3-bromobenzanthrone with pyrazolanthrone followed by treatment with alkali, and the grey dye C.I. Vat Black 8 (6.105), which is made in a similar manner from 3,9-dibromobenzanthrone, pyrazolanthrone and 1-aminoanthraquinone.

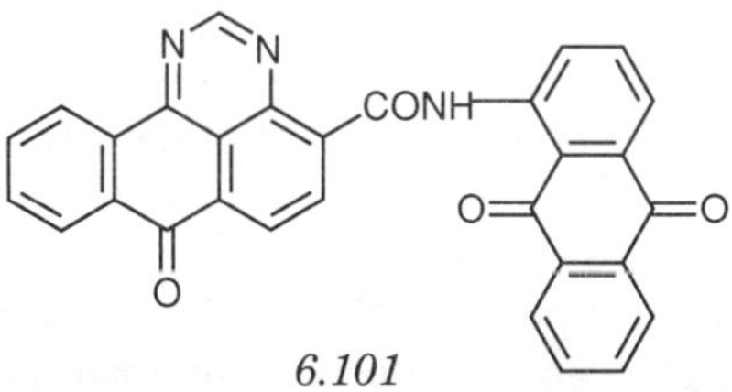

6.101

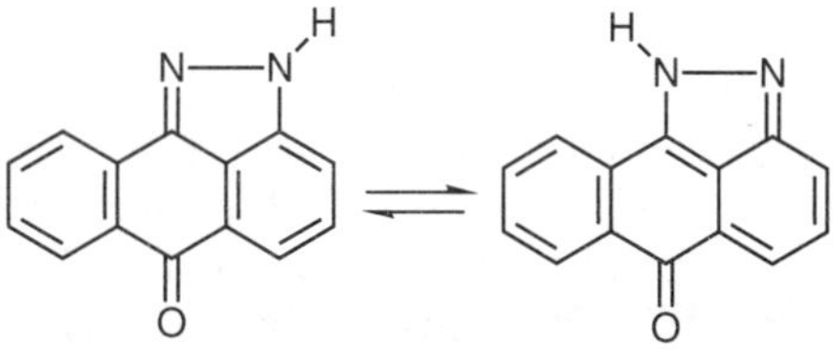

6.102

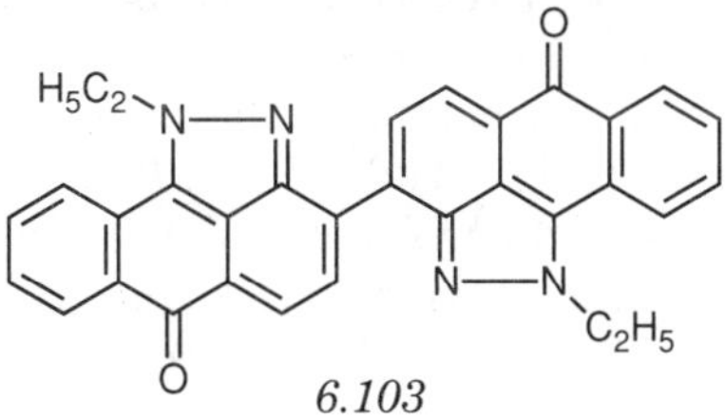

6.103

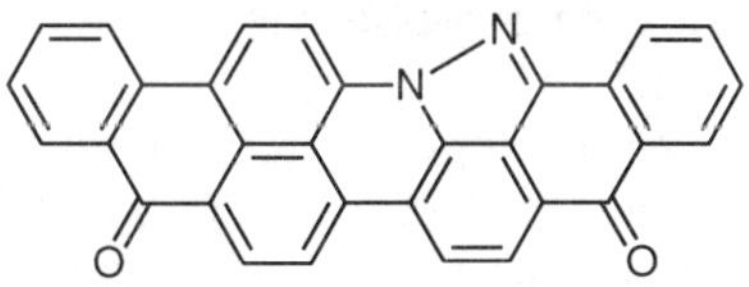

6.104

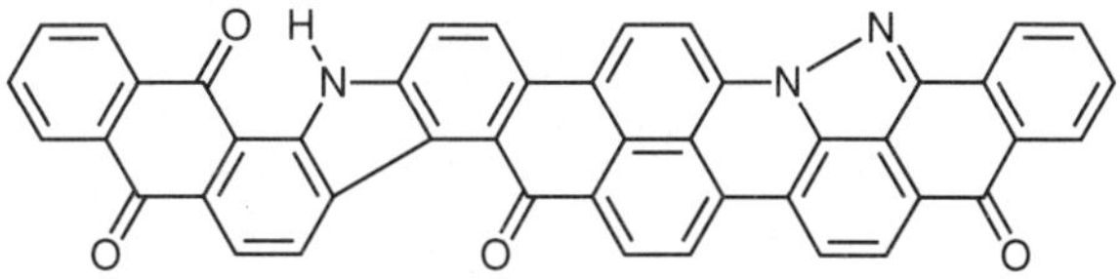

6.105

Heterocyclic ring systems are also used to connect two anthraquinone groups. Typical examples include C.I. Vat Red 10 (6.106), which is an oxazole derivative obtained from 2-amino-3-hydroxyanthraquinone and the appropriate acyl chloride, the similar thiazole derivative C.I. Vat Blue 31 (6.107) and the oxadiazole derivative C.I. Vat Blue 64 (6.108).

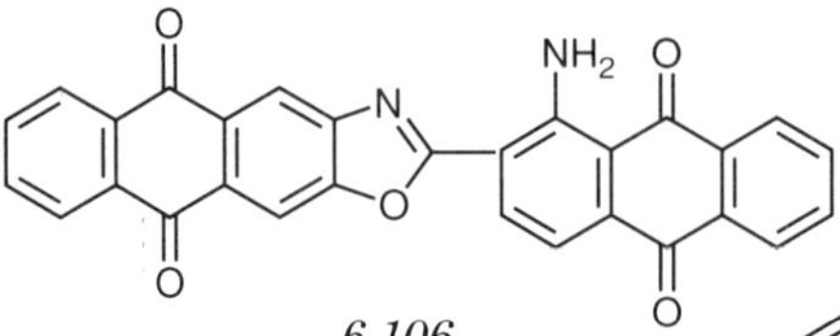

6.106

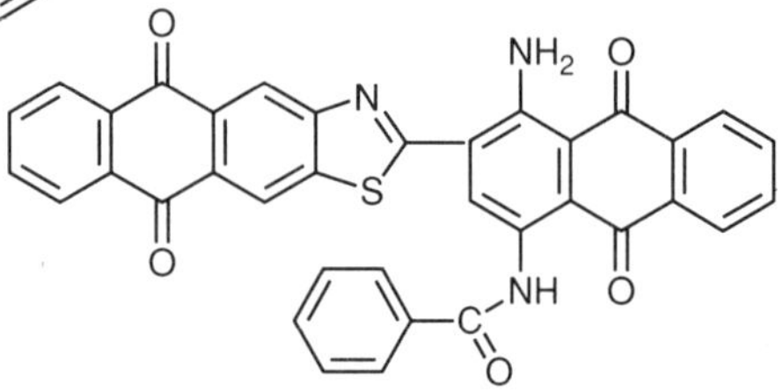

6.107

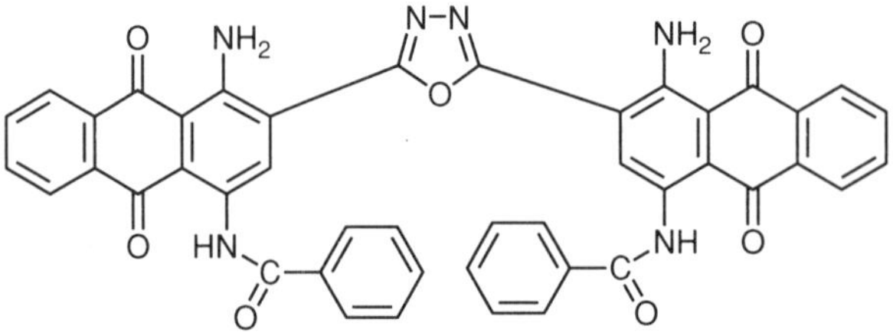

6.108

Although not anthraquinonoid, some related vat dyes derived from perylene-3,4,9,10-tetracarboxylic acid and from naphthalene-1,4,5,8-tetracarboxylic acid are worthy of mention. Thus, for example, when *N*-methylnaphthalimide is fused with alkali the resulting dimer is C.I. Vat Red 23 (6.109). This product is also useful as a pigment (C.I. Pigment Red 179). Dyes of this type also find an application in electrophotographic recording processes. Treatment of naphthalene-1,4,5,8-tetracarboxylic anhydride with *o*-phenylenediamine gives rise to two isomeric diimidazoles, each of which is useful as a vat dye and as a pigment. The products are C.I. Vat Orange 7/C.I. Pigment Orange 43 (6.110) and C.I. Vat Red 15/ C.I. Pigment Red 194 (6.111). The two isomers can be separated chromatographically or by treatment with ethanolic potassium hydroxide. Dichroic pigments of this type are also useful in liquid crystal display devices.

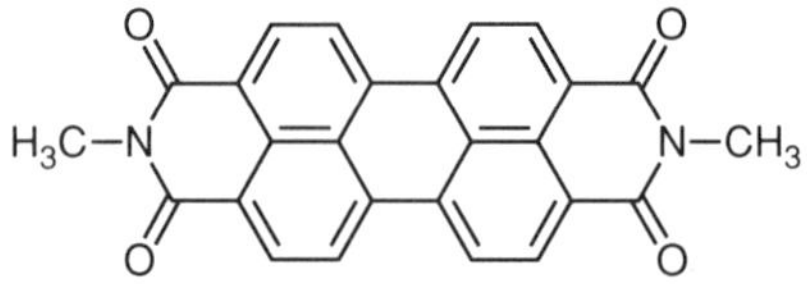

6.109

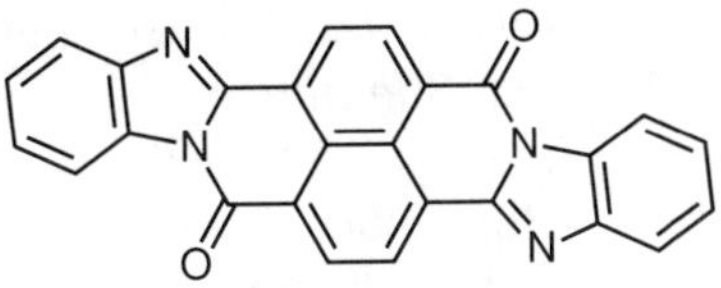

6.110

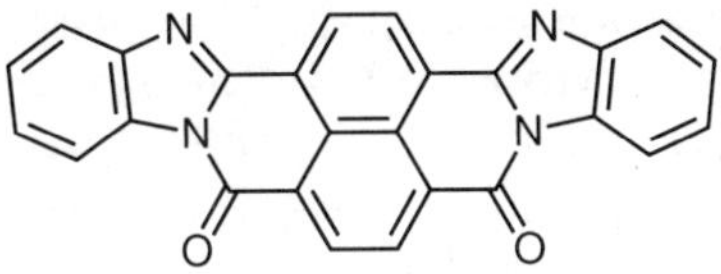

6.111

6.3 INDIGOID AND THIOINDIGOID DYES

The natural blue dye indigo (6.112; C.I. Vat Bluc 1) has been known since antiquity; there are several plants from which indigo can be isolated. Of particular importance is the genus *Indigofera*, which was grown in India and China and in other tropical and subtropical regions [27]. Aqueous extraction of the indigo plant yields indican, the colourless glucoside of indoxyl (3-hydroxyindole). Removal of the sugar residue in a fermentation process and air oxidation of the resulting indoxyl gives indigo. As well as the essential colouring matter, indigotin (6.112), natural indigo contains varying amounts of the isomeric by-product indirubin (6.113) and other compounds. In Europe the woad plant (*Isatis tinctoria*) was cultivated to provide indigo of lower purity in the guise of woad.

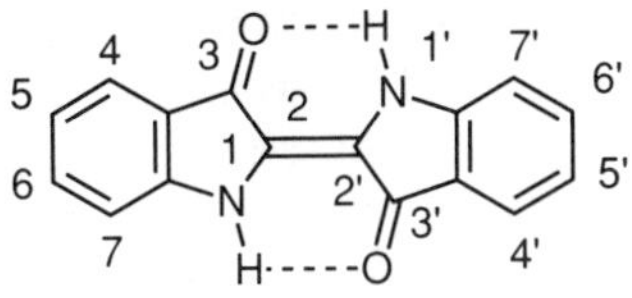

6.112

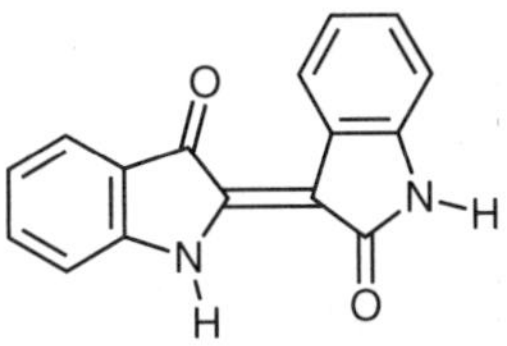

6.113

The constitution of indigo (strictly indigotin) was elucidated by Adolf von Baeyer between 1865 and 1883, on the basis of degradation and synthesis. It was already known that distillation of indigo gave aniline (*anil* is the Portuguese word for indigo). Among other conversions, Baeyer showed that indigo could be reduced to indole and oxidised to isatin (6.114). It should be recognised that the key intermediate indoxyl is a tautomeric system (6.115). Baeyer proposed the *cis* isomer (*Z*-configuration) for indigo.

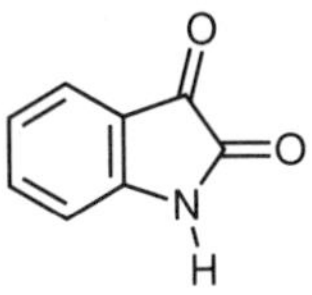

6.114

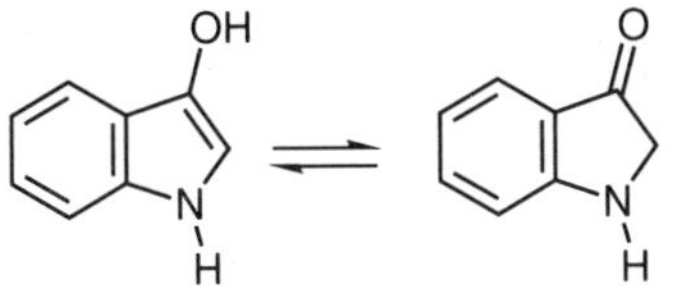

6.115

The correct *E*-configuration (*trans* isomer) was eventually established in 1926 by Posner, using X-ray crystallography [28]. The indigo molecule is almost planar in the solid state and forms a hydrogen-bonded lattice in which each molecule is linked to four others. The observed bond lengths suggest some contribution to the ground state of the molecule from charge-separated structures such as the dipolar species 6.116. Not surprisingly, indigo has a high melting point (390–392°C) and low solubility. Infra-red studies have shown the existence of intermolecular hydrogen bonding in the parent compound and in many derivatives, but intramolecular hydrogen bonding (6.112) cannot be discounted. The *trans* isomer cannot readily be converted into the *cis* form (unlike thioindigo) although derivatives of the *cis* isomer are known, such as compound 6.117.

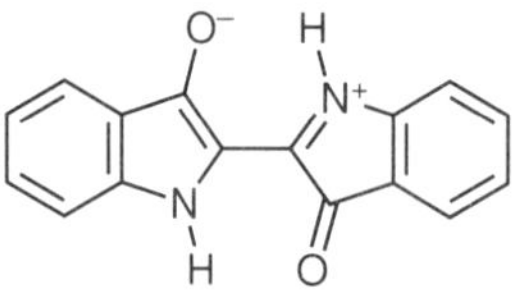

6.116

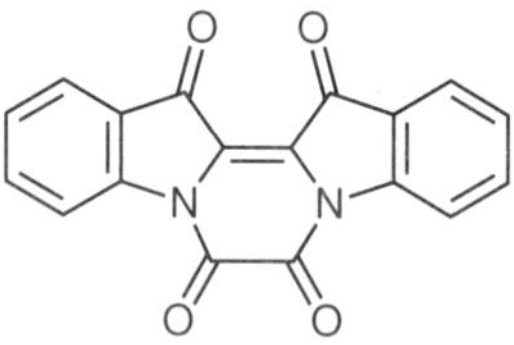

6.117

The colour of indigo depends dramatically upon its physical state and environment; for example, the vapour is red but the colour on the fibre is blue. The marked solvatochromism of indigo (Table 6.4) is attributable mainly to hydrogen bonding. A progressive bathochromic shift of the visible absorption band is observed as the solvent polarity (dielectric constant) increases. *NN′*-Dimethylindigo and thioindigo, in which hydrogen bonding is absent, exhibit only small solvatochromic effects.

TABLE 6.4

Solvatochromism of indigo

Solvent	λ_{max}/nm
Vapour	540
Carbon tetrachloride	588
Xylene	591
Ethanol	606
Dimethyl sulphoxide	642
Solid	660

The isolated chromogen in indigo vapour, like that of thioindigo (λ_{max} 543 nm in DMF), is red. Klessinger and Lüttke have shown that the structural unit responsible for the red colour is the cross-conjugated H chromophore (the basic unit is H-shaped) (6.118). The secondary role of the benzene rings in indigo has been established by the synthesis of simple analogues such as compound 6.119 [29]. PPP–MO calculations support the idea of an H chromophore [30]. Despite the relatively minor role of the benzene rings in the indigo molecule, the type and position of substituents in the chromogen are important for determining the colour of a derivative. In general, electron-donating groups at the 5,5′- and 7,7′-positions bring about bathochromic shifts, whereas electron-withdrawing substituents at these sites cause hypsochromic shifts. Qualitatively, a donor substituent that is *para* or *ortho* to the donor NH group will stabilise the first excited state whereas an acceptor group has a destabilising effect. The reverse situation obtains at 4,4′- and 6,6′-positions that are *ortho* or *para* to the acceptor carbonyl group. The VB explanation is supported by PPP–MO calculations [28].

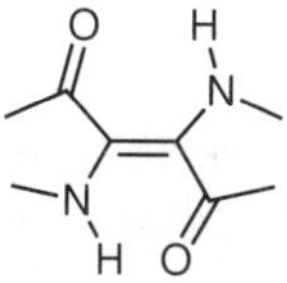

6.118

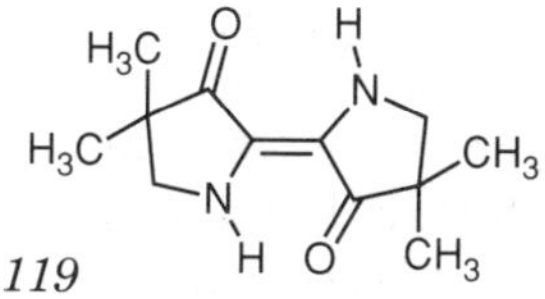

6.119

Indigo itself is still of considerable commercial importance, especially in warp dyeing for woven cotton denim (indigo fades on tone to a pale blue), although indigoid vat dyes have gradually declined in importance relative to anthraquinone derivatives. Indigo also finds use as a pigment (C.I. Pigment Blue 66).

6.3.1 Methods of synthesis

The synthesis of indigo was much more difficult than that of alizarin (6.2) [31]. Baeyer first attempted to obtain indigo in 1865 by reductive dimerisation of isatin (6.114); he finally achieved a seven-step synthesis from phenylglycine via isatin in 1878. Many syntheses have subsequently been developed for indigo, but very few of these have achieved industrial importance.

In 1890 Heumann obtained a low yield of indoxyl (6.115) by fusing phenylglycine with potassium hydroxide at 300°C; indoxyl, once formed, is readily oxidised by atmospheric oxygen to give indigo. An improved yield was obtained at a lower fusion temperature (200°C) from 2-carboxyphenylglycine, obtained from anthranilic acid (Scheme 6.21). The initial cyclisation is like a Dieckmann condensation [32]. This route was adopted by BASF, who first marketed synthetic indigo in 1897. The advent of syn-

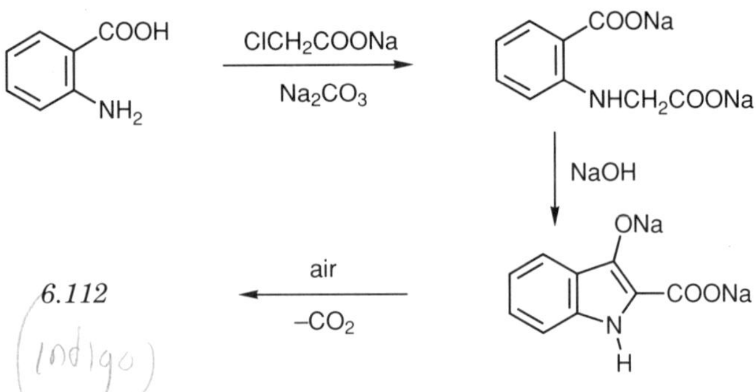

Scheme 6.21

thetic indigo soon led to a rapid decline in demand for the relatively impure natural product. At the time the route to anthranilic acid was difficult and relatively expensive [33]. In 1901 Pfleger found that it is possible to cyclise phenylglycine efficiently by adding sodamide at 190°C to a low-melting eutectic mixture of sodium and potassium hydroxides. This discovery enabled the cheaper aniline to be used as the starting material. Pfleger's route is still used today, except that the required phenylglycine is produced from aniline via the *N*-cyanomethyl derivative obtained by the condensation of aniline with sodium formaldehyde–bisulphite and sodium cyanide; alkaline hydrolysis then gives sodium phenylglycinate.

Derivatives of indigo can be made by synthesis or by direct substitution. Sulphonation of indigo using oleum gives mainly the 5,5′-disulphonic acid (6.120) together with the 5,7′-isomer. This dye is of interest as a food dye (C.I. Food Blue 1) and as an acid dye (C.I. Acid Blue 74); the aluminium salt is a pigment (C.I. Pigment Blue 63). Several halogenated dyes have been marketed, exemplified by C.I. Vat Blue 5 (6.121), which is obtained from indigo by bromination in an organic solvent such as nitrobenzene. The natural dye Tyrian purple (C.I. Natural Violet 1) is 6,6′-dibromoindigo [34].

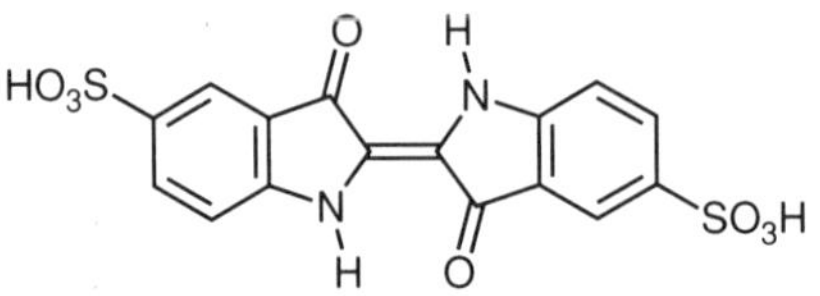

6.120

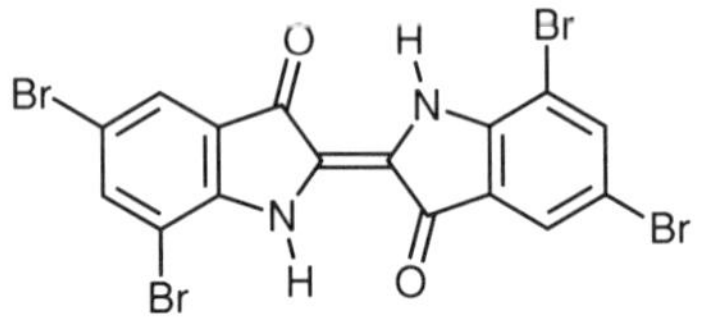

6.121

The sulphur analogue of indigo, thioindigo (6.122), was first synthesised in 1905 by Friedländer from anthranilic acid, using a reaction sequence similar to the earlier Heumann synthesis of indigo (Scheme 6.21); the amino group is first converted into a thiol by diazotisation, treatment with sodium disulphide and reduction of the intermediate diphenyl disulphide.

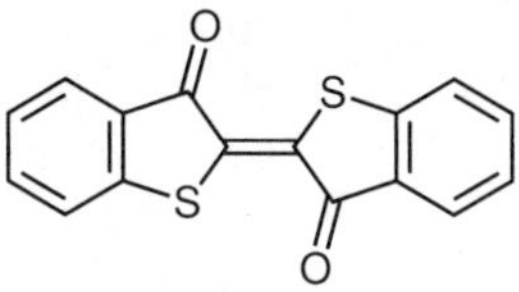

6.122

Reaction of the thiol with chloroacetic acid and subsequent cyclisation gives thioindoxyl, which is then oxidised, usually with sulphur. The final oxidation step, as in the case of indigo, may proceed by a hydride transfer mechanism or by a free-radical coupling reaction [32]. Thioindigo (C.I. Vat Red 41) is a bright red dye that is still available commercially. Unlike indigo, it is photochromic; on exposure to light the *trans* stereoisomer (λ_{max} 546 nm in chloroform) is converted into the more hypsochromic *cis* compound (λ_{max} 490 nm in chloroform) giving a marked change of colour from violet to yellowish-red.

Substituted thioindigoid dyes are usually obtained via the appropriate benzenethiol in a Heumann-type synthesis. The final cyclisation of the phenylthioglycolic acid derivative can often be achieved in concentrated sulphuric acid or by using chlorosulphonic acid. Several routes make use of the Herz reaction, in which a substituted aniline is converted into the corresponding *o*-aminothiophenol by reaction with sulphur monochloride followed by hydrolysis of the intermediate dithiazolium salt (Scheme 6.22) [35]. After reaction between the thiol and chloroacetic acid, the amino group is converted into a nitrile group by a Sandmeyer reaction. Hydrolysis of the nitrile leads to the formation of the required thioindoxyl derivative.

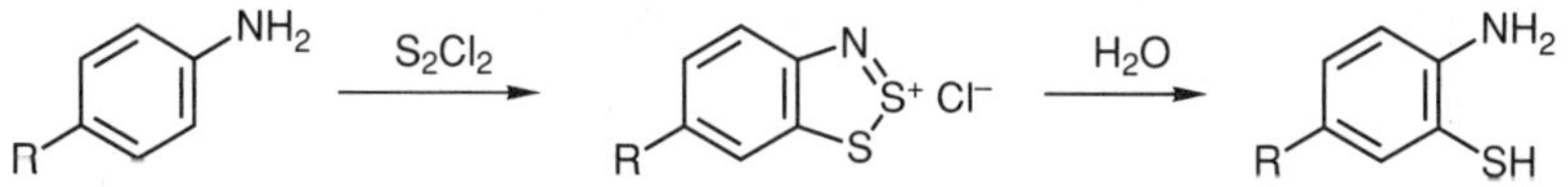

Scheme 6.22

The colour of the thioindigo chromogen is more sensitive to the effect of substituents than that of indigo. Several derivatives of thioindigo are of commercial significance. The presence of donor substituents *para* to the carbonyl groups leads to a pronounced hypsochromic shift, as in the case of C.I. Vat Orange 5 (6.123).

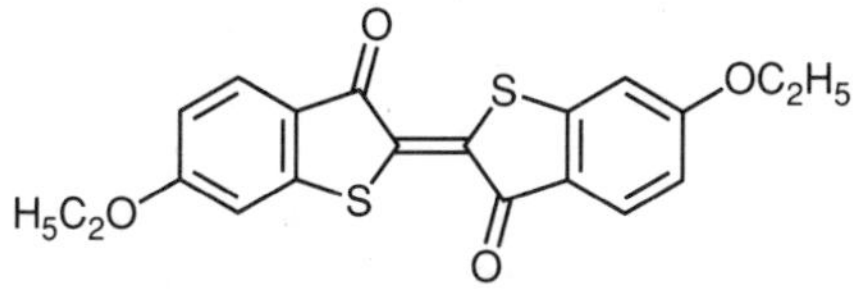

6.123

Dyes containing chloro and methyl substituents are of importance. Typical examples include C.I. Vat Red 1 (6.124) and C.I. Vat Violet 3 (6.125), which also finds use as a pigment (C.I. Pigment Violet 38). 4,4′,7,7′-Tetrachlorothioindigo is widely used as a reddish-violet pigment (C.I. Pigment Red 88). A naphthalene analogue of thioindigo (6.126; C.I. Vat Brown 5) can be prepared from naphthalene-2-thiol and is used for textile printing. A few unsymmetrically substituted thioindigos are also produced; C.I. Vat Red 2 (6.127) is an example.

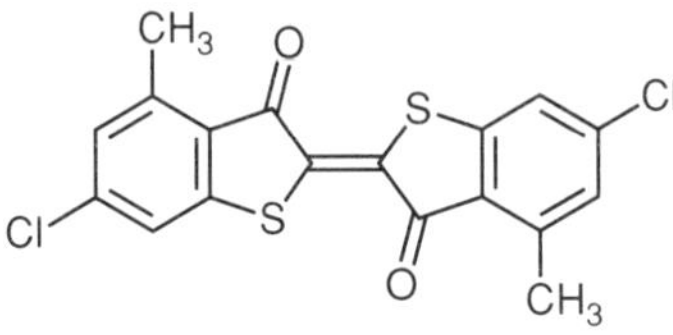

6.124

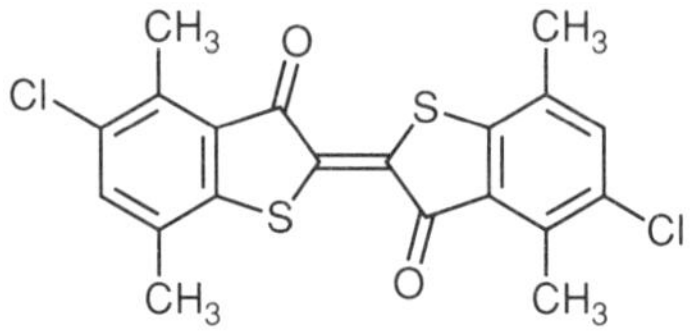

6.125

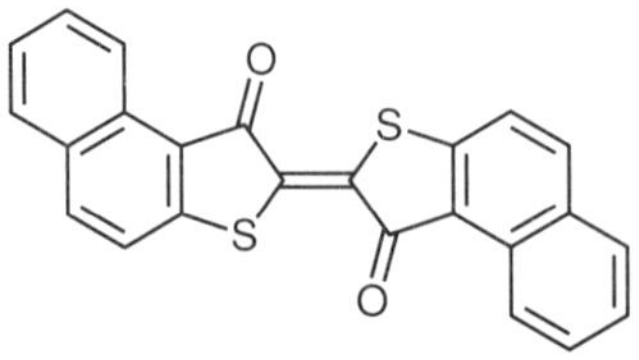

6.126

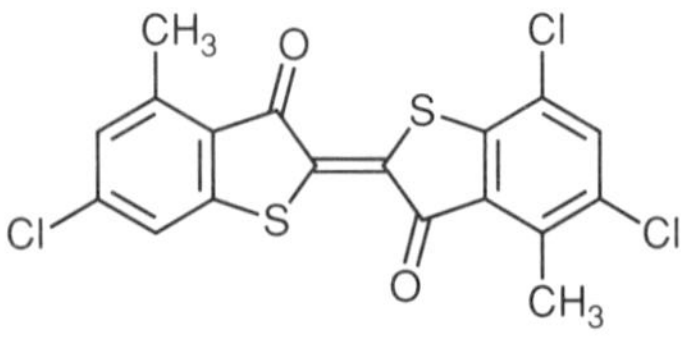

6.127

Dull tertiary hues are available from mixed indigoid–thioindigoid dyes, such as C.I. Vat Brown 42 (6.128) and C.I. Vat Black 1 (6.129). Hybrid dyes of this kind are relatively difficult to prepare. In one method of synthesis a suitable indoxyl or thioindoxyl reacts with 4-nitroso-*NN*-dimethylaniline to form an anil at the 2-position (keto tautomer). Further condensation between the anil and a different thioindoxyl or indoxyl gives the unsymmetrical dye.

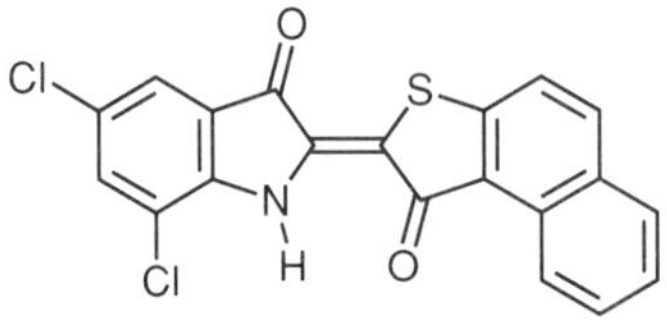

6.128

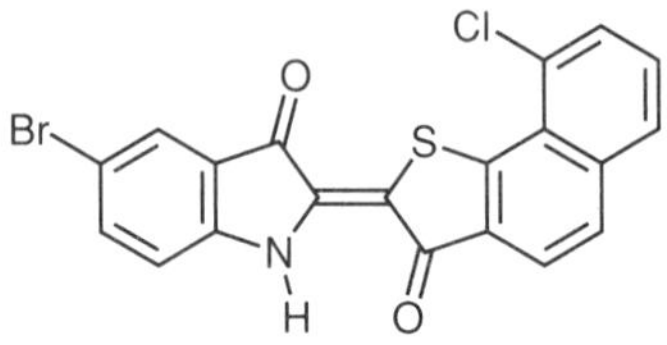

6.129

6.3.2 Application

Vat dyes in general have the disadvantage that reduction to a water-soluble form is necessary before application to cellulosic textiles. Traditionally sodium dithionite in alkaline solution serves as the reducing agent. An alternative sometimes used in vat printing is thiourea dioxide, which is converted into formamidine sulphinate in alkaline solution (see Volume 2).

Indigo and its derivatives are readily reduced under weakly alkaline conditions to the water-soluble leuco form. The disodium salt indigo white (6.130) is classified in the *Colour Index* as C.I. Reduced Vat Blue 1 and is used for cotton dyeing. Acidification of indigo white gives the more stable acid leuco derivative. The related leuco disulphuric acid ester (6.131) was originally marketed as Indigosol O (C.I. Solubilised Vat Blue 1) and was used for wool dyeing. As mentioned in section 6.2, reduction and esterification can be carried out without isolating the leuco compound by using pyridine and iron together with chlorosulphonic acid.

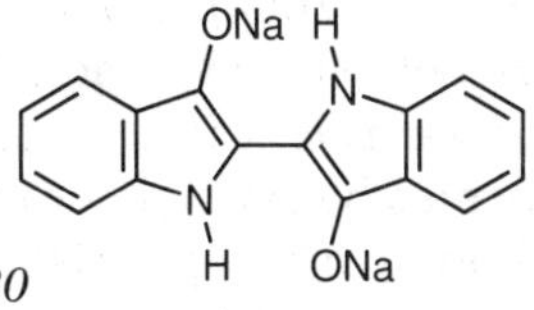

6.130

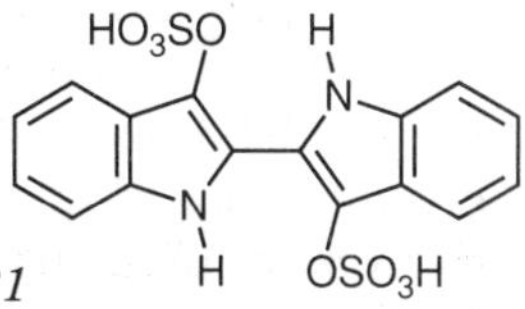

6.131

Nearly all the commercially produced indigoid and thioindigoid vat dyes are also marketed in the solubilised form. Thus, for example, C.I. Vat Blue 5 is also available as C.I. Solubilised Vat Blue 5. In addition, several dyes are sold only as the solubilised derivatives.

6.4 SULPHUR AND THIAZOLE DYES

Although many types of dye contain sulphur other than in sulphonic acid groups, sulphur dyes are usually considered to be those dyes that are best applied from a sodium sulphide dyebath. Like vat dyes, sulphur dyes are water-insoluble before and after application to cellulosic fibres. Disulphide linkages in the dye molecules are readily reduced by sodium sulphide to give alkali-soluble thiol groups. The resulting leuco dye has substantivity for cellulose. Subsequent oxidation within the fibre by exposure to air and oxidising agents such as hydrogen peroxide leads to the regeneration of di- and poly-sulphide linkages between aromatic ring systems. The dyeings generally possess good wet fastness and satisfactory light fastness properties. The fastness to bleaching of sulphur dyes is poor and the dyes are often decomposed by acids with the evolution of hydrogen sulphide. Nevertheless they are relatively cheap to produce and offer a full range of hues from yellow to black, albeit rather dull colours. The blues and blacks are of greatest importance, especially C.I. Sulphur Black 1, which is manufactured on a large scale [36].

The first commercial sulphur dye was discovered accidentally in 1873 by Croissant and Bretonnière who heated lignin-containing organic waste, such as sawdust, with sodium polysulphide at about 300°C; the product was sold under the name *Cachou de Laval* [37]. Even today an equivalent dye (C.I. Sulphur Brown 1) is derived from lignin sulphonate, which is readily available from waste liquors from wood pulp manufacture. The real pioneer of sulphur dyes was Vidal, the first chemist to obtain dyes of this type from specific organic compounds; in particular, Sulphur Black T (C.I. Sulphur Black 1) was made from 2,4-dinitrophenol in 1899. At the turn of the century many of the intermediates available were subjected to sulphurisation (thionation), that is, treatment with sulphur, sodium sulphide or sodium polysulphide to introduce sulphur linkages.

6.4.1 Methods of formation

Two basic methods are used for the manufacture of sulphur dyes. In one, the starting materials are baked either with sulphur alone or with sulphur and sodium sulphide at a temperature between 180 and 350°C. Alternatively the intermediates are heated under reflux in aqueous or alcoholic sodium polysulphide; this process may also be carried out under pressure at temperatures up to about 130°C. Following sulphurisation the dye is precipitated by means of air or chemical oxidation, acidification or a combination of these methods. The sulphurisation process results in the evolution of hydrogen sulphide, which is usually absorbed in aqueous sodium hydroxide for use elsewhere – in the reduction of nitro compounds, for example.

The commercially available sulphur dyes can be divided into three categories. The conventional C.I. Sulphur dyes are water-insoluble, polymeric products, which can also be produced in a finely dispersed form [36]. C.I. Leuco Sulphur dyes are obtained by treating sulphur dyes with, typically, an alkaline mixture of sodium sulphide and sodium hydrosulphide. These dyes are suitable for direct application and are available in concentrated solutions (liquid dyes). Liquid dyes are also produced directly from the thionation mixture without isolation; powder or granular products can be made by drying a slurry of the dye together with a reducing agent. C.I. Solubilised Sulphur dyes, which are thiosulphonic acid derivatives of sulphur dyes, are obtained by the action of sodium sulphite or bisulphite on the parent dye. These products are highly water-soluble but possess only low substantivity for cotton until they have been converted into the leuco form with sodium sulphide. After application an oxidative treatment converts the thiol groups into disulphide bonds, leading to dimeric or polymeric dyes [38]. The thiosulphate dyes ($ArSSO_3Na$) belong to a group of compounds known as Bunte salts [39].

In the *Colour Index* both conventional sulphur dyes and their leuco counterparts are allocated the same C.I. constitution number; a different number is given to the related solubilised version. Thus, for example, C.I. Sulphur Black 1 and C.I. Leuco Sulphur Black 1 have the reference C.I. 53185 whereas C.I. Solubilised Sulphur Black 1 appears under C.I. 53186.

Because of the complexity of the final products, sulphur dyes are classified according to the chemical structure of the organic starting material that predominates in the manufacturing process. Typical intermediates include aromatic amines, with or without nitro and phenolic groups, and diphenylamine derivatives.

6.4.2 Structural features

The dry baking procedure generally results in yellow, orange or brown sulphur dyes, many of which are known to contain benzothiazole groups. For example, Immedial Yellow GG (C.I. Sulphur Yellow 4) is obtained from a melt of 2-(4′-aminophenyl)-6-methylbenzothiazole (dehydrothio-*p*-toluidine) (6.132), 4,4′-diaminobiphenyl (benzidine) (6.133) and sulphur. The thiazole intermediate is obtained when *p*-toluidine is heated with sulphur. Further sulphurisation gives the dithiazole derivative primuline base, sulphonation of which led to the discovery of primuline (C.I. Direct Yellow 59) by Green in 1887. Benzidine (a carcinogen) is no longer used by European manufacturers but sulphur dyes derived from this intermediate are still available commercially in other parts of the world. The structure of C.I. Sulphur Yellow 4 was investigated in 1948 by Zerweck, Ritter and Schubert [40]. These workers degraded the dye in a potassium hydroxide melt and obtained various *o*-aminothiophenols, which were condensed with chloroacetic acid to produce identifiable lactams. From the ratio of lactams obtained, it was deduced that C.I. Sulphur Yellow 4 was a mixture of four main components, exemplified by structure 6.134; these were all synthesised and gave, on admixture, a product that was effectively identical with the commercial dye.

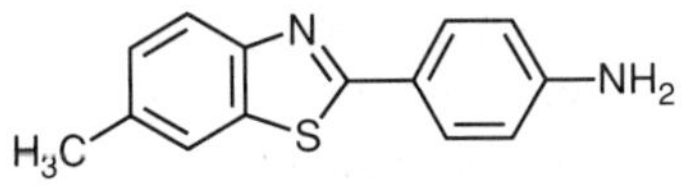

6.132

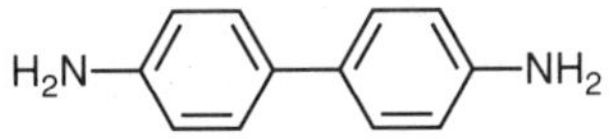

6.133

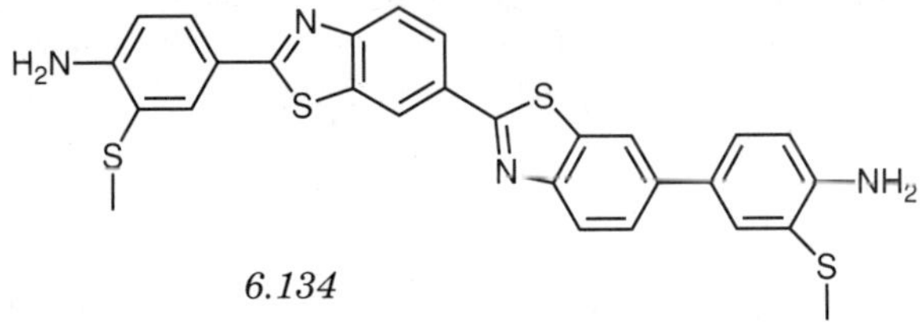

6.134

A similar study of the dye Immedial Orange C (C.I. Sulphur Orange 1), obtained by heating 2,4-diaminotoluene with sulphur, led to the isolation of compound 6.135 after reaction of the alkali melt of the dye with chloroacetic acid. It can be concluded that the commercial dye is a polybenzothiazole. The final hue of the product depends on the temperature and the duration of the sulphurisation; the use of a smaller proportion of sulphur produces C.I. Sulphur Brown 10. A similar product (C.I. Sulphur Brown 8) is obtained when 2,4-dinitrotoluene is baked with sodium polysulphide. A brown sulphur dye of high fastness to light (C.I. Sulphur Brown 52) results when the polycyclic hydrocarbon decacyclene (6.136) is baked with sulphur at 350°C [41].

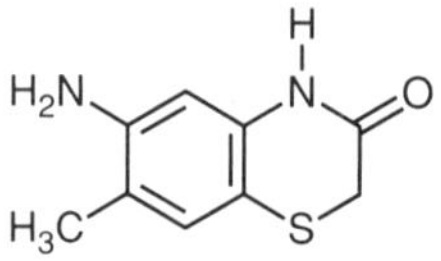

6.135

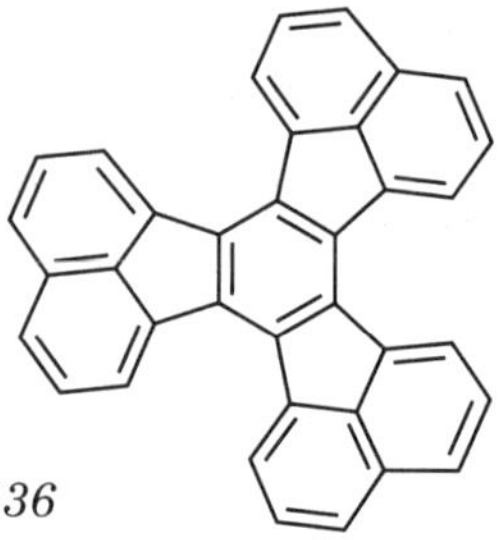

6.136

The use of intermediates capable of forming quinonimine ring systems leads to the synthesis of red, blue, green and black sulphur dyes on treatment with aqueous or alcoholic sodium polysulphide. Thus, for example, condensation of *NN*-dimethyl-*p*-phenylenediamine with phenol gives a diphenylamine intermediate (6.137), which on sulphurisation in alcoholic sodium polysulphide first yields an indoaniline oxidation product (6.138). This quinonimine in turn undergoes thionation to form a phenothiazone chromophore (6.139). Further reaction leads to the synthesis of C.I. Sulphur Blue 9. A similar sequence can be envisaged using indophenol derivatives. For example, the simple indophenol 6.140, which is made by oxidising a mixture of 4-aminophenol and phenol with sodium hypochlorite, gives C.I. Sulphur Blue 14 on sulphurisation with aqueous sodium polysulphide. Similar blue dyes are available from related intermediates. For example, condensation of diphenylamine with 4-nitrosophenol in sulphuric acid gives compound 6.141; this intermediate reacts with aqueous

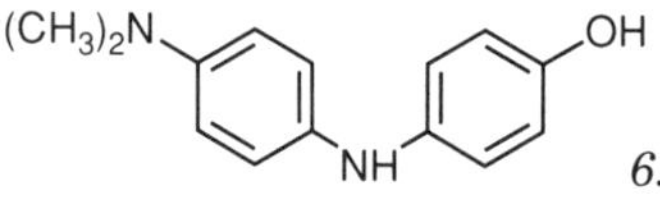

6.137

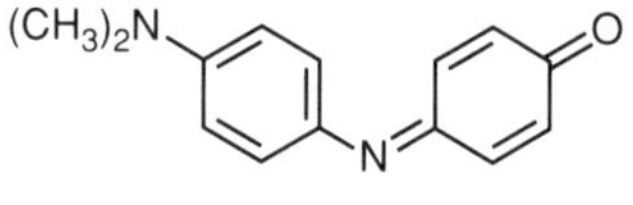

6.138

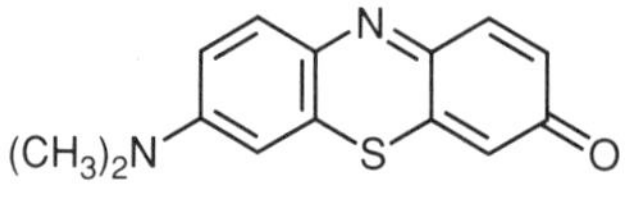

6.139

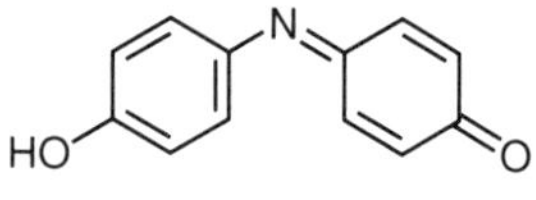

6.140

6.141

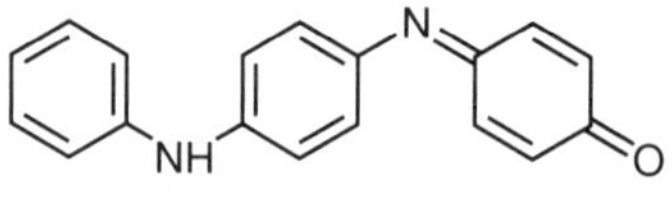

sodium polysulphide to form C.I. Sulphur Blue 13. The incorporation of a naphthalene ring system enables green dyes to be produced. The important dye C.I. Sulphur Green 3, for example, is made by heating compound 6.142 with aqueous sodium polysulphide under reflux in the presence of copper(II) sulphate. C.I. Sulphur Green 14 is based on copper phthalocyanine [36].

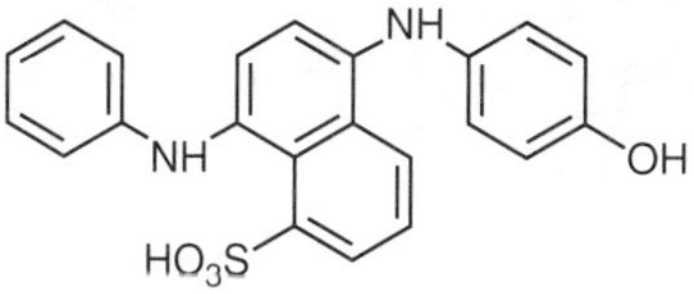

6.142

Sulphurisation of 4-hydroxydiphenylamine in a glycol solvent gives C.I. Sulphur Red 10. A useful red dye (C.I. Sulphur Red 6) is prepared by heating 3-amino-6-hydroxy-2-methylphenazine (6.143) with aqueous sodium polysulphide. The phenazine derivative can be made by oxidising the condensation product of 2,4-diaminotoluene and 4-aminophenol. The substituted phenothiazone 6.144 gives C.I. Sulphur Red 5 on treatment with aqueous sodium polysulphide in the presence of glycerol. Zerweck and his collaborators [40] showed that the chlorine atoms in compound 6.144 could be displaced stepwise by mercapto groups. The resulting polymeric dye was found to be essentially the same as that generated by sulphurisation of the parent 4-hydroxy-4′-methyldiphenylamine. It can therefore be deduced that sulphur dyes of this type contain structural units similar to that shown in structure 6.145, which possesses a central thianthrene ring system isosteric with indanthrone (6.65).

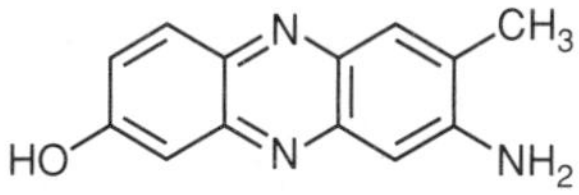

6.143

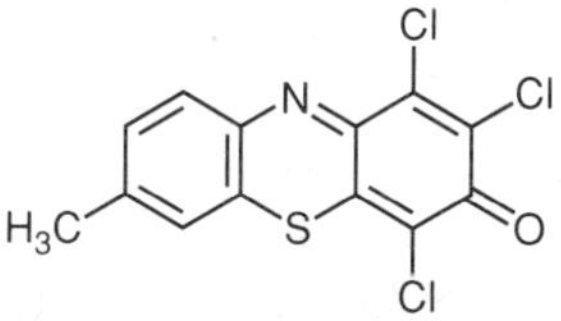

6.144

Only one violet sulphur dye is commercially available: C.I. Sulphur Violet 1, obtained from the substituted indoaniline 6.146. Several black sulphur dyes are currently in use, especially C.I. Sulphur Black 1 which is produced from the relatively simple intermediate 2,4-dinitrophenol and aqueous sodium polysulphide. A similar product (C.I. Sulphur Black 2) is obtained from a mixture of 2,4-dinitrophenol and either picric acid (6.147; $X = NO_2$) or picramic acid (6.147; $X = NH_2$). A black dye possessing superior fastness to chlorine when on the fibre (C.I. Sulphur Black 11) can be made from the naphthalene intermediate 6.148 by heating it in a solution of sodium polysulphide in butanol. An equivalent reaction using the carbazole intermediate 6.149 gives rise to the reddish-blue Hydron Blue R (CAS, C.I. Vat Blue 43). This important compound, which also possesses superior

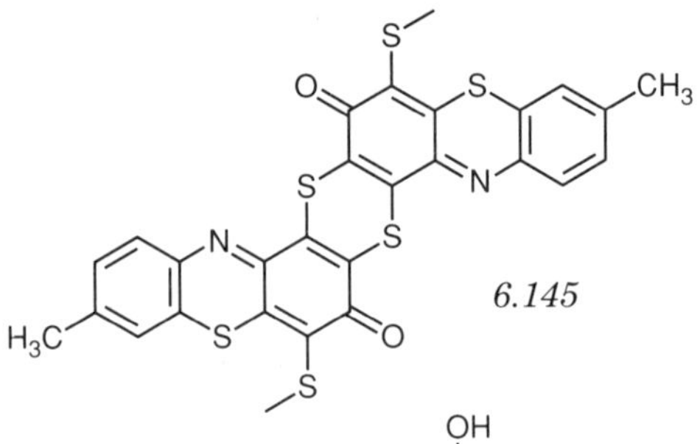

6.145

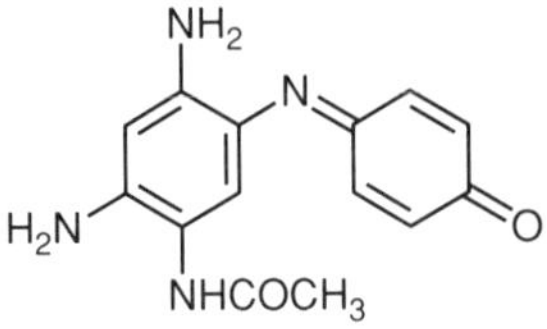

6.146

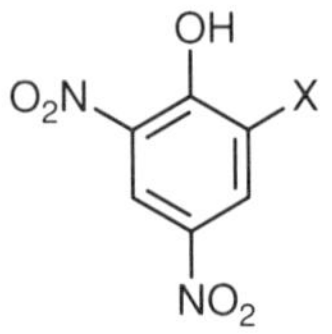

6.147

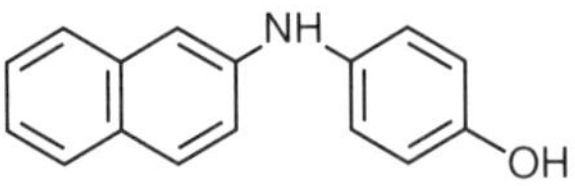

6.148

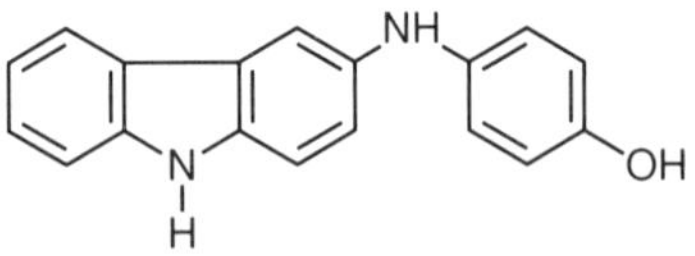

6.149

fastness properties, is classified as a vat dye because it is usually applied from an alkaline sodium dithionite bath; dyes of this type can, however, be applied from a sodium sulphide bath. Interestingly, inclusion of copper(II) sulphate in the sulphurisation of intermediate 6.149 leads to the formation of the bluish-black C.I. Sulphur Black 4.

The many sulphur dyes synthesised via quinonimine intermediates are polymeric products containing numerous disulphide cross-linkages that can be broken by reduction in aqueous sodium sulphide; thioether groups survive the reduction process. The smaller thiolate-containing molecules produced show affinity for cellulose. Although the actual structures of such dyes are complex, their essential features can be illustrated in an idealised model (Scheme 6.23), in which X = >S, >NH or >O, and R indicates substituents or annelated rings.

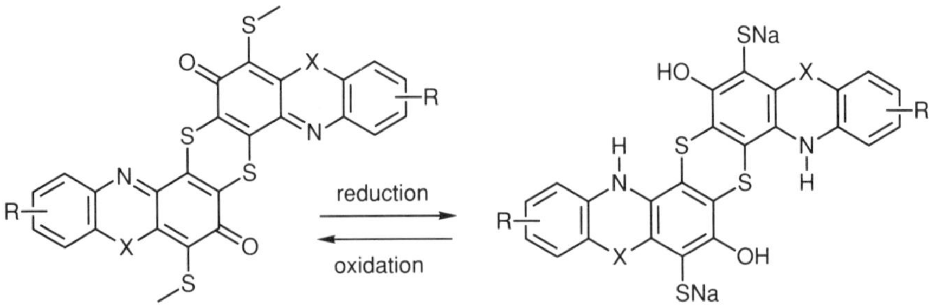

Scheme 6.23

6.4.3 Sulphurised vat dyes and thiazole derivatives

C.I. Vat Blue 43 and the equivalent dye derived from *N*-ethylcarbazole (C.I. Vat Blue 42) are sometimes referred to as sulphurised vat dyes. The term is also applied to some anthraquinonoid thiazole derivatives, such as C.I. Vat Yellow 2 (6.150). This compound can be made by the reaction of 2,6-diamino-1,5-dichloroanthraquinone with benzaldehyde and sulphur. Another anthraquinonoid vat dye containing a thiazole ring system (6.107) has already been mentioned.

The thiazole ring system is found in many types of dye. Thiazole-containing sulphur dyes and primuline were considered in section 6.4.2. Quaternised dehydrothio-*p*-toluidine (6.132) is available as C.I. Basic Yellow 1 (6.151). Other derivatives of this intermediate are used as direct dyes, such as C.I. Direct Yellow 8 (4.58). The benzothiazole ring appears in various azo disperse dyes [12], quaternisation of which gives useful cationic dyes, an important example being C.I. Basic Blue 41 (4.94). Another example containing a quaternised thiazole ring is C.I. Basic Red 29 (4.97).

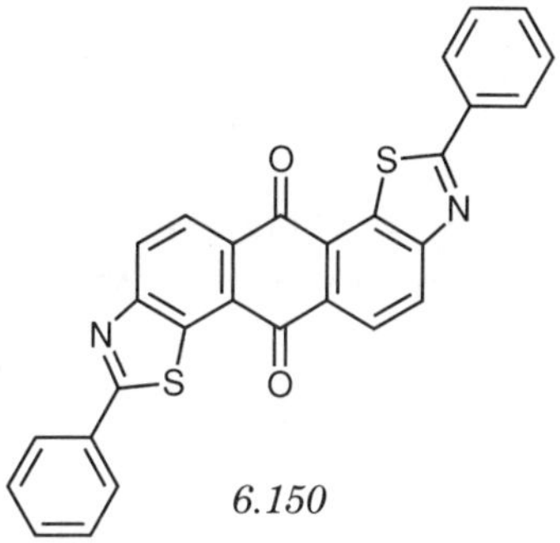

6.150

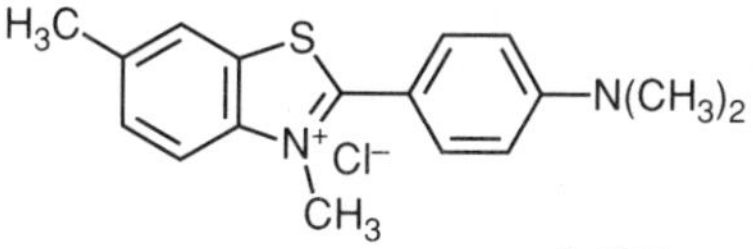

6.151

6.5 DIARYLMETHANE AND TRIARYLMETHANE DYES

Derivatives of triphenylmethane were among the earliest synthetic colorants, and are still in demand where bright, intense colours are needed without the necessity for outstanding fastness to light and chemical reagents. Basic dyes of this type, as well as other cationic dyes, are suitable for dyeing conventional acrylic fibres, on which they show better fastness properties than on natural fibres.

Traditionally dyes in this category are referred to as triarylmethane dyes; carbocyclic ring systems other than benzene, particularly naphthalene, and various heterocyclic systems are also encountered. Few diarylmethane dyes are of commercial significance, although some heterocyclic derivatives and analogues are of interest. Di- and tri-arylmethane dyes are mesomeric cations (charge-resonance systems) in which the positive charge is localised mainly on the terminal nitrogen atoms, as in the case of Michler's Hydrol Blue (Scheme 6.24). One of the resonance canonicals

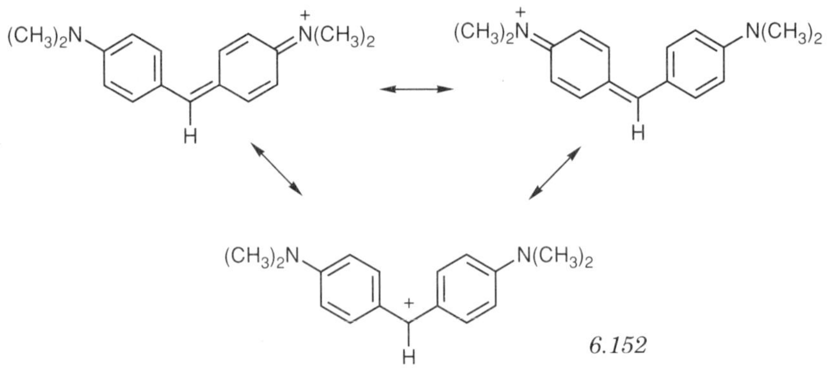

6.152

Scheme 6.24

(6.152) is a carbonium ion, so that dyes of this kind are sometimes referred to as di- or tri-arylcarbonium ions [42]. Since the central carbon atom in these systems is sp^2 hybridised, Zollinger suggests that the terms di- and tri-arylmethine dyes are more appropriate [43]; this class of dyes is structurally similar to the polymethine dyes.

Michler's Hydrol Blue absorbs at 607.5 nm (ε 147 500) in 98% acetic acid [44]. This bright blue dye and related diphenylmethane derivatives have poor fastness properties and are not of commercial importance. Attachment of an amino group to the central carbon atom of Michler's Hydrol Blue gives the yellow dye Auramine O (6.153; C.I. Basic Yellow 2), which absorbs at 434 nm in ethanol and is used in the dyeing of paper and in other applications. The electron-donating amino group produces a very marked hypsochromic shift. Auramine O is somewhat more stable than Michler's Hydrol Blue but, being a ketimine, it is hydrolysed in boiling water to form Michler's ketone (6.154) and ammonia. The dye was first obtained in 1883 by Kern and Caro who heated Michler's ketone with ammonium chloride and zinc chloride. The synthesis first requires the formation of the important ketone intermediate by a Friedel–Crafts acylation (Scheme 6.25). An improved route was devised in 1889 by Sandmeyer, involving the corresponding thioketone (6.156) (Scheme 6.26). Condensation of *NN*-dimethylaniline with formaldehyde gives 4,4′-bis(dimethylamino)diphenylmethane (6.155), which is then heated with sulphur and ammonium chloride, with sodium chloride as diluent, in an atmosphere of ammonia. The resulting Auramine Base (6.157) gives Auramine O when treated with hydrochloric acid. The ethyl analogue of Auramine O is C.I. Basic Yellow 37.

Reduction of Michler's ketone gives Michler's hydrol (6.158), which forms Michler's Hydrol Blue in the presence of acid (Scheme 6.27). Michler's hydrol is produced industrially by the oxidation of the diphenyl-

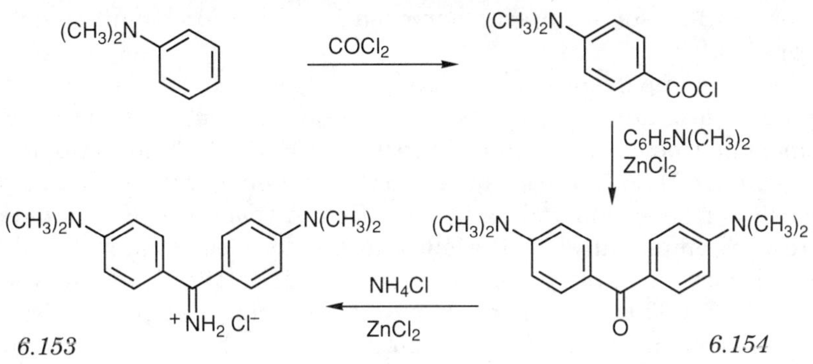

Scheme 6.25

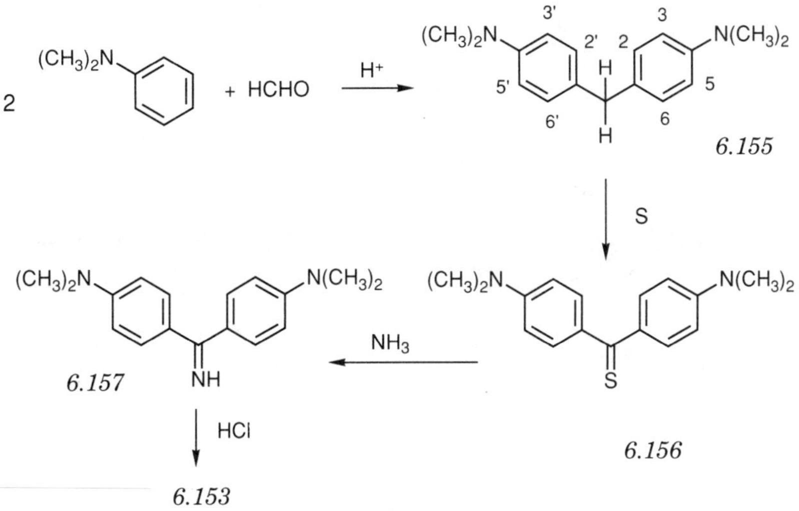

Scheme 6.26

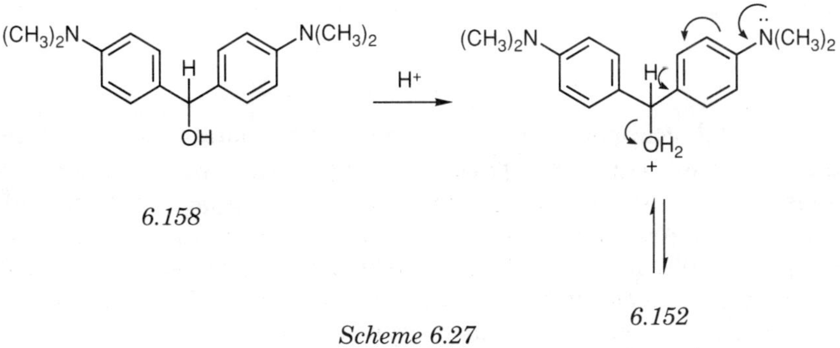

Scheme 6.27

methane precursor (6.155); further oxidation to give Michler's ketone takes place readily.

The more important triphenylmethane dyes probably date from 1834, when Runge obtained red compounds of unknown constitution from crude phenol and from crude aniline. Shortly after Perkin's discovery of mauveine (6.205) in 1856 Natanson obtained the triphenylmethane dye magenta (fuchsine) from crude aniline by oxidation. Although many related dyes were made empirically over the following two decades, it was not until 1880 that Fischer was able to establish structural relationships within this group of dyes. The essential triphenylmethane ring system was identified by its conversion into 4,4′,4″-triaminotriphenylmethane (6.159), the source of pararosaniline (6.160; C.I. Basic Red 9) (Scheme 6.28).

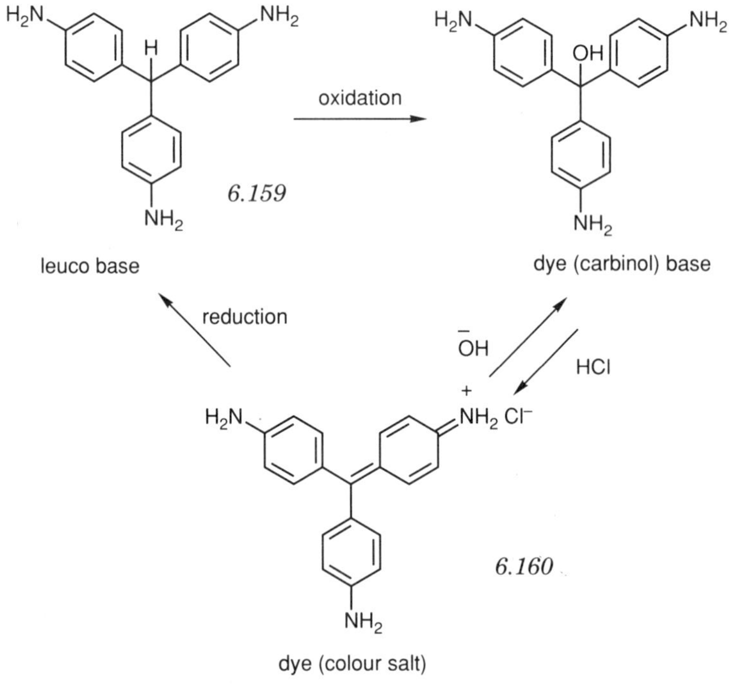

Scheme 6.28

Nearly all commercial triarylmethane dyes contain two or three electron-donor substituents at *para* sites. In general, dyes containing two *p*-amino groups (the malachite green series) are green or greenish-blue, whereas those with three *p*-amino groups (the crystal violet series) are reddish- to bluish-violet. Of less importance are monoamino structures containing alkoxy groups, although these orange to red dyes have very good light fastness on acrylic and modacrylic fibres [16].

6.5.1 Methods of synthesis

In the synthesis of the triarylmethane skeleton the central carbon atom is provided either by a simple compound, such as formaldehyde or phosgene, or a part of an aromatic intermediate. Depending on the oxidation state of the component used, the resulting dye is obtained directly or as the leuco base. The various reactions involved are essentially electrophilic aromatic substitutions in which, usually, the *para* position is activated by a powerful electron-donor substituent such as a dialkylamino group.

The leuco base of the dye is produced by condensation of one mole of an aromatic aldehyde with two moles of an aromatic amine having an unsubstituted *para* position (the aldehyde synthesis) in the presence of a mineral acid or a Friedel–Crafts catalyst. Schiff bases are obtained with primary aromatic amines devoid of crowding substituents adjacent to the amino group. The aldehyde method is illustrated by the classical synthesis of malachite green (6.161; C.I. Basic Green 4), first carried out in 1877 by Fischer (Scheme 6.29). The resulting leuco base is readily oxidised to the dye base and thence to the colour salt. For many years lead(IV) oxide has been used as the oxidising agent but more recently, for environmental reasons, lead-free processes have been used. Of particular interest is oxidation by chloranil (tetrachloro-*p*-benzoquinone) or by air in the presence of catalysts. Mechanistically it is likely that the conversion involves hydride abstraction in the case of chloranil, so that the dye is formed directly from the leuco base [45].

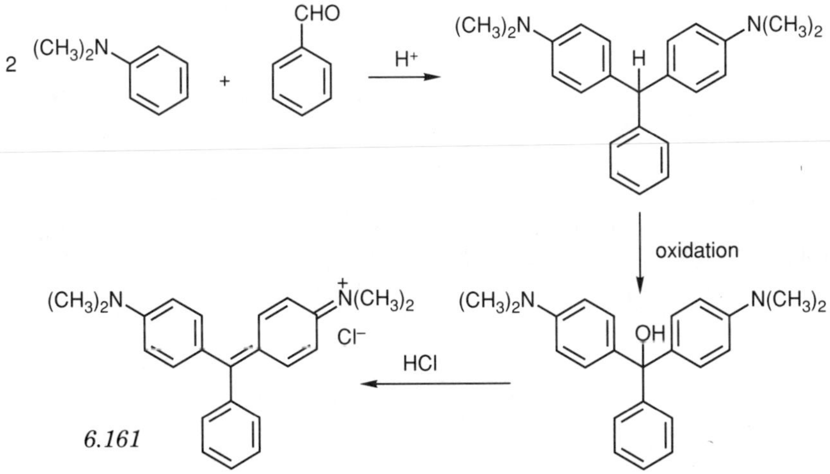

6.161

Scheme 6.29

Condensation of a diphenylmethanol derivative, such as Michler's hydrol (6.158), with a reactive aryl component under acid conditions (the hydrol synthesis) also provides a leuco base. The acid dye Wool Green S

(6.162; C.I. Acid Green 50) is made by reacting Michler's hydrol with R acid (2-naphthol-3,6-disulphonic acid) and oxidising the resulting leuco compound (Scheme 6.30).

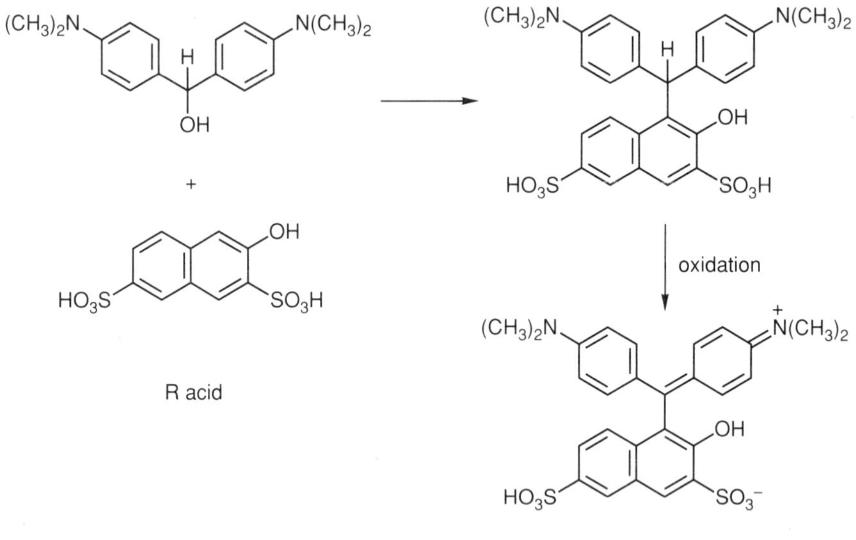

Scheme 6.30

Michler's ketone (6.154) and related carbonyl compounds can be used to obtain colour salts directly by reaction with a reactive intermediate such as an aromatic amine (the ketone synthesis), as in the case of crystal violet (6.163; C.I. Basic Violet 3) (Scheme 6.31).

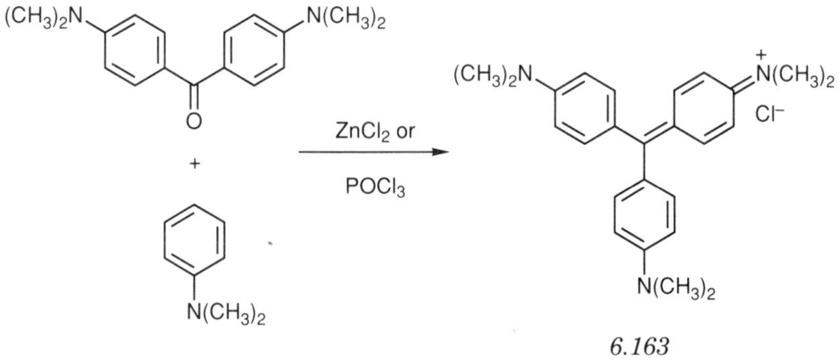

Scheme 6.31

A few triphenylmethane dyes are still obtained by empirical methods first used in the early processes of dye manufacture. For example, in the production of magenta (C.I. Basic Violet 14) a mixture of aniline, toluidines and nitrotoluenes is heated with zinc and iron(II) chlorides for several hours. It is likely that the reaction involves initially the oxidation of a methyl group to an aldehyde group. Condensation to form triphenylmethane derivatives, not surprisingly, leads to mixtures. The main components of magenta are rosaniline (6.164) and pararosaniline (6.160). The yellowish-brown by-product chrysaniline (6.165; C.I. Basic Orange 15), an acridine dye, is present in the mother liquors of magenta manufacture. In the manufacture of methyl violet (C.I. Basic Violet 1), air oxidation of *NN*-dimethylaniline in the presence of a copper(II) salt gives a mixture of the more highly methylated pararosanilines.

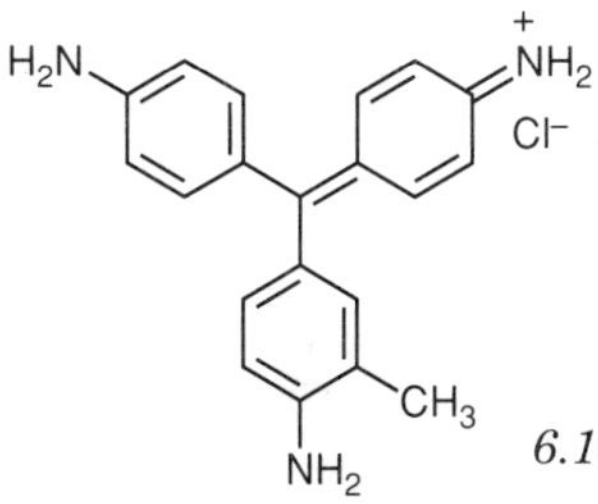

6.164

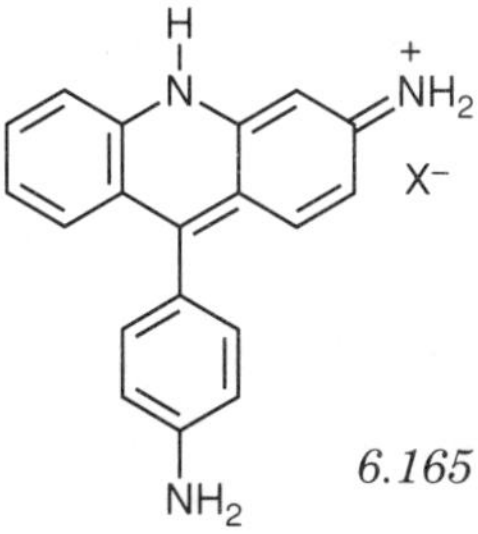

6.165

In the laboratory pure dye bases can be prepared by the reaction of diaryl ketones, such as Michler's ketone (6.154), with appropriate Grignard reagents or organolithium compounds. These tertiary alcohols are also available from related reactions between aryl esters and organometallic reagents.

The basic dyes are usually marketed as chlorides, oxalates or zinc chloride double salts; in the case of malachite green (6.161), the last-named derivative has the formula $3[Dye]Cl.2ZnCl_2.2H_2O$. Acid dyes of this type are often isolated as the sodium salt.

6.5.2 Acid, basic and mordant dyes

Alkylation of aminotriphenylmethane dyes has a bathochromic effect, which is even more pronounced on arylation, as illustrated in Table 6.5.

Unlike Michler's Hydrol Blue (Scheme 6.24) and crystal violet (6.166; $R = CH_3$), malachite green (6.167; $R = CH_3$) shows two absorption bands in the visible region of the electromagnetic spectrum (Figure 6.1). In 98% acetic acid the long-wavelength absorption band (x-band) of malachite green, which is associated with polarisation along the x-axis, appears at 621 nm (ε 104 000) whereas the less significant y-band, which is polarised along the y-axis, is found at 427.5 nm (ε 20 000) [46]: the green colour of the dye arises from these two components. Consequently the presence of

TABLE 6.5

Spectral data for some triphenylmethane dyes in 98% acetic acid

Compound	λ_{max}/nm
6.166; R = H	538
6.166; R = CH_3	589
6.166; R = C_2H_5	592.5
6.167; R = H	570
6.167; R = CH_3	621
6.167; R = C_2H_5	629.5

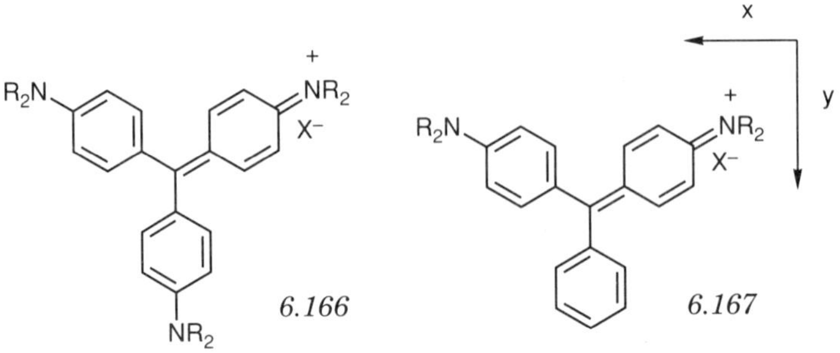

6.166 *6.167*

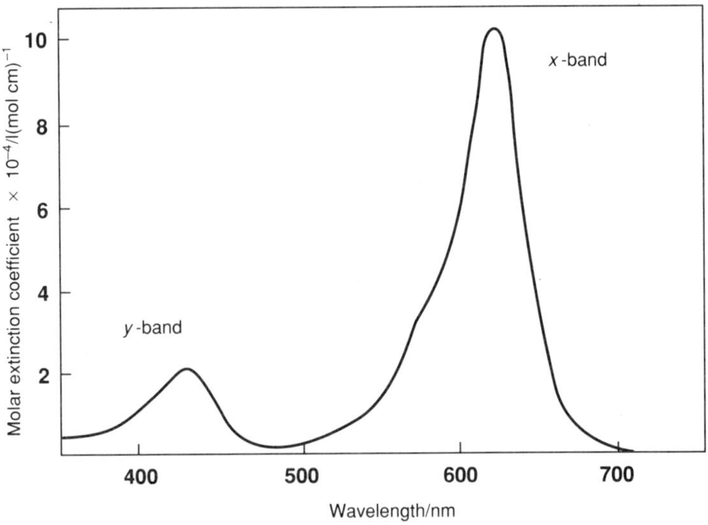

Figure 6.1 – Absorption spectrum of malachite green

an *ortho* substituent in the phenyl ring of malachite green gives rise to a significant crowding effect, which decreases the importance of the *y*-band by partial deconjugation; at the same time the *x*-band is enhanced. Dyes of this type are therefore blue; for example, the 2-chloro derivative of malachite green (6.168; C.I. Basic Blue 1) displays absorption bands at 635 nm (ε 121 000) and 415.5 nm (ε 13 000) [46].

As shown by X-ray crystallography, triphenylmethane dyes are unable to adopt a planar conformation and thus the three aromatic rings are twisted out of the molecular plane by approximately 30°. For example, the molecule of pararosaniline perchlorate is shaped like a three-bladed propeller [47]. Steric effects are therefore significant for *o*-substituted analogues of malachite green but not for those with only *m*- or *p*-substitution (6.169). In accordance with theory [48], crowding substituents adjacent to the central carbon atom in triarylmethane dyes give rise to bathochromic shifts of the long-wavelength absorption band.

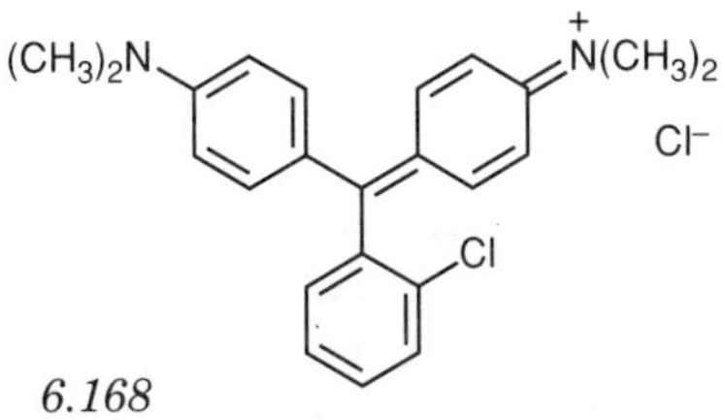

6.168

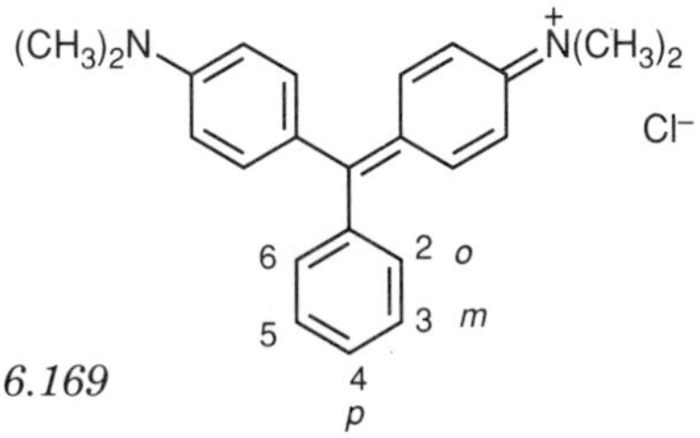

6.169

The *x*-band in malachite green arises from an NBMO→π* transition, so that 3- and 4-substituents affect the energy of the excited state only and bring about spectral shifts of the first absorption band which vary linearly with the appropriate Hammett substituent constants. Thus, electron-withdrawing groups cause bathochromic shifts of the *x*-band whereas donor substituents cause hypsochromic shifts (Table 6.6) [46,49]. The *y*-band arises from a π→π* transition [50] so that substituent effects are less predictable. As the donor strength of the 4-substituent increases, however, the *y*-band moves bathochromically and eventually coalesces with the *x*-band – at 589 nm in the case of crystal violet (6.163), which possesses two NBMOs that are necessarily degenerate [51].

Acid dyes

The presence of at least two sulphonic acid groups in the triarylmethane ring system permits the derivatives to be applied as acid dyes. Although sulphonic acid groups are often present in the intermediates used, acid dyes can also be obtained by direct sulphonation at the leuco stage or of the basic dye itself.

Typically the aldehyde synthesis is used to make the 2,4-disulphonic acid of Brilliant Green (6.170; C.I. Acid Blue 1) from *NN*-diethylaniline and

TABLE 6.6

Spectral data for some derivatives of malachite green

	x-band		y-band	
Substituent	λ_{max}/nm	ε_{max}	λ_{max}/nm	ε_{max}
None	621	104 000	427.5	20 000
3-Cl	630	103 000	426	17 000
4-Cl	627.5	104 000	433	22 000
3-CF_3	634	106 000	424	16 000
4-CF_3	637	104 000	424	15 000
3-CN	637	90 000	426	15 000
4-CN	643	88 000	429	16 000
3-NO_2	637.5	87 000	425	14 000
4-NO_2	645	83 000	425	17 000
3-CH_3	618.5	106 000	433	22 000
4-CH_3	616.5	106 000	437.5	25 000
3-OCH_3	622.5	107 000	435	18 000
4-OCH_3	608	106 000	465	34 000
4-C_6H_5	626	100 000	457	30 000
4-$N(CH_3)_2$	589	117 000	589	117 000

benzaldehyde-2,4-disulphonic acid. The closely related dye C.I. Acid Blue 3 (6.171) is obtained by sulphonating the 3-hydroxy derivative of Brilliant Green at the leuco base stage. After oxidation the dye is isolated as the calcium salt, which also finds use as a food colorant (C.I. Food Blue 5). Direct sulphonation of the colour salt is used to prepare C.I. Acid Blue 93 from diphenylamine blue (6.172; C.I. Solvent Blue 23). The latter dye is made by phenylation of pararosaniline (6.160) with an excess of aniline in the presence of benzoic acid at about 180°C. The main product of sulphonation is the trisulphonic acid. Wool Green S (6.162) has already been

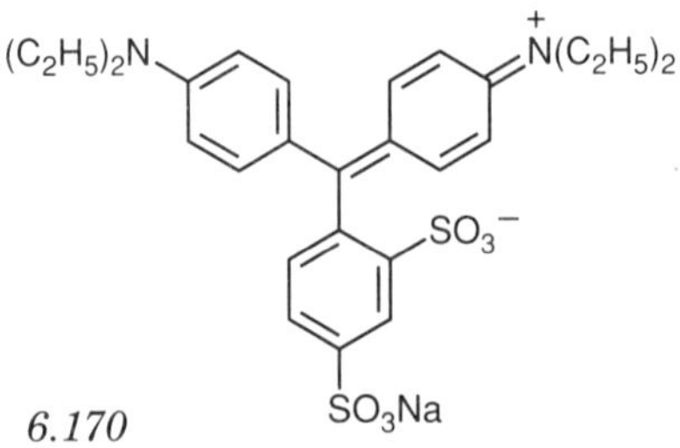

6.170

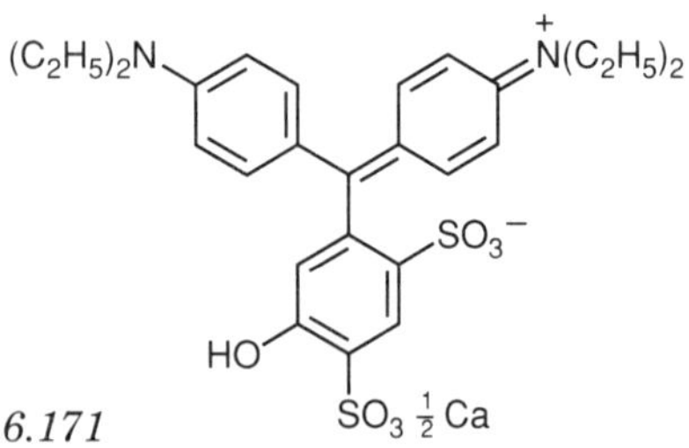

6.171

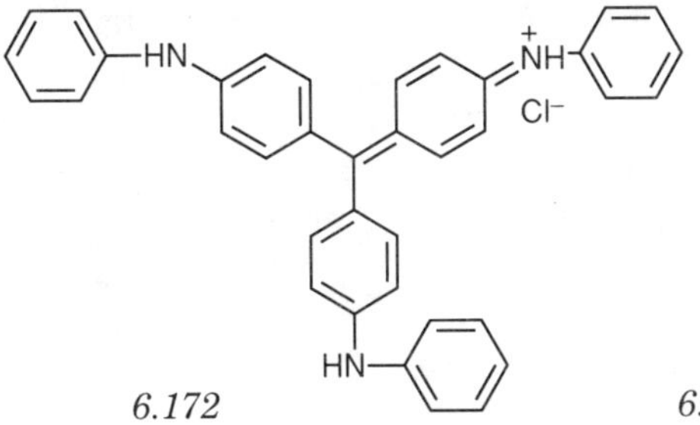

6.172

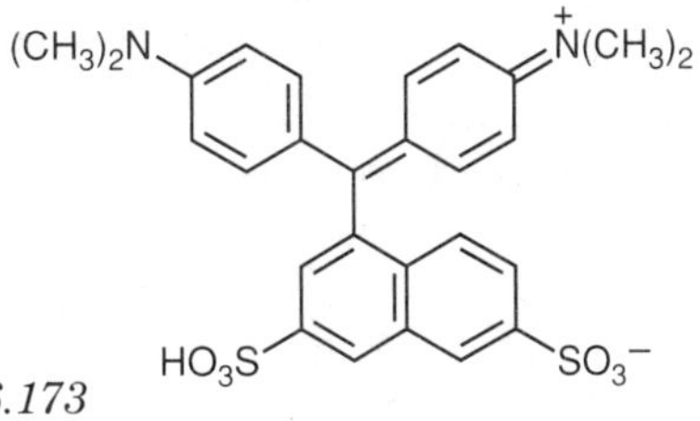

6.173

mentioned as an example of the hydrol synthesis. The very similar Naphthalene Green V (6.173; C.I. Acid Green 16) is obtained by condensing Michler's hydrol with naphthalene-2,7-disulphonic acid followed by oxidation.

Several acid dyes are derived from *N*-sulphobenzyl-substituted amines. Thus in the production of C.I. Acid Violet 17 (6.174) *N*-ethyl-*N*-(3-sulphobenzyl)aniline is condensed with formaldehyde to produce the substituted diphenylmethane, which is then oxidised to the hydrol and condensed with *NN*-diethylaniline. Further oxidation of the leuco base gives the dye. The presence of an *o*-methyl group in the diethylaminophenyl ring gives a bright blue dye (C.I. Acid Blue 15). Condensation of benzaldehyde-2-sulphonic acid with the same intermediate followed by oxidation of the leuco base gives C.I. Acid Blue 9. This dye is also a food colorant (C.I. Food Blue 2) and the barium salt is used as a pigment (C.I. Pigment Blue 24). The corresponding 2-chloro derivative is C.I. Acid Green 9.

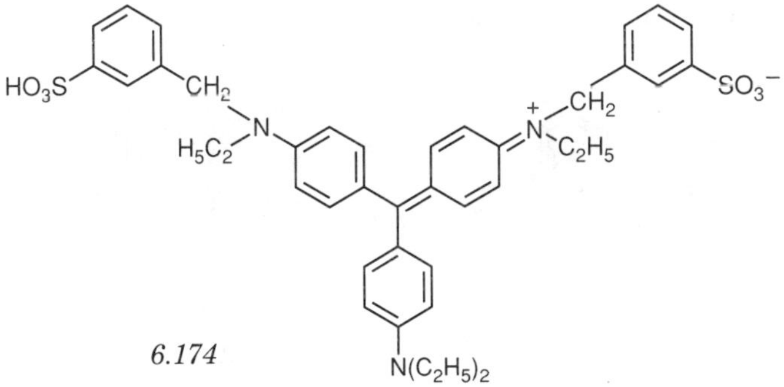

6.174

The nucleophilic displacement of suitable *para* substituents in analogues of malachite green is used as a means of synthesis of some important acid dyes. For example, Brilliant Indocyanine 6B (6.175; R = H; FDN, C.I. Acid Blue 83) is made by the aldehyde method, using 4-chlorobenzaldehyde and *N*-ethyl-*N*-(3-sulphobenzyl)aniline. The resulting leuco base is oxidised to the colour salt, which is then condensed with *p*-phenetidine to give the required product. The sterically hindered dimethyl derivative of this dye, C.I. Acid Blue 90 (6.175; R = CH_3) is also of interest. It is synthesised by first condensing benzaldehyde with *N*-benzyl-*N*-ethyl-*m*-toluidine to give a leuco base, which is then trisulphonated; after oxidation, the 4-sulphonic acid group is displaced by reaction with *p*-phenetidine.

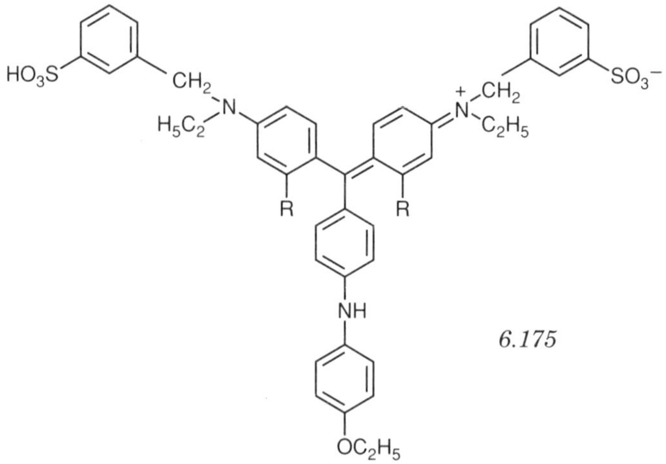

6.175

The incorporation of an indole ring system often leads to an improvement in the light fastness. A suitable example is C.I. Acid Blue 123 (6.176), which is derived from 4,4′-dichlorobenzophenone. Condensation with 1-methyl-2-phenylindole in the presence of phosphorus oxychloride produces the triarylmethane ring system. Replacement of the chlorine atoms with *p*-phenetidine, followed by sulphonation, gives the dye.

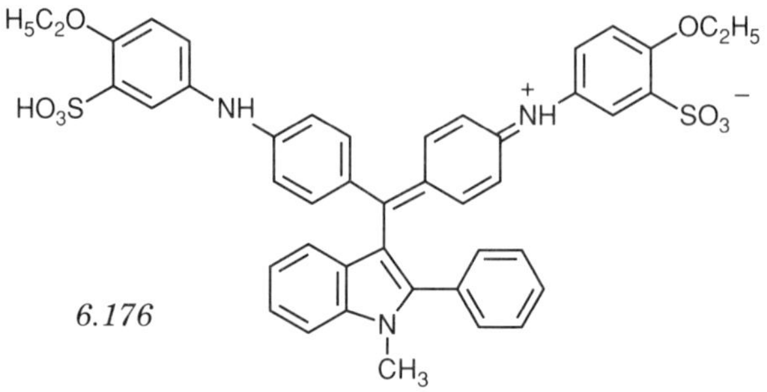

6.176

Basic dyes

Several examples of typical triarylmethane dyes have already been mentioned, in particular, pararosaniline (6.160), malachite green (6.161) and its *o*-chloro derivative (6.168), crystal violet (6.163), rosaniline (6.164) and diphenylamine blue (6.172). Brilliant Green (C.I. Basic Green 1) (6.167;

$R = C_2H_5$), the ethyl analogue of malachite green, is prepared by the aldehyde route and is isolated as the sulphate. The ethyl analogue of crystal violet is ethyl violet (C.I. Basic Violet 4) (6.166; $R = C_2H_5$) and is obtained by the ketone route.

In general the presence of substituents adjacent to terminal dialkylamino groups leads to partial deconjugation and to complex spectral shifts [52]. Steric effects are minimised, however, in the case of primary and secondary *p*-amino groups. Thus rosaniline (6.164) contains one *o*-toluidine residue and three such groupings are present in new magenta (6.177; C.I. Basic Violet 2), which is effectively obtained by the hydrol route, involving the reaction of two moles of *o*-toluidine with one of formaldehyde, oxidation to the hydrol, condensation of this intermediate with *o*-toluidine and oxidation of the leuco base. Methyl groups are also sited in two rings of C.I. Basic Blue 5 (6.178), made by the aldehyde route using 2-chlorobenzaldehyde and *N*-ethyl-*o*-toluidine.

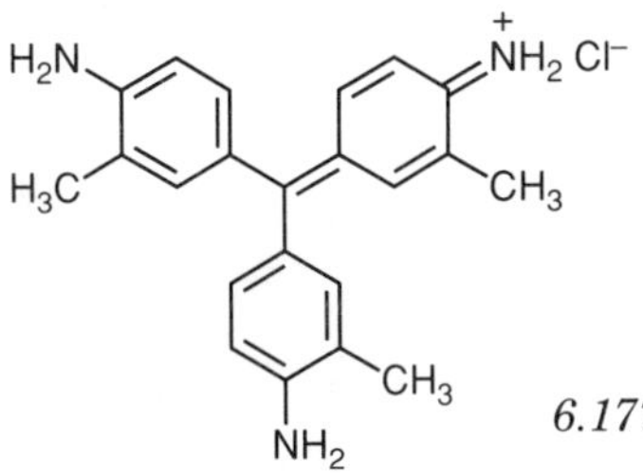

6.177

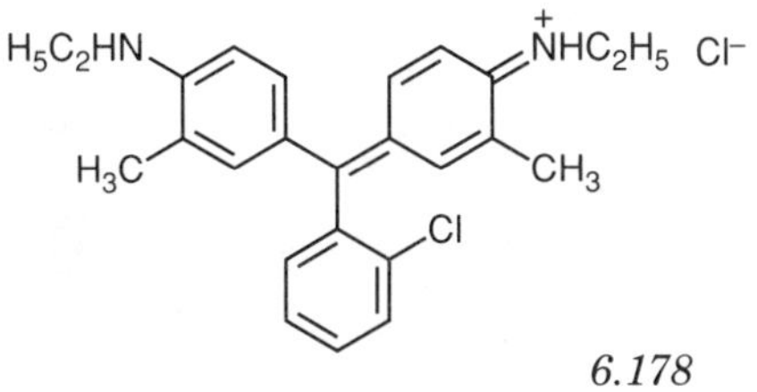

6.178

Various derivatives of naphthyldiphenylmethane are of value as blue dyes; the naphthyl substituent brings about bathochromic shifts by extending the conjugation and also by exerting a steric effect [53]. The Victoria Blues have been used commercially for many years. Typical examples include Victoria Blue B (6.179; C.I. Basic Blue 26), which was discovered in 1883 by Caro and Kern. The dye can be prepared by the ketone route using Michler's ketone (6.154) and 1-phenylaminonaphthalene. A similar condensation between the ethyl analogue of Michler's ketone and 1-ethylaminonaphthalene gives C.I. Basic Blue 7 (6.180).

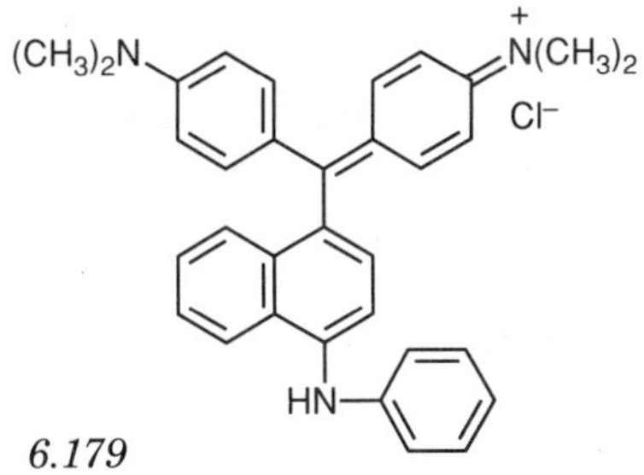

6.179

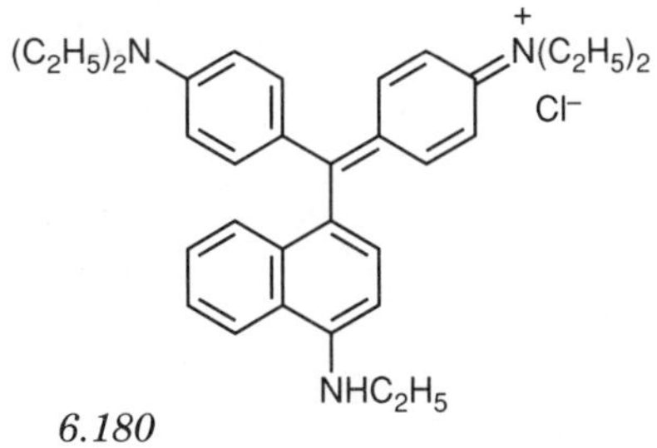

6.180

The phosphotungstomolybdic acid salts (PTMA lakes) of many basic dyes are used as violet, blue and green pigments. Examples include methyl violet (C.I. Pigment Violet 3), Victoria Pure Blue BO (C.I. Pigment Blue 1) and Brilliant Green (C.I. Pigment Green 1). The copper hexacyanoferrate(II) complex of methyl violet is C.I. Pigment Violet 27. The free bases of various triarylmethane derivatives are used as solvent dyes. Illustrative examples include methyl violet (C.I. Solvent Violet 8), Victoria Blue B (C.I. Solvent Blue 4) and malachite green (C.I. Solvent Green 1).

Certain derivatives of basic triarylmethane dyes are used in pressure-sensitive papers. Carbonless copy papers that make use of the so-called crystal violet lactone (6.181) are of particular importance. The reverse side of the top sheet is coated with microcapsules containing a solution of the colourless lactone (colour former) in a non-polar solvent. The bottom sheet is coated with an acidic clay or an acidic phenolic resin. Application of pressure ruptures the minute capsules and interaction between the released lactone and the lower acidic surface leads to instant dye formation (structure 6.182) at the point of contact. The resulting colour gradually fades; slower-acting colour formers that give rise to longer-lasting images, such as benzoylated leuco methylene blue, are therefore usually incorporated into the capsules.

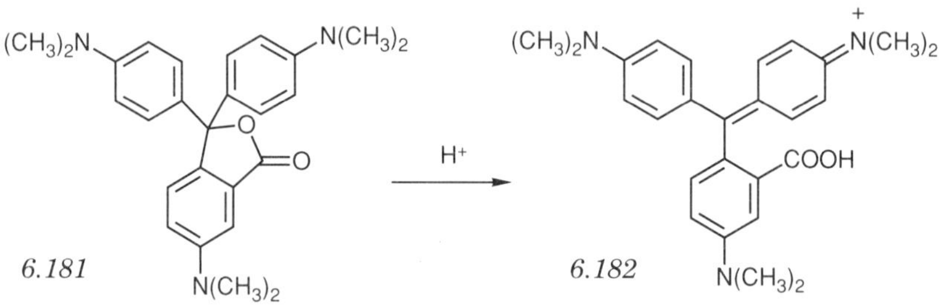

6.181 *6.182*

Mordant dyes

Hydroxytriphenylmethane derivatives give rise to anionic charge-resonance systems that are isoconjugate with the corresponding amino-substituted dyes. Thus, for example, the oxonol analogue of malachite green is benzaurine (Scheme 6.32). The quinonoid neutral form (6.183), which is pale yellow, produces a violet anion (6.184) on the addition of alkali; in strongly acidic solution the neutral form is protonated to give a cationic charge-resonance system (6.185) which is also violet [54]. The analogous trihydroxytriphenylmethane system is aurine.

Technically important dyes are salicylic acid derivatives that function as chrome mordant dyes for wool. Thus Eriochrome Azurol B (6.186; CGY, C.I. Mordant Blue 1) is made by the aldehyde synthesis from 2,6-dichlorobenzaldehyde and 2-hydroxy-3-methylbenzoic (*o*-cresotinic) acid in con-

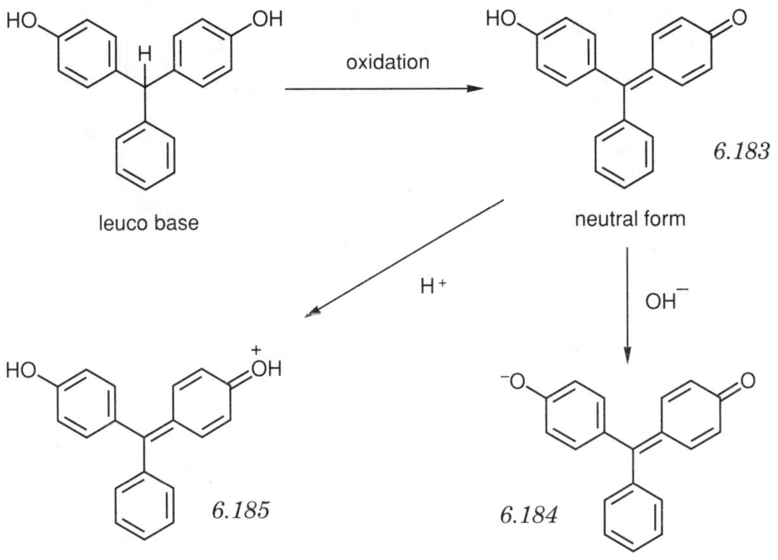

Scheme 6.32

centrated sulphuric acid. Oxidation of the leuco base is achieved by the addition of sodium nitrite. On wool the product, which is isolated as the sodium salt, is a dull maroon colour, changing to a bright blue on treatment with a chromium salt. Some dyes of this type, such as C.I. Mordant Violet 1 (6.187), also contain a basic group. This compound is also prepared by the aldehyde route.

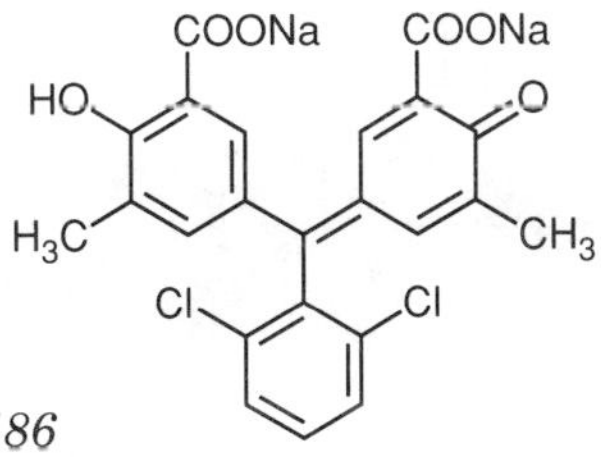

6.186

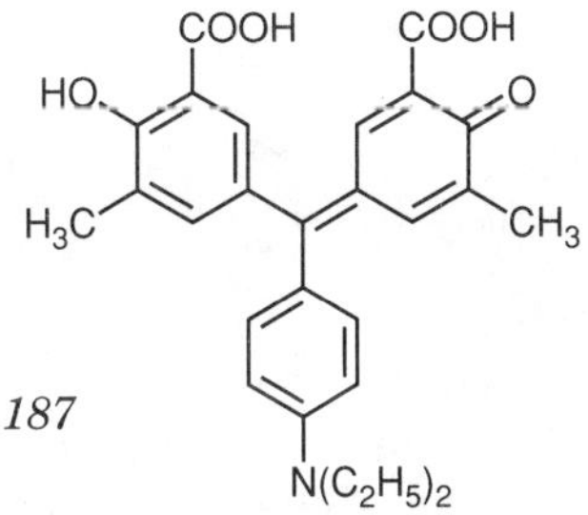

6.187

The acid–base indicator phenolphthalein is a derivative of benzaurine (Scheme 6.33). Condensation of phthalic anhydride with phenol generates the colourless lactone form (6.188), which ionises in alkaline solution and undergoes ring opening to produce the red dianion (6.189) within the pH range 8.3–10.0. The charge-resonance system is destroyed in an excess of alkali by the formation of the colourless triphenylmethanol derivative (6.190). Several derivatives of phenolphthalein are also used as indicators.

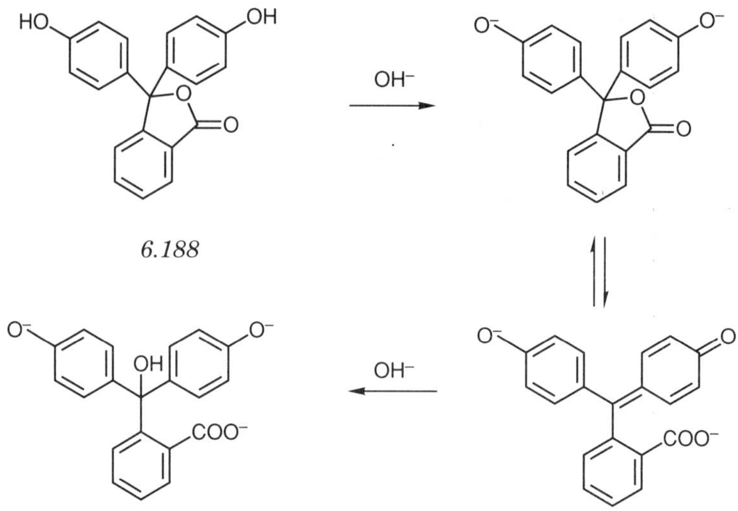

6.188

6.190 *6.189*

Scheme 6.33

The related phenolsulphonephthaleins are particularly versatile as acid–base indicators, covering a wide pH range. Typical examples include bromophenol blue (6.191) (pH range 3.0–4.6) and bromocresol purple (6.192) (pH range 5.2–6.8).

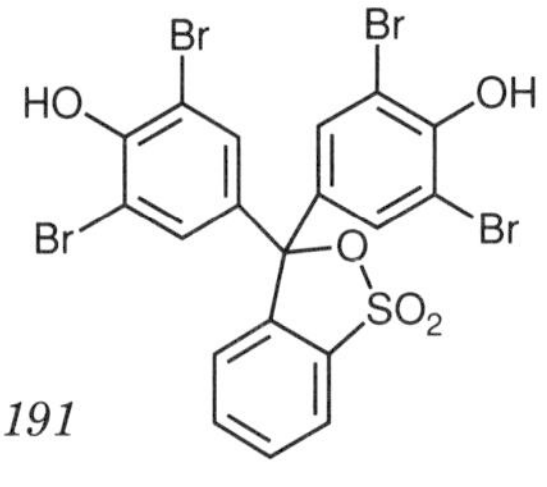

6.191

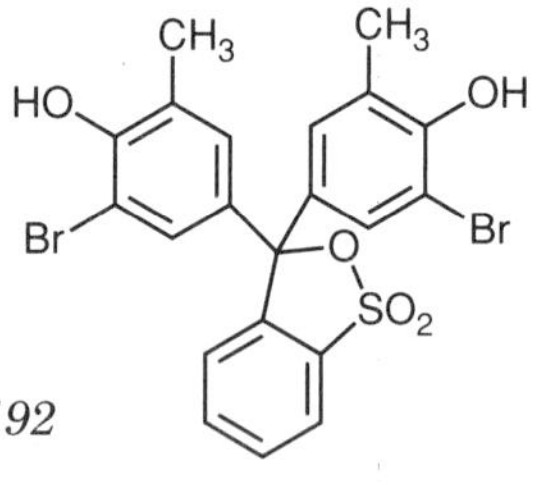

6.192

6.5.3 Xanthene dyes

The most important heterocyclic analogues of the di- and tri-arylmethane dyes are xanthene derivatives containing an oxygen bridge. These dyes produce red, pink and greenish-yellow hues that are strongly fluorescent. The oxygen bridge confers greater rigidity on the system, leading to fluorescence and, in accordance with PMO theory [55], the electron donation results in hypsochromic shifts of the long-wavelength absorption band relative to that of the parent dyes.

The xanthene analogue of Michler's Hydrol Blue (6.152) is Pyronine G (6.193), which is red. Condensation of 3-dimethylaminophenol with formaldehyde followed by dehydration in concentrated sulphuric acid gives the leuco base. Dyes of commercial significance contain three aromatic rings. The bright reddish-violet dye Rhodamine B (6.194; C.I. Basic Violet 10), discovered in 1887 by Cérésole, is a good example. Condensation of 3-diethylaminophenol with phthalic anhydride yields the dye. The free base is used as a solvent dye (C.I. Solvent Red 49) and the phosphotungstomolybdic acid lake finds use as a pigment (C.I. Pigment Violet 1); the ethyl ester of Rhodamine B is C.I. Basic Violet 11. A similar reaction between 3-ethylamino-*p*-cresol and phthalic anhydride gives a redder product on esterification (6.195; C.I. Basic Red 1). The PTMA lake is a valuable pink pigment (C.I. Pigment Red 81), as is the copper hexacyanoferrate(II) complex (C.I. Pigment Red 169).

Several xanthene derivatives are applied as acid dyes. For example, condensation of 3-diethylaminophenol with benzaldehyde-2,4-disulphonic acid, followed by cyclisation and oxidation, gives Acid Rhodamine B (6.196; C.I. Acid Red 52).

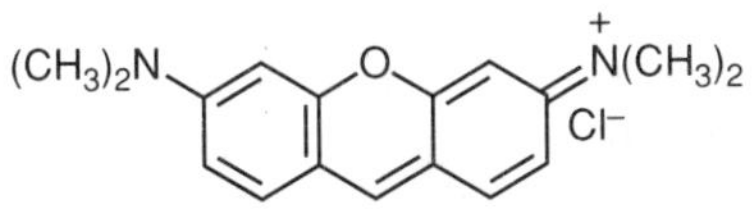

6.193

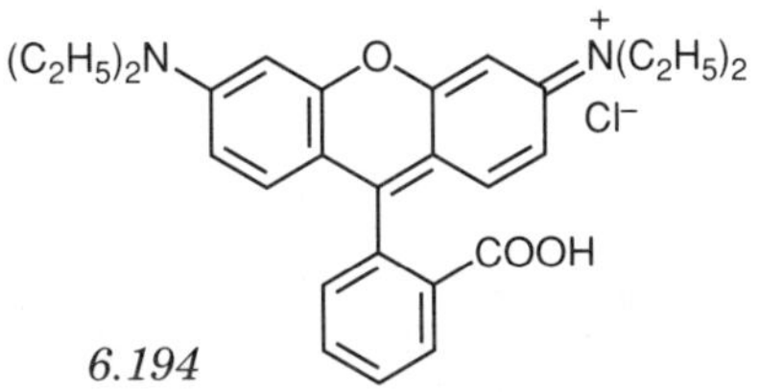

6.194

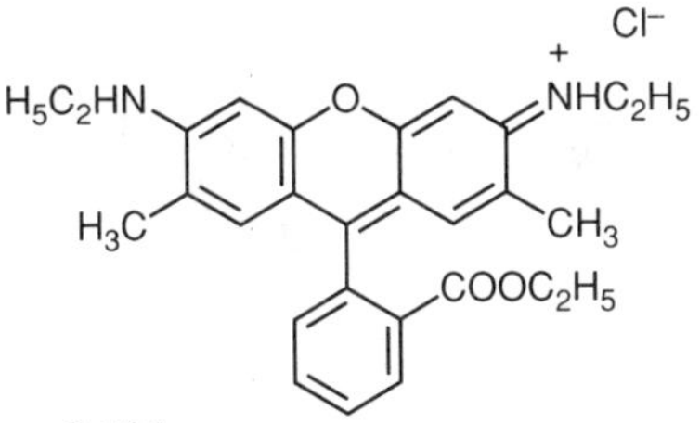

6.195

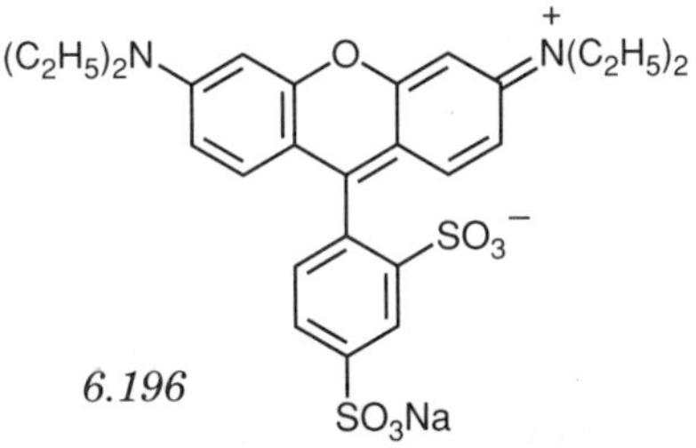

6.196

Fluorescein (6.197; X = H) is the xanthene analogue of phenolphthalein (6.189) and gives rise to an anionic charge-resonance system. This compound, the first xanthene dye (C.I. Acid Yellow 73), was originally made in 1871 by Baeyer from phthalic anhydride and resorcinol. The dye, which has a strong green fluorescence in solution even at very great dilution, is widely used as a marker. Various halogenated derivatives are of interest, especially the bright red tetrabromo compound eosine (6.197; X = Br; C.I. Acid Red 87), and the equivalent tetraiodo compound erythrosine (6.197; X = I; C.I. Acid Red 51). Eosine is widely used for colouring paper, inks and crayons; C.I. Pigment Red 90:1 is the aluminium salt. Purified erythrosine is a permitted food colorant (C.I. Food Red 14); the aluminium salt is also used for this purpose (C.I. Food Red 14:1) and as a pigment (C.I. Pigment

Red 172). Condensation of tetrachlorophthalic anhydride with resorcinol, followed by tetrabromination of the product, gives the bright pink Phloxine B (6.198; C.I. Acid Red 92); the free acid is used as a solvent dye (C.I. Solvent Red 48) and the aluminium salt is a pigment (C.I. Pigment Red 174).

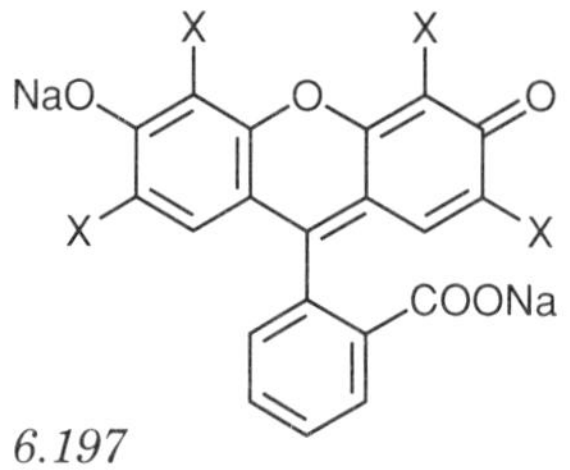

6.197

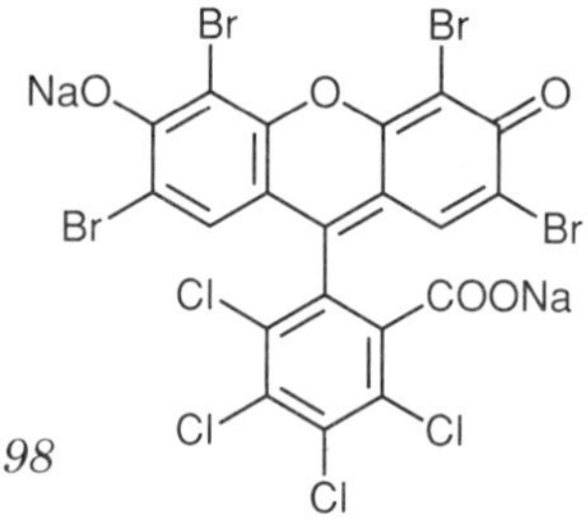

6.198

Xanthene dyes are used as colour formers. These so-called fluorans usually contain amino groups sited *para* and *meta* to the central carbon atom. Such a substitution pattern gives rise to broad absorption and leads to almost black colour production; the lactone 6.199 is a typical example.

Various xanthene derivatives, including fluorescein, are also used as laser dyes to cover the spectral region from 500 to 700 nm. A modern example is the julolidine-based dye Rhodamine 101 (6.200), which absorbs at 576 nm and lases at 648 nm [56].

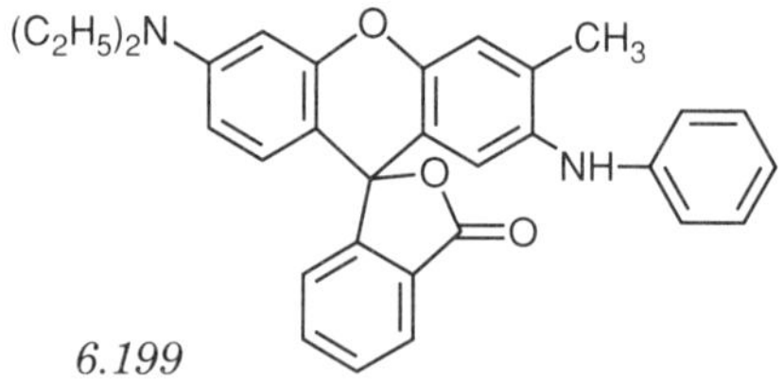

6.199

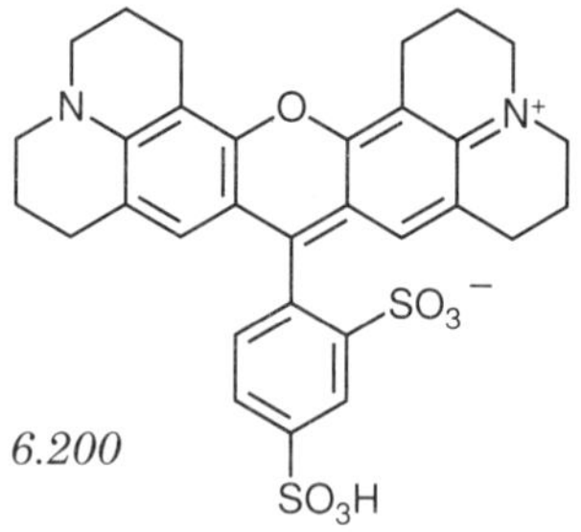

6.200

6.5.4 Acridine dyes

The presence of a nitrogen bridge in di- and tri-arylmethane dyes gives rise to the less significant acridine dyes; the parent heterocyclic ring system is so called because of its irritant action in the nose and throat. The acridine analogue of Michler's Hydrol Blue is acridine orange (6.201; C.I. Basic Orange 14). The dye is made by nitration of 4,4′-bis(dimethylamino)-diphenylmethane (6.155), reduction of the resulting dinitro derivative and oxidation of the cyclised diamino intermediate.

The well-known antiseptic acriflavine (6.202) is obtained by condensation of *m*-phenylenediamine with glycerol and oxalic acid, followed by methylation of the product.

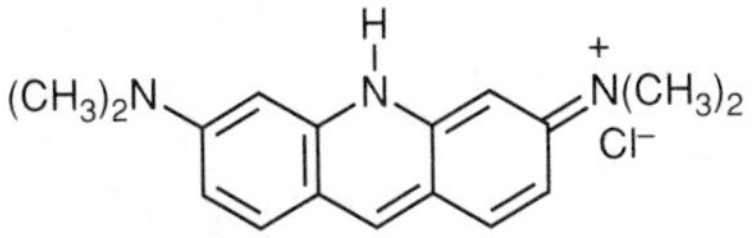

6.201

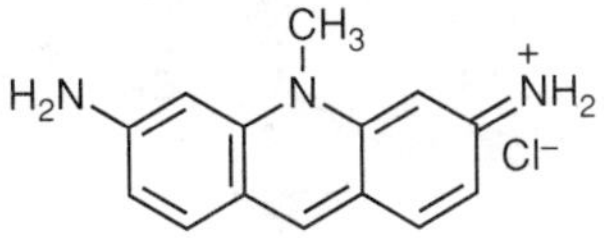

6.202

The triphenylmethane acridine analogue chrysaniline (6.165) has already been mentioned as a by-product in the synthesis of magenta.

6.6 MISCELLANEOUS COLORANTS

6.6.1 Azines, oxazines and thiazines

Heterocyclic analogues of diarylmethane dyes in which the central carbon atom has been replaced by nitrogen are of some importance. Such a replacement leads to significant bathochromic shifts of the long-wavelength absorption band in accordance with PMO theory [55]. Thus Bindschedler's Green (6.203), the aza analogue of Michler's Hydrol Blue, absorbs at 725 nm. This indamine dye is only of historical interest but a related indoaniline derivative is used as a solvent dye (6.204; C.I. Solvent Blue 22). Of more importance are the azines, oxazines and thiazines which contain a heteroatom bridge.

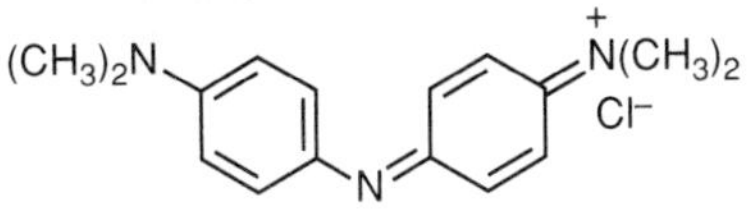

6.203

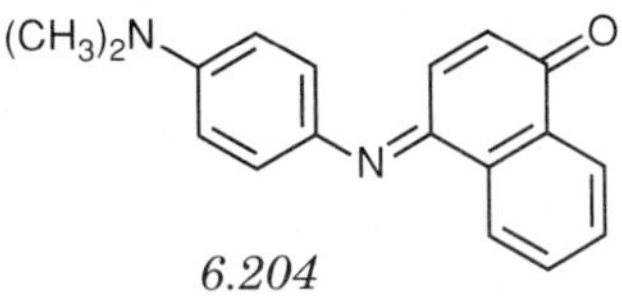

6.204

Azines

The first synthetic dye, mauveine (6.205), belongs to this group. Few dyes of this type remain of industrial significance, an exception being Safranine T (6.206; C.I. Basic Red 2), discovered in 1859 by Greville Williams. The oxidative sequence leading to the formation of Safranine T (Scheme 6.34)

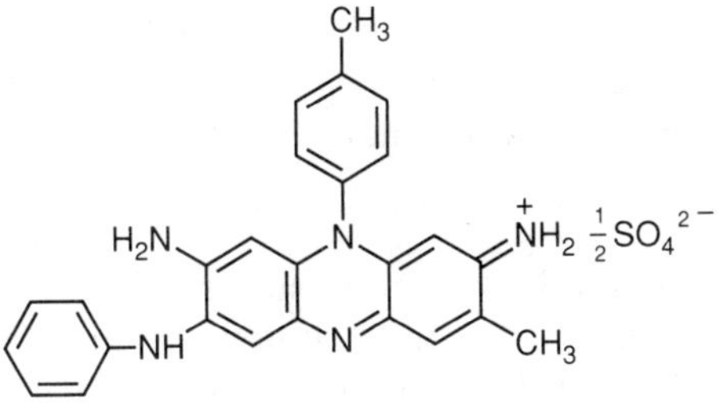

6.205

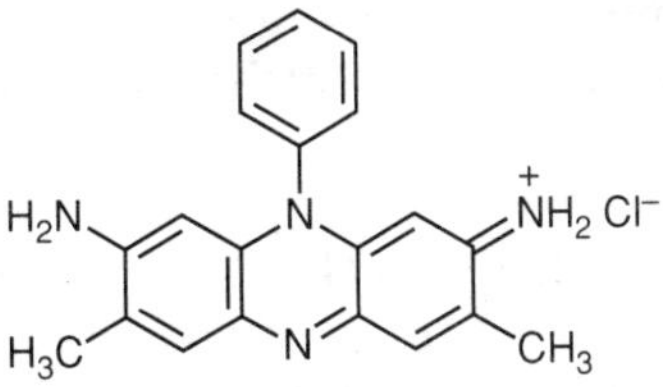

6.206

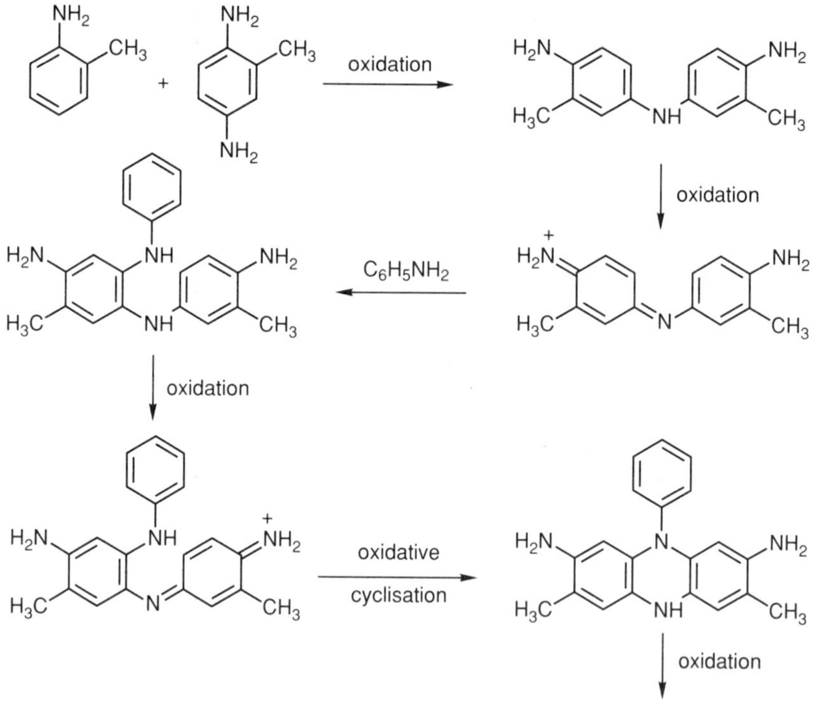

Scheme 6.34

was only elucidated in recent times [57]. The 1:1 mixture of *o*-toluidine and 2-methyl-1,4-diaminobenzene required for the first oxidative step can be obtained by reductive cleavage of the monoazo dye formed by coupling diazotised *o*-toluidine with *o*-toluidine. Further oxidation of the intermediate diphenylamine derivative gives an electrophilic quinonediimine (indamine), which then reacts with aniline. Oxidative cyclisation produces a dihydroazine that yields the dye in a final oxidative step.

Aniline black (C.I. Oxidation Base 1) is a complex polymeric phenazine that can be produced on cotton fabric by impregnation with aniline hydrochloride and suitable inorganic oxidants, such as sodium chlorate, ammonium vanadate and copper hexacyanoferrate(II). Aniline black is also made directly for use as a pigment (C.I. Pigment Black 1).

Oxazines

Bindschedler's Green (6.203) can be made by condensing *p*-nitroso-*NN*-dimethylaniline with *NN*-dimethylaniline. A similar condensation with 2-naphthol gives Meldola's Blue (6.207; C.I. Basic Blue 6), the first oxazine dye, discovered in 1879. The symmetrical C.I. Basic Blue 3 (6.208) is of more commercial significance. It is synthesised by nitrosation of *NN*-diethyl-*m*-anisidine followed by condensation with *NN*-diethyl-*m*-aminophenol, and is used for dyeing acrylic fibres.

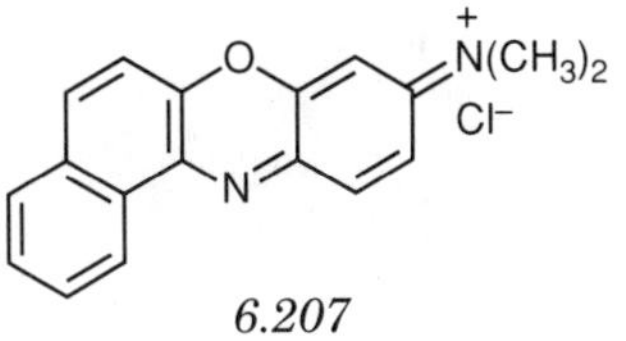

6.207

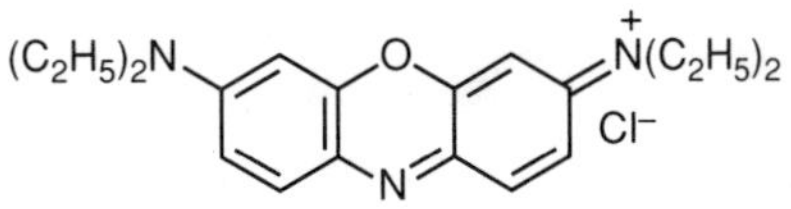

6.208

The dioxazine ring system is the source of some valuable violet pigments, such as C.I. Pigment Violet 23 (6.209). This colorant is obtained by condensing 3-amino-9-ethylcarbazole with chloranil. Sulphonation of the pigment gives the dye C.I. Direct Blue 108. Triphenodioxazines have recently been the source of some blue reactive dyes [17].

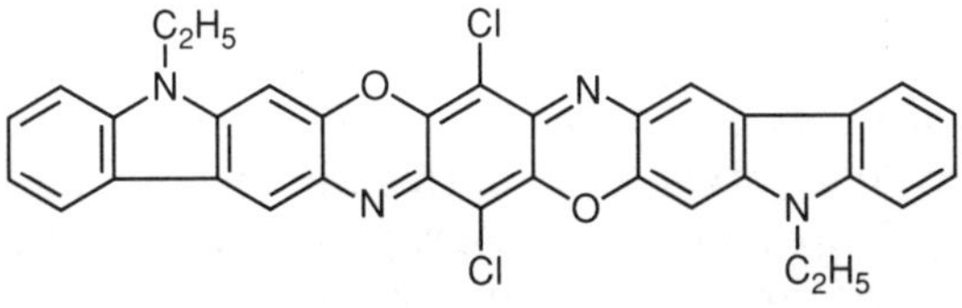

6.209

Thiazines

The most important thiazine dye is methylene blue (6.210; C.I. Basic Blue 9), which is also widely used as a biological stain and as a redox indicator. Its synthesis (Scheme 6.35) involves the oxidative thiosulphonation of *p*-amino-*NN*-dimethylaniline. Condensation of the product with *NN*-dimethylaniline then gives an intermediate indamine that subsequently undergoes oxidative ring closure. Nitration of methylene blue produces methylene green (6.211; C.I. Basic Green 5), which shows better light fastness on acrylic fibres. The *N*-benzoyl derivative of leuco methylene blue is used as a colour former in carbonless copy papers (see section 6.5.2).

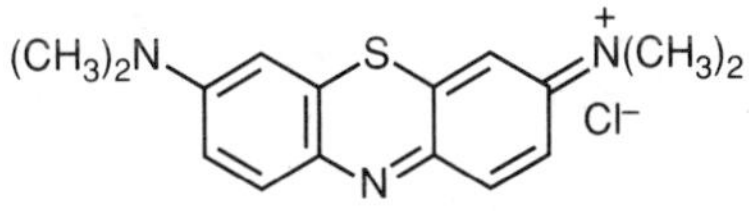

6.210

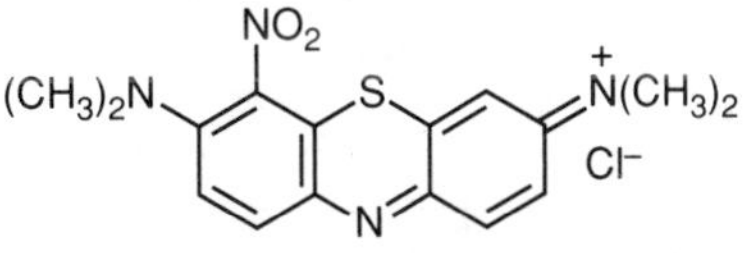

6.211

6.6.2 Polymethine dyes

Cyanine dyes fall within the more general category of polymethine dyes, in which a chain of methine groups is terminated with a donor group and an acceptor group respectively [58].

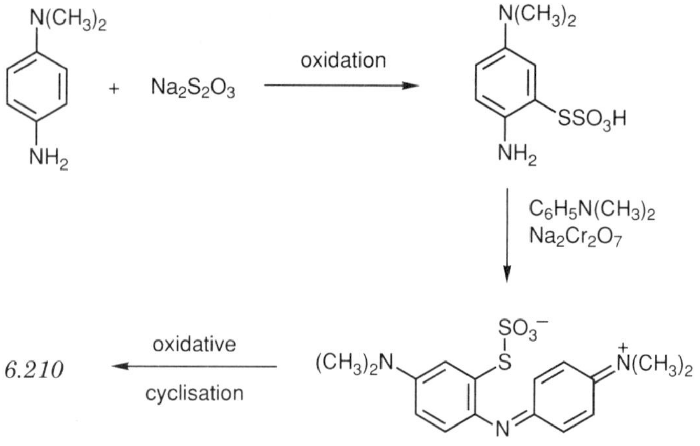

Scheme 6.35

The first cyanine dye was made in 1856 by Greville Williams. Thus the blue charge-resonance system 6.212 was produced when an oxidative coupling took place between *N*-pentylquinolinium iodide and the corresponding 4-methyl derivative. Dyes of this kind were later found to have colour-sensitising properties in photographic films and hundreds of related compounds have now been synthesised, making use of a great variety of heterocyclic terminal groups [59]. Long-chain cyanines absorb beyond the visible region and are useful as infra-red sensitisers in photography. Few cyanine colorants have found use as textile dyes, however, owing to inferior fastness properties. Astra Phloxine (6.213; C.I. Basic Red 12) is an example of a carbocyanine (trimethine) dye derived from 1,3,3-trimethylindoline. Structurally unsymmetrical dyes of this type are exemplified by C.I. Basic Orange 21 (6.214), which is obtained by condensing Fischer's aldehyde (6.215) with 2-methylindole.

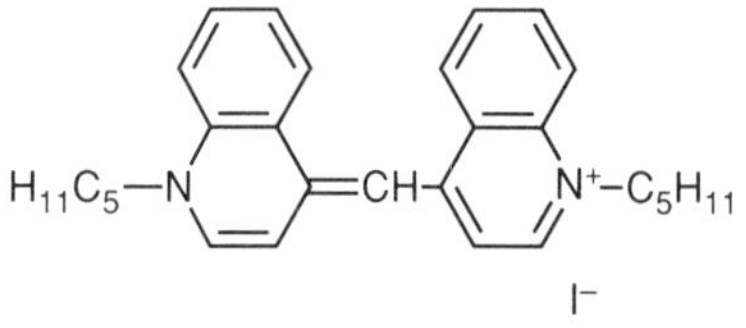

6.212

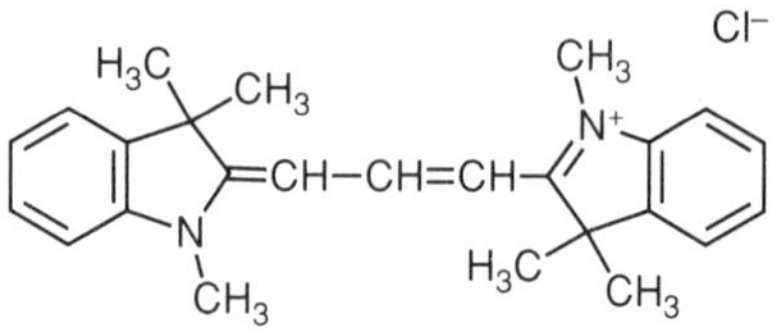

6.213

6.214

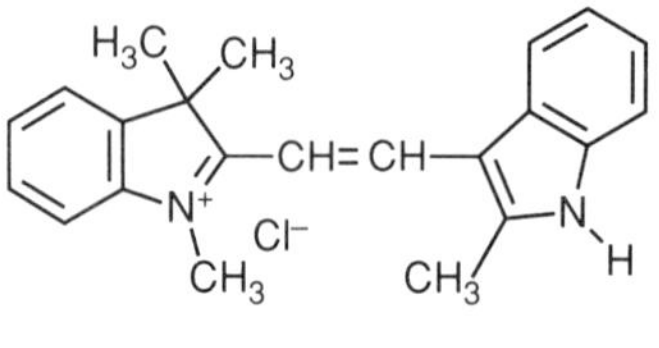

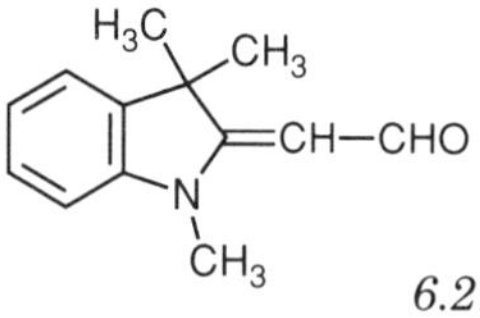

6.215

Of some importance as textile dyes are aza analogues of polymethine (cyanine) dyes. Azacarbocyanines result when Fischer's aldehyde is heated with primary aromatic amines. Thus C.I. Basic Yellow 11 (6.216) is obtained when Fischer's aldehyde is condensed with 2,4-dimethoxyaniline. The equivalent reaction with 2-methylindoline gives C.I. Basic Yellow 21 (6.217), which has superior light fastness. The tinctorially strong, golden-yellow diazacarbocyanine dye C.I. Basic Yellow 28 (6.218) is prepared by coupling diazotised *p*-anisidine with Fischer's base (6.219), followed by quaternisation with dimethyl sulphate. Some triazacarbocyanine dyes are also used commercially.

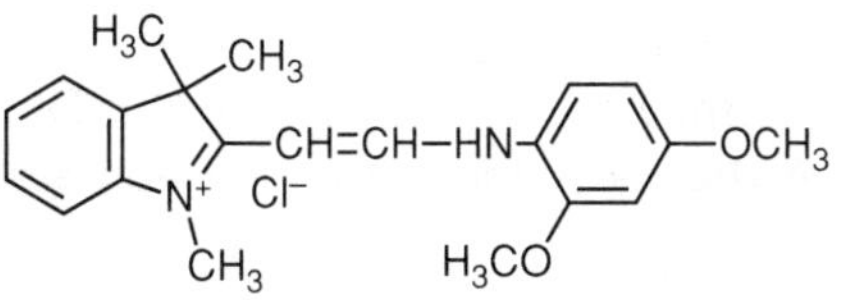

6.216

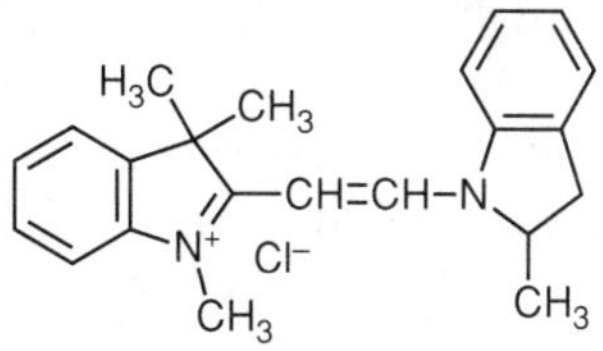

6.217

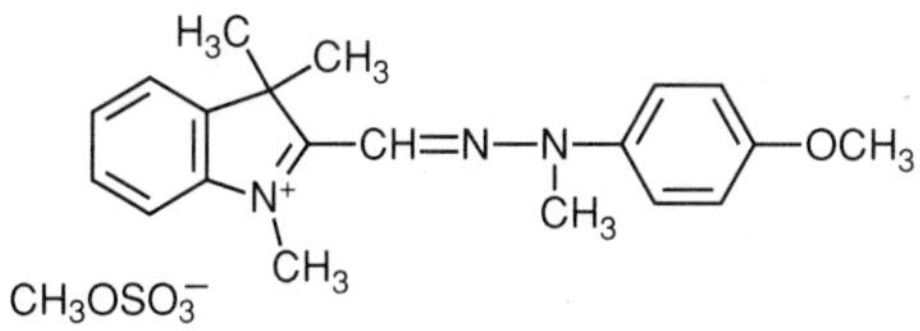

6.218

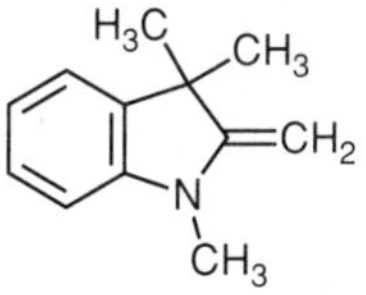

6.219

Hemicyanine (styryl) dyes are readily obtained by heating Fischer's base with an appropriate aldehyde. Typical products include C.I. Basic Red 14 (6.220) and C.I. Basic Violet 16 (6.221).

Cationic monoazo dyes can be classified as diazahemicyanines; examples of such dyes are considered in section 4.10.

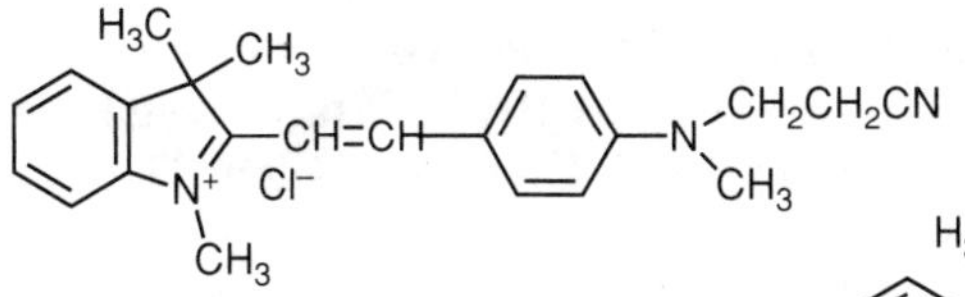

6.220

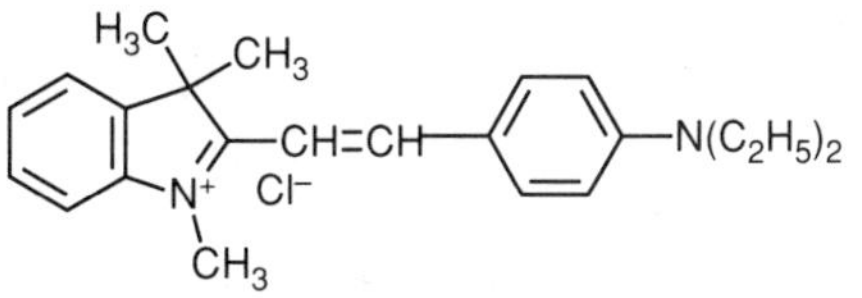

6.221

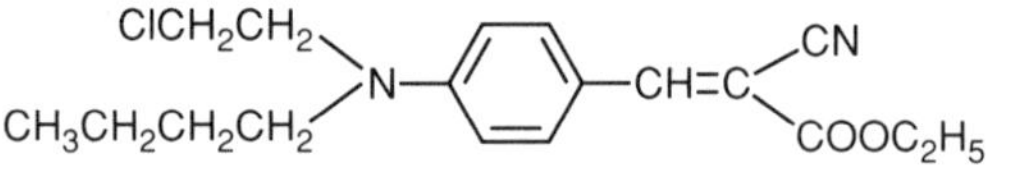

6.222

Uncharged styryl (methine) disperse dyes were originally introduced to provide greenish-yellow colours on cellulose acetate fibres. One such dye still in use is C.I. Disperse Yellow 31 (6.222), which is made by condensing 4-(*N*-butyl-*N*-chloroethylamino)benzaldehyde with ethyl cyanoacetate. Suitable compounds for polyester usually contain the electron-accepting dicyanovinyl group, introduced with the aid of malononitrile. An increased molecular size leads to improved sublimation fastness, as in the case of Terasil Brilliant Yellow 6G (6.223; CGY, C.I. Disperse Yellow 99). A novel polymethine-type structure of great interest is present in the recently introduced disperse dye Foron Brilliant Blue S-R (6.224; S, C.I. Disperse Blue 354), which is claimed to be the most brilliant blue disperse dye currently available [60].

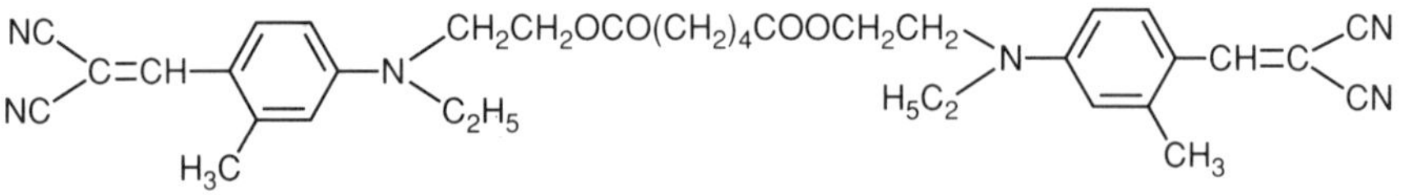

6.223

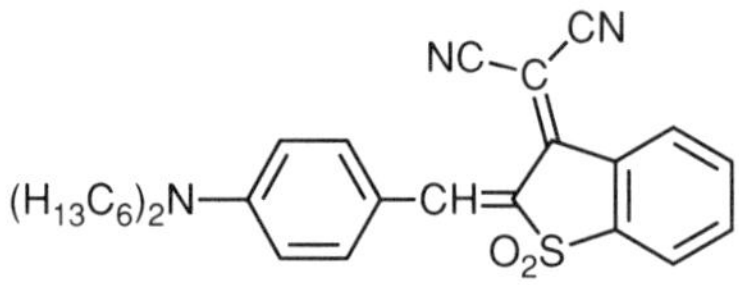

6.224

Quinophthalone (6.225) and its derivatives [61] also fall into the methine category, although they appear in the *Colour Index* under quinoline colouring matters. The parent compound was discovered in 1882 by Jacobsen, who condensed 2-methylquinoline (quinaldine) with phthalic anhydride. The product, quinoline yellow, is used as a solvent dye (C.I. Solvent Yellow 33). The light fastness is improved by the presence of a hydroxy group in the quinoline ring system. Derivatives of this type provide greenish-yellow disperse dyes for polyester. The moderate sublimation fastness of C.I. Disperse Yellow 54 (6.226; R = H) is improved by the introduction of an adjacent bromine atom in C.I. Disperse Yellow 64 (6.226; R = Br).

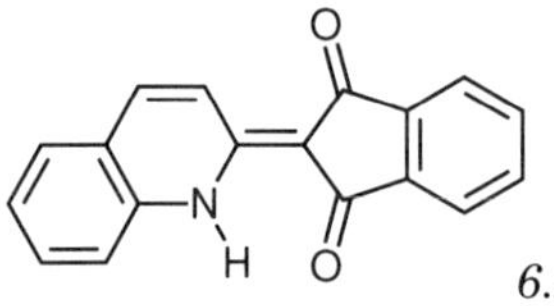

6.225

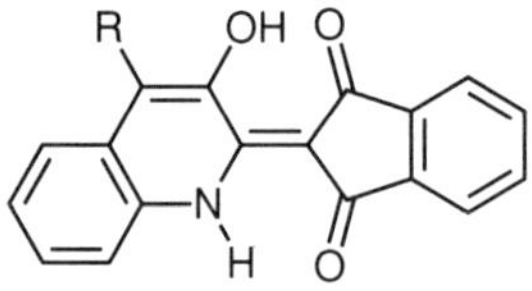

6.226

The chromium complexes of some azomethine derivatives are used as solvent dyes. C.I. Solvent Yellow 32 (6.227) serves as an example and is obtained by condensing one mole of salicylaldehyde with the appropriate amine, followed by treatment with one equivalent of chromium.

A new cross-conjugated methine-type chromogen was introduced in 1984 by ICI. The dye Dispersol Red C-BN (6.228; C.I. Disperse Red 356) exemplifies this system, which contains two α,ω-donor–acceptor dienone segments.

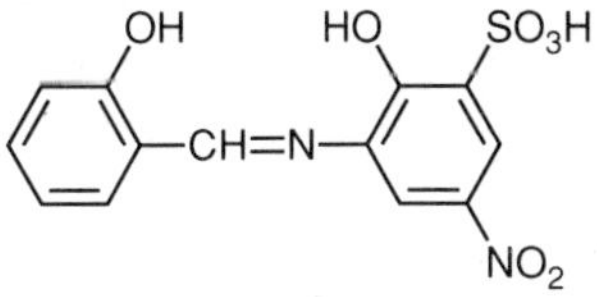

6.227

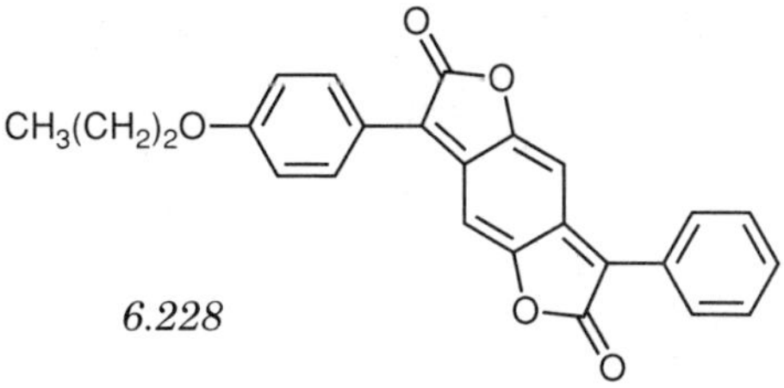

6.228

6.6.3 Nitro and nitroso dyes

Yellow to brown compounds are obtained when one or more nitro groups are conjugated with electron-donor substituents such as hydroxy or, of more importance, amino groups. Such dyes are relatively cheap. Picric acid, 2,4,6-trinitrophenol, can be regarded as the oldest synthetic dye; it was produced in 1771 by Woulfe by treating indigo with nitric acid.

The hydroxy-nitro dye Naphthol Yellow S (6.229; C.I. Acid Yellow 1) was discovered in 1879 by Caro and is still manufactured. It is produced by sulphonation of 1-naphthol to give 1-naphthol-2,4,7-trisulphonic acid, followed by replacement of the 2- and 4-sulpho groups in nitric acid medium. Nucleophilic substitution of 1-chloro-2,4-dinitrobenzene with 4-aminodiphenylamine-2-sulphonic acid gives C.I. Acid Orange 3 (6.230).

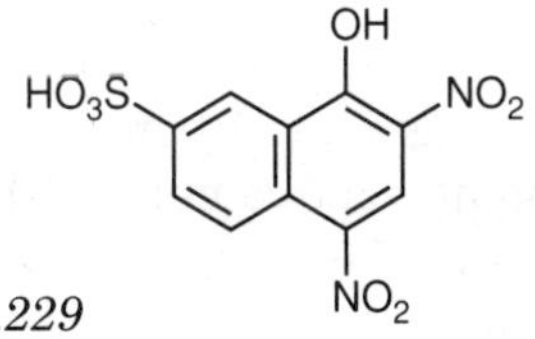

6.229

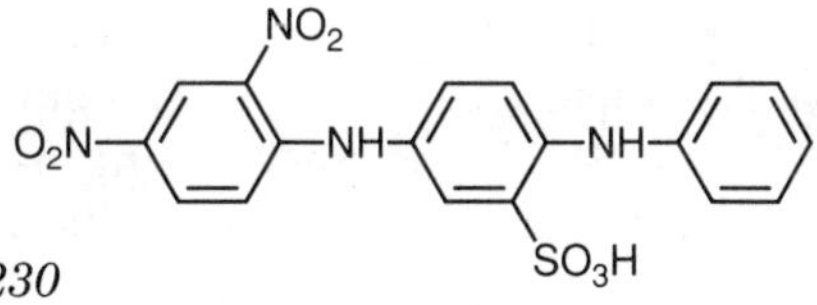

6.230

Of particular importance are disperse dyes based on *o*-nitrodiphenylamine. The light fastness of such compounds is better than that of their *p*-substituted isomers, owing to intramolecular hydrogen bonding. The *o*-nitro derivatives are more bathochromic than the *p*-nitro analogues but the latter possess greater tinctorial strength. Dyes such as C.I. Disperse Yellow 1 (6.231; X = OH) and C.I. Disperse Yellow 9 (6.231; X = NH_2) are examples of simple commercial products used on cellulose acetate. Such dyes generally have a lower substantivity for polyester, but this can be improved by minor structural changes. Thus C.I. Disperse Yellow 42

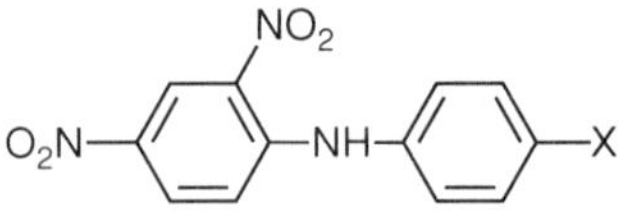

6.231

(6.232; X = C_6H_5) has higher substantivity for polyester than has the parent dye (6.232; X = H; C.I. Disperse Yellow 33) [15]. The incorporation of an azo chromophore into the diphenylamine system also improves sublimation fastness and colour strength; for example, dye 6.231 (X = NH_2) is the source of C.I. Disperse Yellow 70 (6.233).

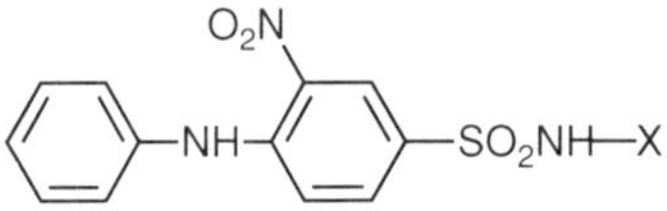

6.232

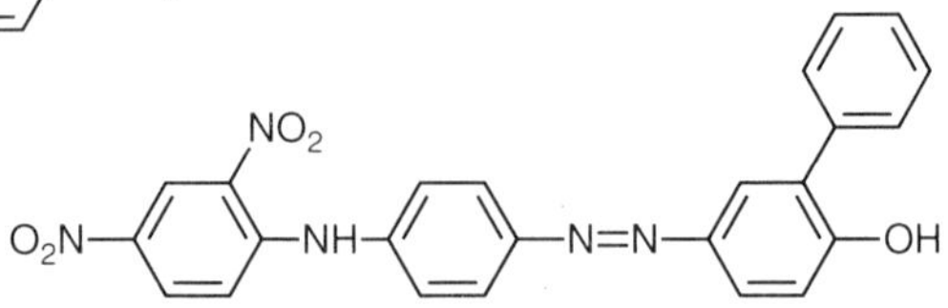

6.233

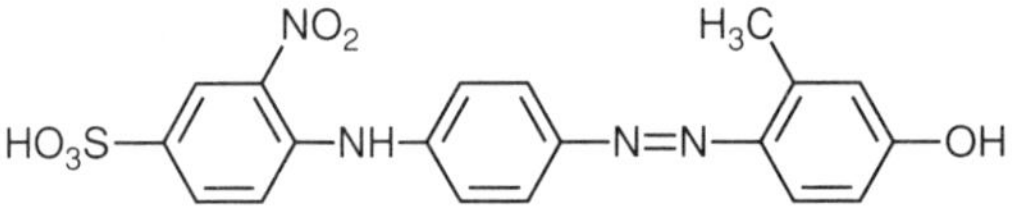

6.234

The introduction of sulphonic acid groups into dyes of this kind provides acid dyes suitable for nylon, such as C.I. Acid Yellow 199 (6.234).

The so-called nitroso dyes are metal complexes of 1-nitroso-2-naphthol. In the parent system a tautomeric equilibrium exists between hydroxy-nitroso (6.235) and quinoneoxime (6.236) forms. Only the green iron(II) complexes, which have good light fastness, have been exploited commercially. The unsubstituted iron(II) complex is Pigment Green B (C.I. Pigment Green 8). The corresponding 6-sulphonic acid complex, Naphthol Green B (C.I. Acid Green 1), was discovered in 1885 by Hoffmann.

6.235

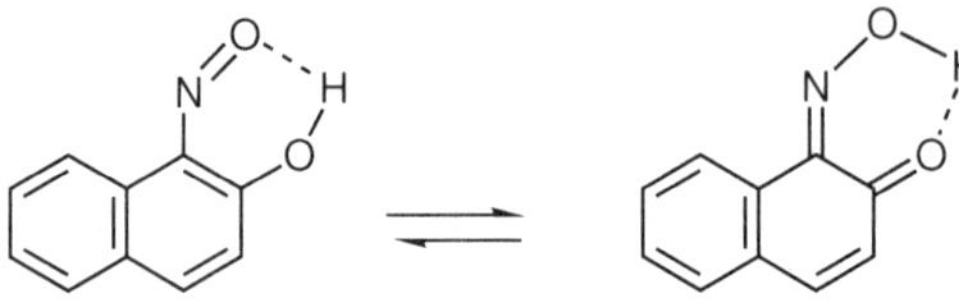

6.236

REFERENCES

1. Y Kogo, H Kikuchi, M Matsuoka and T Kitao, *J.S.D.C.*, **96** (1980) 475.
2. C Graebe and C Liebermann, *Ber.*, **1** (1868) 49, 104.
3. M R Fox, *Dye-makers of Great Britain 1856–1976* (Manchester: ICI, 1987) 97.

4. M Titz and M Nepras, *Coll. Czech. Chem. Comm.*, **37** (1972) 2674.
5. H Zollinger, *Color chemistry* (Weinheim: VCH, 1987) chapter 8.4.
6. J Winkler and W Jenny, *Helv. Chim. Acta*, **48** (1965) 119.
7. M Kikuchi, T Yamagishi and M Hida, *Dyes and Pigments*, **2** (1981) 143.
8. M Nepras, J Fabian, M Titz and B Gas, *Coll. Czech. Chem. Comm.*, **47** (1982) 2569.
9. P F Gordon and P Gregory, *Organic chemistry in colour* (Berlin: Springer-Verlag, 1983) chapter 4.6.
10. C F H Allen, C V Wilson and G F Frame, *J. Org. Chem.*, **7** (1942) 169.
11. *Colour Index International*, 3rd Edn (3rd revision) Vol. 5 (Bradford: SDC, 1987).
12. G Hallas in *Developments in the chemistry and technology of organic dyes*, Ed. J Griffiths (Oxford: Blackwell Scientific Publications, 1984) 31.
13. O Annen, R Egli, R Hasler, B Henzi, H Jakob and P Matzinger, *Rev. Prog. Coloration*, **17** (1987) 72.
14. J F Dawson, *Rev. Prog. Coloration*, **14** (1984) 90.
15. J F Dawson, *J.S.D.C.*, **99** (1983) 183.
16. R Raue, *Rev. Prog. Coloration*, **14** (1984) 187.
17. A H M Renfrew, *Rev. Prog. Coloration*, **15** (1985) 15.
18. H Zollinger, *Color chemistry* (Weinheim: VCH, 1987) chapter 8.11.
19. U Baumgarte, *Rev. Prog. Coloration*, **17** (1987) 29.
20. M Schellenberg and R Steinmetz, *Helv. Chim. Acta*, **52** (1969) 431.
21. J Aoki, *Bull. Chem. Soc. Japan*, **41** (1968) 1017.
22. H Zollinger, *Color chemistry* (Weinheim: VCH, 1987) chapter 8.9.
23. K Venkataraman and V N Iyer in *The chemistry of synthetic dyes*, Vol. 5, Ed. K. Venkataraman (New York: Academic Press, 1971) chapter 3.
24. P F Gordon and P Gregory, *Organic chemistry in colour* (Berlin: Springer-Verlag, 1983) chapter 5.2.
25. J Griffiths, *Colour and constitution of organic molecules* (New York: Academic Press, 1976) chapter 4.6.
26. G A Clowes, *J.C.S.*, *C*, (1968) 2519.
27. K Venkataraman, *The chemistry of synthetic dyes*, Vol. 2 (New York: Academic Press, 1952) 1003.
28. P F Gordon and P Gregory, *Organic chemistry in colour* (Berlin: Springer-Verlag, 1983) chapter 5.3.
29. H Zollinger, *Color chemistry* (Weinheim: VCH, 1987) chapter 2.5.
30. M Klessinger, *Dyes and Pigments*, **3** (1982) 235.
31. J Shore, *Textilveredlung*, **12** (1986) 207.
32. H Zollinger, *Color chemistry* (Weinheim: VCH, 1987) chapter 8.3.
33. P F Gordon and P Gregory, *Organic chemistry in colour* (Berlin: Springer-Verlag, 1983) chapter 1.3.
34. K McLaren, *The colour science of dyes and pigments*, 2nd Edn (Bristol: Adam Hilger, 1986) chapter 1.2.
35. S Geyer and R Meyer, *Chem. Abs.*, **95** (1981) 203782r.
36. W E Wood, *Rev. Prog. Coloration*, **7** (1976) 80.
37. C Heid, K Holoubek and R Klein, *Melliand Textilber.*, **54** (1973) 1314.
38. H Zollinger, *Color chemistry* (Weinheim: VCH, 1987) chapter 9.1.
39. C D Weston in *The chemistry of synthetic dyes*, Vol. 7, Ed. K Venkataraman (New York: Academic Press, 1974) chapter 2.
40. W Zerweck, H Ritter and M Schubert, *Angew. Chem.*, **60** (1948) 141.
41. D G Orton in *The chemistry of synthetic dyes*, Vol. 7, Ed. K. Venkataraman (New York: Academic Press, 1974) chapter 1.
42. P Rys and H Zollinger, *Fundamentals of the chemistry and application of dyes* (New York: Wiley Interscience, 1972) chapter 8.
43. H Zollinger, *Color chemistry* (Weinheim: VCH, 1987) chapter 4.1.
44. C C Barker, G Hallas and A Stamp, *J.C.S.*, (1960) 3790.
45. N Nishimura and T Motoyama, *Bull. Chem. Soc. Japan*, **57** (1984) 1.
46. C C Barker, M H Bride, G Hallas and A Stamp, *J.C.S.*, (1961) 1285.

47. L L Koh and K Eriks, *Acta Cryst.*, *B*, **27** (1971) 1405.
48. M J S Dewar in *Steric effects in conjugated systems*, Ed. G W Gray (London: Butterworths, 1958) 46.
49. A S Ferguson and G Hallas, *J.S.D.C.*, **87** (1971) 187.
50. A C Hopkinson and P A H Wyatt, *J.C.S.*, *B*, (1970) 530.
51. J Griffiths, *Colour and constitution of organic molecules* (New York: Academic Press, 1976) chapter 9.3.
52. G Hallas, *J.S.D.C.*, **83** (1967) 368.
53. G Hallas and D R Waring, *J.C.S.*, *B*, (1970) 979.
54. J Griffiths, *Colour and constitution of organic molecules* (New York: Academic Press, 1976) chapter 9.4.
55. M J S Dewar in *Recent advances in the chemistry of colouring matters* (London: Chemical Society, 1956) 79.
56. P F Gordon and P Gregory in *Developments in the chemistry and technology of organic dyes*, Ed. J Griffiths (Oxford: Blackwell Scientific Publications, 1984) 66.
57. J F Corbett, *J.S.D.C.*, **88** (1972) 438.
58. H Zollinger, *Color chemistry* (Weinheim: VCH, 1987) chapter 3.1.
59. L G S Brooker in *The theory of the photographic process*, Ed. T H James (New York: Macmillan, 1971) 205.
60. J Hu, P Skrabal and H Zollinger, *Dyes and Pigments*, **8** (1987) 189.
61. F Kehrer, P Niklaus and B K Manukian, *Helv. Chim. Acta*, **50** (1967) 2200.

CHAPTER 7

Chemistry of reactive dyes

C Vivian Stead

7.1 INTRODUCTION

Until 1956 dyes for cellulosic fibres relied upon either hydrogen bonding, as in the case of direct dyes, or formation of an insoluble colouring material on the fibre, as in the case of azoic, sulphur and vat dyes, to achieve a satisfactory level of fastness to washing. For many years an attractive alternative had been perceived to lie in designing a dye that would covalently react with the fibre. Removal of such a dye would then entail breaking a chemical bond, and since this would require a high input of energy excellent wash fastness would ensue.

Several attempts to develop such a dye are to be found in the early literature. At the end of the nineteenth century Cross and Bevan [1] described the production of coloured cellulose derivatives by treating the carbohydrate with sodium hydroxide to form 'soda cellulose', benzoylating and then nitrating the resultant benzoyl cellulose; subsequent reduction, diazotisation and coupling produced a coloured derivative. This, and other similarly complicated procedures, could never achieve commercial significance. In the intervening years only an attempt [2] to develop fibre-reactive dyes for cellulose containing an isatoic anhydride residue, such as compound 7.1, was simple enough to warrant mention, but here weakness of the resulting dye–fibre bond precluded practical utility.

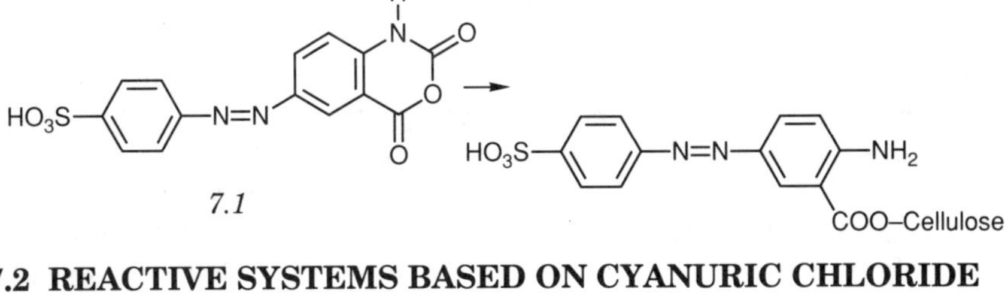

7.1

7.2 REACTIVE SYSTEMS BASED ON CYANURIC CHLORIDE

7.2.1 Dichloro-*s*-triazine dyes

The first reactive dyes for cellulose were launched by ICI in 1956 under the trade name Procion. They resulted from the pioneering work of Stephen and Rattee [3], and the earliest members relied for their effect on the presence of a dichloro-*s*-triazinyl group in the molecule [4]. Currently these types are designated Procion MX dyes. A key intermediate in their preparation is cyanuric chloride (7.2), in which molecule the contribution of canonical forms (such as 7.2a) results in a build-up of positive charge on the ring carbon atoms. This, together with the polarisation of the carbon–chlorine bonds caused by the greater electronegativity of the chlorine atom (7.3), makes the three chlorine atoms labile and readily susceptible to stepwise nucleophilic displacement. If the attacking nucleophile is a water-soluble amine, the reaction can be conveniently carried out on cyanuric chloride suspended in an aqueous solution of the amine at a temperature of about 5°C. Maintaining the pH at about neutrality by the addition of an alkali as required allows one of the chlorine atoms to be smoothly displaced, yielding a dichloro-*s*-triazinylamino compound (7.4) (Scheme 7.1).

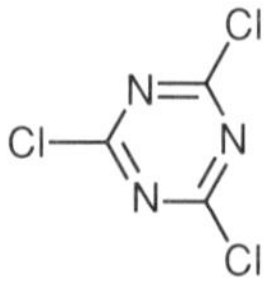

7.2

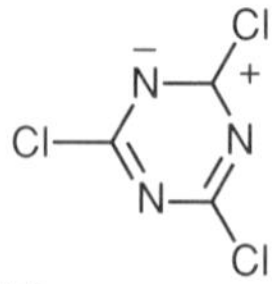

7.2a

7.3

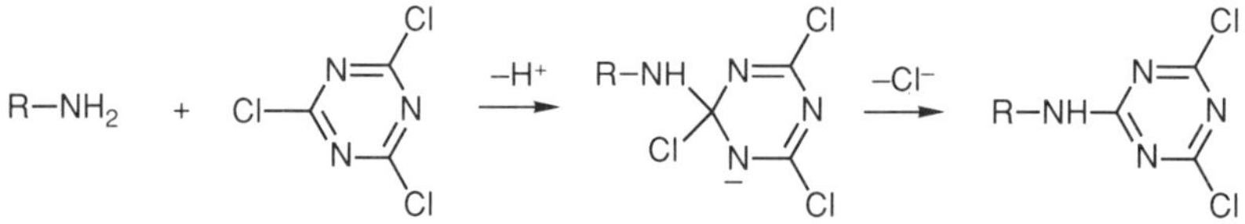

7.4

Scheme 7.1

Selecting a sulphonated dye containing an amino group as nucleophile leads to a dichloro-*s*-triazinyl dye. In certain cases a suitable intermediate may be employed at the condensation stage, subsequent processing then leading to the dye. Both these methods are illustrated for the preparation of C.I. Reactive Red 1 (7.5) in Scheme 7.2. In these dyes the electronic influences responsible for the lability of the chlorine atoms are muted by feedback of electrons from the linking amino group (7.6), making the dye quite stable in neutral solution although susceptible to attack by hydroxide ions on the alkaline side and subject to autocatalytic hydrolysis on the acid side. To guard against this hydrolysis a buffer is added to the solution to ensure stability during isolation and further buffer is added to the dyestuff paste before drying [5].

The dichloro-*s*-triazine dyes are highly reactive and can be applied to cellulosic textiles at temperatures of only 30–40°C. The basic chemical

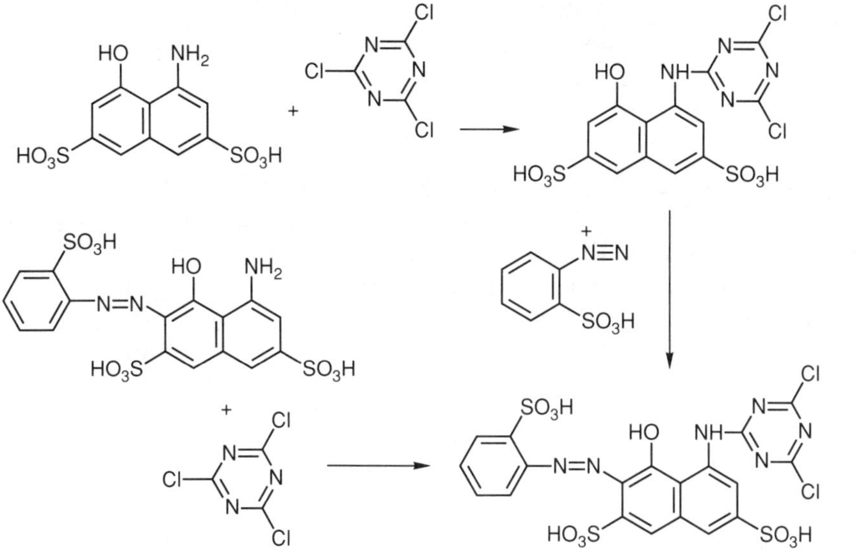

7.5

Scheme 7.2

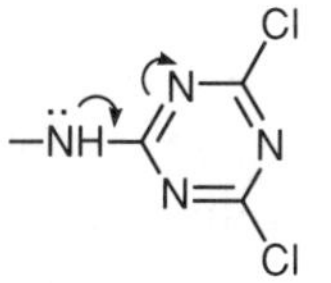

7.6

principles behind their application can be illustrated by considering their use in the exhaust dyeing process [6]. Here the dye is dissolved in water and the cellulosic material immersed in the solution. Salt is then added, causing the dye to be exhausted on to the fabric. Since these purely physical steps take place at neutrality, at which pH virtually no nucleophiles are present, chemical reaction does not occur. When satisfactory exhaustion has been achieved, alkali is added to raise the pH to about 10.5, causing the nucleophiles necessary for reaction to be generated and to attack the reactive system (Scheme 7.3). The attacking nucleophile can be either a cellulosate anion or a hydroxide ion, the former resulting in fixation of the dye to the fabric (7.7) whilst the latter merely leads to hydrolysis of the dye (7.8) and its loss from the dyeing process.

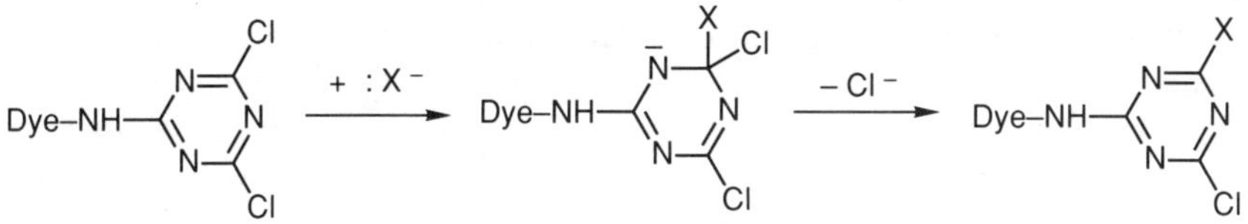

Scheme 7.3

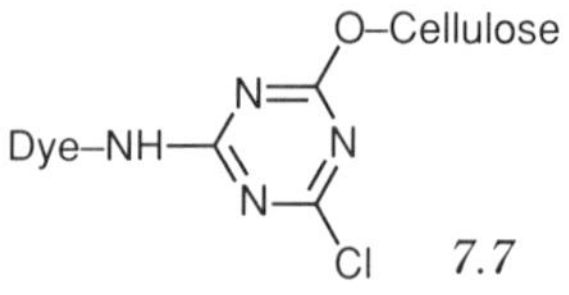

7.7

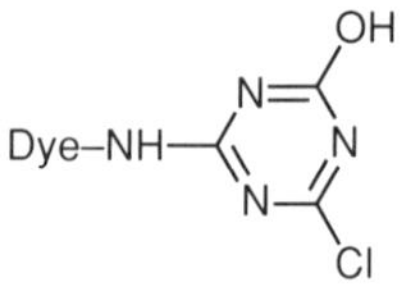

7.8

The vast excess of water in the system would, at first sight, suggest that hydrolysis should be the overwhelming reaction, but two factors oppose this. Firstly, most of the dye molecules have been physically adsorbed on to the fibre and are thus ideally sited for reaction with it; secondly, the ready ionisation of hydroxy groups on the cellulose molecule sets up an internal pH within the fibre that is primarily maintained by cellulosate anions, the resultant charge on the fibre surface expelling hydroxide ions from the fibre interior. These two effects combine to ensure that fixation predominates over hydrolysis, allowing fixation efficiencies of up to about 70% to be achieved in ideal cases. The dyeing process is finally completed by washing out unfixed dye and inorganic materials.

It is worth noting that when hydrolysis occurs to produce species 7.8 the dye does not have a further chance to achieve fixation by virtue of its remaining chlorine atom since, in the alkaline conditions of the fixation stage, ionisation of the acidic hydroxy group on the triazine ring leads to a massive feedback of negative charge into the ring, causing total deactivation of the remaining chlorine atom. The third chlorine atom in the 'fixed' species (7.7) may remain unchanged, or it may be hydrolysed, producing form 7.9, or it may react further with the fibre to produce the crosslinked form 7.10 [7,8].

7.9

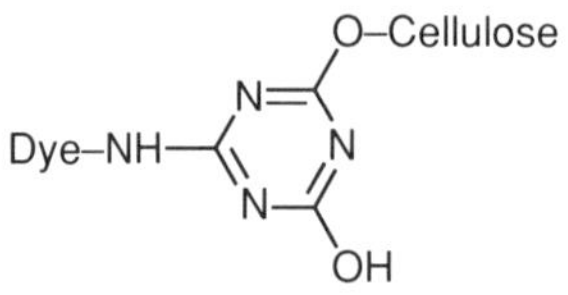

7.10

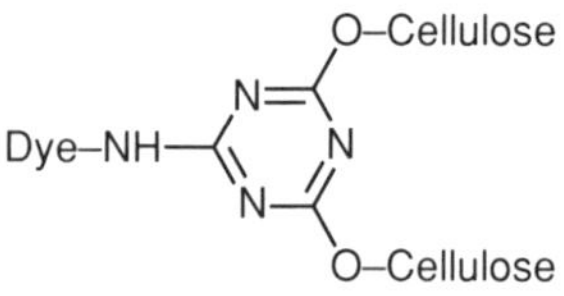

Numerous variations on this exhaustion technique have been designed for the application of dichloro-*s*-triazine reactive dyes to cellulose; all rely on the basic principles outlined above for their efficacy. In the pad–batch process a weakly alkaline pad liquor is prepared immediately prior to use by raising the pH of a concentrated solution of dye to a level at which only slow reaction will occur. This liquor is padded on to the fabric, which is then stored for several hours. The small volume of dye liquor used for the amount of fabric being dyed enables the dye to be readily adsorbed on to the fibre surface without the need for a separate and prolonged exhaustion stage. The dye slowly reacts with the fabric and again, at the end of the process, any hydrolysed dye is washed out together with inorganic materials.

With the dichloro-*s*-triazine dyes the operation of dyeing processes at low temperatures means that fairly small chromogens are required to ensure adequate mobility of dye on the fibre during the exhaustion stage. This requirement makes the dichloro-*s*-triazine dyes eminently suitable for producing bright dyeings but leads to difficulties in obtaining satisfactory deep tertiary shades, since the polyazo chromogens used in this area are often too large to allow dyes of acceptable performance to be prepared. A weakness with certain of these dyes, particularly red dyes based on the use of H acid as coupling component, is that under conditions of low pH the dye–fibre bond is broken by acid-catalysed hydrolysis, leading to wash fastness deficiencies.

7.2.2 Monochloro-*s*-triazine dyes

Reaction of a dichloro-*s*-triazine dye with an amine at a temperature of about 25–40°C results in one of the chlorine atoms being displaced, producing a less reactive monochloro-*s*-triazine dye (7.11). Dyes having this reactive system were launched by ICI (as Procion H dyes) and Ciba (as Cibacron dyes [9]) shortly after the marketing of the dichloro-*s*-triazines. The monochloro-*s*-triazine dyes react with cellulose in dyeing applications in exactly the same manner as the dichloro dyes, except that a higher temperature (usually about 80°C) and slightly higher pH (about 11) is required. Closely related are the rather more reactive monoalkoxy-monochloro-*s*-triazines (7.12), which are ideally applied at about 70°C. These dyes are prepared by condensing a dye base with a 2-alkoxy-4,6-dichloro-*s*-triazine, often 2-methoxy-4,6-dichloro-*s*-triazine, which is itself prepared by reaction of cyanuric chloride with the appropriate alcohol.

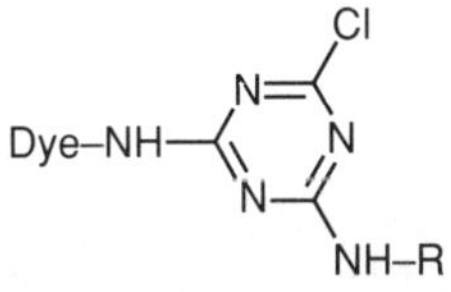

7.11

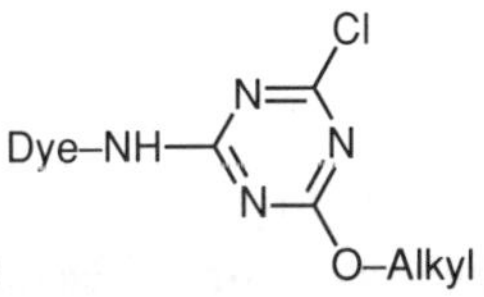

7.12

Although these simple monochloro-*s*-triazines are used in dyeing applications, their use in textile printing is of greater interest. In this application the relative inertness of the dye towards alkali at room temperature allows the satisfactory formulation of a stable print paste containing both the dye and an alkali. This can then be printed on to the fabric and formation of the dye–fibre bond is initiated by raising the temperature above 100°C, normally by steaming. In the printing process the dye does not need to have high substantivity; indeed the substantivity is generally minimised in order to facilitate easy washing out of any unfixed dye at the end of the process. This was done in the Cibacron Pront range of dyes, which utilised the higher reactivity of the alkoxy-*s*-triazine type to shorten the steaming time required for the dye–fibre reaction to take place. In this range a bulky alkoxy group, such as the isopropoxy group, was chosen for the triazine ring in order to disrupt the planarity of the molecule and thus speed removal of unfixed dye from the fabric in the washing-off process.

7.2.3 Bis(monochloro-*s*-triazine) dyes[1]

If two chloro-*s*-triazine residues are incorporated into the dye molecule then, since reaction of either group with the fibre will result in fixation of the dye, a higher fixation efficiency would be expected. In the case of bis(dichloro-*s*-triazinyl) dyes, however, only the arrangement 7.13 is available. The difficulties of introducing sufficient sulphonic acid groups into these rather large dye molecules in order to achieve adequate solubility and dye mobility at low application temperatures precludes their commercial utility.

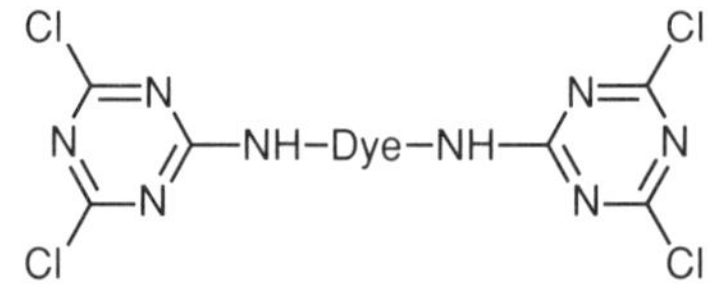

7.13

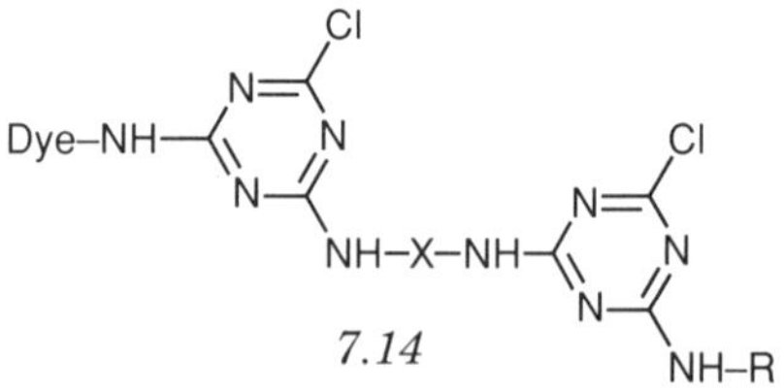

7.14

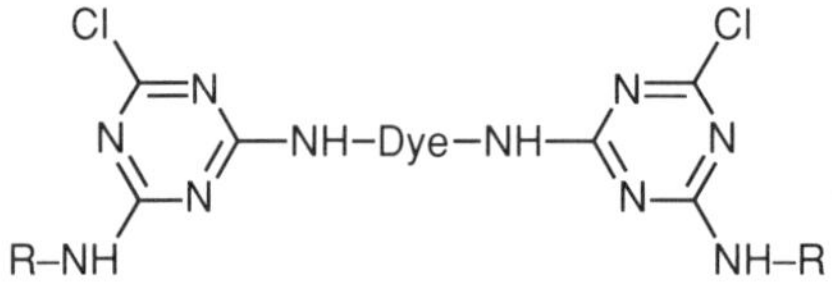

7.15

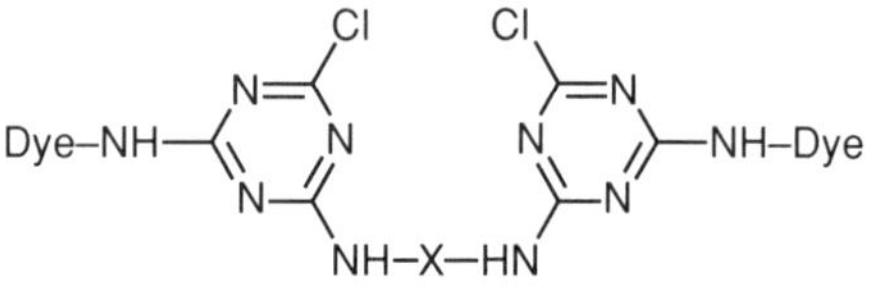

7.16

The monochloro-*s*-triazines present a different picture, with three possible arrangements (7.14–7.16). All of these types have been marketed, the earliest exploitation being in the Procion Supra range of high-efficiency printing dyes (now withdrawn), which included members having the pattern 7.14; these, however, suffer from the economic disadvantage of requiring two reactive groupings to fix a single chromogen to the fibre. Type 7.15, in which the dye is a cheap disazo chromogen, and type 7.16 do not have this drawback, since here fixation of only one of the reactive systems leads to attachment to the substrate of either one disazo chromogen (in the case of type 7.15) or two chromogens (in the case of type 7.16). These dyes, having about twice the molecular size of simple chloro-*s*-triazinyl dyes, are

---

[1] Symmetrical 1:2 chromium and cobalt complexes with a reactive group on each azo residue fall within the definition of this class but are discounted in this discussion.

much more substantive and this allows them to achieve excellent exhaustion on to the fabric at the preferred dyeing temperature of about 80°C, leading to fixation values of about 70–80%. Dyes of these two types make up the Procion H–E range of high-efficiency exhaust dyes.

7.3 OTHER CHLOROHETEROCYCLIC SYSTEMS

The success of the chloro-*s*-triazine reactive dyes initiated a wide-ranging investigation of other halogenoheterocyclic compounds that could be used as the basis for reactive dyes, and a host of patent specifications claiming the utility of many such compounds has appeared [10]. Here mention is made only of systems that have been commercially exploited.

Tetrachloropyrimidine, which is prepared by chlorination of barbituric acid (7.17; X = H) and treatment of the resulting 5-chlorobarbituric acid (7.17; X = Cl) with phosphorus oxychloride and dimethylaniline, is used in the Drimarene X (S) dyes. Two isomers (7.18a and 7.18b) are possible with these pyrimidine-based dyes.

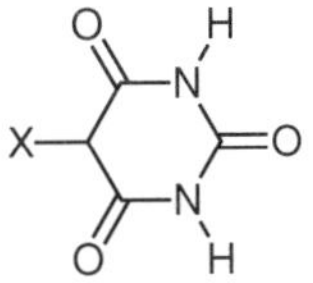

7.17

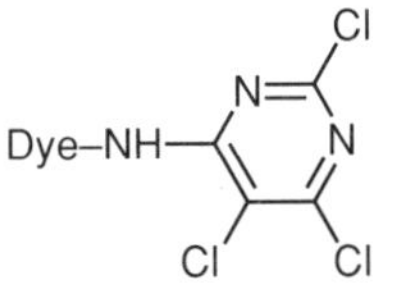

7.18a

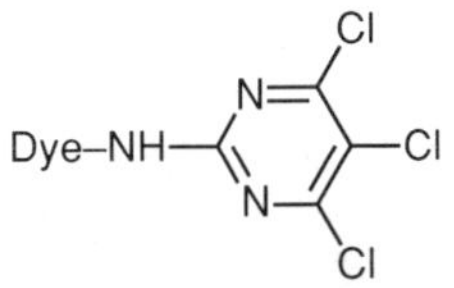

7.18b

A variation in the mode of attachment of the reactive system 7.19 to the chromogen is used in the Levafix E (BAY) range, in which the dye base is condensed with 2,3-dichloroquinoxaline-6-carbonyl chloride, thus preserving both chlorine atoms carried by the heterocyclic ring [11]. This intermediate is prepared from 3,4-diaminobenzoic acid by condensation with oxalic acid to yield 2,3-dihydroxyquinoxaline-6-carboxylic acid, which is then treated with phosphorus oxychloride and phosphorus pentachloride. In similar fashion members of the now withdrawn Elisiane (Fran) (7.20) range were prepared from 1,4-dichlorophthalazine-6-carbonyl chloride, which was made from trimellitic acid (benzene-1,2,4-tricarboxylic acid) by condensation with hydrazine and treatment of the resulting 1,4-dihydroxyphthalazine-6-carboxylic acid with phosphorus oxychloride and phosphorus pentachloride [12].

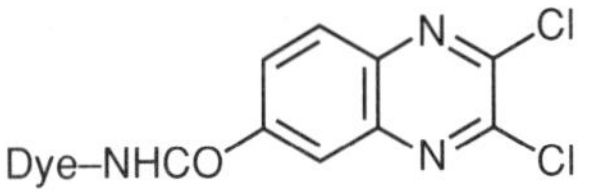

7.19

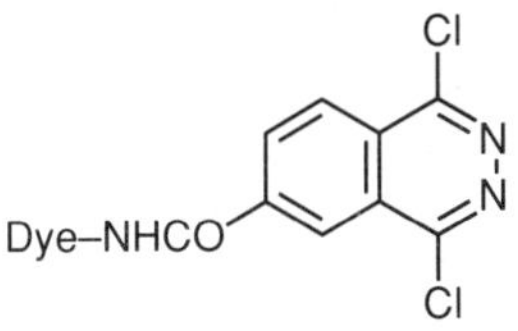

7.20

The *s*-triazine ring is, however, unique amongst the six-membered nitrogen heterocycles in possessing three nitrogen atoms ideally placed to provide the necessary activation of the chlorine atoms. Calculations of charge distribution over the various nitrogen heterocycles show that the largest positive charge recorded in the series resides on the carbon atoms of the *s*-triazine ring (Figure 7.1) [13]. As a result, despite possessing more than one chlorine atom attached to the heterocyclic ring, dyes derived from alternative halogenoheterocycles are generally less reactive than the dichloro-*s*-triazine dyes.

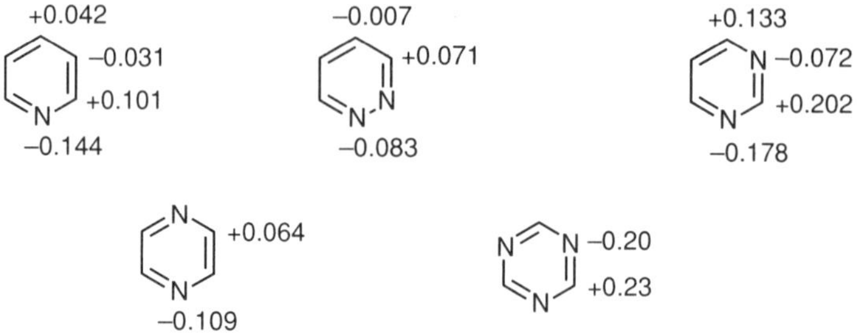

Figure 7.1 – Levels of reactivity of various heterocyclic rings

The Levafix E dyes (7.19) are the most reactive of this group, requiring a reaction temperature of about 50°C in dyeing applications, whilst the others resemble the monochloro-*s*-triazine dyes in reactivity. In the pyrimidine series the deficiency in reactivity can conveniently be overcome by incorporating an electron-attracting substituent in the 5-position of the pyrimidine ring and one compound of this type that has been commercially exploited is 5-cyano-2,4-6-trichloropyrimidine (7.21). The preparation of this compound starts from barbituric acid (7.17; X = H), which with sodium cyanate and hydrochloric acid gives the 5-carbonamido derivative (7.22). Treatment of this with phosphorus oxychloride and dimethylaniline results in simultaneous dehydration of the carbonamido group and replacement of the hydroxy groups by chlorine atoms, yielding 5-cyano-2,4-6-trichloropyrimidine. This compound has a level of reactivity comparable with that of cyanuric chloride and, like tetrachloropyrimidine, it forms two isomers on condensation with a dye base. One 5-cyanodichloropyrimidinyl

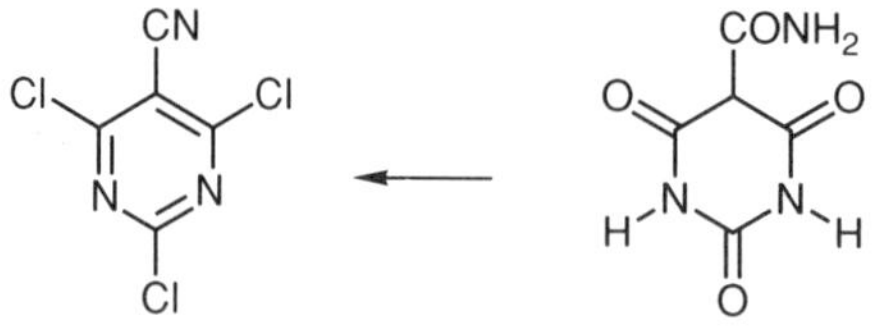

7.21 7.22

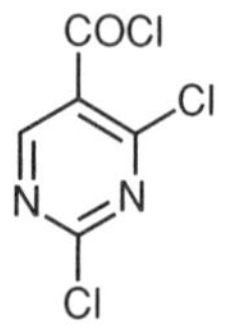

7.23

dye, C.I. Reactive Red 219, figures alongside the dichloro-*s*-triazines in the Procion MX range. In the Reactofil (Gy) dyes, which again displayed near cold-dyeing properties, 2,4-dichloro-5-chlorocarbonylpyrimidine (7.23) was used as the source of the reactive system. Here the 5-carboxylic acid chloride function served both to provide a means of attachment to the dye base and to activate the pyrimidine ring chlorine atoms. This pyrimidine was prepared from 2,4-dihydroxypyrimidine-5-carboxylic acid (the condensation product of urea and formylenemalonic ester) with, for example, phosphorus oxychloride.

A different arrangement is to be found in the 4,5-dichloropyridazone system, upon which the Primazin P (BASF) range [14] of printing dyes depended. Here the reactive 4-chlorine atom is not activated by attachment to a fully aromatic nitrogen heterocycle but the >C=C–CO– grouping in the pyridazone ring forms a vinylogue of a carboxylic acid chloride. Several 4,5-dichloropyridazone derivatives have been described as intermediates for the production of dyes of this type. One such is 3-(4,5-dichloropyridaz-6-on-1-yl)propionyl chloride (7.26); this is prepared from 2-butyne-1,4-diol, which is chlorinated to yield dichloromaleic acid (7.24). Cyclisation of this with β-hydrazinopropionitrile (7.25) followed by hydrolysis of the cyano group with sulphuric acid and treatment with thionyl chloride gives the dichloropyridazone (Scheme 7.4).

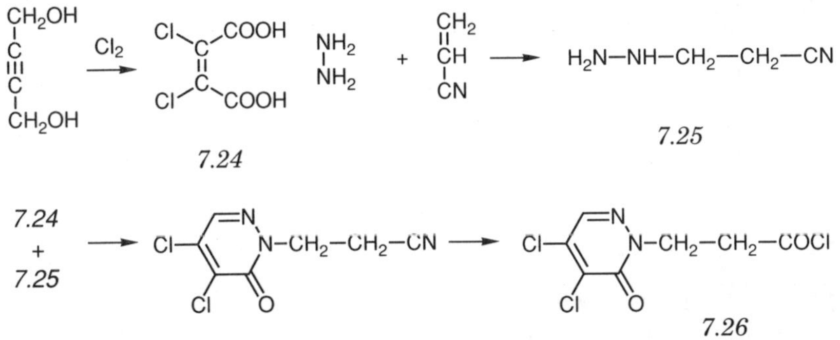

Scheme 7.4

Sulphur, as well as nitrogen, was used as a ring hetero atom in certain Reatex (Fran) dyes containing a 2-chlorobenzothiazole derivative as reactive system (7.27) [15]. Typically this group was introduced by reaction of a dye base with 2-chlorobenzothiazole-6-sulphonyl chloride, prepared by chlorosulphonation of 2-chlorobenzothiazole.

7.27

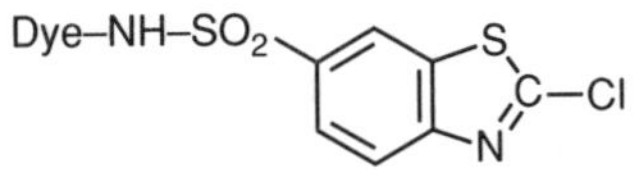

7.4 ALTERNATIVE LEAVING GROUPS ON HETEROCYCLIC SYSTEMS

7.4.1 General

Whilst chlorine is the most common choice as displaceable substituent on a heterocyclic reactive system, a large number of other leaving atoms and groups have been described in the patent literature. Most of these have been mentioned as substituents on an *s*-triazine ring and include bromine atoms, cyano, azido and trichloromethyl [16] groups, 2-nitro-4-sulphophenoxy [17] and 2,4-disulphophenoxy groups, 2,4-dinitrothiophenoxy [18], thiocyanato [19] and sulphonic acid [20] groups and a variety of other sulphur, carbon and phosphorus [21] derivatives (Figure 7.2).

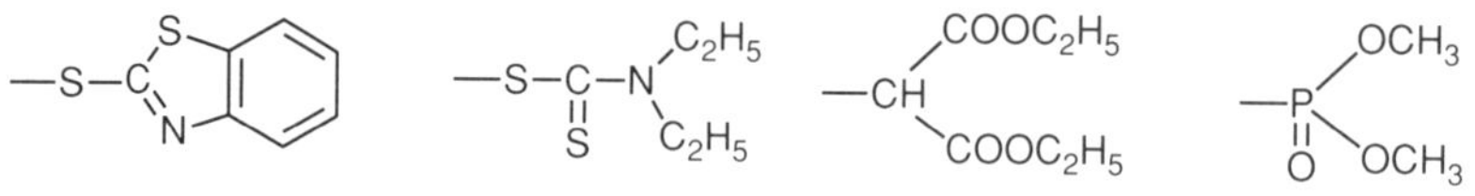

Figure 7.2 – Some leaving groups on heterocyclic systems

These alternatives are occasionally made by condensation of a dye base with the appropriately substituted triazine, such as cyanuric bromide or 2,4-bis(trichloromethyl)-6-chloro-*s*-triazine, but are more generally made by replacement of the chlorine atom in a chloro-*s*-triazinyl dye by reaction with the sodium salt of the appropriate nucleophile. They offer no advantage over the chloro compounds and have not been commercially exploited, the majority having been disclosed by ICI in order to prevent competitors evading their patents by marketing such variations. Three other displaceable entities have achieved commercial importance, however: the quaternary ammonium and sulphone groupings and the fluorine atom.

7.4.2 Quaternary groups

Monochloro-*s*-triazines react with tertiary amines possessing a sterically accessible nitrogen atom to produce a quaternary ammonium derivative (7.28) (Scheme 7.5) [22]. Suitable tertiary amines are trimethylamine (7.29), *NN*-dimethylhydrazine (7.30), pyridine (7.31; R = H) and substituted pyridines such as nicotinic acid (7.31; R = COOH) and 1,4-diazabicyclo[2,2,2]octane (7.32).

In structure 7.28 the quaternary ammonium group is made particularly labile by the full positive charge that it carries. This increases the polarisation of the C–N bond, making these compounds much more reactive than the parent chloro-*s*-triazines. In the case of the quaternary derivatives formed from the non-planar aliphatic amines 7.29, 7.30 and 7.32, steric strains further destabilise the bond so that reaction with cellu-

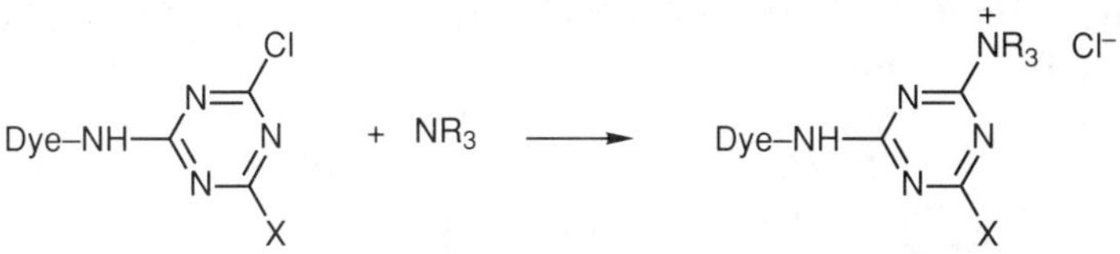

(X = NH_2, NHR, OR)

7.28

Scheme 7.5

$N(CH_3)_3$

7.29

$(CH_3)_2N{-}NH_2$

7.30

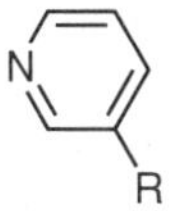

7.31

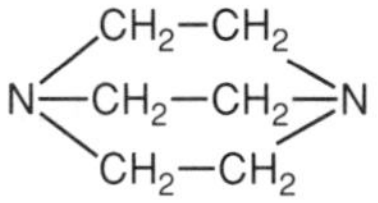

7.32

lose occurs under alkaline conditions at about 30°C, whereas temperatures of about 40–50°C are required for the pyridinium derivatives. Although the properties of these quaternary salts make them in the main very attractive as reactive dyes, they have one major drawback: the objectionable smell of many of the parent tertiary amines, which are liberated on fixation. This has largely restricted their use, but no such problem arises with the odour-free nicotinic acid. Quaternary salts based on this amine are marketed in the Kayacelon React (Nippon Kayaku) range (C.I. Reactive Red 221, for example) in which advantage is taken of the ability of quaternary derivatives to react with cellulose under neutral conditions, allowing them to be used in conjunction with disperse dyes for the dyeing of polyester/cotton blends [23].

7.4.3 Fluorine atoms

A fluorine atom is used as leaving group in the Cibacron F (CGY) range (7.33), which embodies a monofluoro-*s*-triazine reactive system, and in the Levafix E–A (BAY) and Drimarene K (S) ranges of 5-chlorodifluoropyrimidinyl dyes (mixtures of the isomers 7.34a and 7.34b). The reactive intermediates required for the preparation of these dyes, cyanuric fluoride and 5-chloro-2,4-6-trifluoropyrimidine, are both made from the corresponding chloro compounds by an exchange reaction with potassium fluoride.

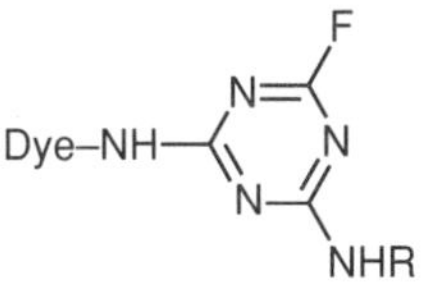

7.33

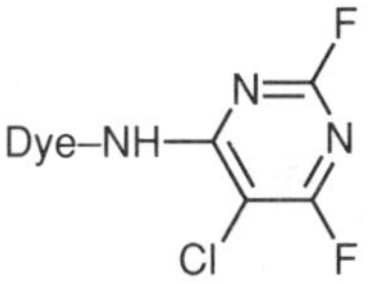

7.34a

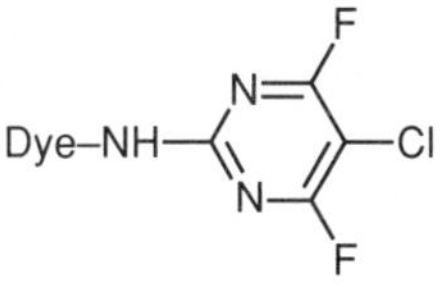

7.34b

In these compounds the greater electronegativity of fluorine compared with that of chlorine again results in a higher level of reactivity than that of the chloro compounds, and dyeing temperatures of about 60°C for the fluoro-*s*-triazines and 40–50°C for the 5-chlorodifluoropyrimidines are appropriate. The 'hard' character of the fluorine atom, resulting from the tight binding of the electron shell around the atomic nucleus, makes reaction with dye bases carrying aliphatic amino groups preferable; this has prompted the preparation and use of novel dye intermediates carrying such substituents. An interesting intermediate of this type is 5-aminomethyl-1-naphthylamine-2-sulphonic acid (7.35), prepared by condensing 1-acetylaminonaphthalene-2-sulphonic acid with *N*-hydroxymethylphthalimide (Tscherniac–Einhorn reaction) followed by hydrolysis of the resulting 5-substituted naphthalene derivative (7.36).

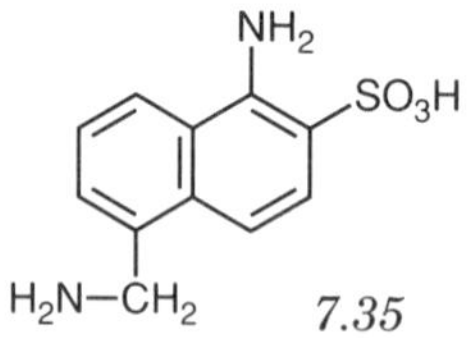

7.35

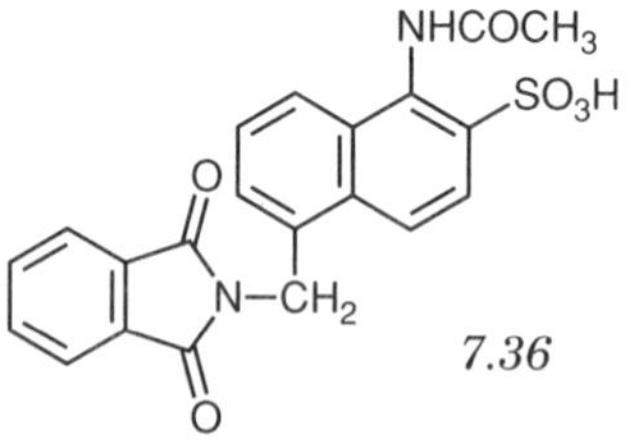

7.36

The dye–fibre bond formed by the 5-chlorodifluoropyrimidinyl dyes is more stable to acid hydrolysis than that formed by the dichloro-*s*-triazines, but it does tend to undergo oxidative cleavage under the influence of light in the presence of perborates.

7.4.4 Methylsulphonyl groups

In the Levafix P (BAY) (7.37) range of rapid-fixing printing dyes the leaving group is a methylsulphonyl group [24]. The preparation of the required reactive intermediate, 4,5-dichloro-6-methyl-2-methylsulphonylpyrimidine (7.38), starts with a pyrimidine ring synthesis involving the condensation of *S*-methylthiourea and ethyl acetoacetate; the resulting 4-hydroxy-6-methyl-2-methylthiopyrimidine is chlorinated, the thiol group oxidised to a sulphone and the hydroxy group then exchanged for a chlorine atom by the action of phosphorus oxychloride (Scheme 7.6). The 4-chlorine atom of structure 7.38 is preferentially eliminated on condensation with the dye

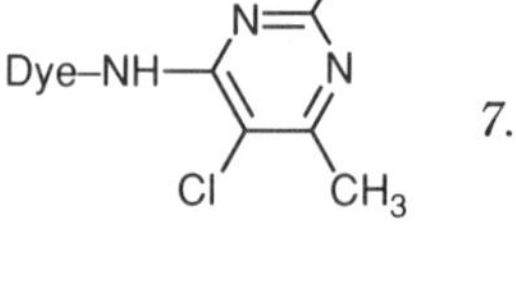

7.37

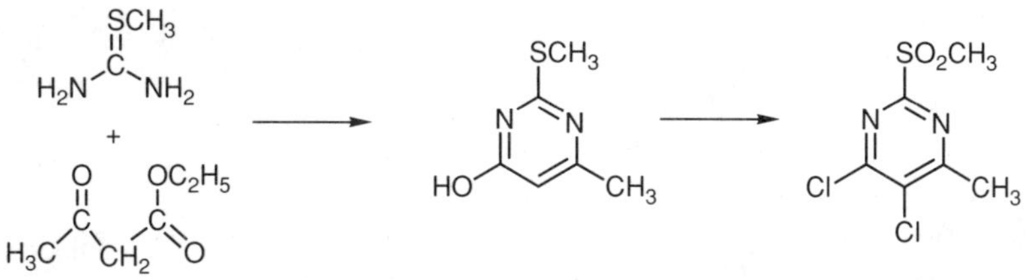

7.38

Scheme 7.6

base, leading to dyes of structure 7.37. This would appear to be inconsistent with the behaviour in the dye–fibre reaction stage, where the methylsulphonyl group is a particularly effective leaving group and leads to rapid fixation. The reason for this difference in behaviour is somewhat complex and is related to the electron cloud around the methylsulphonyl group, which can be readily distorted. This allows easier formation of the transition state for nucleophilic displacement when the 'hard' cellulosate nucleophile approaches, thus facilitating displacement and speeding reaction.

7.5 CHROMOGENS USED IN HALOGENOHETEROCYCLIC REACTIVE DYES

The demands placed upon the chromogen in the design of a reactive dye are small. The only structural features necessary are a sufficiency of sulphonic acid groups to ensure adequate water-solubility and an amino group to which the reactive system can be attached. It is thus possible to use any chromogen as a basis for a reactive dye and most ranges incorporate monoazo, disazo, metallised mono- and dis-azo, anthraquinone and phthalocyanine chromogens, and perhaps also formazans and triphenodioxazines. Moving across the shade range a general pattern emerges for the most commonly used dye bases (dyes carrying an amino group that give reactive dyes on condensation with a halogenoheterocycle), which is largely dictated by accessibility of intermediates. This is summarised below, illustrated by actual dye structures wherever possible. Where these have not been disclosed skeletal structures indicating the salient features of the dye molecule are given.

7.5.1 Greenish-yellow

These are invariably of the monoazo type, using pyridone, pyrazolone and, less frequently, acetoacetarylamide and barbituric acid coupling components. The ready availability of phenylenediaminesulphonic acids, together with the general lack of coupling components containing amino groups, dictates that the reactive system is most conveniently attached to the diazo component in structures such as C.I. Reactive Yellow 86 (7.39) and C.I. Reactive Yellow 1 (7.40).

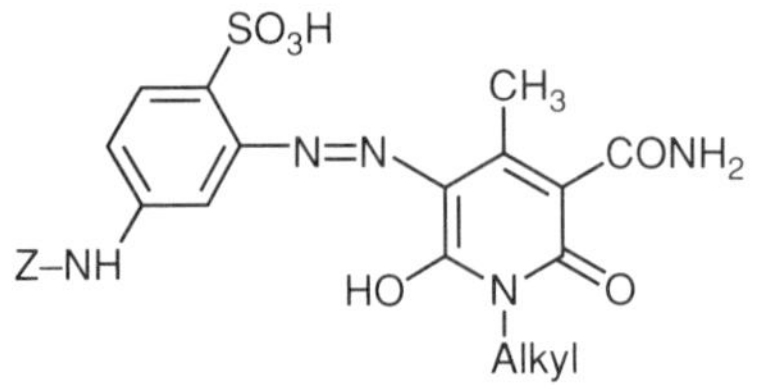

7.39

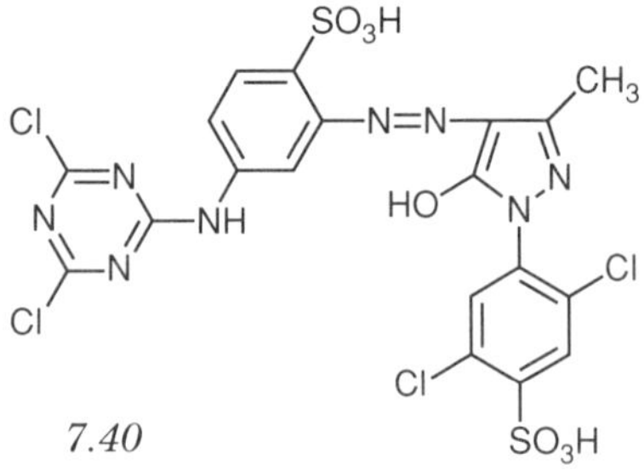

7.40

7.5.2 Reddish-yellow

Here diazotised arylamine sulphonic acids coupled with *p*-coupling anilines present a convenient source of dye bases that give reddish-yellow hues on reaction with a halogenoheterocycle. Typical is C.I. Reactive Yellow 3 (7.41).

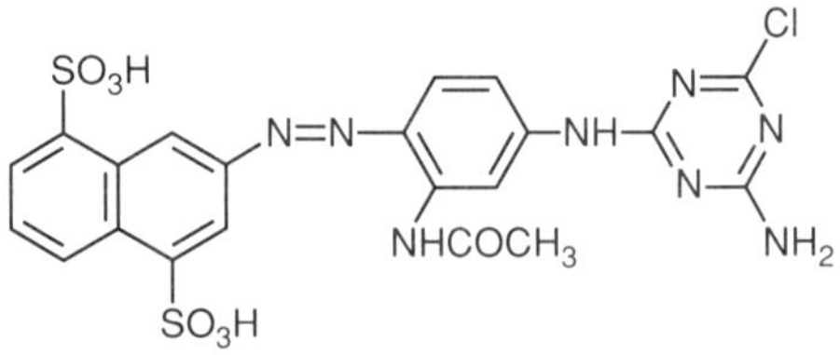

7.41

7.5.3 Orange and scarlet

Dye bases for these dyes are usually obtained using either J acid or its *N*-methyl derivative as coupling component; *N*-methylation reduces the rather high substantivity conferred by J acid, which can make it difficult to wash off any unfixed dye. Aniline diazo components are used for yellowish-orange hues, typically C.I. Reactive Orange 1 (7.42; X = H). Naphthylaminesulphonic acids give redder hues, as in C.I. Reactive Orange 4 (7.43), and 4-methoxyanilinesulphonic acids give scarlets, such as C.I. Reactive Red 8 (7.42; X = CH_3O).

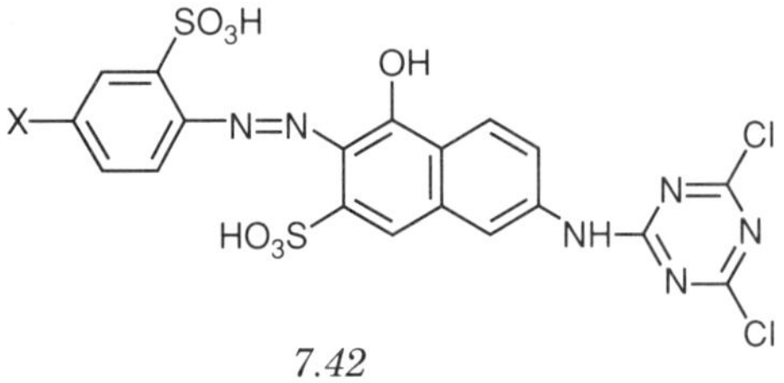

7.42

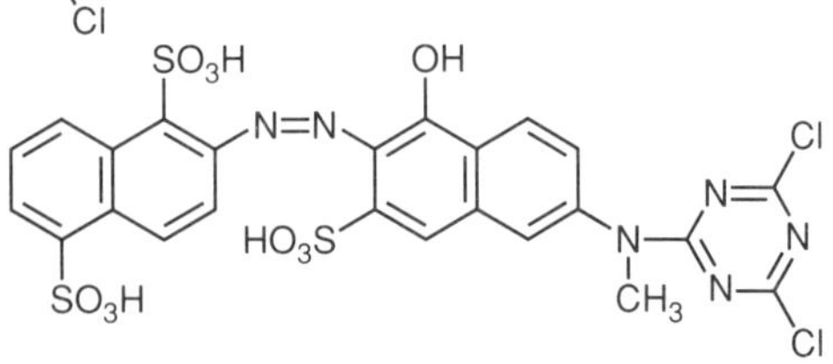

7.43

7.5.4 Red

The all-important coupling component here is H acid used, typically, in C.I. Reactive Red 17 (7.44).

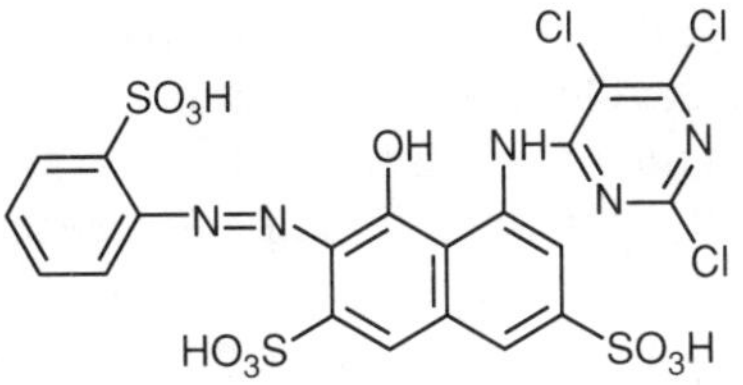

7.44

7.5.5 Rubine and violet

Copper complexes of 2-aminophenolsulphonic acids, diazotised and coupled with J acid for rubines and with H acid for violets, dominate this part of the colour gamut. C.I. Reactive Red 6 (7.45) and C.I. Reactive Violet 1 (7.46) are examples.

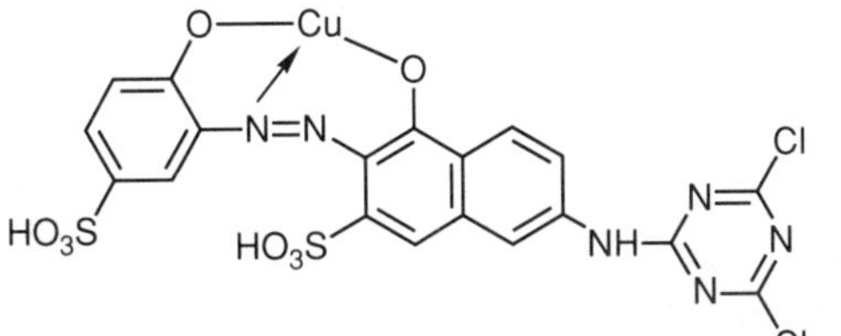

7.45

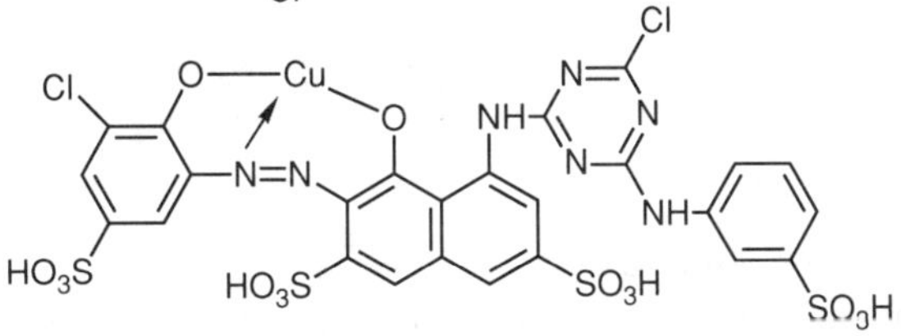

7.46

7.5.6 Dull blue

The use of copper complexes of aminonaphtholsulphonic acid diazo components with H acid, such as C.I. Reactive Blue 13 (7.47), leads to blue chromogens. Often these dyes are more easily prepared by using a naphthylaminesulphonic acid as diazo component and oxidatively coppering the resulting monoazo dye.

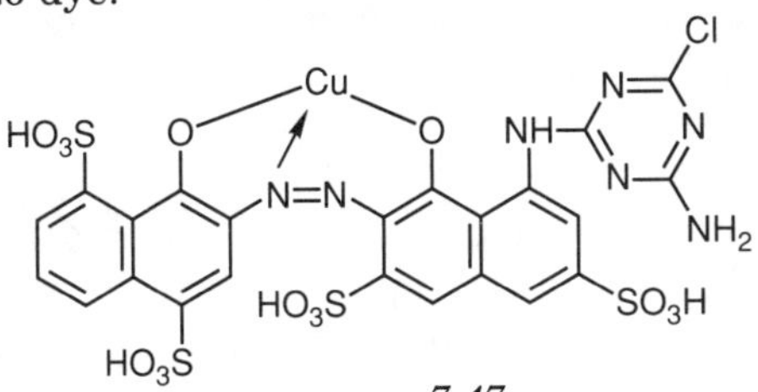

7.47

7.5.7 Bright blue

Two different types of chromogen dominate this area. Initially it was the sole province of anthraquinone dyes, the bases for which are prepared by condensing bromamine acid with a phenylenediaminesulphonic acid. C.I. Reactive Blue 2 (7.48) is typical of these dyes. A major feature of recent years has been the development of triphenodioxazine dyes, such as C.I. Reactive Blue 198, which are tinctorially much stronger and economically advantageous. No actual triphenodioxazine structures have been disclosed but these dyes are of the general pattern shown by structure 7.49, where Z represents a halogenoheterocyclic reactive group and $H_2N–R–NH_2$ represents a diamine.

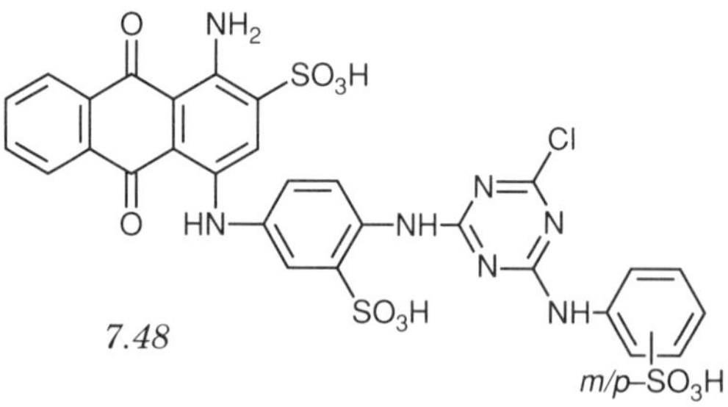

7.48

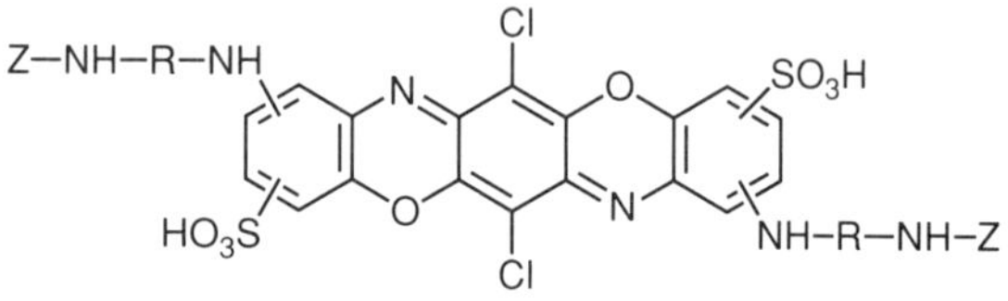

7.49

7.5.8 Navy blue

Disazo dyes based on twice-coupled H acid chromogens [such as C.I. Reactive Blue 171 (7.50)], formazans [such as C.I. Reactive Blue 160 (7.51)] and copper complexes of A→M→E disazo compounds [such as C.I. Reactive Blue 40 (7.52)] supply this hue sector. Again few structures have been disclosed, and in the skeleton structures shown irrelevant substituents have been omitted and Z used to indicate a halogenoheterocyclic reactive group. The twice-coupled H acid types dominate this hue sector because of their economic advantages, whilst the formazans have the best fastness properties.

7.5.9 Brown

Unmetallised A→M→E disazo dyes such as C.I. Reactive Brown 1 (7.53) dominate this sector, aniline M components giving yellowish-browns and naphthylamine M components reddish-browns; 1:2 chromium complexes have also been used.

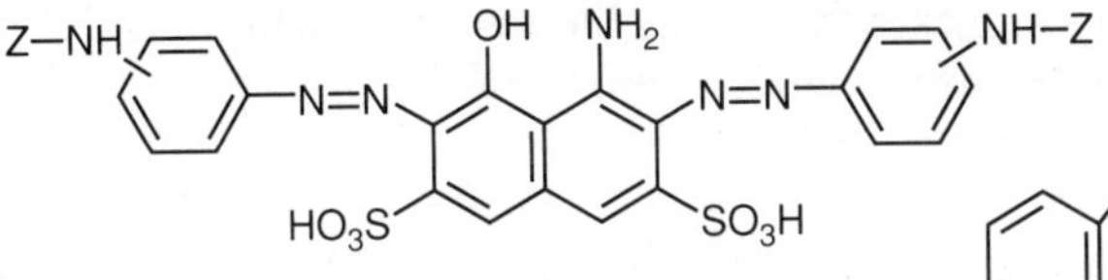

7.50

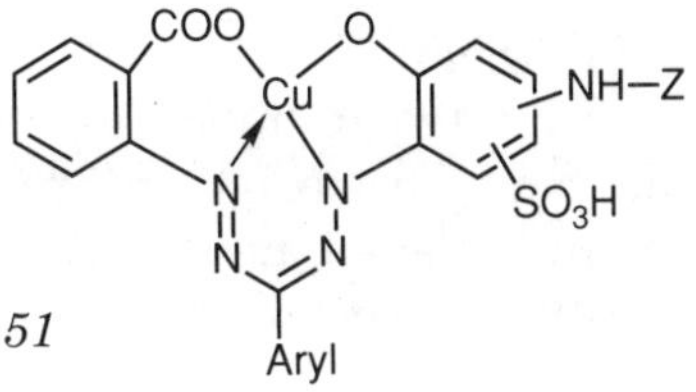

7.51

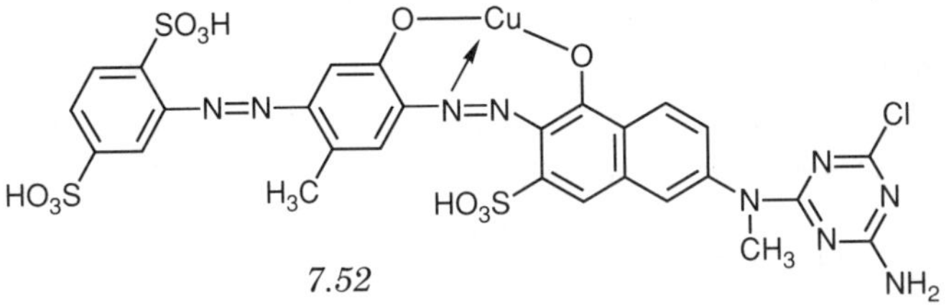

7.52

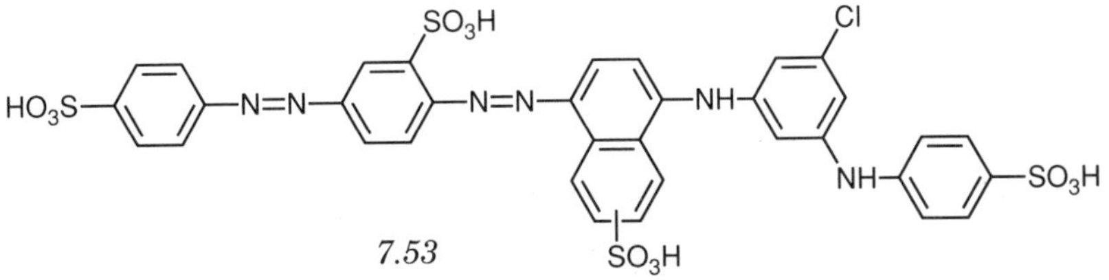

7.53

7.5.10 Turquoise

This sector is totally dominated by phthalocyanine dyes. They are prepared by chlorosulphonating copper phthalocyanine, partially reacting the tetrasulphonyl chloride with a suitable diamine and solubilising by hydrolysing the remaining sulphonyl chloride groups. The reactive system is then attached to the pendent amino-carrying chain. The products are not homogeneous but are statistical mixtures that are assigned approximate constitutions, such as structure 7.54 for C.I. Reactive Blue 7, CuPc representing the copper phthalocyanine nucleus.

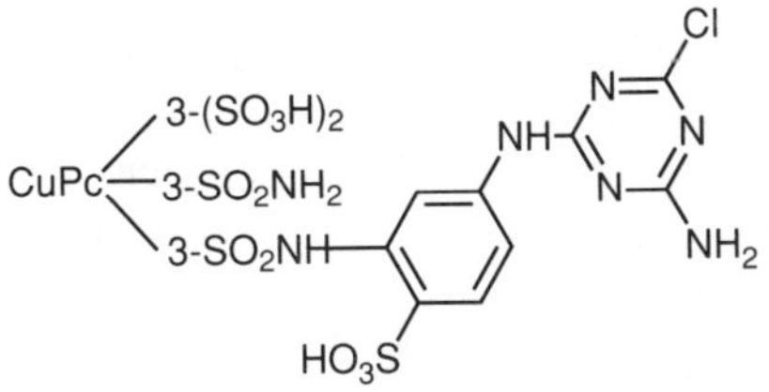

7.54

The use of nickel phthalocyanine as starting material results in the chlorosulphonyl groups entering the 4- rather than the 3-positions of the peripheral benzene rings and, more importantly, prevents the catalytic fading of azo yellow components, which can be observed when applying green colours with phthalocyanine/azo dye mixtures.

7.5.11 Green

Only a limited number of green reactive dyes has been marketed. Of these, some combine azo and anthraquinone or azo and phthalocyanine chromogens in the same molecule. Twice-coupled H acid structures of the type described in section 7.5.8 can be used to give dull greens if the diazo components are correctly chosen, as in the case of C.I. Reactive Green 19.

7.5.12 Black

Black dyeings are often based on twice-coupled H acid navy dyes with minor amounts of red and yellow; A→M→E disazo navy dyes and their copper complexes may be used similarly. Mixed and homogeneous 1:2 metal complexes may also be used, especially the homogeneous type (7.55) formed from 2-aminophenols diazotised and coupled with H acid, such as C.I. Reactive Black 4.

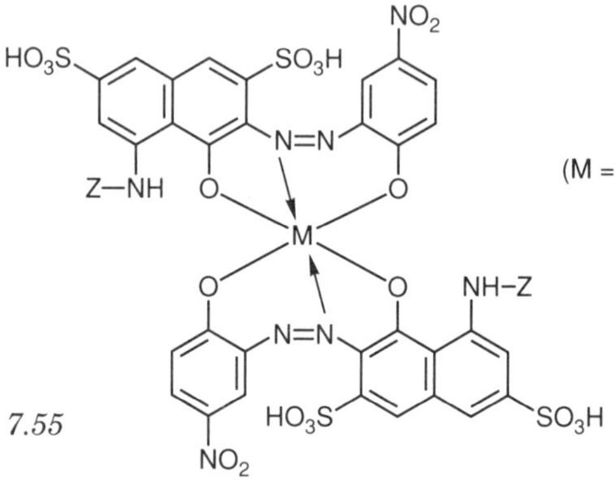

(M = Cr or Co)

7.55

7.6 VINYL SULPHONE AND VINYL SULPHONAMIDE DYES

7.6.1 Reaction mechanism and properties of vinyl sulphone dyes

A second major group of reactive dyes, based on a vinyl sulphone reactive system, differs from the halogenoheterocyclic reactive dyes in that it functions by virtue of a nucleophilic addition mechanism rather than a nucleophilic substitution mechanism. These dyes were first introduced by Hoechst, which markets them under the trade name Remazol. They contain a β-sulphatoethylsulphone grouping which, under the action of alkali, eliminates sulphuric acid producing a vinyl sulphone group [25]. In this group the carbon–carbon double bond is polarised by the powerfully electron-attracting sulphone group. This polarisation imparts a positive charge to the terminal carbon atom, allowing nucleophilic addition of ionised cellulose to occur. Addition of a proton completes the dye–fibre reaction (Scheme 7.7).

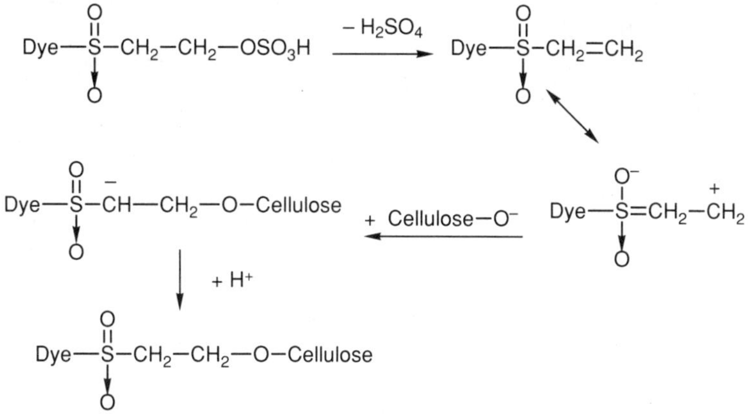

Scheme 7.7

These dyes are intermediate in reactivity between the mono- and dichloro-*s*-triazine dyes, requiring an exhaust dyeing temperature of 50–60°C, depending on pH. They are applicable by a wide variety of dyeing and printing techniques, similar to those outlined for the halogenoheterocyclic type but modified to cater for the difference in properties; for example, in printing applications the greater reactivity compared with the monochloro-*s*-triazines makes two-stage processes preferable, the dye and alkali being applied to the fabric separately. The substantivity is markedly lower than that of the halogenoheterocyclic dyes since, unlike the heterocyclic ring, the vinyl sulphone grouping has little if any affinity for the fibre. In exhaust dyeing procedures this can be overcome by the drop in solubility that occurs at the initial acid-elimination stage of the dye reaction sequence. As any unfixed dye is low in substantivity it is readily washed off. Unlike the chloro-*s*-triazines, the bond formed with the fibre is at its weakest at alkaline pH [26].

7.6.2 Intermediates for vinyl sulphone dyes

In the halogenoheterocyclic dyes the reactive system can be attached to any suitable chromogen that contains an amino group, but preparation of dyes of the vinyl sulphone series requires the synthesis of special intermediates containing the reactive system. Several such intermediates are used in the Remazol range, all being anilines or naphthylamines carrying a β-sulphatoethylsulphone group introduced by a standard procedure. The synthesis of a typical intermediate, 4-β-sulphatoethylsulphonylaniline (7.56), starts from acetanilide which with chlorosulphonic acid yields 4-chlorosulphonylacetanilide. This is reduced with sodium sulphite and the sulphinic acid produced condensed with 2-chloroethanol to give the 4-hydroxyethylsulphone. Hydrolysis of the acetylamino group and esterification of the hydroxy group by treatment with concentrated sulphuric acid complete the process (Scheme 7.8).

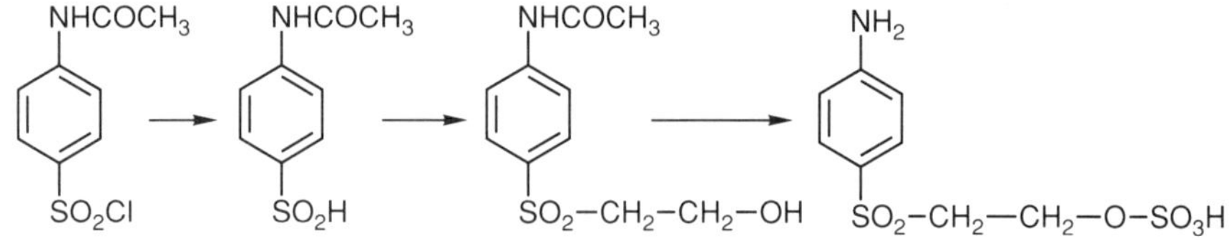

7.56

Scheme 7.8

The isomeric 3-β-sulphatoethylsulphonylaniline is similarly prepared from nitrobenzene, chlorosulphonation of which yields 3-nitrobenzenesulphonyl chloride, the synthesis then involving a reduction rather than a hydrolysis stage after hydroxyethylsulphone formation.

7.6.3 Chromogens used in vinyl sulphone dyes

Ranges of vinyl sulphone dyes are entirely constructed using anilines and naphthylamines containing the β-sulphatoethylsulphone reactive system. These intermediates have wide applicability in the azo class, since they are conveniently used as diazo components with any of the gamut of coupling components available. Typical examples are C.I. Reactive Yellow 17 (7.57) and C.I. Reactive Red 22 (7.58).

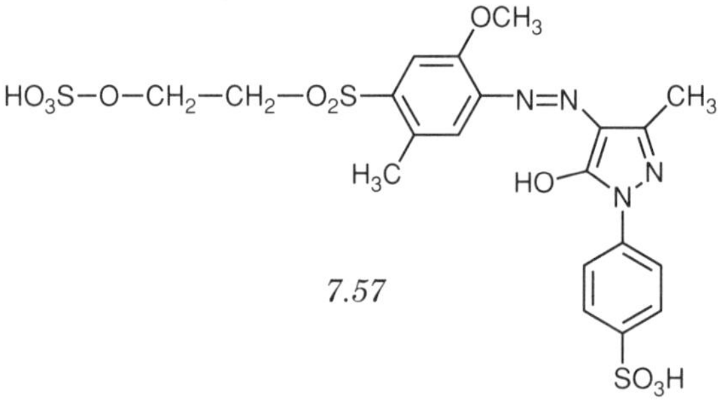

7.57

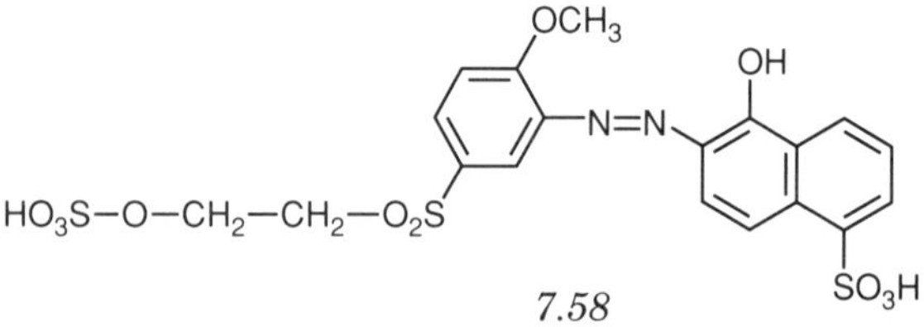

7.58

No general discussion of vinyl sulphone dyes is called for, but two dyes do deserve special mention. C.I. Reactive Black 5 (7.59) is applicable by a wide variety of dyeing and printing techniques. It has a much lower level of substantivity than typical halogenoheterocyclic dyes, yet achieves good uptake on to the fibre, leading to excellent fixation, by the progressive loss of two solubilising groups as the β-sulphatoethylsulphone groupings are converted into vinyl sulphone groups. After fixation the inherently low substantivity of any unfixed dye makes washing-off easy in a region of the colour gamut where this is normally notoriously difficult. Despite the fastness properties being only moderate, the ease of application coupled with a low selling price has made this the largest-selling reactive dye.

C.I. Reactive Blue 19 (7.60) possesses good fastness (but poor solubility and levelling) and an extremely attractive bright blue hue which for many years could not be challenged by other reactive dyes. The success of the Remazol range owes much to skilful marketing of these two dyes.

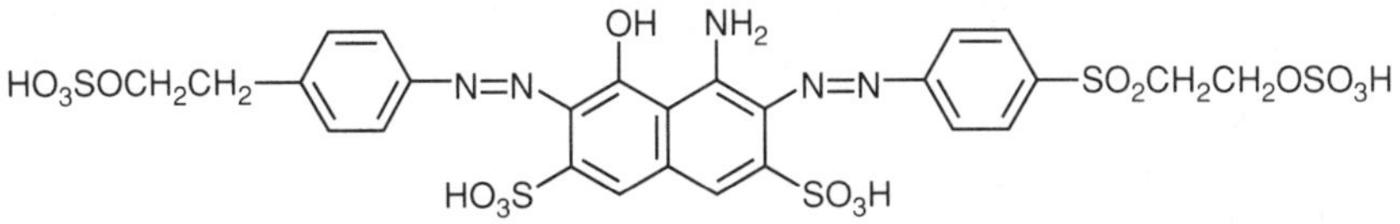

7.59

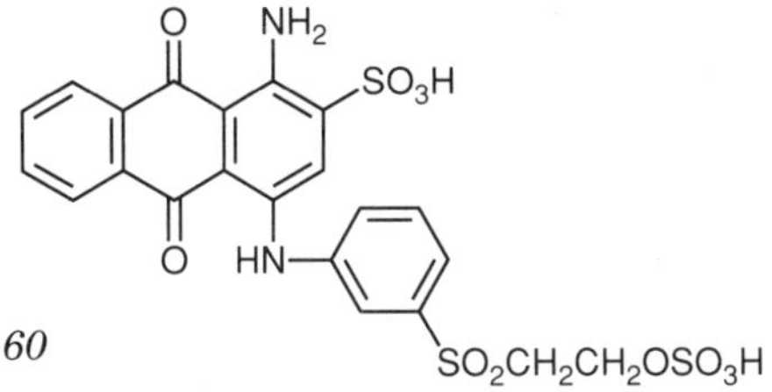

7.60

7.6.4 Vinyl sulphonamide dyes

A neat alternative to the vinyl sulphone type is presented by the vinyl sulphonamide structures, which have been exploited by Hoechst in its Remazol D dyes. These utilise carbyl sulphate (7.61) as a key intermediate. This interesting compound is prepared by reaction of sulphur trioxide with ethylene [27], and reacts with a dye base or an intermediate containing a suitable amino group to introduce a β-sulphatoethylsulphonamide group into the molecule [28]. Where an intermediate has been used the product is subsequently converted into a dyestuff. Like the β-sulphatoethylsulphone precursors of vinyl sulphone dyes, the β-sulphatoethylsulphonamide group eliminates sulphuric acid, yielding a vinyl sulphonamide. Reaction of this with the fibre then parallels that of the vinyl sulphone dyes (Scheme 7.9).

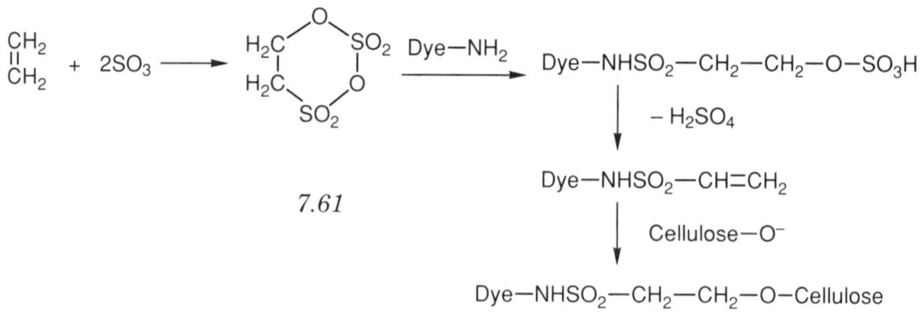

7.61

Scheme 7.9

7.7 REACTIVE SYSTEMS BASED ON OTHER ADDITION MECHANISMS

7.7.1 β-Sulphatoethylsulphamoyl dyes

The β-sulphatoethylsulphamoyl reactive system (7.62; X = OSO_3H) has a level of reactivity roughly equivalent to that of a monochloro-*s*-triazine system. The mechanism of reaction is said to involve elimination of the elements of sulphuric acid yielding an ethylenimine derivative (7.63) that can then be attacked by ionised cellulose leading to fixation (Scheme 7.10) [29].

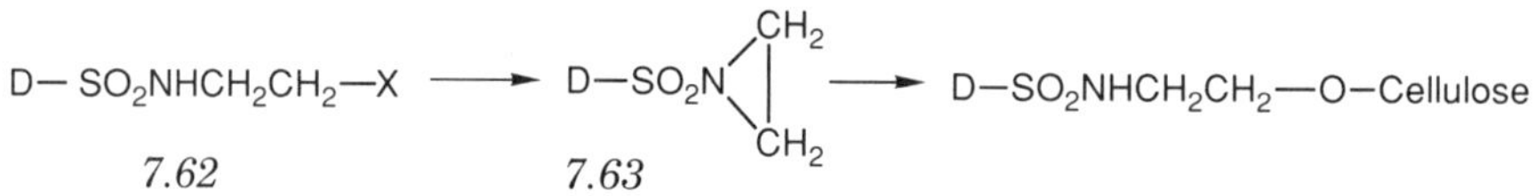

7.62 7.63

Scheme 7.10

Dyes containing this system were used in the original Levafix (BAY) range (not to be confused with Levafix E, E–A and P ranges, which contain heterocyclic reactive groups). The members of this range contained a minimum of two such groups, a typical structure being 7.64. Since they were otherwise devoid of water-solubilising groups, it is quite possible that the true function of the β-sulphatoethylsulphamoyl groups was to tempo-

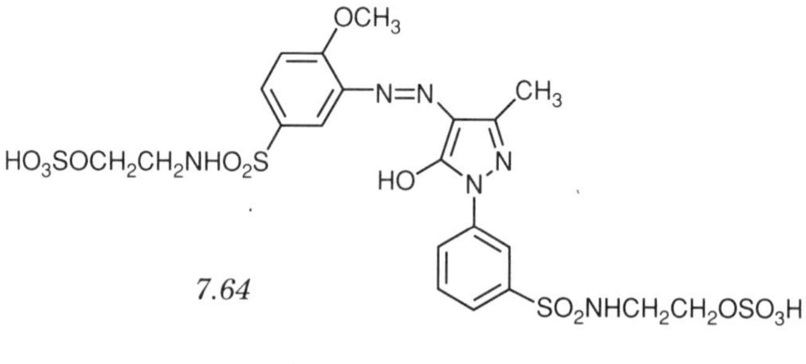

7.64

rarily solubilise the dye molecule rather than to cause dye–fibre reaction, the reactive group merely regenerating (7.62; X = OH) under the reaction conditions.

7.7.2 β-Chloroethylsulphamoyl dyes

A similar fixation mechanism has been proposed for the β-chloroethylsulphamoyl dyes (7.62; X = Cl) [30]. The low level of reactivity of this group makes dyes containing it more suitable for application to polyamide fibres than to cellulose and has led to their principal use being as disperse reactive dyes for nylon (section 7.10.3).

Dyes in which the β-chloroethylsulphamoyl group is attached to a phthalocyanine nucleus have a somewhat higher level of reactivity, and these dyes are useful for the printing and dyeing of cellulosic fibres. The reactive group can be conveniently introduced into this chromogen by chlorosulphonation of copper phthalocyanine and partial reaction of the resulting tetra(sulphonylchloride) with β-chloroethylamine, the remaining sulphonyl chloride groups being either hydrolysed to sulphonic acid groups or converted to sulphonamide groups by reaction with ammonia. This process leads to a mixture of water-soluble compounds generally formulated as structure 7.65. C.I. Reactive Blue 3 and C.I. Reactive Blue 25 are of this type.

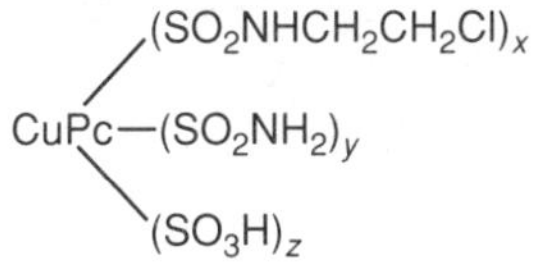

7.65

$x + y + z = 4$

7.7.3 β-Sulphatopropionylamino dyes

Early members of the Primazin (BASF) range contained a β-sulphatopropionylamino group [31], a typical structure being 7.66. This group may be viewed as a precursor of the feebly reactive acrylamido group. Again, the fact that these dyes contain two solubilising groups as well as the sulphato group on the reactive system suggests that temporary solubilisation may not have been the main factor in their performance.

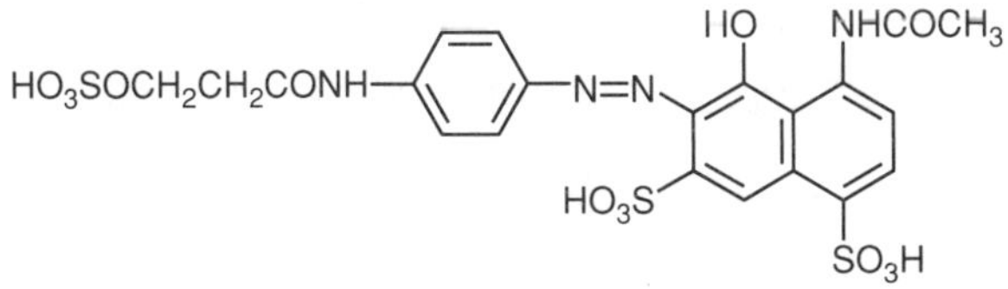

7.66

7.7.4 Tetrafluorocyclobutane derivatives

Among the many reactive systems that have been the subject of patent claims but not commercially exploited, an interesting series of dyes containing a tetrafluorocyclobutane system disclosed by Hoechst [32] merit mention for the unusual chemistry involved. They are prepared by acylation of an amino-containing dye base with either 2,2,3,3-tetrafluorocyclobutane-1-carbonyl chloride (7.67; $n = 0$) or β-(2,2,3,3-tetrafluorocyclobutyl)acryloyl chloride (7.67; $n = 1$). These novel acid chlorides are obtained by addition of tetrafluoroethylene and acrylonitrile or α-cyanobutadiene, hydrolysis of the cyano group to a carboxylic acid group and conversion to the acid chloride (Scheme 7.11).

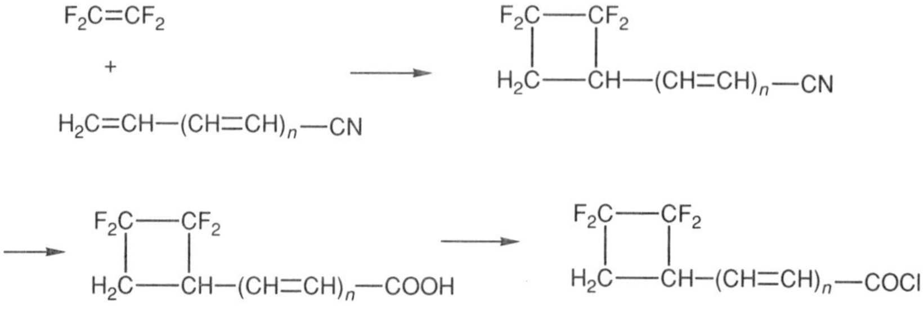

7.67

Scheme 7.11

The probable mode of reaction involves introduction of an activated double bond by elimination of hydrogen fluoride followed by addition of a cellulose hydroxy group (Scheme 7.12).

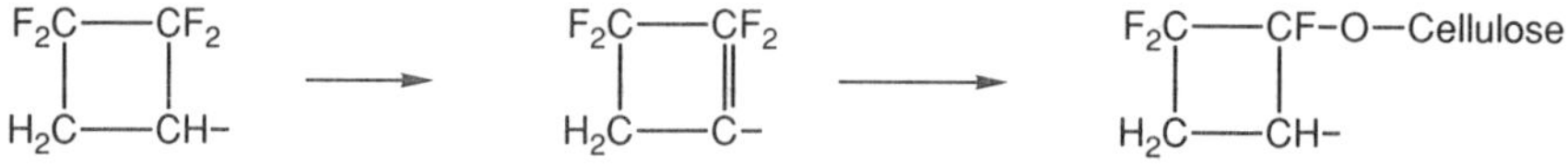

Scheme 7.12

7.8 DYES CONTAINING BOTH CHLOROTRIAZINE AND VINYL SULPHONE REACTIVE SYSTEMS

Condensation of a dichloro-*s*-triazine dye with an intermediate containing a β-sulphatoethylsulphone group allows the ready preparation of dyes of structure 7.68, containing both monochloro-*s*-triazine and β-sulphatoethylsulphone groups. Such dyes form the basis of the Sumifix Supra (Sumitomo) range [33].

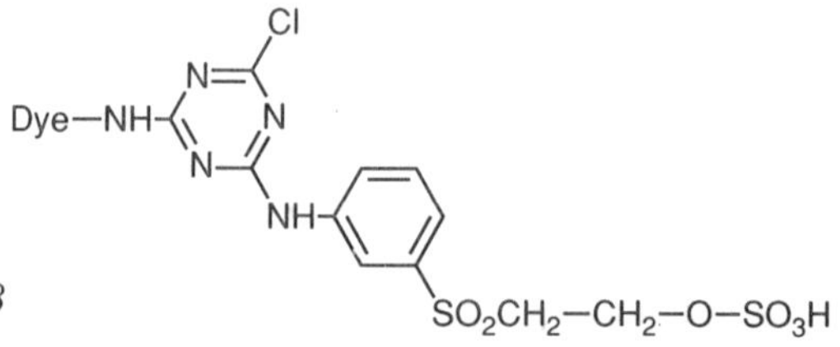

7.68

The presence of the two reactive groups in the molecule has been claimed to give improved fixation. However, whilst both reactive systems can contribute to the fixation process, the greater reactivity of the β-sulphatoethylsulphone system relative to the monochloro-*s*-triazinyl system will ensure that most of the fixation arises from reaction of the vinyl sulphone group with the fibre. The main usefulness of this class lies in the convenience with which the β-sulphatoethylsulphone grouping can be introduced into a wide range of chromogens and, possibly, the higher substantivity that arises because of the presence of the triazine ring in the molecule.

7.9 APPLICATION OF REACTIVE DYES TO POLYESTER/COTTON BLENDS

In order to dye a polyester/cotton blend it is necessary to use a disperse dye to colour the polyester together with a suitable dye for the cotton component, and it is obviously economically desirable to apply both dyes simultaneously in a single process rather than separately. In the single dyeing process the application temperature is fixed by the need to apply the disperse dye at an elevated temperature, and when the chosen cotton dye is a reactive dye this limits the choice to those dyes designed for hot application. An immediate difficulty arises because satisfactory application of the disperse dye calls for the dyeing process to be operated either at neutrality or under acid conditions, since alkaline conditions lower the dispersion stability, whereas the presence of alkali is normally required for the application of a reactive dye in order to ionise the cellulose and promote dye–fibre reaction.

Three distinct approaches to solving this problem have been adopted. The simplest has been to choose the more reactive of the hot-dyeing dyes and to run the dyeing process with a minimum of alkali (at a pH of about 8). Such a compromise can be achieved by using either the β-sulphatoethylsulphone dyes of the Remazol range, which have relatively high reactivity, or selected monoalkoxymonochloro-*s*-triazines, which have higher reactivity than the conventional monoaminomonochloro-*s*-triazines.

A more novel approach was pursued by ICI with the Procion T dyes [34], which exploited the finding that dyes containing a phosphonic acid group would react under acid conditions at high temperatures (about 220°C) with cotton in the presence of dicyandiamide to form a phosphonic acid ester bond between the dye and the fibre. The proposed mechanism (Scheme 7.13) entails acid-catalysed addition of the phosphonic acid group across the nitrile triple bond to give an intermediate (7.69) which is then split by the cellulose to give the dye–fibre link; this is quite stable to hydrolysis, conferring excellent fastness properties. Whilst the dyeing conditions used in the process are entirely compatible with the requirements for thermofixation of a disperse dye to polyester, they do represent an unusually severe treatment of cotton and result in a considerable loss in strength of

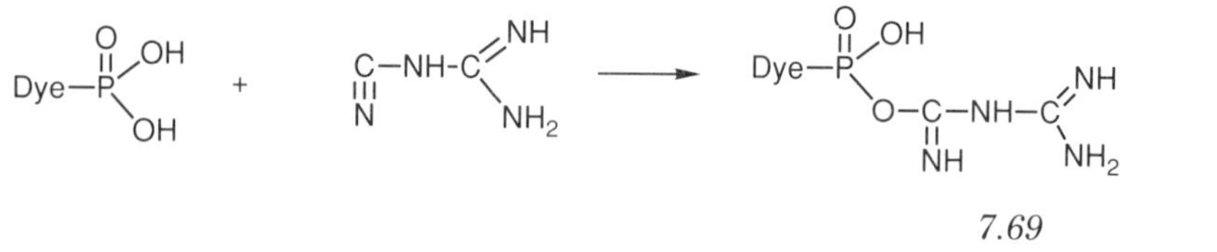

Scheme 7.13

this component of the fibre blend. This is not a matter of concern, however, since the polyester is the major contributor of strength to the blend.

The most recent approach to the problem stems from the finding that quaternary derivatives of monochloro-*s*-triazines and nicotinic acid will react with cellulose at high temperatures under neutral conditions. Dyes of type 7.70 contain two such groups, and form the basis of the Kayacelon React range marketed by Nippon Kayaku for application to the fibre blend in conjunction with a disperse dye without any addition of alkali.

7.70

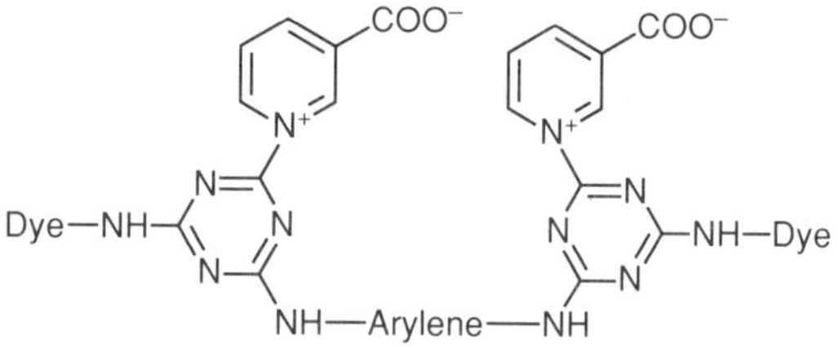

7.10 REACTIVE DYES FOR WOOL AND NYLON

7.10.1 General

Wool and nylon both carry aliphatic amino groups; wool also carries thiol (SH) groups. Due to the presence of these excellent nucleophiles, both fibre types are readily coloured by reactive dyes. Indeed, the first commercially available dye that reacted with the fibre was a wool dye: Supramine Orange R, introduced by IG Farbenindustrie in the 1930s. This dye contained a chloroacetylamino group from which the labile chlorine atom can be readily displaced under acid or neutral dyeing conditions by a fibre amino group, although it does not seem to have been realised at the time that the high fastness to washing was a consequence of dye–fibre reaction.

In late 1952 the first dyes containing β-sulphatoethylsulphone groups were marketed, under the trade name Remalan (HOE), for the dyeing of wool and nylon. These dyes were applied under neutral to slightly acid conditions and functioned by nucleophilic addition across the activated double bond of the generated vinyl sulphone. The chemistry that had been elucidated for these dyes provided a springboard for Hoechst to market the Remazol dyes rapidly when the launch of the Procion dyes alerted it to the possibility of dye–fibre reaction in the case of cellulose.

Despite the ideal chemistry of wool and nylon, reactive dyes have made their major impact in the dyeing of cellulosic fibres. This is because there were other anionic dyes that could meet the fastness requirements on wool and nylon, but not on the cellulosic fibres. Nevertheless, reactive dyes have found a niche in the dyeing of these nitrogenous fibres. Some of the dyes used are similar to cellulosic dyes, but with the number of sulphonic acid groups often reduced to conform with the different dyeing requirements. Thus the Cibalan (CGY) dyes were simple monochloro-*s*-triazines, as were later unmetallised members of the Procilan (ICI) range. Similarly the Drimalan F (S) and Verofix (BAY) dyes contain a 5-chloro-2,4-difluoropyrimidinyl group.

7.10.2 Reactive dyes for wool

In addition to the aforementioned dyes, which may be regarded merely as modifications of cellulose reactive dyes, several ranges have been introduced with reactive groups designed specifically to fit into the dyeing processes commonly applied to wool [35]. Conventional wool dyeing often involves a rapid unlevel initial strike at low temperature, followed by a prolonged migration treatment at the boil to attain optimum levelness. To observe these requirements and satisfactorily colour the wool any reactive group used on a wool dye must be slow to react, and this implies groups of low intrinsic reactivity. One such group is the acrylamido group ($-NHCOCH{=}CH_2$), which was used in the metal-complex members of the original Procilan (ICI) range. These dyes were unsymmetrical 1:2 chromium complexes containing a single sulphonic acid group and a single acrylamido group; they were designed to be applied at the boil, about an hour being required for the slow addition of the nucleophile across the activated double bond of the reactive group.

The Lanasol (CGY) dyes rely on an α-bromoacrylamido group (7.71), which has been shown to react initially with wool by an elimination rather than an addition mechanism, although the latter is not precluded [36]. Patents covering these dyes usually describe dyes having an α,β-dibromoacrylamido group in the molecule, hydrogen bromide being readily eliminated from this group in aqueous solution thus enabling it to function as a precursor of the α-bromoacrylamido group [37] (Scheme 7.14).

In the discontinued Hostalan (HOE) dyes [38], now marketed as Procilan

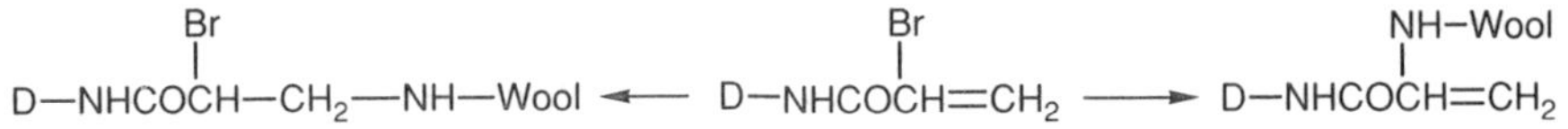

7.71

Scheme 7.14

E (ICI) brands, a variation on the Remazol type is exploited. These dyes utilise the *N*-methyltaurine derivative of the vinyl sulphone system (7.72); this slowly eliminates *N*-methyltaurine under mildly acid conditions at the boil, thus allowing the dye to level before it reacts with the fibre (Scheme 7.15).

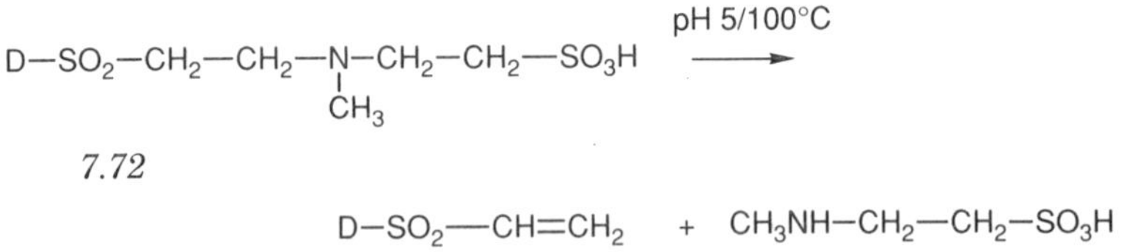

Scheme 7.15

7.10.3 Reactive dyes for nylon

One range of dyes has been marketed specifically for the dyeing of nylon. This was the Procinyl (ICI) range of disperse reactive dyes, most members of which carried a β-chloroethylsulphonamide group in the diazo component (7.73). The objective of these dyes was to combine the excellent levelling properties of disperse dyes on nylon with improved wet fastness. Unfortunately, few dyers have been willing to pay the higher price of Procinyl dyes for the modest gain in wet fastness offered.

7.73

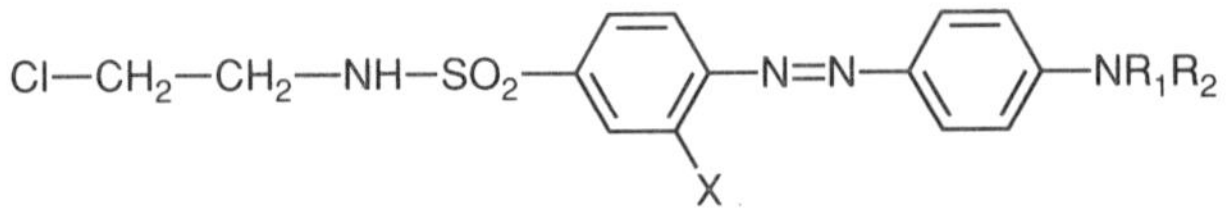

7.11 NOVEL DYEING PROCEDURES

In addition to straightforward reaction of the dye with the fibre, several processes have been evolved over the years for exploiting chemical reactivity of a dyestuff in the dyeing process. These have generally failed to make a significant impact in textile coloration, however.

An early example was the Procion Resin process [39]. In this method a chlorotriazinyl reactive dye was condensed with an amino sugar, such as *N*-methylglucosamine, and the carbohydrate residue on the resulting dye crosslinked with the fibre by application of a suitable resin. In the same vein, the Calcobond dyes of American Cyanamid (7.74) relied on cross-linking between the fibre and an *N*-methylol group on the dye, this group being introduced by aminating a chlorotriazine dye and allowing the diaminotriazine so formed to react with formaldehyde [40] (Scheme 7.16).

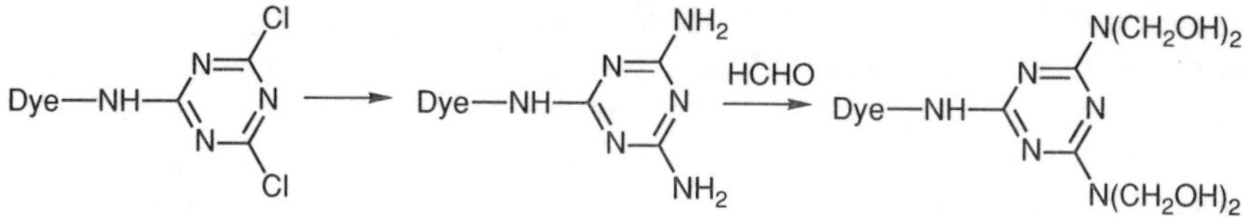

7.74

Scheme 7.16

A somewhat different principle was used in the case of the Basozol dyes (BASF) [41]. Here a 1:2 metal-complex dye containing an amino group in each of the azo chromogens was applied to the substrate in conjunction with tris(acryloyl)hexahydro-*s*-triazine, also known as Fixing Agent P (7.75), which reacted with amino nucleophiles in the Basozol dye and cellulosate nucleophiles in the fibre. Despite a lack of commercial success, the current patent literature shows that these types of approach are still being studied.

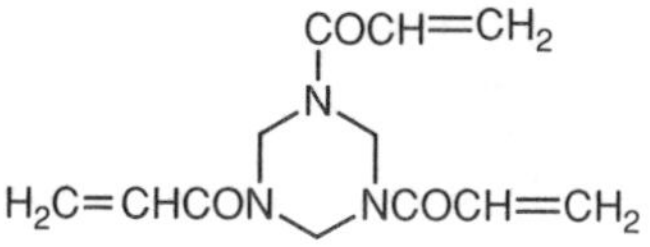

7.75

7.12 FUTURE PROSPECTS

It is over thirty years since reactive dyes for cellulose were introduced and they now account for about 25% of the total dye consumption on that fibre. They emerged after the results of the work on the mechanisms of organic reactions was in place, and this enabled their development to be characterised by the study and application of reaction mechanisms involved in the dye–fibre reaction. This factor has paid handsome dividends. The work continues, increasingly gaining cost-effectiveness by enhancing fixation at the expense of hydrolysis. Efforts to exploit novel alternative reaction mechanisms, such as polymerisation, have met with little success and the simple nucleophilic substitution and addition mechanisms of dye fixation remain totally dominant.

Reactive dyes offer bright shades and good properties combined with relatively easy application at a moderate cost. These attributes have enabled them to gain their market share chiefly at the expense of azoic dyes. They have achieved more limited market penetration against vats, which offer excellent fastness at a high price, and against directs, which generally offer cheapness and easy application but duller shades and only poor fastness properties. It is difficult to see any encroachment on the vat dye area on account of the superb fastness properties of vats, but development of simplified dyeing methods entailing reduced dyeing times and less washing-off could lead to some market share being gained at the expense of direct dyes. Enhancing efficiency by the incorporation of two reactive systems offers promise in this direction but current interest in using dissimilar reactive systems, rather than choosing to duplicate the more effective of the systems, seems unusual. It is perhaps also surprising that more attention has not been paid to the development of simple and quick aftertreatment processes to minimise washing-off.

Reactive dyes still offer ample scope for further progress in the dyeing of cellulose, with less opportunity being apparent for improvements in printing since improved pigment printing systems now offer serious competition to reactive dye systems. It is to be expected that reactive dyes will continue to attract a considerable research effort aimed at more efficient dyeing systems for many years and that this effort will not be unrewarded.

REFERENCES

1. C F Cross and E J Bevan, *Researches on cellulose* (1895–1900) 34.
2. BASF, BP 259 634.
3. W E Stephen and I D Rattee, *J.S.D.C.*, **76** (1960) 6.
4. W E Stephen, *Chimia*, **19** (1965) 261.
5. ICI, BP 838 337.
6. W F Beech, *Fibre reactive dyes* (London: Logos Press 1970) 25.
7. T L Dawson and A S Fern, *J.S.D.C.*, **76** (1960) 210.
8. C Preston and A S Fern, *Chimia*, **15** (1961) 177.
9. J Wegmann, *J.S.D.C.*, **76** (1970) 205.
10. C V Stead, *Dyes and Pigments*, **3** (1982) 161.
11. K G Kleb, E Siegel and K Sasse, *Angew. Chem.*, Internat. Edn, **3** (1964) 312.
12. CFMC, BP 1 002 648.
13. T E Peacock, *Electronic properties of aromatic and heterocyclic molecules* (London: Academic Press, 1965) 103.
14. H R Hensel and G Lützel, *Angew. Chem.*, Internat. Edn, **4** (1965) 408
15. M A Avsac, *Teintex*, **30** (1965), 163.
16. W Möck, *Melliand Textilber.*, **45** (1964) 652.
17. ICI, BP 846 765; 925 994; 927 773.
18. ICI, BP 923 068.
19. ICI, BP 852 911.
20. ICI, BP 849 772.
21. ICI, BP 954 992; 970 585.
22. ICI, BP 937 182; 946 998.
23. N Morimura and M Ojima, *Amer. Dyestuff Rep.*, **74** (1985) 28.
24. K H Schündehutte and K Trautner, *Chimia* Supplement, **22** (1968) 155.
25. E Bohnert, *J.S.D.C.*, **75** (1959) 581.
26. O Thumm and J Benz, *J.S.D.C.*, **77** (1961) 734.
27. Monsanto, BP 666 154.
28. HOE, BP 985 604.

29. K G Kleb, K Sasse and E Siegel, *Angew. Chem.*, Internat. Edn, **3** (1964) 408.
30. ICI, BP 823 098; 826 689; 921 425.
31. G Meyer and H Wirth, *Melliand Textilber.*, **43** (1962) 850.
32. HOE, BP 1 087 033.
33. S Fujioka and S Abeta, *Dyes and Pigments*, **3** (1982) 281.
34. G Alberghina, M E Amato, S Fisichella and I D Rattee, *Dyes and Pigments*, **7** (1986) 397.
35. D M Lewis, *Rev. Prog. Coloration*, **8** (1977) 10.
36. E Mäusezahl, *Textilveredlung*, **5** (1965) 839.
37. CGY, BP 1 223 304; 1 309 294.
38. K Osterloh, *Textil Praxis*, **26** (1971) 164.
39. I D Rattee, *J.S.D.C.*, **78** (1962) 69.
40. H R McCleary, A L Cate, F Fordemwalt and F F Loffelman, *Amer. Dyestuff Rep.*, **56** (1967) 46.
41. BASF, BP 923 162; 968 258; 1 118 174.

Colorants index

The colorants index is subdivided into two parts. In Part 1 colorants are listed in alphabetical order according to their *Colour Index* generic names (for completeness, all entries should be preceded by the letters C.I., but these are omitted here for ease of reference). The numbers in square brackets are the structure numbers as used in the text. Where possible each colorant has also been identified by a classical or a commercial name in the index, and these are given in Part 2 with a cross-reference to Part 1 in each case. Not all the commercial names appear in the text. The code letters following the commercial names identify the manufacturer; further details are given in Volume 5 of the *Colour Index International*.

Part 1

Part 2

Chemicals index

The numbers in square brackets are the structure numbers as used in the text.

Subject index

This index refers to both Volumes 1 and 2 of this book.